VOLUME NINETY

CURRENT TOPICS IN DEVELOPMENTAL BIOLOGY

Organogenesis in Development

VOLUME NINETY

Current Topics in DEVELOPMENTAL BIOLOGY

Organogenesis in Development

Edited by

PETER KOOPMAN
Institute for Molecular Bioscience
University of Queensland
Brisbane, Queensland
Australia

ELSEVIER

AMSTERDAM • BOSTON • HEIDELBERG • LONDON
NEW YORK • OXFORD • PARIS • SAN DIEGO
SAN FRANCISCO • SINGAPORE • SYDNEY • TOKYO
Academic Press is an imprint of Elsevier

Academic Press is an imprint of Elsevier
525 B Street, Suite 1900, San Diego, CA 92101-4495, USA
30 Corporate Drive, Suite 400, Burlington, MA 01803, USA
32, Jamestown Road, London NW1 7BY, UK
Linacre House, Jordan Hill, Oxford OX2 8DP, UK

First edition 2010

ISBN: 978-0-12-380912-4
ISSN: 0070-2153

For information on all Academic Press publications
visit our website at elsevierdirect.com

Printed and bound in the USA

10 11 12 13 9 8 7 6 5 4 3 2 1

Contents

Contributors ix
Preface xiii

1. How to Make a Heart: The Origin and Regulation of Cardiac Progenitor Cells

Stéphane D. Vincent and Margaret E. Buckingham

1. Introduction 2
2. The Origin of the Heart Fields and Cardiac Progenitor Cell Behavior 5
3. Markers of Cardiac Progenitors and the Distinction Between First and Second Heart Fields (FHF and SHF) 7
4. Cardiac Progenitor Contributions to the Cell Types of the Heart 10
5. Subdomains of the SHF 14
6. Molecular Mechanisms that Govern SHF Cell Behavior—Transcriptional Regulators and Signaling Pathways 18
7. Conclusion 29

Acknowledgments 30
References 30

2. Vascular Development—Genetic Mechanisms and Links to Vascular Disease

John C. Chappell and Victoria L. Bautch

1. Introduction 44
2. VEGF in Vascular Development 49
3. BMP in Vascular Development 53
4. Notch/Delta/Jagged in Vascular Development 55
5. Perspectives 60

Acknowledgments 62
References 62

3. **Lung Organogenesis**

David Warburton, Ahmed El-Hashash, Gianni Carraro, Caterina Tiozzo, Frederic Sala, Orquidea Rogers, Stijn De Langhe, Paul J. Kemp, Daniela Riccardi, John Torday, Saverio Bellusci, Wei Shi, Sharon R Lubkin, and Edwin Jesudason

1. Introduction 74
2. Developmental Anatomy of the Lung 75
3. Molecular Embryology of the Lung 81
4. Mechanobiology of the Developing Lung 111
5. Stem/Progenitor Cell Biology of the Lung 118
6. Postnatal and Adult Lung 128
7. Conclusions 130

Acknowledgments 131
References 131

4. **Transcriptional Networks and Signaling Pathways that Govern Vertebrate Intestinal Development**

Joan K. Heath

1. Introduction 160
2. Formation of the Definitive Endoderm 162
3. The Formation and Regionalization of the Primitive Gut Tube 165
4. Establishment of the Crypt–Villus Axis 175
5. Establishing the Stem Cell Niche and Homeostasis in the Intestinal Epithelium 179
6. Role of Intestinal Development Pathways in Cancer 185

Acknowledgments 186
References 186

5. **Kidney Development: Two Tales of Tubulogenesis**

Melissa Little, Kylie Georgas, David Pennisi, and Lorine Wilkinson

1. Introduction: How You Get a Kidney 194
2. The First Tale of Tubulogenesis: a Branching Tree 197
3. Tubulogenesis *Via* MET—Tube One Induces Tube Two 203
4. Patterning the Resulting Tubules 207
5. Moving From Structure to a Functional Filter 213
6. Disruptions to Kidney Tubulogenesis in Human Disease 217
7. Conclusion 220

Acknowledgments 221
References 221

6. The Game Plan: Cellular and Molecular Mechanisms of Mammalian Testis Development

Elanor N. Wainwright and Dagmar Wilhelm

1. Introduction 232
2. Introducing the Players: Cell Biology and Morphology of the Gonads 233
3. Origin of Sertoli Cells 235
4. Kickoff in Testis Determination: *Sry* and Sertoli Cell Specification 235
5. The Goalkeeper: *Sox9* and Sertoli Cell Differentiation 237
6. Forward Players: Beyond *Sox9* 241
7. The Sweepers: Peritubular Myoid Cells 243
8. Interplay Between PM and Other Testicular Cells 245
9. The Midfielders: Leydig Cells 245
10. The White Lines: Endothelial Cells 249
11. What All the Fuss Is About: Germ Cells 251
12. Concluding Remarks 253

Acknowledgments 254
References 254

7. Building Pathways for Ovary Organogenesis in the Mouse Embryo

Chia-Feng Liu, Chang Liu, and Humphrey H-C Yao

1. Evolution of the Hypotheses for Ovary Organogenesis in Mammals 264
2. Building the Foundation: Morphogenesis of the Ovary 266
3. Making Eggs: Establishment of the Female Germline 268
4. Determination of the Ovarian Identity: Differentiation of Granulosa Cells 272
5. Emerging Pathways for the Establishment of Female Somatic Environment and Maintenance of Female Germ Cells 276
6. Conclusion and Perspectives 278

References 283

8. Vertebrate Skeletogenesis

Véronique Lefebvre and Pallavi Bhattaram

1. Introduction 292
2. Structural Organization and Advantages of the Vertebrate Skeleton 293
3. Development of Skeletogenic Cells 294
4. Development of Cartilage Anlagen 298
5. Development of Cartilage Growth Plates 300
6. Bone Development 302

7. Synovial Joint Formation 306
8. Skeleton Variation 309
9. Perspectives 310
Acknowledgments 311
References 311

9. The Molecular Regulation of Vertebrate Limb Patterning
Natalie C. Butterfield, Edwina McGlinn, and Carol Wicking

1. Introduction 320
2. The Early Limb 322
3. Patterning Along the AP Axis 325
4. Proximal–Distal Patterning and Outgrowth 328
5. Interaction Between the ZPA and the AER 332
6. Towards a Systems Biology Approach to Limb Patterning 334
Acknowledgments 335
References 335

10. Eye Development
Jochen Graw

1. Introduction 344
2. Overview of Eye Development 346
3. Early Stage: The Eye Field 346
4. Lens Development 354
5. The Cornea 363
6. The Iris and the Ciliary Body 366
7. The Retina 367
8. The Optic Nerve 374
9. Conclusion and Perspectives 377
Acknowledgments 378
References 378

Index *387*
Contents of Previous Volumes *409*

Contributors

Victoria L. Bautch
Department of Biology, McAllister Heart Institute, and Lineberger Comprehensive Cancer Center, The University of North Carolina at Chapel Hill, Chapel Hill, North Carolina, USA

Saverio Bellusci
Developmental Biology and Regenerative Medicine Program, California Institute for Regenerative Medicine Training Program, The Saban Research Institute, Childrens Hospital Los Angeles, Los Angeles, California, USA, and Excellence Cluster in Cardio-Pulmonary Systems, University of Giessen, ECCPS/Medical Clinic II Klinkstr, Giessen, Germany

Pallavi Bhattaram
Department of Cell Biology, and Orthopaedic and Rheumatologic Research Center, Lerner Research Institute, Cleveland Clinic, Euclid Avenue, Cleveland, Ohio, USA

Margaret E. Buckingham
Institut Pasteur, Département de Biologie du Développement, CNRS URA 2578, Paris, France

Natalie C. Butterfield*
Institute for Molecular Bioscience, The University of Queensland, Brisbane, Queensland, Australia

Gianni Carraro
Developmental Biology and Regenerative Medicine Program, California Institute for Regenerative Medicine Training Program, The Saban Research Institute, Childrens Hospital Los Angeles, Los Angeles, California, USA

John C. Chappell
Department of Biology, and McAllister Heart Institute, The University of North Carolina at Chapel Hill, Chapel Hill, North Carolina, USA

Ahmed El-Hashash
Developmental Biology and Regenerative Medicine Program, California Institute for Regenerative Medicine Training Program, The Saban Research Institute, Childrens Hospital Los Angeles, Los Angeles, California, USA

* Present address: Division of Developmental Biology, MRC-National Institute for Medical Research, Mill Hill London, United Kingdom

Kylie Georgas
Institute for Molecular Bioscience, The University of Queensland, St. Lucia, Australia

Jochen Graw
Helmholtz Center Munich—German Research Center for Environmental Health, Institute of Developmental Genetics, Neuherberg, Germany

Joan K. Heath
Ludwig Institute for Cancer Research, Royal Melbourne Hospital, and Department of Surgery, University of Melbourne, Parkville, Victoria, Australia

Edwin Jesudason
Alder Hey Children's Hospital & Division of Child Health, University of Liverpool, Liverpool, Merseyside, UK

Paul J. Kemp
School of Biosciences, Cardiff University, Cardiff, South Glamorgan, UK

Stijn De Langhe
National Jewish Health, Denver, Colorado, USA

Véronique Lefebvre
Department of Cell Biology, and Orthopaedic and Rheumatologic Research Center, Lerner Research Institute, Cleveland Clinic, Euclid Avenue, Cleveland, Ohio, USA

Melissa Little
Institute for Molecular Bioscience, The University of Queensland, St. Lucia, Australia

Chang Liu
Department of Animal Science, University of Illinois at Urbana-Champaign, Illinois, USA

Chia-Feng Liu
Department of Veterinary Biosciences, University of Illinois at Urbana-Champaign, Illinois, USA

Sharon R Lubkin
Department of Mathematics and Department of Biomedical Engineering, North Carolina State University, Raleigh, North Carolina, USA

Edwina McGlinn
Department of Genetics, Harvard Medical School, Boston, Massachusetts, USA

David Pennisi
Institute for Molecular Bioscience, The University of Queensland, St. Lucia, Australia

Daniela Riccardi
School of Biosciences, Cardiff University, Cardiff, South Glamorgan, UK

Orquidea Rogers
Developmental Biology and Regenerative Medicine Program, California Institute for Regenerative Medicine Training Program, The Saban Research Institute, Childrens Hospital Los Angeles, Los Angeles, California, USA

Frederic Sala
Developmental Biology and Regenerative Medicine Program, California Institute for Regenerative Medicine Training Program, The Saban Research Institute, Childrens Hospital Los Angeles, Los Angeles, California, USA

Wei Shi
Developmental Biology and Regenerative Medicine Program, California Institute for Regenerative Medicine Training Program, The Saban Research Institute, Childrens Hospital Los Angeles, and Systems Biology and Disease Graduate Program, Keck School of Medicine and School of Dentistry, University of Southern California, Los Angeles, California, USA

Caterina Tiozzo
Developmental Biology and Regenerative Medicine Program, California Institute for Regenerative Medicine Training Program, The Saban Research Institute, Childrens Hospital Los Angeles, Los Angeles, California, USA

John Torday
Los Angeles County Harbor-UCLA Medical Center, Los Angeles, California, USA

Stéphane D. Vincent
Institut Pasteur, Département de Biologie du Développement, CNRS URA 2578, Paris, France

Elanor N. Wainwright
Division of Molecular Genetics and Development, Institute for Molecular Biosciences, The University of Queensland, Brisbane, Queensland, Australia

David Warburton
Developmental Biology and Regenerative Medicine Program, California Institute for Regenerative Medicine Training Program, The Saban Research Institute, Childrens Hospital Los Angeles, Los Angeles, California, USA

Carol Wicking
Institute for Molecular Bioscience, The University of Queensland, Brisbane, Queensland, Australia

Dagmar Wilhelm
Division of Molecular Genetics and Development, Institute for Molecular Biosciences, The University of Queensland, Brisbane, Queensland, Australia

Lorine Wilkinson
Institute for Molecular Bioscience, The University of Queensland, St. Lucia, Australia

Humphrey H-C Yao
Department of Veterinary Biosciences, University of Illinois at Urbana-Champaign, Illinois, USA

PREFACE

I often liken the development of an embryo to the building of a skyscraper, or a spacecraft. In all cases, the structures are enormously complex, and bring together vast numbers of components, arranged in exactly the right place, assembled in a logistically correct time sequence, and connected with the other components in such a way as to make a functional whole. The two critical differences are that embryogenesis is often completed in a matter of days, as opposed to years needed to build complex man-made structures, and that embryogenesis does not depend on large teams of architects, engineers, or construction workers. Remarkably, an embryo builds itself.

This volume focuses on how organs develop in the vertebrate embryo. Assembly of each organ requires a detailed schedule of cell differentiation, migration, proliferation, elimination, communication and cooperation—a microcosm of the embryo as a whole. These processes, and their coordination in time and space, call for accurate reading of the genetic blueprint, and the deployment of a raft of signaling molecules, receptors, signal transduction systems, and transcription factors to execute the blueprint.

The following chapters explore the cellular events that contribute to the proper development of a number of key organ systems and the molecular genetic mechanisms that regulate and coordinate these events. The focus of this volume is vertebrate development, with an emphasis on information gleaned from genetically tractable organisms such as mice and zebrafish. Authors have explored the experimental underpinnings of the knowledge base, and so these chapters deal in considerable detail with the study of genetic mutants in which organogenesis is perturbed. This, in turn, leads to a consideration of human developmental diseases involving defects in organogenesis, and to an improved understanding of the etiology of these diseases.

A theme that emerges from these pages is the recurrent use of a limited set of regulatory molecules in different developmental contexts. The signaling events that direct the development of the many different organs typically involve members of the hedgehog, transforming growth factor (TGF)-β, delta-notch, Eph, and Wnt families and their receptors. Also, wherever cells are differentiating and organs developing, members of the Hox, Sox, Pax, and forkhead transcription factor families can be found. How so many different functions can be carried out by so few molecules remains something of a mystery, although combinatorial action is likely to be involved, given that a large number of combinations can be generated from a relatively small number of individual elements.

These chapters also emphasize how much we still do not know and have yet to discover. What regulatory components are we missing by focusing only on protein-coding genes? How do organs determine their appropriate size and shape? How do the subcompartments of any organ arrange themselves so accurately to achieve their coordinated function and at the same time incorporate appropriate wiring and plumbing in the form of neural, vascular, and lymphatic networks? How come some organs can recover from damage, wholly or partially, whereas others cannot? Clearly these questions are fascinating in themselves, as well as representing some of the key challenges in regenerative medicine.

Any monograph is only as good as its authors. A critical reader will always ask who is providing the information, what are their credentials, and can they be trusted to give reliable data and interpretation. I am fortunate to have been able to cajole several of the world's leading experts in organogenesis into writing for this volume. These authors are not only working at the cutting edge on their organ of interest, but also have a demonstrated flair for organization, communication, and in many instances, drawing great summary diagrams. The result is an authoritative, timely, and hopefully useful volume, and I express my gratitude to the authors for the care and gusto with which they have approached their tasks

This volume covers the development of 10 different organ systems in 10 chapters. These represent some of the more obvious organs in the body, organs that have attracted a large amount of research attention, either through their experimental accessibility or involvement in human disease, or both. Of course, given the dozens of organs in the vertebrate body between the hair follicles on our head and the nails on our toes, there is much more to cover. I hope that future volumes will address some of the organ systems that we have not had the opportunity to deal with here. Each one is different and each a remarkable feat of engineering.

Peter Koopman
June 2010

CHAPTER ONE

How to Make a Heart: The Origin and Regulation of Cardiac Progenitor Cells

Stéphane D. Vincent *and* Margaret E. Buckingham

Contents

1. Introduction 2
2. The Origin of the Heart Fields and Cardiac Progenitor Cell Behavior 5
3. Markers of Cardiac Progenitors and the Distinction Between First and Second Heart Fields (FHF and SHF) 7
   3.1. Cardiac progenitors and cell fate determination 7
   3.2. Markers of the heart fields 8
4. Cardiac Progenitor Contributions to the Cell Types of the Heart 10
   4.1. Cell types derived from the SHF 10
   4.2. Neural crest 11
   4.3. The proepicardial organ 13
5. Subdomains of the SHF 14
   5.1. The anterior SHF and contributions to the arterial pole of the heart 16
   5.2. The posterior SHF and formation of the venous pole 17
6. Molecular Mechanisms that Govern SHF Cell Behavior—Transcriptional Regulators and Signaling Pathways 18
   6.1. Anterior/posterior patterning of the SHF: retinoic acid signaling 20
   6.2. Left/right patterning: Nodal signaling through Pitx2c 21
   6.3. Maintenance of proliferation in the SHF: FGF, hedgehog, and canonical Wnt signaling 22
   6.4. Interactions with neural crest in the SHF: FGF, Notch, and semaphorin signaling 24
   6.5. Prevention of differentiation in the SHF: canonical Wnt signaling and transcriptional repression 25
   6.6. Regulation of SHF differentiation potential, recruitment to the heart tube, and differentiation: Shh, Notch, BMP, and non-canonical Wnt signaling 26
7. Conclusion 29
Acknowledgments 30
References 30

Institut Pasteur, Département de Biologie du Développement, CNRS URA 2578, Paris, France

Current Topics in Developmental Biology, Volume 90
ISSN 0070-2153, DOI 10.1016/S0070-2153(10)90001-X

Abstract

The formation of the heart is a complex morphogenetic process that depends on the spatiotemporally regulated contribution of cardiac progenitor cells. These mainly derive from the splanchnic mesoderm of the first and second heart field (SHF), with an additional contribution of neurectodermally derived neural crest cells that are critical for the maturation of the arterial pole of the heart. The origin and distinguishing characteristics of the two heart fields, as well as the relation of the SHF to the proepicardial organ and to a proposed third heart field are still subjects of debate. In the last ten years many genes that function in the SHF have been identified, leading to the establishment of a gene regulatory network in the mouse embryo. It is becoming increasingly evident that distinct gene networks control subdomains of the SHF that contribute to different parts of the heart. Although there is now extensive information about mutant phenotypes that reflect problems in the integration of progenitor cells into the developing heart, relatively little is known about the mechanisms that regulate SHF cell behavior. This important source of cardiac progenitor cells must be maintained as a proliferative, undifferentiated cell population. Selected subpopulations, at different development stages, are directed to myocardial, and also to smooth muscle and endothelial cell fates, as they integrate into the heart. Analysis of signaling pathways that impact the SHF, as well as regulatory factors, is beginning to reveal mechanisms that control cardiac progenitor cell behavior.

1. Introduction

The heart occupies an important place in the popular imagination. In medieval times, in Europe, it was regarded as the seat of courage. King Richard I of England, renowned for his bravery, was called Richard the Lion Heart and the great queen Elizabeth I proudly claimed, in terms now politically incorrect, "I know that I have the mind and body of a weak and feeble woman, but I have the heart and stomach of a king, and of a king of England too!". After this period of male bravure, the heart, also regarded as the site of the soul and hence a holy relic, is now mainly a symbol of love, much evident in the commercialization of romance on St Valentine's Day. Fixation on the heart reflects its obvious role as a vital organ. The beating of the heart as it pumps blood around the body is synonymous of life and until very recently the official definition of death was arrest of the heartbeat.

The heart is the first organ to form in the embryo where its early function is essential for the circulation of nutrients and removal of waste, as soon as the number of cells reaches a point where diffusion is no longer efficient. Early heart defects are a frequent source of lethality when genes are mutated in mouse models. In the human population almost 1% of newborn children have some form of congenital heart defects and cardiac malformations probably account for as many as 30% of embryos/foetuses lost before birth (Bruneau, 2008). These figures indicate the vital requirement for a fully functional heart and also reflect the degree of precision required during cardiogenesis. The construction of the heart is a complex process, involving the integration of different cell populations at distinct sites as development proceeds. In this review, we discuss the origins of cardiac progenitor cells and the regulation of their contributions to the heart. The characterization of the second heart field (SHF) as a major contributor, with increasingly detailed genetic information about the regulatory factors and signaling pathways that affect the behavior of cells that transit through this field has considerably advanced our understanding of cardiogenesis. Knowledge about the progenitor cells that form the heart is also of importance for potential stem cell therapies in the context of the failing adult heart.

We shall discuss the formation of the heart from the stand point of mammalian cardiogenesis. In this context, the mouse is the best studied model. We shall particularly focus on the properties of the SHF because it constitutes a major source of cardiac progenitor cells as the primitive heart tube grows. However, the formation of the tube and the contribution of other sources of cells will also be considered in the context of cardiogenesis, briefly summarized as follows (see Fig. 1.1).

The first differentiated myocardial cells are detected in the cardiac crescent, in splanchnic mesoderm underlying the head folds. As the embryo grows, the crescent fuses at the midline to form the primitive cardiac tube which rapidly begins to pump blood. It is now established that cardiac progenitor cells mainly lie medially and posteriorly to the crescent and then are located behind the heart tube, extending posteriorly and also anteriorly into pharyngeal mesoderm, as a result of morphogenetic movements as the embryo develops. These progenitor cells constitute the SHF, in contrast to the region of the crescent referred to as the first heart field (FHF). In the mouse embryo, the early heart tube has a mainly left ventricular identity and its expansion depends on contributions from the SHF (Buckingham *et al.*, 2005). The early FHF-derived cardiac tube thus provides a scaffold for subsequent growth.

Other cell populations also contribute to the formation of the heart, in addition to the splanchnic mesoderm of the heart fields. The proepicardial organ (PEO) is a transitory mesenchymal structure that forms at the posterior end of the heart tube. Cells from the PEO grow over the myocardium of the tube to form the outer layer of the epicardium. Some of these epicardial cells undergo an epithelial/mesenchymal transition (EMT) and enter the heart where they contribute the smooth muscle of the coronary blood vessels and also constitute the population of cardiac fibroblasts/interstitial cells.

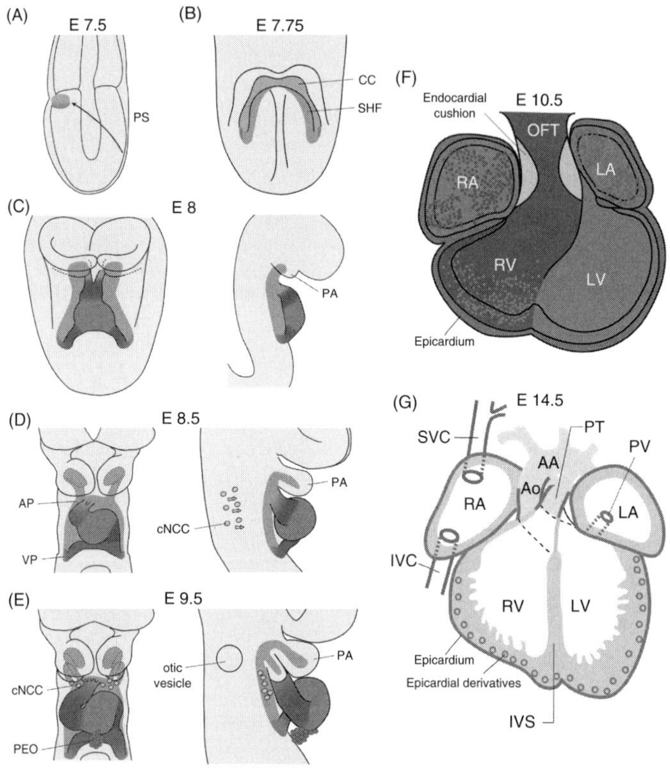

Figure 1.1 (A) Migration of cells anteriorly from the primitive streak (PS). (B) Formation of the cardiac crescent (CC), with the second heart field (SHF) lying medial to it. (C–E) Front (left) and lateral (right) views of the heart tube as it begins to loop with contributions of cardiac neural crest cells (cNCC), which migrate from the pharyngeal arches (PA) to the arterial pole (AP). The proepicardial organ (PEO) forms in the vicinity of the venous pole (VP). (F) The looped heart tube, with the cardiac compartments—OFT, outflow tract; RA, right atrium; LA, left atrium; RV, right ventricle; LV, left ventricle. (G) The mature heart which has undergone septation—IVS, interventricular septum; AA, aortic arch; Ao, aorta; PT, pulmonary trunk; PV, pulmonary vein; SVC, superior caval vein; IVC, inferior caval vein. The first heart field (FHF) and its myocardial contribution are shown in red, the SHF and its derivatives in dark green (myocardium) and pale green (vascular endothelial cells), cNCC in yellow (vascular smooth muscle of the AA, endocardial cushions), and PEO derivatives in blue. (See Color Insert.)

The functional form of the mature arterial pole of the heart depends on neural crest cells. These are of neurectodermal origin and migrate from the dorsal neural tube. Cardiac neural crest (Hutson and Kirby, 2007) transits through the posterior pharyngeal arches and invades the anterior domain of the SHF before entering the anterior part of the heart tube.

This constitutes the outflow tract of the heart and neural crest plays a major role in the remodeling of this region, contributing to septum and valve formation, which results in the separation of the myocardial base of the pulmonary trunk and aorta. These great arteries connect with right and left ventricles, respectively, and ensure blood flow to the lungs and body.

The venous pole of the heart, initially located at the posterior end of the heart tube, moves anteriorly as the tube undergoes looping. The inflow region of the tube develops with the addition of myocardium that will form the atria and then the base of the pulmonary and caval veins that recycle blood from the lungs and body to the left and right atria, respectively. The development of the cardiac chambers results from expansion of regions of the heart tube and subsequent septation and valve formation to give the mature heart (Fig. 1.1).

2. The Origin of the Heart Fields and Cardiac Progenitor Cell Behavior

Cardiac mesoderm derives from the anterior part of the primitive streak (PS), as shown previously by cell labeling and grafting experiments in the mouse (Lawson *et al.*, 1991; Tam *et al.*, 1997) as well as for the chick embryo (see Kirby, 2007). Since in both amniote models, the outflow tract of the heart tube has been shown to derive from the SHF, the question of whether there is pre-patterning of progenitor cells already in the streak can be addressed using the avian embryo, which is more amenable to experimental manipulations.

This question is intimately linked to the mode of migration of cells from the streak to the heart-forming region of splanchnic mesoderm (Fig. 1.1). Earlier experiments suggested that cells that give rise to outflow tract myocardium are situated more anteriorly in the cardiogenic region of the streak (Garcia-Martinez and Schoenwolf, 1993); however, these and other fate mapping experiments were limited by the technical methods of the time (see Abu-Issa and Kirby, 2007). This was also the case for studies at later developmental stages when it was difficult to ensure that only cardiac progenitors were labeled. Thus Stalsberg and DeHaan (1969) had concluded that migrating cells behaved as cohesive groups, whereas Redkar *et al.* (2001) suggested that there was considerable dispersion. In a recent fate mapping analysis, using sophisticated time lapse imaging microscopy after marker electroporation in the quail embryo, Cui *et al.* (2009) showed that cardiac progenitor cells, originating from similar anterior/posterior levels in the streak to those previously identified, change their relative positions as they migrate. This results in medial/lateral repositioning as the region where cardiogenic mesoderm is located undergoes morphological changes driven by endodermal folding. This would be consistent with the medial/lateral location of first and SHFs, respectively. They concluded that cardiac progenitors can move as cohorts of cells that will contribute to specific regions of the heart tube.

The results of retrospective clonal analysis in the mouse embryo had shown that a period of dispersive progenitor cell growth precedes coherent growth that accompanies cardiogenesis (Meilhac *et al.*, 2003). This retrospective approach does not throw any light on the spatial location of progenitors; however, it provides important temporal insights. Notably it distinguishes two myocardial cell lineages that segregate early, around the onset of gastrulation (Meilhac *et al.*, 2004). These two lineages can be equated with the contribution of first and SHFs, in that the first is the exclusive source of the early left ventricle whereas the second is the exclusive source of outflow tract myocardium. Both lineages contribute to other parts of the heart. Prospective clonal analysis in the mouse embryo will be required to address spatial issues of myocardial lineage segregation, as well as the timing, at or before gastrulation.

The issue of cell behavior in the SHF is linked to that of proliferation. Two recent studies in the chick embryo use three-dimensional reconstructions to interpret BrdU data on cell proliferation (Soufan *et al.*, 2006; Van den Berg *et al.*, 2009). They conclude that the early heart tube has a low level of proliferation and that changes in cell size play a significant role in its expansion. Growth of the heart tube also depends on the addition of progenitor cells. This mainly occurs in a proliferative center which is located in a dorsal/medial position, in the posterior SHF. Cell tracing experiments suggest that cells move anteriorly from this proliferative zone to contribute to the arterial as well as the venous pole of the cardiac tube and indeed impairment of proliferation in this posterior zone affects both poles of the heart. This finding for the chick embryo has to be equated with observations on clonal growth and cell fate determination for the mouse embryo. Retrospective clonal analysis suggests that the mouse heart tube is more proliferative (Meilhac *et al.*, 2004) and the anterior as well as the posterior region of the SHF is clearly proliferating, as indicated by mutants that affect proliferation in the anterior SHF, with consequences for arterial pole formation (see Section 6.3).

Explant experiments and dye-tracing of cells show that the anterior part of the SHF is programmed to make outflow tract and right ventricular myocardium and contributes to this part of the heart (Zaffran *et al.*, 2004). In contrast, the posterior part of the SHF is programmed to make atrial myocardium and indeed cells in this domain contribute to the atria (Galli *et al.*, 2008). These observations do not preclude that there may be posterior/anterior movement of cardiac progenitor cells, but would suggest that such a phenomenon may be limited. However, more detailed spatiotemporal fate mapping in the mouse SHF is required to clarify this issue.

3. Markers of Cardiac Progenitors and the Distinction Between First and Second Heart Fields (FHF and SHF)

3.1. Cardiac progenitors and cell fate determination

One of the first markers of cardiac progenitor cells is Mesp1, which is required for the delamination of these cells from the primitive streak. A *Mesp1*$^{Cre/+}$ line crossed to the *Rosa-26* conditional reporter marks all cardiac cells in the heart of mesodermal origin (Saga *et al.*, 1999) and has proved to be very useful for genetic experiments in which mutations are targeted to cardiac mesoderm. Experiments with embryonic stem (ES) cells have led to the proposal that this transcription factor may act as a master regulator of cardiovascular cell fates (Bondue *et al.*, 2008), down-regulating pluripotency genes and early mesodermal genes and up-regulating genes for key cardiac transcription factors such as Gata4 or Nkx2-5. In keeping with this, injection of *Mesp1* RNA into *Xenopus* embryos leads to ectopic heart formation (David *et al.*, 2008); however, these authors emphasize an indirect role for Mesp1 in the induction of the Wnt inhibitor, Dkk1, and effects on endodermal induction of cardiogenesis. Another series of experiments in ES cells also points to the role of Mesp1 in promoting cardiovascular cell fate in the presence of Dkk1 (Lindsley *et al.*, 2008). These authors also demonstrate that Mesp1 triggers an epithelial/mesenchymal transition (EMT) in ES cell embroid bodies. In the mouse embryo, Mesp1 is implicated in EMT, required for exit of cells from the streak, but the gene is rapidly down-regulated thereafter and it is not clear whether it directly activates transcriptional regulators of the cardiac program, which are detected later.

Genes such as *Gata4* or *Nkx2-5* are expressed in the cardiac crescent where myocardial cell differentiation first takes place. T-box transcription factors, such as Tbx5 and Hand1/2 (basic helix–loop–helix), as well as Mef2c (MADS-box) factors, are also implicated in the differentiation of cells in the crescent, as well as in the heart. In the context of master regulators, it was shown recently that Gata4 and Tbx5, in the presence of the chromatin remodeling component, Baf60c/Smarcd3 specifically associated with cardiogenesis, can induce beating myocardial tissue when ectopically expressed in mesoderm (Takeuchi and Bruneau, 2009). Gata4 and Baf60c induce Nkx2-5 expression which acts with Gata4 to initiate the cardiac program; Tbx5 is required for full differentiation. The concept of a master regulator gene comes from the demonstration that transcription factors of the MyoD family can play this role for skeletal myogenesis (Weintraub *et al.*, 1991). In this case, these factors have intrinsic chromatin remodeling activity (Tapscott, 2005), which cardiac regulatory factors appear to lack; hence the requirement for Baf60c. Skeletal myogenesis is also exceptional in its dependence on a single determination factor, whereas most tissue programs depend on combinations of factors. These may also be interchangeable, which would explain why no single mutation in a cardiac regulatory gene completely abolishes myocardial cell differentiation.

3.2. Markers of the heart fields

Expression of *Islet1* in the SHF first led to an appreciation of the full extent of this field and its contribution to the venous, as well as the arterial, pole of the heart (Cai *et al.*, 2003). Islet1 has been regarded as a marker of the SHF. However, more recently Islet1 protein has been detected in the cardiac crescent and, in the absence of Nkx2-5, Islet1 expression is maintained in differentiating myocardial cells of the crescent as well as the heart at later stages, suggesting that this regulation operates in both heart fields (Prall *et al.*, 2007). *Islet1-Cre* activation of a highly sensitive conditional *FLAP* reporter in the *Gata4* gene resulted in expression in the FHF as well as the SHF (Ma *et al.*, 2008). Furthermore, a conditional *Rosa26* reporter when activated by a new *Islet1-Cre* line is more broadly expressed in the heart than was seen previously, although part of the left ventricle remains negative (Sun *et al.*, 2007). This observation, also reported with other *SHF-Cre* driver lines, such as *Tbx1-Cre* (Brown *et al.*, 2004) or *Mef2c-Cre* (Verzi *et al.*, 2005), may also reflect subsequent expansion of initially right ventricular myocardium into the left ventricular compartment as the heart tube matures. Such expansion of genetically marked cells is also seen for *Tbx2*, a marker of the atrioventricular canal (AVC); cells that had expressed *Tbx2* subsequently expand into part of the left ventricular myocardium (Aanhaanen *et al.*, 2009). In this case, as AVC myocardium is derived from both first and second lineages, this does not necessarily imply that this is now a second lineage contribution (Harvey *et al.*, 2009). Despite detection of Islet1 in the FHF, it is notable that *Islet1* null mutants still form the primitive cardiac tube and the mutant phenotype primarily reflects a problem with the SHF contribution to the heart (Cai *et al.*, 2003). Islet1 positive cardiac progenitors have now been identified in *Xenopus*, where they co-localize initially with Nkx2-5 positive cells but subsequently appear to constitute a progenitor cell field; knock-down of Islet1 suggests that it is not essential for early heart formation (Brade *et al.*, 2007). Thus there is some indication of an equivalent to the SHF in *Xenopus*, and there is evidence for early segregation of two cardiogenic lineages, such that the second lineage, as in the chick, appears to mainly contribute to the outflow tract (Gessert and Kühl, 2009). In zebrafish, there is evidence for two phases of myocardial differentiation, a first Islet1-dependent contribution to the venous pole, and a second Fgf8-dependent addition of cardiomyocytes to the arterial pole (de Pater *et al.*, 2009).

In addition to *Islet1*, a number of genes expressed in the SHF have now been characterized (Fig. 1.2). In many cases, such as *Foxc1/Foxc2* (Seo and Kume, 2006), there is no evidence that they are also expressed in the FHF. Furthermore, as in the case of *Islet1*, when mutated they give rise to typical

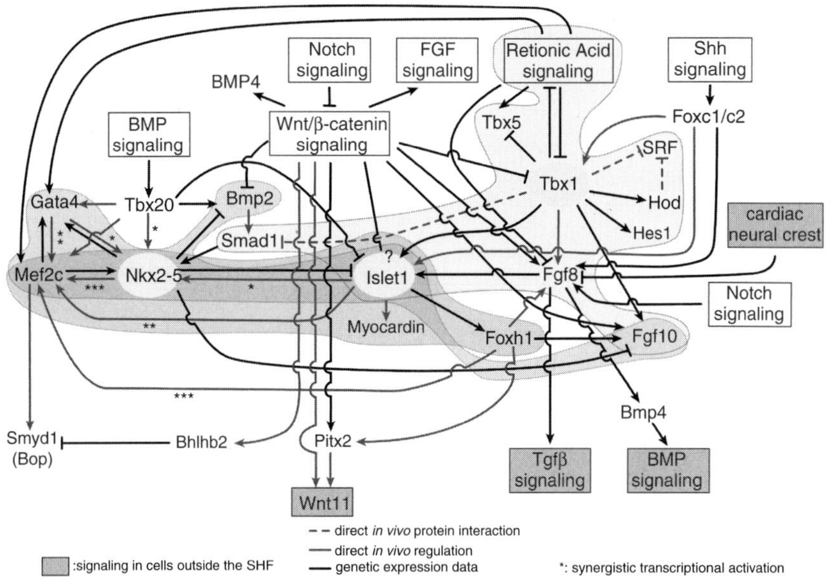

Figure 1.2 Regulatory network in the SHF. Three nodes of regulation are highlighted: Tbx1 (light grey), Islet1 (dark grey), and Nkx2-5 (mid-grey). Direct *in vivo* regulations are indicated by dark grey lines, direct protein–protein interactions are indicated by dashed lines, and genetic expression data are indicated by dark lines. Asterisks highlight synergistic transcriptional activation of enhancers. Grey boxes indicate effects outside of the SHF. The question mark above Islet1 is discussed in Section 6.3: several groups reported that *Islet1* is activated by the Wnt/β-catenin pathway, but a recent report suggested that down-regulation of *Islet1* by Wnt/β-catenin is required for SHF progenitor proliferation.

SHF phenotypes. The question of whether there is a continuum between FHF and SHF to the point where they are regarded as more or less differentiated regions of the same field is partly semantic. They are clearly juxtaposed in the mouse at E7.75 and contiguous in the chick at later stages (Abu-Issa and Kirby, 2008), where Islet1 is also characteristically expressed in undifferentiated cells (Yuan and Schoenwolf, 2000). Distinct first and second myocardial cell lineages exist in the mouse (Meilhac *et al.*, 2004) and mutant phenotypes indicate that many cardiac regulators are mainly functional in the progenitor cell population of the SHF, where a gene regulatory network, based on analysis of mutants and transcriptional regulatory elements, is observed (Fig. 1.2). A classic example of this is provided by two enhancers of the *Mef2c* gene, which are targets of Islet1 and Gata factors (Dodou *et al.*, 2004) and of Foxh1 and Nkx2-5 (von Both *et al.*, 2004), respectively, to drive SHF expression. Another example is provided by the recent identification of an *Islet1* enhancer active in the SHF that requires Foxc2 binding sites for activity, with an implication of Gata4 as well as Fox factors (Kang *et al.*, 2009; Kappen and Salbaum, 2009). Furthermore, factors present in the SHF may interact with each other, as shown for Tbx1 and SRF (Chen *et al.*, 2009), adding a further level of complexity to a network in which Tbx1 activates *Hod* and Hod represses *SRF* (Liao *et al.*, 2008). Another level of factor interaction is illustrated by Tbx20 which synergizes with Islet1 and Gata4 to activate the *Mef2c* enhancer and an *Nkx2-5* cardiac enhancer (Takeuchi *et al.*, 2005).

4. Cardiac Progenitor Contributions to the Cell Types of the Heart

4.1. Cell types derived from the SHF

As stated in Section 1, the FHF is the major contributor to early left ventricular myocardium, whereas the SHF contributes to myocardium of other regions of the heart and most notably to that of the outflow tract. However, myocardium is not the only derivative of the SHF. The origin of the endocardium has been controversial. This endothelium forms the inner sheath of the cardiac tube and compartments of the heart as they develop. It plays a critical role in trabeculation of chamber myocardium and in valve formation, initiated by delamination of endocardial cells to form the cushions (Kirby, 2007). Retroviral tracing in the avian embryo (Wei and Mikawa, 2000) had suggested that myocardial and endocardial progenitors are already distinct at gastrulation. More recently, lineage studies in the zebrafish suggest that endocardial cells are derived from a hematopoietic/vascular lineage (Bussmann *et al.*, 2007). However, genetic tracing in the mouse embryo suggests that cells that had expressed *Islet1* (Cai *et al.*, 2003; Moretti *et al.*, 2006), *Nkx2-5* (Stanley *et al.*, 2002), or activated the *Mef2c SHF* enhancer (Verzi *et al.*, 2005) contribute to endocardium and myocardium. Furthermore *Flk1* expressing progenitors contribute to both tissues (Motoike *et al.*, 2003). *Nfatc1* now provides a marker for endocardium that distinguishes these cells from other endothelial cells. Using this marker in the ES cell model system, it has been shown that multipotent Flk1 positive cardiac progenitors give rise to both endocardial and myocardial derivatives (Misfeldt *et al.*, 2009). This is, therefore, in keeping with the prevailing view that the endocardium is an SHF derivative. Interestingly, a recent paper (Ferdous *et al.*, 2009) showed that the endothelial/endocardial fate in the developing embryo depends on an Ets-related protein, Etsrp17. This is a direct activator of the endothelial *Tie2* gene. Upstream of endocardial formation, Nkx2-5 transactivates the *Etsrp17* gene. Etsrp17 is first detected in the cardiac crescent, suggesting that FHF derivatives may also contribute to endocardium.

The mesoderm of the arches can be regarded as an extension of the SHF, which becomes incorporated into the pharyngeal arches as these transitory structures bulge out as pouches on either side of the pharyngeal region (Fig. 1.1). SHF marker genes such as *Fgf10* (Kelly *et al.*, 2001), *Islet1* (Cai *et al.*, 2003), and *Tbx1* (Xu *et al.*, 2004) are expressed in the mesodermal core of all the arches. In the chick embryo, dye-labeling experiments have demonstrated the contribution of the mesodermal core of arches 1 and 2 to outflow tract myocardium (Kirby, 2007) and this has also been shown in the mouse embryo for arch 2 (Kelly *et al.*, 2001). Endothelial cells of the derivatives of the pharyngeal arch arteries, at the arterial pole of the heart (Fig. 1.3) derive from SHF mesoderm of the posterior arches (3–6). In the mouse embryo, this is supported by genetic tracing experiments showing that these endothelial cells derive from progenitors that have expressed *Mesp1* and *Islet1* (Sun *et al.*, 2007) and activated the *Mef2c SHF* enhancer (Verzi *et al.*, 2005) and also by phenotypes such as that of the *Tbx1* mutant (Zhang *et al.*, 2005) or of *Fgf8/Fgf10* double mutants targeted to the mesoderm of the SHF (Watanabe *et al.*, 2010).

The ES cell system also shows that smooth muscle cells derive from the multipotent cardiac progenitors that give rise to myocardium and endothelial derivatives (Moretti *et al.*, 2006; Wu *et al.*, 2006). In the chick embryo, cell tracing and ablation experiments in the SHF have established that this is the source of smooth muscle cells in the outflow tract that will contribute to the sub-pulmonary and aortic smooth muscle at the base of these great arteries. This SHF contribution follows that of myocardium, which is contributed over a first 24-h period (HH14-18) (Waldo *et al.*, 2005b; Ward *et al.*, 2005). In the mouse embryo, cells that have expressed *Mesp1* (Saga *et al.*, 2000) and *Islet1* (Moretti *et al.*, 2006; Sun *et al.*, 2007) also contribute to smooth muscle at the arterial pole of the heart and *Tbx1* regulates this derivative, as well as myocardium (Chen *et al.*, 2009). It is not yet clear whether SHF cells are multipotent for smooth, myocardial, and endothelial lineages as shown by clonal analysis for ES cells. However, interestingly, the rare Islet1 positive cells detected in foetal hearts (Laugwitz *et al.*, 2005), which probably represent undifferentiated SHF progenitors, are multipotent in this respect (Bu *et al.*, 2009).

4.2. Neural crest

Cardiac neural crest, that invades the mesodermal core of the posterior pharyngeal arches, contributes the smooth muscle of the pharyngeal arch arteries and their derivatives at the arterial pole of the heart (Fig. 1.3). Neural crest cells from the arches also migrate through the anterior SHF into the outflow tract where they form the endocardial cushions (Fig. 1.1). In the absence of neural crest, outflow tract septation and arterial pole maturation are compromised (Hutson and Kirby, 2007).

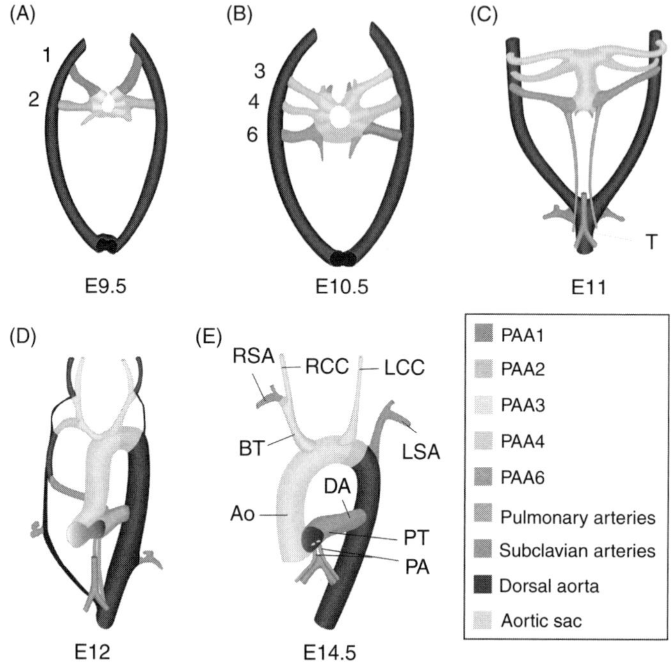

Figure 1.3 Pharyngeal arch artery remodeling. Ventral representation of the arterial network connected to the heart as development proceeds. The arterial pole of the heart (not represented here) feeds into the aortic sac (grey). The heart is connected to the paired dorsal aortas (black) via the artery of each of the five pharyngeal arches; the pharyngeal arch arteries (PAAs). During development, the pharyngeal arches and their arteries are symmetrically formed following a rostro-caudal, temporal gradient. (A) At E9.5, only the PAA1 (orange) and PAA2 (blue) are connected to the aortic sac. (B) At E10.5, the PAA1 and PAA2 are no longer connected directly to the heart but form the capillary beds of their respective pharyngeal arches. The heart is now connected via PAA3 (light green), PAA4 (dark green), and PAA6 (purple). (C) At E11, the network is still symmetric and the pulmonary arteries (dark yellow) are clearly visible. (E) From E11.5, remodeling of the PAAs is initiated via the increase of blood flow in the left PAA6. This results in the stabilization of the aortic arch on the left side at the expense of the right side. Segments of the dorsal aortas are degenerating, leading to the individualization of the future common carotid arteries (RCC, right common carotid and LCC, left common carotid). (D) At E14.5, the left PAA4 contributes to the segment of the aortic arch between the LCC and the left subclavian artery (in bronze), whereas the right PAA4 will form a segment that connects the RSA to the brachycephalic trunk (BT—remains of the right aortic arch). The BT is also connected to the derivative of the right PAA3, the right common carotid (RCC). The right PAA6 is not maintained, whereas the left PAA6 contributes to the *ductus arteriosus* (DA). The DA is an embryonic shunt that connects the (left) dorsal aorta to the pulmonary trunk. At birth, the DA closes, allowing the establishment of the pulmonary and systemic blood circulations (adapted from Kaufman and Bard (1999); Ao, aorta; PT, pulmonary trunk; T, trachea; PA, pulmonary arteries). (See Color Insert.)

Neural crest ablation not only affects septation, but also results in a reduction in outflow tract myocardium and consequent arterial pole defects (Waldo *et al.*, 2005a). The effect of neural crest on outflow tract development, demonstrated by experiments in the chick embryo, is also shown for the mouse, for example in the phenotype of mutants for *Tbx3*, expressed in neural crest (Mesbah *et al.*, 2008), as well as at other sites in the heart. Pax3 is a key regulator of neural crest and in the *Splotch*^{2H} mouse, where Pax3 function is affected, neural crest migration is reduced, again resulting in outflow tract defects, including ectopic myocardial differentiation with abnormal distribution of Islet1 positive cells of the SHF (Bradshaw *et al.*, 2009). Interactions between neural crest and the anterior SHF affect the behavior of both cell populations and their contributions to the heart (see Section 6.4).

4.3. The proepicardial organ

The PEO is a transitory structure, which forms as a group of cells close to the venous pole of the heart tube (Fig. 1.1). It has been thought to be derived from coelomic mesenchyme of the *septum transversum* (Männer *et al.*, 2001) and not from the SHF (Wessels and Pérez-Pomares, 2004); however, its relation to the SHF is not clear (see Section 5). In the early embryo this mesenchyme expresses *Islet1* (Ma *et al.*, 2008), so that the PEO and subsequent epicardium and coronary blood vessels are marked by *Islet1-Cre* genetic tracing (Moretti *et al.*, 2006; Sun *et al.*, 2007; Zhou *et al.*, 2008b). *Nkx2-5-Cre* tracing also marks these cells (Zhou *et al.*, 2008a, b) and, contrary to Islet1, Nkx2-5 is required for PEO development (Zhou *et al.*, 2008b). *Wt1* and *Tbx18* are expressed in the PEO and in the epicardium. *Wt1* mutant mice have coronary vascular defects and it has now been shown that this is due to a direct activation by Wt1 of the *Snail* gene, required for EMT of epicardial cells (Martínez-Estrada *et al.*, 2010). ES cells mutant for *Wt1*, fail to form cardiomyocytes, as well as other mesodermal derivatives, but this probably results from a failure of EMT required for embroid body "gastrulation."

The smooth muscle cells of the coronary vasculature arise from the epicardium and until recently it was thought that the endothelial cells also came from this source. However, genetic tracing experiments with fluorescent markers (Red-Horse *et al.*, 2010) now show that endothelial cells in the coronary vessels and capillaries derive from the venous plexus, at the *sinus venosus*, which invades the heart after formation of the epicardium. The question of whether the PEO can give rise to myocardium is a subject of debate. Myocardial differentiation of PEO-derived cells had been reported *in vitro* but only when levels of FGF versus BMP signaling were manipulated (Kruithof *et al.*, 2006; Van Wijk *et al.*, 2009). Previous fate mapping studies in the chick or mouse had not shown any proepicardial contribution to myocardium (Winter and Gittenberger-de Groot, 2007). Using a *Wt1* $^{GFP\text{-}Cre/+}$ line, the fate of PEO cells has been followed and, surprisingly, in addition to the expected derivatives, cardiomyocytes were detected in the walls of the cardiac chambers and in the interventricular septum (Zhou *et al.*, 2008a). Furthermore a subset of GFP-positive non-myocardial cells, isolated from foetal hearts, differentiated into myocardium in culture. In a parallel experiment, a *Tbx18-Cre* line was used to trace cells that had expressed this gene, and also resulted in the labeling of cardiomyocytes (Cai *et al.*, 2008). Labeling of the smooth muscle of coronary blood vessels and of cardiac fibroblasts was consistent with the observation that a subset of PEO cells express *Tbx18*. A caveat for all such genetic tracing experiments is that the interpretation depends on the expression of the *Cre* driver—whether this precisely reflects expression of the endogenous gene and whether the expression of the gene in question is restricted to the postulated progenitor source (see Christoffels *et al.*, 2009). In the case of *Tbx18*, there is later expression in the myocardium and notably in the interventricular septum, where many labeled cells were present in the genetic tracing experiment, which complicates the interpretation. However, as in the case of *Wt1*, dye-labeled epicardium resulted in labeled cardiomyocytes after culture, consistent with some myocardial contribution (Cai *et al.*, 2008; Zhou *et al.*, 2008a). The suggestion that PEO-derived cells can contribute to myocardium is important in a therapeutic context also, since the fibroblasts of the heart, which come from the epicardium, are a potential endogenous source of cardiac stem cells. Unlike the rare Islet1 positive cells present in the foetal heart (Laugwitz *et al.*, 2005), which are no longer present in the adult, the fibroblast population is maintained. An association between the epicardium and regeneration is demonstrated in the zebrafish heart where up-regulation of *Raldh2* in the epicardium, triggered by injury, leads to a retinoic acid activated cascade which results in the extensive cardiac repair that characterizes this organism (Lepilina *et al.*, 2006), although this is probably not due to direct formation of myocardium from epicardially derived cells (Jopling *et al.*, 2010; Kikuchi *et al.*, 2010).

5. Subdomains of the SHF

Genes that function within the genetic network of the SHF (Fig. 1.2) are not all expressed throughout this heart field. Cellular level resolution is lacking and even *Islet1* may not be expressed in all progenitors. The results of *Cre* tracing experiments tend to be interpreted in terms of onset of expression of the *Cre* driver, but heterogeneity between cells in the SHF may also affect the results and this is, of course, also the case for mutant phenotypes. If information on expression/co-expression of SHF genes at the cellular level is still lacking, there is increasing evidence for the existence of subdomains of the SHF, distinguishable by gene expression and by their contribution to the heart (Fig. 1.4).

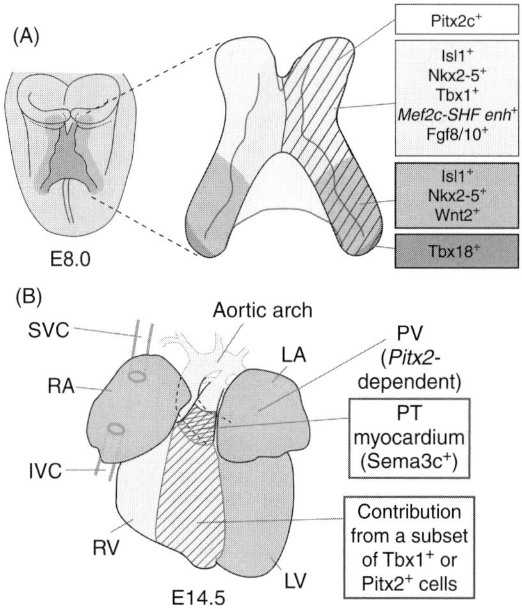

Figure 1.4 Subdomains within the SHF. (A) The SHF is characterized by the expression of Nkx2-5 and Islet1 shown at mouse embryonic day (E) 8.0. The anterior part of the SHF (light grey) is composed of cells that also express Tbx1, Fgf8, or Fgf10, and activate the *Mef2c*-SHF enhancer, whereas Wnt2 is expressed in addition to Nkx2-5 and Islet1 in the posterior SHF (mid-grey). At the most posterior and lateral side of the SHF resides a domain that is characterized by the expression of Tbx18 but not Nkx2-5 (dark grey). Pitx2c is expressed only on the left side of the SHF (diagonal stripes). (B) These different subdomains contribute to specific domains within the four-chambered heart shown at E14.5. Atrial myocardium contains cells that have expressed Islet1 and Nkx2-5 from the posterior SHF (mid-grey). The myocardium of the caval veins (IVC and SVC) is derived from cells that had expressed *Tbx18* (dark grey), whereas pulmonary vein (PV) myocardium is derived from cells of the posterior SHF (mid-grey) that also expressed Pitx2. Cells of the anterior part of the SHF contribute to the right ventricle (RV) and outflow tract myocardium. Interestingly, a subset of Tbx1$^+$ and Pitx2$^+$ cells contribute to a more centrally located region of the RV (stipled stripes), encompassing the interventricular septum domain, including the myocardium at the base of pulmonary trunk (PT) (darker diagonal stripes). This pulmonary trunk (PT) myocardium is characterized by the expression of *Sema3c* and is constituted from SHF progenitors that had responded to Shh. In *Tbx1* mutants, this domain is affected (see Section 6.3). RA, LA; right and left atria, respectively; RV, LV; right and left ventricles, respectively.

5.1. The anterior SHF and contributions to the arterial pole of the heart

The anterior part of the SHF is marked by expression of *Fgf8, Fgf10* (Kelly *et al.*, 2001), and *Tbx1* (Xu *et al.*, 2004) and cells that have transcribed these genes form the arterial pole of the heart (Fig. 1.4). The *Mef2c SHF* enhancer also functions here (Dodou *et al.*, 2004). Within this anterior region, subpopulations can be distinguished. Tbx1 controls the addition of cells that will constitute pulmonary trunk myocardium at the outlet of the right ventricle (Maeda *et al.*, 2006) and this is severely reduced in *Tbx1* mutant hearts (Théveniau-Ruissy *et al.*, 2008) where the SHF contribution to smooth muscle is also affected (Chen *et al.*, 2009). In human congenital heart disease, outflow tract alignment defects, such as tetralogy of Fallot, are frequent and probably result from a deficit in sub-pulmonary myocardium (Van Praagh, 2009). Further evidence that different progenitor cell populations form sub-pulmonary versus sub-aortic myocardium comes from retrospective clonal analysis and transgene expression profiles, which distinguish superior/inferior regions of the outflow tract and myocardium at the base of each artery (Bajolle *et al.*, 2008). The transgenes that mark different SHF domains and arterial pole derivatives reflect insertion site effects now shown to correspond to the *Sema3c* gene that marks sub-pulmonary myocardium (Théveniau-Ruissy *et al.*, 2008) and to the *Hes1* gene, also expressed in sub-aortic myocardium. *Hes1* mutants have outflow tract defects, leading to over-riding aorta and ventricular septal defects (Rochais *et al.*, 2009; Van Bueren *et al.*, 2010). It has been suggested, as a result of experiments in the chick embryo, that the SHF contribution to smooth muscle, and probably also to myocardium at the base of the pulmonary trunk, spirals into the outflow tract from the right part of the SHF (Ward *et al.*, 2005). However, genetic tracing experiments in the mouse (see Section 6) suggest that cells that had expressed *Pitx2*, that marks the left side of the SHF, contribute to pulmonary trunk myocardium (Ai *et al.*, 2006). Interestingly cells that have expressed both *Pitx2* and *Tbx1* (Huynh *et al.*, 2007; Maeda *et al.*, 2006) also contribute to a central subdomain of the right ventricle extending into the interventricular septal region. The pulmonary trunk is further distinguished in fate mapping experiments which show it is derived from SHF cells that had received a sonic hedgehog (Shh) signal (Hoffmann *et al.*, 2009).

The anterior part of the SHF extends into the mesodermal core of the pharyngeal arches which gives rise to arterial pole myocardium or endothelial cells of the pharyngeal arch arteries (Section 4.1). In addition to myocardial versus endothelial cells derived from the mesodermal core of anterior compared to posterior pharyngeal arches, there are differences between progenitor populations that assume the same cell fate, but derive from different arches. This probably reflects differences in the pharyngeal environment which in turn reflects the progressive development of the pharyngeal arches on the anterior/posterior axis. Thus, for example, the arterial derivatives of the fourth pharyngeal arch are particularly subject to interruption or hypoplasia. In mutants in which *Tbx1* is deleted in mesoderm, with consequent loss of pharyngeal arches 3–6 and failure of pharyngeal arch artery development, restoration of mesodermal *Tbx1* expression rescues most of these defects, but not the fourth pharyngeal arch phenotype (Zhang *et al.*, 2006), now shown to be due to *Gbx2* regulation by Tbx1 in the pharyngeal ectoderm (Calmont *et al.*, 2009). Gain of function *Tbx1* expression also specifically affects the fourth pharyngeal arch arteries as well as pulmonary trunk myocardium (Vitelli *et al.*, 2009). The mesodermal core of the anterior pharyngeal arches (1–3) not only contributes to arterial pole myocardium (1–2) or endothelial cells of pharyngeal arch arteries (3), but also to skeletal muscles of the head (Noden and Francis-West, 2006). These contributions have now been mapped for the first two arches in the chick embryo and an overlapping gradient of gene expression for skeletal myogenic versus SHF markers demonstrated, following a proximal/distal gradient within the core mesoderm (Nathan *et al.*, 2008; Tirosh-Finkel *et al.*, 2006). Genetic tracing experiments in the mouse indicate that cells that had expressed *Islet1*, for example, populate a subset of head muscles as well as giving rise to myocardium (Harel *et al.*, 2009; Nathan *et al.*, 2008). Again these results point to subdomains of the SHF, with different potential cell fates.

5.2. The posterior SHF and formation of the venous pole

Cells that contribute to the atria, at the venous pole of the heart, express Islet1, but not the anterior SHF markers (Cai *et al.*, 2003; Galli *et al.*, 2008) (Fig. 1.4). This posterior contribution of Islet1 positive cells to the atria and atrio-ventricular canal depends on Wnt2 signaling leading to *Gata6* activation in a feed-forward regulatory loop that is specific to this domain (Tian *et al.*, 2010). Furthermore, explant experiments suggest that this region of the SHF is programmed to assume an atrial fate (Galli *et al.*, 2008), whereas explants from the anterior region form right ventricular or outflow tract myocardium (Zaffran *et al.*, 2004). In the posterior domain, there is also evidence of regional heterogeneity, during the complex development of the venous pole (Anderson *et al.*, 2006). The dorsal mesenchymal protrusion (or *spina vestibuli*) is a morphologically distinct structure that contributes to the atrio-ventricular septum and undergoes a later myocardial transition (Mommersteeg *et al.*, 2006; Snarr *et al.*, 2007b). The mesodermal cells that form this structure are Islet1 positive and the dorsal mesenchymal protrusion, formed from the dorsal mesocardium is thought to be a SHF derivative (Meilhac *et al.*, 2004; Snarr *et al.*, 2007a). Its formation is particularly dependent on Shh signaling (Goddeeris *et al.*, 2008) (see Section 6.6).

As the venous pole develops, after formation of atrial myocardium (by E9.5), the *sinus venosus* develops with formation of myocardium around the caval veins. This is marked by the expression of *Tbx18*, and the absence of expression of *Nkx2-5*, both in the surrounding mesenchyme of the SHF, as well as the caval vein myocardium (Christoffels *et al.*, 2006). Tbx18 is required for the correct development of this part of the *sinus venosus* into caval vein myocardium. Explant experiments now confirm that these Tbx18 positive progenitor cells, that are Nkx2-5, and also Islet1, negative, form Nkx2-5 negative myocardium. In close proximity to these progenitors, Tbx18 negative, Islet1 positive cells contribute to pulmonary vein myocardium, with overlapping expression only at the lateral most border of these domains. Tbx18 positive cells are first detected caudal/lateral to the cardiac crescent (Fig. 1.1) and cell, as well as genetic, tracing experiments show that they subsequently contribute to the venous pole. Labeled cells were also seen in the PEO (Section 4.3). Interestingly, cells in the Tbx18 positive domain of cardiogenic mesoderm had initially expressed *Islet1* and *Nkx2-5*. The question, therefore, arises of whether this constitutes another heart field (Christoffels *et al.*, 2006; Mommersteeg *et al.*, 2010) and also of the relation with the PEO. Formal cell lineage studies will show whether there is an additional early lineage segregation, as seen for the first and second myocardial cell lineages (Meilhac *et al.*, 2004). However, the Tbx18 positive population clearly represents a distinct domain which contributes to a specific part of the venous pole and is characterized by the presence of a different transcriptional network from most of the posterior SHF. Recent results in the chick embryo suggest that *Tbx18* expressing cells of the PEO and subdomain of the second/third heart field derive from a common precursor pool and demonstrate how segregation to epicardial or myocardial lineages is promoted by FGF or BMP signaling, respectively, both in explants and *in vivo* (Van Wijk *et al.*, 2009).

6. Molecular Mechanisms that Govern SHF Cell Behavior–Transcriptional Regulators and Signaling Pathways

Signaling pathways and their interaction with transcriptional regulators underlie the patterning of the SHF, in terms of both its extent and its subdomains. The contribution and future cardiac identity of progenitor cells depends on the signals to which they are exposed. In the SHF, cells are maintained as proliferating, non-differentiated progenitors until they enter the heart tube. Indications of how regulatory networks (Fig. 1.2) and, particularly, signaling pathways affect SHF cell behavior are beginning to emerge (Fig. 1.5).

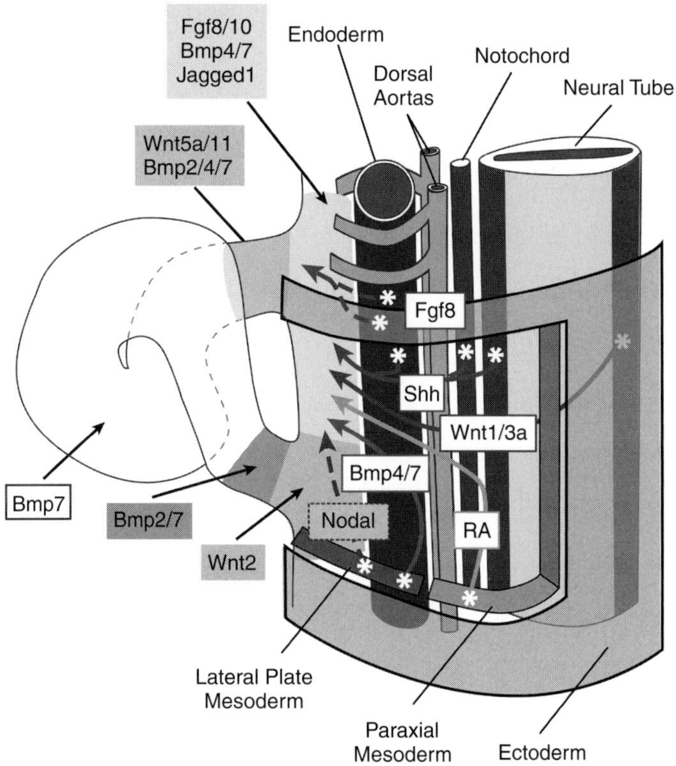

Figure 1.5 The SHF is a target of multiple signals. Schematic representation of a left-sided view of the heart region. The SHF itself is a source of signals: the Notch ligand Jagged1, is involved in the activation of *Fgf8*, but Notch signaling also interferes with canonical Wnt signaling, inhibiting progenitor proliferation. Fgf8 and Fgf10 are important for the promotion of SHF proliferation, whereas Bmp4/7 is required for the survival of the cNCCs. Bmp2, which is repressed by Nkx2-5 and also by canonical Wnt signaling, inhibits SHF proliferation. In the posterior SHF, Wnt2 controls the contribution of venous pole progenitors. Fgf8 also comes from the pharyngeal endoderm and ectoderm (grey arrow). The outflow tract is also a source of signals for the SHF: Bmp2 has been shown to be required for the deployment of SHF cells. Non-canonical Wnt11 and Wnt5a do not affect the SHF, but control outflow tract maturation. The midline is another signaling source. Shh (purple arrow), from the endoderm (in dark grey) (but also from the notochord and the floorplate, in blue), affects arterial pole formation, probably through an effect on SHF proliferation, and is required for the formation of the atrial septum at the venous pole. Also coming from the midline, canonical Wnts (Wnt1, Wnt3a, green arrow) from the dorsal neural tube (green domain) are important for the maintenance of proliferating progenitors within the SHF and inhibition of differentiation. On the contrary, lateral signals such as Bmp4 (red arrow) from the lateral plate mesoderm (red domain) promote cardiac differentiation. The paraxial mesoderm (posterior to the SHF, in brown) is a source of retinoic acid (RA, brown arrow) that is important in the early stages of cardiogenesis for anterior/posterior patterning and limiting the posterior boundary of the SHF. On the left side, and only between E8 and E8.5, Nodal (dashed blue arrow) activates *Pitx2c* asymmetrically on the left side of the SHF. Asymmetric *Pitx2c* expression is maintained independently of Nodal and affects both the arterial and venous poles of the heart (see Section 6.2). (See Color Insert.)

6.1. Anterior/posterior patterning of the SHF: retinoic acid signaling

The patterning of the SHF on the anterior/posterior axis depends on retinoic acid signaling, as indicated by the abnormal posterior expansion in expression of anterior SHF marker genes, such as *Tbx1, Fgf8*, and *Fgf10* in *Raldh2* mutant embryos (Ryckebusch *et al.*, 2008; Sirbu *et al.*, 2008). In avian embryos, vitamin A deficiency, or over-expression, of retinoic acid also up- or down-regulates *Tbx1* expression, respectively (Roberts *et al.*, 2006). Reduction in retinoic acid synthesis results in many of the phenotypic features of DiGeorge syndrome at the arterial pole of the heart, in which *TBX1* is implicated, and analysis of compound *Raldh2/Tbx1* mouse mutants indicates that decrease in levels of retinoic acid accelerates the recovery from arterial growth delay seen in *Tbx1*$^{-/-}$ embryos (Ryckebüsch *et al.*, 2010; Vermot *et al.*, 2003). This genetic interaction also operates in *Tbx1* mutant embryos in which the *Raldh2* expression domain is moved anteriorly and enzymes that degrade retinoic acid are down-regulated (Guris *et al.*, 2006; Ivins *et al.*, 2005; Liao *et al.*, 2008; Roberts *et al.*, 2006). In *Raldh2* mutants the posterior domain of *Islet1* expression in the SHF is also abnormally extended caudally (Ryckebusch *et al.*, 2008; Sirbu *et al.*, 2008), indicating that not only the patterning, but also the limits of the SHF are affected by retinoic acid.

In the zebrafish embryo, retinoic acid has been shown to control the size of the cardiac field (Keegan *et al.*, 2005). Retinoic acid levels regulate anterior/posterior expression of genes in the *Hox* clusters, which are potential effectors of retinoic acid signaling in the SHF. In zebrafish, this scenario is complex since indirect effects of Hox5b, acting in the forelimb field, downstream of retinoic acid signaling, may control signals that affect the number of atrial progenitors (Waxman *et al.*, 2008). In the mouse *Raldh2* mutant, despite the expansion of SHF marker expression, the heart tube fails to grow which may reflect the fact that although more cells express SHF genes, the level of expression is reduced (Ryckebusch *et al.*, 2008). A related phenomenon is seen in retinoic acid receptor mutants where the distal outflow tract fails to form. This late phenotype is associated with a marked reduction in *Mef2c* expression and genetic tracing with the *Mef2c SHF* enhancer (*Cre*) shows that this subpopulation of cells is strongly reduced and does not enter the heart, whereas *Islet1* and *Fgf8* expression in the SHF appears normal (Li *et al.*, 2010). This effect of interference in retinoic acid signaling on activation of the *Mef2c SHF* enhancer, probably mediated through Gata4, with failure to form the distal part of the arterial pole, provides another example of regulatory subdomains of the SHF (see Section 5).

6.2. Left/right patterning: Nodal signaling through Pitx2c

The SHF is also patterned on the left/right axis by the Nodal signaling pathway which leads to the activation of *Pitx2c* in the left side of the SHF. This transcription factor also has a dynamic expression pattern in the heart itself, where it is important for cardiac remodeling during development (Tessari *et al.*, 2008). Interference with this left/right signaling has a striking effect on cardiac morphogenesis, and asymmetry, exemplified by right atrial isomerism (see Franco and Campione, 2003; Tessari *et al.*, 2008), but interestingly *Pitx2* mutants still undergo cardiac looping and indeed signals that drive this striking aspect of asymmetric development remain poorly understood. In the posterior domain of the left SHF, where it is strongly expressed, Pitx2c functions to repress the acquisition of right atrial identity, followed with a transgenic marker in explant cultures, as well as *in vivo* (Galli *et al.*, 2008). Pitx2c also represses proliferation in the left *sinus venosus,* but not detectably in the left SHF. It is not clear whether these are direct effects, since the function of Pitx2 as a transcriptional activator or repressor, and its transcriptional targets, are poorly understood at present. At later stages in the posterior SHF, Pitx2 is required to initiate the formation of pulmonary vein myocardium, but the identity of this myocardium appears to depend on Nkx2-5, since in *Nkx2-5* mutants markers of caval vein myocardium begin to be expressed in pulmonary vein myocardium (Mommersteeg *et al.*, 2007). Here, too, the molecular role of Pitx2 and the nature of its interaction with Nkx2-5 are obscure.

In addition to its role at the venous pole, Pitx2c also plays a role in outflow tract development. By genetic lineage tracing and cell type specific conditional mutation, it is now clear that this is not due to a primary effect on neural crest cells, but due to the function of Pitx2c in the left anterior SHF and its derivatives (Ai *et al.*, 2006). Once the outflow tract is formed in the mouse embryo, it undergoes rotation, essential for the final juxtaposition of the great arteries, and this depends on Pitx2c (Bajolle *et al.*, 2006). Spiraling of the outflow tract structure and consequent rotation of the aortic sac, which affects the initially symmetrical blood flow, has been further investigated in *Pitx2* mutants, where Pitx2 indirectly induced asymmetric signaling, through PDGF and VEGF2 receptors, points to downstream mechanisms (Yashiro *et al.*, 2007). Genetic interaction between *Pitx2* and *Tbx1* has been demonstrated, with the suggestion that in the anterior SHF, *Pitx2* expression requires *Nkx2-5* and *Tbx1*, transitorily expressed on the left side (Nowotschin *et al.*, 2006). Foxh1, another transcription factor implicated in the development of the anterior SHF (von Both *et al.*, 2004), functions in the Nodal signaling pathway. It has now been shown that mutations in *FOXH1* and other mutants of this pathway are linked to arterial pole defects in the human population (Roessler *et al.*, 2008).

Sonic hedgehog signaling is involved in the establishment of left/right asymmetry and characterization of cardiac defects in the *Shh* mutant mouse shows a phenotype at the arterial pole similar to tetralogy of Fallot, with complete pulmonary trunk atresia (Washington Smoak *et al.*, 2005), which may result from an asymmetric effect on the SHF. Furthermore, left atrial isomerism, with bilateral *Pitx2* expression, is seen in the *Shh* mutant embryo, in which the heart remains attached to the dorsal mesocardium (Hildreth *et al.*, 2009).

6.3. Maintenance of proliferation in the SHF: FGF, hedgehog, and canonical Wnt signaling

A number of signaling pathways that impact the SHF have been shown to promote proliferation. This is the case for FGF signaling. Fgf8 is produced by the mesodermal cells of the SHF, and also by endoderm, and ectoderm of the pharyngeal arches. *Fgf8* mutants do not gastrulate, but *Fgf8* hypomorphs have demonstrated its importance in arterial pole development, with defects that ressemble those in *Tbx1* mutants (Abu-Issa *et al.*, 2002; Frank *et al.*, 2002). Mutation of *Fgf8* in the *Tbx1* expressing domains of the SHF and endoderm/ectoderm of the arches results in phenotypes that suggest that *Fgf8* lies downstream of *Tbx1* and indeed a regulatory element on the *Fgf8* locus is directly activated by Tbx1 (Hu *et al.*, 2004). However, targeting of an *Fgf8* coding sequence to the *Tbx1* gene shows that, whereas Fgf8 can replace Tbx1 for many aspects of pharyngeal arch artery formation (Brown *et al.*, 2004), Fgf8 and Tbx1 play independent roles in outflow tract development. Notably Tbx1, but not Fgf8, is crucial in a *Hoxa3* expressing domain of the pharynx which includes the mesoderm of the anterior SHF (Vitelli *et al.*, 2006). Conditional mutants, in which *Fgf8* deletion has been targeted to specific expression domains, clarify its role in pharyngeal arch artery and in outflow tract development (Ilagan *et al.*, 2006; Macatee *et al.*, 2003; Park *et al.*, 2006). In addition to effects on myocardial derivatives, targeting *Fgf8* deletion to *Tbx1* expressing cells with a *Tbx1-Cre* shows its role in the SHF cells that will form smooth muscle (Brown *et al.*, 2004). Mesodermal *Fgf8* expression is clearly important for SHF development. *Fgf10* is also expressed in the SHF, but *Fgf10* mutants do not demonstrate an SHF phenotype (Marguerie *et al.*, 2006). However, mesodermal *Fgf8*/*Fgf10* compound mutants display increasingly severe pharyngeal artery and outflow tract defects, as alleles of *Fgf8* and *Fgf10* are removed. This shows that the SHF contribution to the arterial pole, which depends on maintenance of proliferation is very sensitive to FGF dosage (Watanabe *et al.*, 2010). Conditional deletion in cardiac mesoderm of *Fgfr1* or *Fgfr2*, which encode the principal receptors of Fgf8 and Fgf10, or of *Frs2α* which encodes an adaptor protein that links FGFR to MAPK and P13K signaling cascades within the cell, as well as conditional over-expression of Sprouty that interferes with this intracellular signaling, demonstrates an autocrine requirement for FGF signaling in the SHF (Park *et al.*, 2008; Zhang *et al.*,

2008). In these different genetic situations in the mouse embryo, and in chick, FGF signaling is required for SHF proliferation. In zebrafish embryos, a reduction in FGF signaling reduces the number of cardiomyocytes, particularly in the ventricle (where Fgf8 is expressed), and an ectopic increase in FGF increases the number of ventricular and atrial cardiomyocytes prior to cardiac tube differentiation (Marques *et al.*, 2008). In *Xenopus*, FGF from anterior neural ectoderm increases the extent of *Nkx2-5* expression in mesoderm, rendering it cardiogenic (Keren-Politansky *et al.*, 2009). In these models, FGF signaling, probably through its effects on proliferation, increases the extent of the heart field and consequent numbers of cardiomyocytes.

Hedgehog (Hh) signaling is also often associated with proliferation. A number of defects at the venous as well as at the arterial pole have been documented when Shh signaling from endoderm is abrogated (see Section 6.6; Goddeeris *et al.*, 2007; Hoffmann *et al.*, 2009). In the mouse model it is not clear that this is due to an effect on SHF proliferation; however, in the chick embryo, Shh signaling is clearly important for maintaining progenitor cell proliferation in the critical time frame which precedes addition of cells to the heart tube (Dyer and Kirby, 2009).

Canonical Wnt signaling, in addition to playing earlier roles in mesoderm development and negative modulation of cardiac specification, also promotes progenitor cell proliferation, as evidenced by work with ES cultures and embryo model systems (see Cohen *et al.*, 2008). Wnt/β-catenin signaling is active in the SHF and specific deletion of the gene encoding β-catenin in cardiac mesoderm leads to right ventricular and outflow tract hypoplasia, probably due to a reduction in SHF proliferation (Ai *et al.*, 2007; Cohen *et al.*, 2007; Klaus *et al.*, 2007; Kwon *et al.*, 2007; Lin *et al.*, 2007; Qyang *et al.*, 2007). SHF marker expression is reduced, including transcripts of *Islet1, Fgf10*, and *Shh*, suggesting an upstream role for canonical Wnt signaling, although this also reflects the reduction in cell number. β-Catenin can directly activate transcription from *Islet1* and *Fgf10* promoters (Cohen *et al.*, 2007; Lin *et al.*, 2007), but the significance of this is not demonstrated *in vivo*. Complementary gain of function experiments, by LiCl treatment and conditional expression of stabilized β-catenin, results in expansion of SHF progenitors. Transcriptome analysis of these Islet1 positive cells shows up-regulation of *Fgfs*, which will promote proliferation, when canonical Wnt signaling is increased (Kwon *et al.*, 2009).

A caveat for genetic experiments that depend on manipulation of β-catenin is that this has a second function in cell adhesion, which may also influence the SHF phenotypes. Of Wnt ligands potentially involved in autocrine signaling within the SHF, only Wnt2 expressed in the posterior SHF has been shown, in concert with β-catenin, to directly regulate proliferation of venous pole progenitors (Tian *et al.*, 2010).

A link between Notch signaling and the canonical Wnt/β-catenin pathway has been demonstrated (Kwon *et al.*, 2009). Conditional deletion of *Notch1*, with an *Islet1-Cre* line, promoted proliferation of Islet1 positive progenitors in the SHF at the expense of cardiogenesis, with absence of the arterial pole, including the right ventricle, as had been seen when β-catenin is over-expressed (Cohen *et al.*, 2007). In the absence of Notch1, higher levels of stabilized β-catenin and increased Wnt/β-catenin signaling were observed, suggesting that Notch signaling normally represses cardiac progenitor proliferation by negatively regulating the active form of β-catenin (Kwon *et al.*, 2009).

6.4. Interactions with neural crest in the SHF: FGF, Notch, and semaphorin signaling

FGF signaling in the SHF was thought to directly affect neural crest. However, targeted deletion, with a neural crest (*Pax3*)*Cre* line, of conditional mutants for the *Fgfr1* and *Fgfr2* genes or for *Frs2α* which encodes an FGFR adaptor protein, does not result in any arterial pole defects, in contrast to deletion in the SHF. This is also the case when Sprouty, the intracellular inhibitor of FGF signaling is up-regulated in neural crest cells (Park *et al.*, 2008; Zhang *et al.*, 2008). In mesodermal *Fgf8* mutants, BMP/TGFβ signaling is down-regulated in the SHF and it is proposed that this pathway normally provides a relay that then affects neural crest (Park *et al.*, 2008), which requires Smad4 mediated signaling for survival (Nie *et al.*, 2008). In addition to Fgf8 and Fgf10, Fgf15 is also implicated in outflow tract development. It is expressed, independently of Tbx1, in pharyngeal arch mesoderm (arch 3) and in the anterior SHF. In *Fgf15* mutants, outflow tract morphogenesis is abnormal, and this is probably at least partly due to effects on neural crest (Vincentz *et al.*, 2005). Neural crest also feeds back on FGF signaling in the SHF, since in chick embryos neural crest ablation results in an increase in Fgf8 which perturbs SHF development and this can be rescued when the level of Fgf8 is reduced (Hutson *et al.*, 2006). An increase in FGF signaling might be expected to increase the cardiac progenitor pool at the expense of differentiation; however, in *Splotch*^{2H} mutant embryos, where neural crest migration is reduced, ectopic myocardial differentiation is observed (Bradshaw *et al.*, 2009), suggesting perturbation of additional SHF regulators in this case.

Interference with Notch signaling in neural crest cells results in arterial pole and arch artery phenotypes (High *et al.*, 2008, 2007; Varadkar *et al.*, 2008). Manipulation, using *Islet1-* or *Mef2-Cre* lines of a dominant-negative form of Mastermind-like (MAML), a co-activator, of the Notch transcriptional complex, interferes with Notch signaling in the SHF and results in arterial pole defects (High *et al.*, 2009), probably also partly due to effects on neural crest. Similar effects are seen when the gene encoding for Jagged1 is conditionally deleted, indicating that this is the principal Notch ligand in the SHF. Jagged1 is also produced by endothelial derivatives of the SHF, with important consequences for neural crest derived smooth muscle differentiation (High and Epstein, 2008). Interference with Notch signaling correlates with a decrease in BMP signaling, which in turn can be explained (Park *et al.*, 2008) by an observed down-regulation of *Fgf8* in the pharyngeal region (High *et al.*, 2009). Changes in the expression of other SHF marker genes, such as *Islet1, Fgf10*, or *Tbx1*, are not observed suggesting that this is a specific effect of Notch signaling acting upstream of *Fgf8*. Thus in addition to a direct effect of Notch signaling in the SHF on neural crest, there is also probably an indirect effect through Fgf8 which then affects BMP signaling.

Semaphorin signaling, functioning through Plexin receptors, is important for neural crest migration. Semaphorin3C which is expressed in a subdomain of the SHF and marks pulmonary trunk myocardium (Théveniau-Ruissy *et al.*, 2008), is required for normal neural crest migration into the arterial pole of the heart (Brown *et al.*, 2001; Feiner *et al.*, 2001). *Sema3C* is directly regulated by Gata6 in smooth muscle derived from neural crest, where it is also expressed (Lepore *et al.*, 2006); *GATA6* mutations in humans have been linked to arterial pole malformations (Kodo *et al.*, 2009). In the mouse embryo, PlexinD1, which interacts with Semaphorin3C, and PlexinA2, both expressed in cardiac neural crest, are required for correct outflow tract development (Toyofuku *et al.*, 2008). PlexinD1 is also expressed in endocardial and endothelial cells (Gitler *et al.*, 2004) and its expression in these Tie2 positive cells is essential for arterial pole development (Zhang *et al.*, 2009).

6.5. Prevention of differentiation in the SHF: canonical Wnt signaling and transcriptional repression

In addition to its role in promoting proliferation in the SHF there is evidence that canonical Wnt signaling also prevents the onset of differentiation. This was suggested by experiments in the *Xenopus* embryo where cardiac differentiation is negatively affected, with inhibition of Gata4/6 and Nkx2-5 expression (Lavery *et al.*, 2008). In experiments in which canonical Wnt signaling is increased by conditional expression of stabilized β-catenin (Kwon *et al.*, 2009), maintenance of Islet1 positive cells in the outflow tract would be in keeping with a delay in cardiac differentiation. Further analysis of genes expressed by Islet1 positive cells in which β-catenin is stabilized shows down-regulation of the gene encoding Myocardin which, together with SRF, promotes myocardial and smooth muscle differentiation, and also of the repressor Bhlhb2, shown to be a direct Wnt/β-catenin target (Fig. 1.2). However, surprisingly, down-regulation of the level of *Islet1* expression was also observed. Because of its presence in the SHF, Islet1 had been associated with proliferation; however, it may, at least when expressed at a high level, promote differentiation. Manipulation of Islet1 in the ES cell system suggests that it can promote myocardial differentiation, and furthermore Islet1 directly activates a regulatory element of the *Myocardin* gene. This may explain why maintenance of Islet1 in cardiomyocytes in the *Nkx2-5* mutant (Prall *et al.*, 2007) is compatible with differentiation. The importance of differences in levels of Islet1 in determining its effects on differentiation or proliferation is further supported by manipulation of stabilized β-catenin versus Islet1 in ES-derived cardiac progenitors, leading to the conclusion that a low level of Islet1 is required for Wnt/β-catenin mediated proliferation of these cells (Kwon *et al.*, 2009).

It is possible that differences in function, according to the level of expression, may also hold for factors like Nkx2-5 or Mef2c, present both in the SHF and in the heart where they play a role in activating myocardial genes. In this context, modulation or prevention of expression of genes encoding factors that can promote differentiation will be critical for maintaining the progenitor cell pool. Tbx1 represses *SRF* and *Tbx5* expression in the SHF, since these genes, which are implicated in myocardial differentiation, are up-regulated in *Tbx1* mutants (Liao *et al.*, 2008).

6.6. Regulation of SHF differentiation potential, recruitment to the heart tube, and differentiation: Shh, Notch, BMP, and non-canonical Wnt signaling

A number of signaling pathways affect subdomains of the SHF (Section 5). Whereas they may primarily exert an effect on proliferation, as in the case of FGF signaling in the anterior SHF, in other cases it is beginning to be evident that they exert effects on myocardial differentiation potential, in terms of the SHF contribution to a region of the heart. At the arterial pole, Shh from the endoderm affects pharyngeal arch mesoderm and *Tbx1* expression (Yamagishi *et al.*, 2003), with additional effects on the maintenance and deployment of neural crest (Goddeeris *et al.*, 2007; Washington Smoak *et al.*, 2005). In the absence of Shh, pharyngeal vasculature as well as outflow tract development is affected (Kolesová *et al.*, 2008). The effect of Shh on *Tbx1* expression has particular implications for the formation of pulmnary trunk myocardium (Théveniau-Ruissy *et al.*, 2008). In the chick embryo, manipulation of Shh signaling affects migration of SHF cells into the outflow tract (Dyer and Kirby, 2009) suggesting that it is also important for SHF recruitment. In the absence of Shh signaling in the posterior SHF, there are specific venous pole defects. The source of the signal is the underlying pulmonary endoderm, where Islet1 is also present and is required for Shh expression (Lin *et al.*, 2006). Development of the dorsal mesocardial protrusion and subsequent formation of the primary atrial septum depend on this signaling pathway (Goddeeris *et al.*, 2008; Hoffmann *et al.*, 2009). Observations on the dorsal myocardium suggest that Shh plays a role in the specification of atrial septal precursors and in the recruitment of cells from this part of the SHF. A role for Shh in cardiac cell specification is in keeping with observations in P19 embryonic carcinoma cells. In the absence of primary cilia, which are intimately related to Hh signaling, myocardial derivatives are compromised in these cells. Furthermore mouse mutants that lack cilia have cardiac abnormalities (Clement *et al.*, 2009; Slough *et al.*, 2008). During cardiogenesis in the zebrafish embryo, Hh signaling promotes cardiomyocyte formation as well as regulating the number of cardiac progenitor cells (Thomas *et al.*, 2008), pointing to a dual role.

Notch signaling is an important regulator of many developmental processes, with a spectrum of effects on cell behavior. Modification of this pathway in the heart leads to cardiac phenotypes which reflect the variety of its roles (High and Epstein, 2008). Components of the pathway are present in the SHF. Among Notch functions are regulation of cell fate choices and of the decision to differentiate rather than proliferate. In cells of the anterior SHF, a role in myocardial versus smooth muscle fates, for example, has not yet been demonstrated. However, effects on myocardialization of the outflow tract have been reported after interference with Notch signaling, by manipulation in the SHF of a dominant-negative form of MAML (see Section 6.4) (High *et al.*, 2009). MAML can interact with Mef2c (Shen *et al.*, 2006), which may complicate the interpretation. However, similar AP defects are seen when the gene for *Jagged1* is conditionally deleted, supporting the role of Notch in promoting differentiation.

BMP signaling promotes cardiac specification and myocardial differentiation, as established in classic experiments on the chick embryo (see Schultheiss *et al.*, 1997), and also more recently for mesoderm in the core of the anterior pharyngeal arches which can form cardiac or skeletal muscle (Tirosh-Finkel *et al.*, 2006). The overall picture (Fig. 1.5) in which SHF progenitors lie medially to the differentiating cardiac crescent, at a distance from sources of BMP emanating from lateral mesoderm, is consistent with their maintenance in an undifferentiated state. Medially produced signals, such as canonical Wnts from the neural tube, will promote proliferation (Section 6.1). Mouse mutant phenotypes demonstrate the importance of BMP signaling for heart development. Deletion of *Bmpr1a*, encoding the BMP type 1 receptor, in early cardiac mesoderm results in failure to form a differentiating cardiac tube, although Islet1 positive cardiac progenitors are present (Klaus *et al.*, 2007). Later conditional deletion in *Islet1* expressing cells results in an abnormal right ventricle and outflow tract where increased numbers of Islet1 positive cells suggest a differentiation defect, since *Islet1* is normally down-regulated in cardiomyocytes. This is accompanied by a reduction in Tbx20, required to repress *Islet1* in the outflow tract (Yang *et al.*, 2006). Bmp4 is required for outflow tract development (McCulley *et al.*, 2008), where it probably affects myocardium formation, as well as the survival of neural crest (Nie *et al.*, 2008). Outflow tract elongation is reduced in embryos lacking both Bmp4 and Bmp7 (Liu *et al.*, 2004) and Bmp2, acting with Bmp4, is also implicated in cardiac development (Uchimura *et al.*, 2009). Bmp2 and Bmp4, expressed by the outflow tract myocardium, are candidate molecules for inducing differentiation of SHF progenitors at the arterial pole of the heart tube and it has also been proposed, from experiments in the chick embryo, that they play an additional role in the recruitment of these cells (Somi *et al.*, 2004; Waldo *et al.*, 2001).

Bmp4, together with other genes for BMP/TGFβ family members, are targets of FGF signaling in the SHF (Park *et al.*, 2008), where Wnt/β-catenin signaling also up-regulates *Bmp4* (Ai *et al.*, 2007) (Fig. 1.2). This might, at first sight, seem contradictory, since these signaling pathways promote proliferation in the SHF where maintenance of an undifferentiated state is important. However, the levels of expression are probably critical and different signaling thresholds may be required for an effect on proliferation or myocardial differentiation. A balance between FGF/BMP had been shown to be important in this context (Barron *et al.*, 2000). The importance of regulating the level of TGFβ/BMP signaling within the SHF is demonstrated by the role of Nkx2-5 which represses *Bmp2* expression. In the absence of Nkx2-5, there is cardiac over-specification and a reduction in proliferating SHF progenitors leading to outflow tract truncation. The SHF effect of Bmp2 is mediated by Smad1. In mesodermal *Smad1* mutants SHF proliferation and outflow tract length are increased; furthermore the *Nkx2-5* phenotype is alleviated on loss of *Smad1* alleles, demonstrating the presence of an Nkx2-5/Bmp2/Smad1 negative feedback loop that controls progenitor specification and proliferation in the SHF (Prall *et al.*, 2007). Tbx1 is another SHF transcriptional regulator that can affect BMP/TGFβ signaling, in this case by direct interaction with Smad1 (Fulcoli *et al.*, 2009).

Non-canonical Wnt signaling is associated with cardiac differentiation, as shown when Wnt5a, in combination with the canonical Wnt inhibitor Dkk-1, is added to stromal vascular cells (Palpant *et al.*, 2007). *In vivo* in lower vertebrates, non-canonical Wnt signaling induces cardiac specification as shown for Wnt11 in *Xenopus* embryos (Pandur *et al.*, 2002). In the mouse embryo, *Wnt5a* mutants have arterial pole defects, but these may be due to effects on neural crest, through PlexinA2 down-regulation (Schleiffarth *et al.*, 2007). *Wnt11* null mice have arch artery patterning and outflow tract defects (Zhou *et al.*, 2007) also seen with *Dishevelled* mutants (Etheridge *et al.*, 2008). However, expression of Islet1 and other markers is normal and this is probably not a SHF phenotype. Wnt11, expressed at a high level in the mouse outflow tract, is implicated in polarized cell behavior required for correct arterial pole development (Phillips *et al.*, 2005, 2007). Wnt11 mediated effects on cardiomyocyte organization are also important for ventricular myocardium development (Nagy *et al.*, 2010). *Wnt11* participates in a gene regulatory network in which it is a direct target of Pitx2 and of canonical Wnt/β-catenin signaling, whereas *Tgfβ2* is a target of Wnt11 acting through ATF/CREB. *Wnt11, Vangl2, Scribble*, and *TGFβ2* mutants have a similar outflow tract phenotype suggesting that TGFβ2 is implicated in the cellular polarization required for outflow tract development (Zhou *et al.*, 2007).

7. Conclusion

Studies on the mouse embryo have provided extensive information about mutant phenotypes which affect different parts of the heart. The existence of subdomains within the SHF is now emerging, with potentially important consequences for the diagnosis and prognosis of human congenital heart disease as well as for the fundamental understanding of cardiogenesis. The developmental and regulatory history of these cell populations raises many of the same questions as do the first and second heart fields, in terms of lineage, cell behavior and gene networks. Complex genetic networks for transcriptional regulators and signaling pathways are documented for the SHF, but their spatiotemporal importance is poorly understood. Apparently confusing observations on the interactions between signaling pathways and effects on proliferation or differentiation probably partly reflect the lack of definition in terms of the location of subpopulations and developmental timing. Indications of the importance of the latter come from observations on SHF cell fate choices, illustrated by the addition of smooth muscle to the outflow tract after the myocardial contribution.

Quantitative data are also mainly lacking and are probably very important in determining the impact of a signaling pathway or a transcriptional regulator. The level of BMP/TGFβ signaling in the SHF required for neural crest survival, for example, is probably not the same as the level that promotes myocardial differentiation. The importance of gene dosage can be approached by study of an allelic series, as in the case of *Fgf8/Fgf10*, where progressively more severe phenotypes point to the underlying importance of FGF signaling on SHF proliferation. However, in this example, quantitative effects on interacting factors and signaling pathways in the FGF network have not yet been evaluated. The suggestion that higher levels of Islet1 promote differentiation rather than SHF proliferation illustrates the importance of quantitative considerations. Moving from a qualitative to a quantitative level of description and extending this to the cellular level is a major challenge for developmental biology in general. The cellular and molecular mechanisms that underlie the behavior of cardiac progenitor cells are also poorly understood at present. Here the challenge is to distinguish regulatory circuits, within the cells, that promote proliferation, prevent or promote differentiation, direct cell fate choices and affect cell interactions and cell movement. Dissection of regulatory outcomes, through a pathway like that of canonical Wnt signaling, is beginning to provide insight at this level. In addition to analysis of mutant phenotypes in the embryo, model cell systems, such as cardiogenic progenitors derived from ES cells, facilitate regulatory analysis, into the role of Wnt signaling for example. Of course, such cell culture systems present an artificial situation, with the risk that selection of a progenitor cell type or conjunction of regulatory inputs is a product of the *in vitro* system rather than a major component of the *in vivo* situation. Verification of a hypothesis, in the *in vivo* embryonic context, is often difficult, but essential. Explants, of regions of the SHF, for example, can provide a useful intermediate state for regulatory studies.

Analysis of outcomes at a single cell level in the SHF will depend on resolutive imaging techniques and genetically manipulable fluorescent probes. This technology is developing rapidly and provides a complementary approach to high-throughput analyses of gene expression or regulation which is also becoming practicable at the level of isolated single cells. Integrating this information for cardiac progenitor cell behavior in normal and mutant, or experimentally perturbed, contexts will depend on sophisticated modeling. In the long term, the aim is to transform the current crude level at which cardiogenesis is perceived into a precise spatiotemporal map of cardiac progenitor cell regulation that predicts behavior.

ACKNOWLEDGMENTS

We thank Didier Rocancourt for the artwork. Work on cardiogenesis in the Buckingham lab is supported by the Institut Pasteur, the CNRS (URA 2578) and by the EU through the Heart Repair (FP6 - LSHM-CT-2005-018630) and CardioCell (FP7 - HEALTH-2007-2.4.2-5) projects. Stéphane D. Vincent is an INSERM research fellow.

REFERENCES

Aanhaanen, W. T. J., Brons, J. F., Domínguez, J. N., Rana, M. S., Norden, J., Airik, R., Wakker, V., de Gier-de Vries, C., Brown, N. A., Kispert, A., Moorman, A. F. M. and Christoffels, V. M. (2009). The tbx2+ primary myocardium of the atrioventricular canal forms the atrioventricular node and the base of the left ventricle. *Circ. Res.* **104**, 1267–1274.

Abu-Issa, R. and Kirby, M. L. (2007). Heart field: from mesoderm to heart tube. *Annu. Rev. Cell. Dev. Biol.* **23**, 45–68.

Abu-Issa, R. and Kirby, M. L. (2008). Patterning of the heart field in the chick. *Dev. Biol.* **319**, 223–233.

Abu-Issa, R., Smyth, G., Smoak, I., Yamamura, K. and Meyers, E. N. (2002). Fgf8 is required for pharyngeal arch and cardiovascular development in the mouse. *Development* **129**, 4613–4625.

Ai, D., Fu, X., Wang, J., Lu, M. F., Chen, L., Baldini, A., Klein, W. H. and Martin, J. F. (2007). Canonical wnt signaling functions in second heart field to promote right ventricular growth. *Proc. Natl. Acad. Sci. USA* **104**, 9319–9324.

Ai, D., Liu, W., Ma, L., Dong, F., Lu, M.-F., Wang, D., Verzi, M. P., Cai, C., Gage, P. J., Evans, S., Black, B. L., Brown, N. A. and Martin, J. F. (2006). Pitx2 regulates cardiac left-right asymmetry by patterning second cardiac lineage-derived myocardium. *Dev. Biol.* **296**, 437–449.

Anderson, R. H., Brown, N. A. and Moorman, A. F. M. (2006). Development and structures of the venous pole of the heart. *Dev. Dyn.* **235**, 2–9.

Bajolle, F., Zaffran, S., Kelly, R. G., Hadchouel, J., Bonnet, D., Brown, N. A. and Buckingham, M. E. (2006). Rotation of the myocardial wall of the outflow tract is implicated in the normal positioning of the great arteries. *Circ. Res.* **98**, 421–428.

Bajolle, F., Zaffran, S., Meilhac, S. M., Dandonneau, M., Chang, T., Kelly, R. G. and Buckingham, M. E. (2008). Myocardium at the base of the aorta and pulmonary trunk is prefigured in the outflow tract of the heart and in subdomains of the second heart field. *Dev. Biol.* **313**, 25–34.

Barron, M., Gao, M. and Lough, J. (2000). Requirement for BMP and FGF signaling during cardiogenic induction in non-precardiac mesoderm is specific, transient, and cooperative. *Dev. Dyn.* **218**, 383–393.

Bondue, A., Lapouge, G., Paulissen, C., Semeraro, C., Iacovino, M., Kyba, M. and Blanpain, C. (2008). Mesp1 acts as a master regulator of multipotent cardiovascular progenitor specification. *Cell Stem Cell* **3**, 69–84.

Brade, T., Gessert, S., Kühl, M. and Pandur, P. (2007). The amphibian second heart field: xenopus islet-1 is required for cardiovascular development. *Dev. Biol.* **311**, 297–310.

Bradshaw, L., Chaudhry, B., Hildreth, V., Webb, S. and Henderson, D. J. (2009). Dual role for neural crest cells during outflow tract septation in the neural crest-deficient mutant Splotch(2H). *J. Anat.* **214**, 245–257.

Brown, C. B., Feiner, L., Lu, M. M., Li, J., Ma, X., Webber, A. L., Jia, L., Raper, J. A. and Epstein, J. A. (2001). PlexinA2 and semaphorin signaling during cardiac neural crest development. *Development* **128**, 3071–3080.

Brown, C. B., Wenning, J. M., Lu, M. M., Epstein, D. J., Meyers, E. N. and Epstein, J. A. (2004). Cre-mediated excision of fgf8 in the tbx1 expression domain reveals a critical role for fgf8 in cardiovascular development in the mouse. *Dev. Biol.* **267**, 190–202.

Bruneau, B. G. (2008). The developmental genetics of congenital heart disease. *Nature* **451**, 943–948.

Bu, L., Jiang, X., Martin-Puig, S., Caron, L., Zhu, S., Shao, Y., Roberts, D. J., Huang, P. L., Domian, I. J. and Chien, K. R. (2009). Human ISL1 heart progenitors generate diverse multipotent cardiovascular cell lineages. *Nature* **460**, 113–117.

Buckingham, M., Meilhac, S. and Zaffran, S. (2005). Building the mammalian heart from two sources of myocardial cells. *Nat. Rev. Genet.* **6**, 826–835.

Bussmann, J., Bakkers, J. and Schulte-Merker, S. (2007). Early endocardial morphogenesis requires scl/tal1. *PLoS Genet.* **3**, e140.

Cai, C.-L., Liang, X., Shi, Y., Chu, P.-H., Pfaff, S. L., Chen, J. and Evans, S. (2003). Isl1 identifies a cardiac progenitor population that proliferates prior to differentiation and contributes a majority of cells to the heart. *Dev. Cell* **5**, 877–889.

Cai, C., Martin, J., Sun, Y., Cui, L., Wang, L., Ouyang, K., Yang, L., Bu, L., Liang, X., Zhang, X., Stallcup, W., Denton, C., McCulloch, A., Chen, J. and Evans, S. (2008). A myocardial lineage derives from tbx18 epicardial cells. *Nature* **454**, 104–108.

Calmont, A., Ivins, S., Van Bueren, K. L., Papangeli, I., Kyriakopoulou, V., Andrews, W. D., Martin, J. F., Moon, A. M., Illingworth, E. A., Basson, M. A. and Scambler, P. J. (2009). Tbx1 controls cardiac neural crest cell migration during arch artery development by regulating gbx2 expression in the pharyngeal ectoderm. *Development* **136**, 3173–3183.

Chen, L., Fulcoli, F., Tang, S. and Baldini, A. (2009). Tbx1 regulates proliferation and differentiation of multipotent heart progenitors. *Circ. Res.* **105**, 842–851.

Christoffels, V. M., Grieskamp, T., Norden, J., Mommersteeg, M. T. M., Rudat, C. and Kispert, A. (2009). Tbx18 and the fate of epicardial progenitors. *Nature* **458**, E8–9.

Christoffels, V. M., Mommersteeg, M. T. M., Trowe, M.-O., Prall, O. W. J., de Gier-de Vries, C., Soufan, A. T., Bussen, M., Schuster-Gossler, K., Harvey, R. P., Moorman, A. F. M. and Kispert, A. (2006). Formation of the venous pole of the heart from an Nkx2-5-negative precursor population requires tbx18. *Circ. Res.* **98**, 1555–1563.

Clement, C. A., Kristensen, S. G., Møllgård, K., Pazour, G. J., Yoder, B. K., Larsen, L. A. and Christensen, S. T. (2009). The primary cilium coordinates early cardiogenesis and hedgehog signaling in cardiomyocyte differentiation. *J. Cell. Sci.* **122**, 3070–3082.

Cohen, E. D., Tian, Y. and Morrisey, E. E. (2008). Wnt signaling: an essential regulator of cardiovascular differentiation, morphogenesis and progenitor self-renewal. *Development* **135**, 789–798.

Cohen, E. D., Wang, Z., Lepore, J. J., Lu, M. M., Taketo, M. M., Epstein, D. J. and Morrisey, E. E. (2007). Wnt/beta-catenin signaling promotes expansion of isl-1-positive cardiac progenitor cells through regulation of FGF signaling. *J. Clin. Invest.* **117**, 1794–1804.

Cui, C., Cheuvront, T. J., Lansford, R. D., Moreno-Rodriguez, R. A., Schultheiss, T. M. and Rongish, B. J. (2009). Dynamic positional fate map of the primary heart-forming region. *Dev. Biol.* **332**, 212–222.

David, R., Brenner, C., Stieber, J., Schwarz, F., Brunner, S., Vollmer, M., Mentele, E., Müller-Höcker, J., Kitajima, S., Lickert, H., Rupp, R. and Franz, W.-M. (2008). MesP1 drives vertebrate cardiovascular differentiation through dkk-1-mediated blockade of wnt-signalling. *Nat. Cell Biol.* **10**, 338–345.

de Pater, E., Clijsters, L., Marques, S. R., Lin, Y. F., Garavito-Aguilar, Z. V., Yelon, D. and Bakkers, J. (2009). Distinct phases of cardiomyocyte differentiation regulate growth of the zebrafish heart. *Development* **136**, 1633–1641.

Dodou, E., Verzi, M. P., Anderson, J. P., Xu, S.-M. and Black, B. L. (2004). Mef2c is a direct transcriptional target of ISL1 and GATA factors in the anterior heart field during mouse embryonic development. *Development* **131**, 3931–3942.

Dyer, L. and Kirby, M. (2009). Sonic hedgehog maintains proliferation in secondary heart field progenitors and is required for normal arterial pole formation. *Dev. Biol.* **330**, 305–317.

Etheridge, S. L., Ray, S., Li, S., Hamblet, N. S., Lijam, N., Tsang, M., Greer, J., Kardos, N., Wang, J., Sussman, D. J., Chen, P. and Wynshaw-Boris, A. (2008). Murine dishevelled 3 functions in redundant pathways with dishevelled 1 and 2 in normal cardiac outflow tract, cochlea, and neural tube development. *PLoS Genet.* **4**, e1000259.

Feiner, L., Webber, A. L., Brown, C. B., Lu, M. M., Jia, L., Feinstein, P., Mombaerts, P., Epstein, J. A. and Raper, J. A. (2001). Targeted disruption of semaphorin 3c leads to persistent truncus arteriosus and aortic arch interruption. *Development* **128**, 3061–3070.

Ferdous, A., Caprioli, A., Iacovino, M., Martin, C. M., Morris, J., Richardson, J. A., Latif, S., Hammer, R. E., Harvey, R. P., Olson, E. N., Kyba, M. and Garry, D. J. (2009). Nkx2-5 transactivates the ets-related protein 71 gene and specifies an endothelial/endocardial fate in the developing embryo. *Proc. Natl. Acad. Sci. USA* **106**, 814–819.

Franco, D. and Campione, M. (2003). The role of pitx2 during cardiac development: Linking left–right signaling and congenital heart diseases. *Trends Cardiovasc. Med.* **13**, 157–163.

Frank, D. U., Fotheringham, L. K., Brewer, J. A., Muglia, L. J., Tristani-Firouzi, M., Capecchi, M. R. and Moon, A. M. (2002). An fgf8 mouse mutant phenocopies human 22q11 deletion syndrome. *Development* **129**, 4591–4603.

Fulcoli, F. G., Huynh, T., Scambler, P. J. and Baldini, A. (2009). Tbx1 regulates the BMP-smad1 pathway in a transcription independent manner. *PLoS ONE* **4**, e6049.

Galli, D., Domínguez, J. N., Zaffran, S., Munk, A., Brown, N. A. and Buckingham, M. E. (2008). Atrial myocardium derives from the posterior region of the second heart field, which acquires left–right identity as pitx2c is expressed. *Development* **135**, 1157–1167.

Garcia-Martinez, V. and Schoenwolf, G. C. (1993). Primitive-streak origin of the cardio-vascular system in avian embryos. *Dev. Biol.* **159**, 706–719.

Gessert, S. and Kühl, M. (2009). Comparative gene expression analysis and fate mapping studies suggest an early segregation of cardiogenic lineages in xenopus laevis. *Dev. Biol.* **334**, 395–408.

Gitler, A. D., Lu, M. M. and Epstein, J. A. (2004). PlexinD1 and semaphorin signaling are required in endothelial cells for cardiovascular development. *Dev. Cell* **7**, 107–116.

Goddeeris, M. M., Rho, S., Petiet, A., Davenport, C. L., Johnson, G. A., Meyers, E. N. and Klingensmith, J. (2008). Intracardiac septation requires hedgehog-dependent cellular contributions from outside the heart. *Development* **135**, 1887–1895.

Goddeeris, M. M., Schwartz, R., Klingensmith, J. and Meyers, E. N. (2007). Independent requirements for hedgehog signaling by both the anterior heart field and neural crest cells for outflow tract development. *Development* **134**, 1593–1604.

Guris, D. L., Duester, G., Papaioannou, V. E. and Imamoto, A. (2006). Dose-dependent interaction of tbx1 and crkl and locally aberrant RA signaling in a model of del22q11 syndrome. *Dev. Cell* **10**, 81–92.

Harel, I., Nathan, E., Tirosh-Finkel, L., Zigdon, H., Guimarães-Camboa, N., Evans, S. M. and Tzahor, E. (2009). Distinct origins and genetic programs of head muscle satellite cells. *Dev. Cell* **16**, 822–832.

Harvey, R. P., Meilhac, S. M. and Buckingham, M. E. (2009). Landmarks and lineages in the developing heart. *Circ. Res.* **104**, 1235–1237.

High, F. A. and Epstein, J. A. (2008). The multifaceted role of notch in cardiac development and disease. *Nat. Rev. Genet.* **9**, 49–61.

High, F. A., Jain, R., Stoller, J. Z., Antonucci, N. B., Lu, M. M., Loomes, K. M., Kaestner, K. H., Pear, W. S. and Epstein, J. A. (2009). Murine jagged1/notch signaling in the second heart field orchestrates fgf8 expression and tissue–tissue interactions during outflow tract development. *J. Clin. Invest.* **119**, 1986–1996.

High, F. A., Lu, M. M., Pear, W. S., Loomes, K. M., Kaestner, K. H. and Epstein, J. A. (2008). Endothelial expression of the notch ligand jagged1 is required for vascular smooth muscle development. *Proc. Natl. Acad. Sci. USA* **105**, 1955–1959.

High, F. A., Zhang, M., Proweller, A., Tu, L., Parmacek, M. S., Pear, W. S. and Epstein, J. A. (2007). An essential role for notch in neural crest during cardiovascular development and smooth muscle differentiation. *J. Clin. Invest.* **117**, 353–363.

Hildreth, V., Webb, S., Chaudhry, B., Peat, J. D., Phillips, H. M., Brown, N., Anderson, R. H. and Henderson, D. J. (2009). Left cardiac isomerism in the sonic hedgehog null mouse. *J. Anat.* **214**, 894–904.

Hoffmann, A. D., Peterson, M. A., Friedland-Little, J. M., Anderson, S. A. and Moskowitz, I. P. (2009). Sonic hedgehog is required in pulmonary endoderm for atrial septation. *Development* **136**, 1761–1770.

Hu, T., Yamagishi, H., Maeda, J., McAnally, J., Yamagishi, C. and Srivastava, D. (2004). Tbx1 regulates fibroblast growth factors in the anterior heart field through a reinforcing autoregulatory loop involving forkhead transcription factors. *Development* **131**, 5491–5502.

Hutson, M. R. and Kirby, M. L. (2007). Model systems for the study of heart development and disease. Cardiac neural crest and conotruncal malformations. *Semin. Cell Dev. Biol.* **18**, 101–110.

Hutson, M. R., Zhang, P., Stadt, H. A., Sato, A. K., Li, Y.-X., Burch, J., Creazzo, T. L. and Kirby, M. L. (2006). Cardiac arterial pole alignment is sensitive to FGF8 signaling in the pharynx. *Dev. Biol.* **295**, 486–497.

Huynh, T., Chen, L., Terrell, P. and Baldini, A. (2007). A fate map of tbx1 expressing cells reveals heterogeneity in the second cardiac field. *Genesis* **45**, 470–475.

Ilagan, R., Abu-Issa, R., Brown, D., Yang, Y.-P., Jiao, K., Schwartz, R. J., Klingensmith, J. and Meyers, E. N. (2006). Fgf8 is required for anterior heart field development. *Development* **133**, 2435–2445.

Ivins, S., Lammerts van Beuren, K., Roberts, C., James, C., Lindsay, E., Baldini, A., Ataliotis, P. and Scambler, P. J. (2005). Microarray analysis detects differentially expressed genes in the pharyngeal region of mice lacking tbx1. *Dev. Biol.* **285**, 554–569.

Jopling, C., Sleep, E., Raya, M., Martí, M., Raya, A. and Belmonte, J. C. I. (2010). Zebrafish heart regeneration occurs by cardiomyocyte dedifferentiation and proliferation. *Nature* **464**, 606–609.

Kang, J., Nathan, E., Xu, S., Tzahor, E. and Black, B. (2009). Isl1 is a direct transcriptional target of forkhead transcription factors in second heart field-derived mesoderm. *Dev. Biol.* **334**, 513–522.

Kappen, C. and Salbaum, J. M. (2009). Identification of regulatory elements in the isl1 gene locus. *Int. J. Dev. Biol.* **53**, 935–946.

Kaufman, M. H. and Bard, J. B. (1999). *The Anatomical Basis of Mouse Development.* Academic Press, London.

Keegan, B. R., Feldman, J. L., Begemann, G., Ingham, P. W. and Yelon, D. (2005). Retinoic acid signaling restricts the cardiac progenitor pool. *Science* **307**, 247–249.

Kelly, R. G., Brown, N. A. and Buckingham, M. E. (2001). The arterial pole of the mouse heart forms from fgf10-expressing cells in pharyngeal mesoderm. *Dev. Cell* **1**, 435–440.

Keren-Politansky, A., Keren, A. and Bengal, E. (2009). Neural ectoderm-secreted FGF initiates the expression of Nkx2.5 in cardiac progenitors via a p38 MAPK/CREB pathway. *Dev. Biol.* **335**, 374–384.

Kikuchi, K., Holdway, J. E., Werdich, A. A., Anderson, R. M., Fang, Y., Egnaczyk, G. F., Evans, T., Macrae, C. A., Stainier, D. Y. R. and Poss, K. D. (2010). Primary contribution to zebrafish heart regeneration by gata4+ cardiomyocytes. *Nature* **464**, 601.

Kirby, M. (2007). *Cardiac Development.* Oxford University Press, Oxford.

Klaus, A., Saga, Y., Taketo, M. M., Tzahor, E. and Birchmeier, W. (2007). Distinct roles of wnt/beta-catenin and bmp signaling during early cardiogenesis. *Proc. Natl. Acad. Sci. USA* **104**, 18531–18536.

Kodo, K., Nishizawa, T., Furutani, M., Arai, S., Yamamura, E., Joo, K., Takahashi, T., Matsuoka, R. and Yamagishi, H. (2009). GATA6 mutations cause human cardiac outflow tract defects by disrupting semaphorin–plexin signaling. *Proc. Natl. Acad. Sci. USA* **106**, 13933–13938.

Kolesová, H., Roelink, H. and Grim, M. (2008). Sonic hedgehog is required for the assembly and remodeling of branchial arch blood vessels. *Dev. Dyn.* **237**, 1923–1934.

Kruithof, B. P. T., Van Wijk, B., Somi, S., Kruithof-de Julio, M., Pérez Pomares, J. M., Weesie, F., Wessels, A., Moorman, A. F. M. and van den Hoff, M. J. B. (2006). BMP and FGF regulate the differentiation of multipotential pericardial mesoderm into the myocardial or epicardial lineage. *Dev. Biol.* **295**, 507–522.

Kwon, C., Arnold, J., Hsiao, E. C., Taketo, M. M., Conklin, B. R. and Srivastava, D. (2007). Canonical wnt signaling is a positive regulator of mammalian cardiac progenitors. *Proc. Natl. Acad. Sci. USA* **104**, 10894–10899.

Kwon, C., Qian, L., Cheng, P., Nigam, V., Arnold, J. and Srivastava, D. (2009). A regulatory pathway involving notch1/beta-catenin/isl1 determines cardiac progenitor cell fate. *Nat. Cell Biol.* **11**, 951–957.

Laugwitz, K.-L., Moretti, A., Lam, J., Gruber, P., Chen, Y., Woodard, S., Lin, L.-Z., Cai, C.-L., Lu, M. M., Reth, M., Platoshyn, O., Yuan, J.X.-J., Evans, S. and Chien, K. R. (2005). Postnatal isl1+ cardioblasts enter fully differentiated cardiomyocyte lineages. *Nature* **433**, 647–653.

Lavery, D., Martin, J., Turnbull, Y. and Hoppler, S. (2008). Wnt6 signaling regulates heart muscle development during organogenesis. *Dev. Biol.* **323**, 177–188.

Lawson, K. A., Meneses, J. J. and Pedersen, R. A. (1991). Clonal analysis of epiblast fate during germ layer formation in the mouse embryo. *Development* **113**, 891–911.

Lepilina, A., Coon, A. N., Kikuchi, K., Holdway, J. E., Roberts, R. W., Burns, C. G. and Poss, K. D. (2006). A dynamic epicardial injury response supports progenitor cell activity during zebrafish heart regeneration. *Cell* **127**, 607–619.

Lepore, J. J., Mericko, P. A., Cheng, L., Lu, M. M., Morrisey, E. E. and Parmacek, M. S. (2006). GATA-6 regulates semaphorin 3c and is required in cardiac neural crest for cardiovascular morphogenesis. *J. Clin. Invest.* **116**, 929–939.

Li, P., Pashmforoush, M. and Sucov, H. M. (2010). Retinoic acid regulates differentiation of the secondary heart field and TGFbeta-mediated outflow tract septation. *Dev. Cell* **18**, 480–485.

Liao, J., Aggarwal, V. S., Nowotschin, S., Bondarev, A., Lipner, S. and Morrow, B. E. (2008). Identification of downstream genetic pathways of tbx1 in the second heart field. *Dev. Biol.* **316**, 524–537.

Lin, L., Bu, L., Cai, C.-L., Zhang, X. and Evans, S. (2006). Isl1 is upstream of sonic hedgehog in a pathway required for cardiac morphogenesis. *Dev. Biol.* **295**, 756–763.

Lin, L., Cui, L., Zhou, W., Dufort, D., Zhang, X., Cai, C.-L., Bu, L., Yang, L., Martin, J., Kemler, R., Rosenfeld, M. G., Chen, J. and Evans, S. M. (2007). Beta-catenin directly regulates islet1 expression in cardiovascular progenitors and is required for multiple aspects of cardiogenesis. *Proc. Natl. Acad. Sci. USA* **104**, 9313–9318.

Lindsley, R. C., Gill, J. G., Murphy, T. L., Langer, E. M., Cai, M., Mashayekhi, M., Wang, W., Niwa, N., Nerbonne, J. M., Kyba, M. and Murphy, K. M. (2008). Mesp1 coordinately regulates cardiovascular fate restriction and epithelial–mesenchymal transition in differentiating ESCs. *Cell Stem Cell* **3**, 55–68.

Liu, W., Selever, J., Wang, D., Lu, M.-F., Moses, K. A., Schwartz, R. J. and Martin, J. F. (2004). Bmp4 signaling is required for outflow-tract septation and branchial-arch artery remodeling. *Proc. Natl. Acad. Sci. USA* **101**, 4489–4494.

Ma, Q., Zhou, B. and Pu, W. T. (2008). Reassessment of isl1 and Nkx2-5 cardiac fate maps using a gata4-based reporter of cre activity. *Dev. Biol.* **323**, 98–104.

Macatee, T. L., Hammond, B. P., Arenkiel, B. R., Francis, L., Frank, D. U. and Moon, A. M. (2003). Ablation of specific expression domains reveals discrete functions of ectoderm- and endoderm-derived FGF8 during cardiovascular and pharyngeal development. *Development* **130**, 6361–6374.

Maeda, J., Yamagishi, H., McAnally, J., Yamagishi, C. and Srivastava, D. (2006). Tbx1 is regulated by forkhead proteins in the secondary heart field. *Dev. Dyn.* **235**, 701–710.

Männer, J., Pérez-Pomares, J. M., Macías, D. and Muñoz-Chápuli, R. (2001). The origin, formation and developmental significance of the epicardium: a review. *Cells Tissues Organs (Print)* **169**, 89–103.

Marguerie, A., Bajolle, F., Zaffran, S., Brown, N. A., Dickson, C., Buckingham, M. E. and Kelly, R. G. (2006). Congenital heart defects in fgfr2-IIIb and fgf10 mutant mice. *Cardiovasc. Res.* **71**, 50–60.

Marques, S. R., Lee, Y., Poss, K. D. and Yelon, D. (2008). Reiterative roles for FGF signaling in the establishment of size and proportion of the zebrafish heart. *Dev. Biol.* **321**, 397–406.

Martínez-Estrada, O. M., Lettice, L. A., Essafi, A., Guadix, J. A., Slight, J., Velecela, V., Hall, E., Reichmann, J., Devenney, P. S., Hohenstein, P., Hosen, N., Hill, R. E., Muñoz-Chapuli, R. and Hastie, N. D. (2010). Wt1 is required for cardiovascular progenitor cell formation through transcriptional control of snail and E-cadherin. *Nat. Genet.* **42**, 89–93.

McCulley, D., Kang, J., Martin, J. and Black, B. (2008). BMP4 is required in the anterior heart field and its derivatives for endocardial cushion remodeling, outflow tract septation, and semilunar valve development. *Dev. Dyn.* **237**, 3200–3209.

Meilhac, S. M., Esner, M., Kelly, R. G., Nicolas, J.-F. and Buckingham, M. E. (2004). The clonal origin of myocardial cells in different regions of the embryonic mouse heart. *Dev. Cell* **6**, 685–698.

Meilhac, S. M., Kelly, R. G., Rocancourt, D., Eloy-Trinquet, S., Nicolas, J.-F. and Buckingham, M. E. (2003). A retrospective clonal analysis of the myocardium reveals two phases of clonal growth in the developing mouse heart. *Development* **130**, 3877–3889.

Mesbah, K., Harrelson, Z., Théveniau-Ruissy, M., Papaioannou, V. E. and Kelly, R. G. (2008). Tbx3 is required for outflow tract development. *Circ. Res.* **103**, 743–750.

Misfeldt, A., Boyle, S., Tompkins, K., Bautch, V., Labosky, P. and Baldwin, H. (2009). Endocardial cells are a distinct endothelial lineage derived from flk1+ multipotent cardiovascular progenitors. *Dev. Biol.* **333**, 78–89.

Mommersteeg, M. T. M., Brown, N. A., Prall, O. W. J., de Gier-de Vries, C., Harvey, R. P., Moorman, A. F. M. and Christoffels, V. M. (2007). Pitx2c and Nkx2-5 are required for the formation and identity of the pulmonary myocardium. *Circ. Res.* **101**, 902–909.

Mommersteeg, M.T.M., Domínguez, J. N., Wiese, C., Norden, J., de Gier-de Vries, C., Burch, J. B. E., Kispert, A., Brown, N. A., Moorman, A. F. M. and Christoffels, V. M. (2010). The sinus venosus progenitors separate and diversify from the first and second heart fields early in development. *Cardiovasc. Res.* **87**, 92–101.

Mommersteeg, M. T. M., Soufan, A. T., de Lange, F. J., van den Hoff, M. J. B., Anderson, R. H., Christoffels, V. M. and Moorman, A. F. M. (2006). Two distinct pools of mesenchyme contribute to the development of the atrial septum. *Circ. Res.* **99**, 351–353.

Moretti, A., Caron, L., Nakano, A., Lam, J. T., Bernshausen, A., Chen, Y., Qyang, Y., Bu, L., Sasaki, M., Martin-Puig, S., Sun, Y., Evans, S. M., Laugwitz, K.-L. and Chien, K. R. (2006). Multipotent embryonic isl1+ progenitor cells lead to cardiac, smooth muscle, and endothelial cell diversification. *Cell* **127**, 1151–1165.

Motoike, T., Markham, D. W., Rossant, J. and Sato, T. N. (2003). Evidence for novel fate of flk1+ progenitor: contribution to muscle lineage. *Genesis* **35**, 153–159.

Nagy, I. I., Railo, A., Rapila, R., Hast, T., Sormunen, R., Tavi, P., Räsänen, J. and Vainio, S. J. (2010). Wnt-11 signalling controls ventricular myocardium development by patterning N-cadherin and beta-catenin expression. *Cardiovasc. Res.* **85**, 100–109.

Nathan, E., Monovich, A., Tirosh-Finkel, L., Harrelson, Z., Rousso, T., Rinon, A., Harel, I., Evans, S. M. and Tzahor, E. (2008). The contribution of islet1-expressing splanchnic mesoderm cells to distinct branchiomeric muscles reveals significant heterogeneity in head muscle development. *Development* **135**, 647–657.

Nie, X., Deng, C.-X., Wang, Q. and Jiao, K. (2008). Disruption of smad4 in neural crest cells leads to mid-gestation death with pharyngeal arch, craniofacial and cardiac defects. *Dev. Biol.* **316**, 417–430.

Noden, D. M. and Francis-West, P. (2006). The differentiation and morphogenesis of craniofacial muscles. *Dev. Dyn.* **235**, 1194–1218.

Nowotschin, S., Liao, J., Gage, P. J., Epstein, J. A., Campione, M. and Morrow, B. E. (2006). Tbx1 affects asymmetric cardiac morphogenesis by regulating pitx2 in the secondary heart field. *Development* **133**, 1565–1573.

Palpant, N. J., Yasuda, S., MacDougald, O. and Metzger, J. M. (2007). Non-canonical wnt signaling enhances differentiation of sca1+/c-kit+ adipose-derived murine stromal vascular cells into spontaneously beating cardiac myocytes. *J. Mol. Cell. Cardiol.* **43**, 362–370.

Pandur, P., Läsche, M., Eisenberg, L. M. and Kühl, M. (2002). Wnt-11 activation of a non-canonical wnt signalling pathway is required for cardiogenesis. *Nature* **418**, 636–641.

Park, E. J., Ogden, L. A., Talbot, A., Evans, S., Cai, C.-L., Black, B. L., Frank, D. U. and Moon, A. M. (2006). Required, tissue-specific roles for fgf8 in outflow tract formation and remodeling. *Development* **133**, 2419–2433.

Park, E., Watanabe, Y., Smyth, G., Miyagawa-Tomita, S., Meyers, E., Klingensmith, J., Camenisch, T., Buckingham, M. and Moon, A. (2008). An FGF autocrine loop initiated in second heart field mesoderm regulates morphogenesis at the arterial pole of the heart. *Development* **135**, 3599–3610.

Phillips, H. M., Murdoch, J. N., Chaudhry, B., Copp, A. J. and Henderson, D. J. (2005). Vangl2 acts via RhoA signaling to regulate polarized cell movements during development of the proximal outflow tract. *Circ. Res.* **96**, 292–299.

Phillips, H. M., Rhee, H. J., Murdoch, J. N., Hildreth, V., Peat, J. D., Anderson, R. H., Copp, A. J., Chaudhry, B. and Henderson, D. J. (2007). Disruption of planar cell polarity signaling results in congenital heart defects and cardiomyopathy attributable to early cardiomyocyte disorganization. *Circ. Res.* **101**, 137–145.

Prall, O. W. J., Menon, M. K., Solloway, M. J., Watanabe, Y., Zaffran, S., Bajolle, F., Biben, C., McBride, J. J., Robertson, B. R., Chaulet, H., Stennard, F. A., Wise, N., Schaft, D., Wolstein, O., Furtado, M. B., Shiratori, H., Chien, K. R., Hamada, H., Black, B. L., Saga, Y., Robertson, E. J., Buckingham, M. E. and Harvey, R. P. (2007). An Nkx2-5/bmp2/smad1 negative feedback loop controls heart progenitor specification and proliferation. *Cell* **128**, 947–959.

Qyang, Y., Martin-Puig, S., Chiravuri, M., Chen, S., Xu, H., Bu, L., Jiang, X., Lin, L., Granger, A., Moretti, A., Caron, L., Wu, X., Clarke, J., Taketo, M. M., Laugwitz, K.-L., Moon, R. T., Gruber, P., Evans, S. M., Ding, S. and Chien, K. R. (2007). The renewal and differentiation of isl1+ cardiovascular progenitors are controlled by a wnt/beta-catenin pathway. *Cell Stem Cell* **1**, 165–179.

Red-Horse, K., Ueno, H., Weissman, I. L. and Krasnow, M. A. (2010). Coronary arteries form by developmental reprogramming of venous cells. *Nature* **464**, 549–553.

Redkar, A., Montgomery, M. and Litvin, J. (2001). Fate map of early avian cardiac progenitor cells. *Development* **128**, 2269–2279.

Roberts, C., Ivins, S., Cook, A. C., Baldini, A. and Scambler, P. J. (2006). Cyp26 genes a1, b1 and c1 are down-regulated in tbx1 null mice and inhibition of cyp26 enzyme function produces a phenocopy of DiGeorge syndrome in the chick. *Hum. Mol. Genet.* **15**, 3394–3410.

Rochais, F., Dandonneau, M., Mesbah, K., Jarry, T., Mattei, M.-G. and Kelly, R. G. (2009). Hes1 is expressed in the second heart field and is required for outflow tract development. *PLoS ONE* **4**, e6267.

Roessler, E., Ouspenskaia, M. V., Karkera, J. D., Vélez, J. I., Kantipong, A., Lacbawan, F., Bowers, P., Belmont, J. W., Towbin, J. A., Goldmuntz, E., Feldman, B. and Muenke, M. (2008). Reduced NODAL signaling strength via mutation of several pathway members including FOXH1 is linked to human heart defects and holoprosencephaly. *Am. J. Hum. Genet.* **83**, 18–29.

Ryckebüsch, L., Bertrand, N., Mesbah, K., Bajolle, F., Niederreither, K., Kelly, R. G. and Zaffran, S. (2010). Decreased levels of embryonic retinoic acid synthesis accelerate recovery from arterial growth delay in a mouse model of DiGeorge syndrome. *Circ. Res.* **106**, 686–694.

Ryckebusch, L., Wang, Z., Bertrand, N., Lin, S.-C., Chi, X., Schwartz, R., Zaffran, S. and Niederreither, K. (2008). Retinoic acid deficiency alters second heart field formation. *Proc. Natl. Acad. Sci. USA* **105**, 2913–2918.

Saga, Y., Kitajima, S. and Miyagawa-Tomita, S. (2000). Mesp1 expression is the earliest sign of cardiovascular development. *Trends Cardiovasc. Med.* **10**, 345–352.

Saga, Y., Miyagawa-Tomita, S., Takagi, A., Kitajima, S., Miyazaki, J. and Inoue, T. (1999). MesP1 is expressed in the heart precursor cells and required for the formation of a single heart tube. *Development* **126**, 3437–3447.

Schleiffarth, J. R., Person, A. D., Martinsen, B. J., Sukovich, D. J., Neumann, A., Baker, C. V. H., Lohr, J. L., Cornfield, D. N., Ekker, S. C. and Petryk, A. (2007). Wnt5a is required for cardiac outflow tract septation in mice. *Pediatr. Res.* **61**, 386–391.

Schultheiss, T. M., Burch, J. B. and Lassar, A. B. (1997). A role for bone morphogenetic proteins in the induction of cardiac myogenesis. *Genes Dev.* **11**, 451–462.

Seo, S. and Kume, T. (2006). Forkhead transcription factors, foxc1 and foxc2, are required for the morphogenesis of the cardiac outflow tract. *Dev. Biol.* **296**, 421–436.

Shen, H., McElhinny, A. S., Cao, Y., Gao, P., Liu, J., Bronson, R., Griffin, J. D. and Wu, L. (2006). The notch coactivator, MAML1, functions as a novel coactivator for MEF2C-mediated transcription and is required for normal myogenesis. *Genes Dev.* **20**, 675–688.

Sirbu, I. O., Zhao, X. and Duester, G. (2008). Retinoic acid controls heart anteroposterior patterning by down-regulating isl1 through the fgf8 pathway. *Dev. Dyn.* **237**, 1627–1635.
Slough, J., Cooney, L. and Brueckner, M. (2008). Monocilia in the embryonic mouse heart suggest a direct role for cilia in cardiac morphogenesis. *Dev. Dyn.* **237**, 2304–2314.
Snarr, B. S., O'Neal, J. L., Chintalapudi, M. R., Wirrig, E. E., Phelps, A. L., Kubalak, S. W. and Wessels, A. (2007a). Isl1 expression at the venous pole identifies a novel role for the second heart field in cardiac development. *Circ. Res.* **101**, 971–974.
Snarr, B. S., Wirrig, E. E., Phelps, A. L., Trusk, T. C. and Wessels, A. (2007b). A spatiotemporal evaluation of the contribution of the dorsal mesenchymal protrusion to cardiac development. *Dev. Dyn.* **236**, 1287–1294.
Somi, S., Buffing, A. A. M., Moorman, A. F. M. and van den Hoff, M. J. B. (2004). Dynamic patterns of expression of BMP isoforms 2, 4, 5, 6, and 7 during chicken heart development. *Anat. Rec. A Discov. Mol. Cell Evol. Biol.* **279**, 636–651.
Soufan, A. T., van den Berg, G., Ruijter, J. M., de Boer, P. A. J., van den Hoff, M. J. B. and Moorman, A. F. M. (2006). Regionalized sequence of myocardial cell growth and proliferation characterizes early chamber formation. *Circ. Res.* **99**, 545–552.
Stalsberg, H. and DeHaan, R. L. (1969). The precardiac areas and formation of the tubular heart in the chick embryo. *Dev. Biol.* **19**, 128–159.
Stanley, E. G., Biben, C., Elefanty, A., Barnett, L., Koentgen, F., Robb, L. and Harvey, R. P. (2002). Efficient cre-mediated deletion in cardiac progenitor cells conferred by a 3'UTR-ires-cre allele of the homeobox gene Nkx2-5. *Int. J. Dev. Biol.* **46**, 431–439.
Sun, Y., Liang, X., Najafi, N., Cass, M., Lin, L., Cai, C.-L., Chen, J. and Evans, S. M. (2007). Islet 1 is expressed in distinct cardiovascular lineages, including pacemaker and coronary vascular cells. *Dev. Biol.* **304**, 286–296.
Takeuchi, J. and Bruneau, B. (2009). Directed transdifferentiation of mouse mesoderm to heart tissue by defined factors. *Nature* **459**, 708–711.
Takeuchi, J. K., Mileikovskaia, M., Koshiba-Takeuchi, K., Heidt, A. B., Mori, A. D., Arruda, E. P., Gertsenstein, M., Georges, R., Davidson, L., Mo, R., Hui, C.-C., Henkelman, R. M., Nemer, M., Black, B. L., Nagy, A. and Bruneau, B. G. (2005). Tbx20 dose-dependently regulates transcription factor networks required for mouse heart and motoneuron development. *Development* **132**, 2463–2474.
Tam, P. P., Parameswaran, M., Kinder, S. J. and Weinberger, R. P. (1997). The allocation of epiblast cells to the embryonic heart and other mesodermal lineages: the role of ingression and tissue movement during gastrulation. *Development* **124**, 1631–1642.
Tapscott, S. J. (2005). The circuitry of a master switch: myod and the regulation of skeletal muscle gene transcription. *Development* **132**, 2685–2695.
Tessari, A., Pietrobon, M., Notte, A., Cifelli, G., Gage, P. J., Schneider, M. D., Lembo, G. and Campione, M. (2008). Myocardial pitx2 differentially regulates the left atrial identity and ventricular asymmetric remodeling programs. *Circ. Res.* **102**, 813–822.
Théveniau-Ruissy, M., Dandonneau, M., Mesbah, K., Ghez, O., Mattei, M.-G., Miquerol, L. and Kelly, R. G. (2008). The del22q11.2 candidate gene tbx1 controls regional outflow tract identity and coronary artery patterning. *Circ. Res.* **103**, 142–148.
Thomas, N. A., Koudijs, M., van Eeden, F. J. M., Joyner, A. L. and Yelon, D. (2008). Hedgehog signaling plays a cell-autonomous role in maximizing cardiac developmental potential. *Development* **135**, 3789–3799.
Tian, Y., Yuan, L., Goss, A. M., Wang, T., Yang, J., Lepore, J. J., Zhou, D., Schwartz, R. J., Patel, V., Cohen, E. D. and Morrisey, E. E. (2010). Characterization and in vivo pharmacological rescue of a wnt2-gata6 pathway required for cardiac inflow tract development. *Dev. Cell* **18**, 275–287.
Tirosh-Finkel, L., Elhanany, H., Rinon, A. and Tzahor, E. (2006). Mesoderm progenitor cells of common origin contribute to the head musculature and the cardiac outflow tract. *Development* **133**, 1943–1953.

Toyofuku, T., Yoshida, J., Sugimoto, T., Yamamoto, M., Makino, N., Takamatsu, H., Takegahara, N., Suto, F., Hori, M., Fujisawa, H., Kumanogoh, A. and Kikutani, H. (2008). Repulsive and attractive semaphorins cooperate to direct the navigation of cardiac neural crest cells. *Dev. Biol.* **321**, 251–262.

Uchimura, T., Komatsu, Y., Tanaka, M., McCann, K. and Mishina, Y. (2009). Bmp2 and bmp4 genetically interact to support multiple aspects of mouse development including functional heart development. *Genesis* **47**, 374–384.

Van Bueren, K. L., Papangeli, I., Rochais, F., Pearce, K., Roberts, C., Calmont, A., Szumska, D., Kelly, R. G., Bhattacharya, S. and Scambler, P. J. (2010). Hes1 expression is reduced in tbx1 null cells and is required for the development of structures affected in 22q11 deletion syndrome. *Dev. Biol.* **340**, 369–380.

Van den Berg, G., Abu-Issa, R., de Boer, B. A., Hutson, M. R., de Boer, P. A. J., Soufan, A. T., Ruijter, J. M., Kirby, M. L., van den Hoff, M. J. B. and Moorman, A. F. M. (2009). A caudal proliferating growth center contributes to both poles of the forming heart tube. *Circ. Res.* **104**, 179–188.

Van Praagh, R. (2009). The first Stella Van Praagh memorial lecture: the history and anatomy of tetralogy of fallot. *Semin. Thorac. Cardiovasc. Surg. Pediatr. Card. Surg. Annu.* **12**, 19–38.

Van Wijk, B., van den Berg, G., Abu-Issa, R., Barnett, P., Van Der Velden, S., Schmidt, M., Ruijter, J. M., Kirby, M. L., Moorman, A. F. M. and van den Hoff, M. J. B. (2009). Epicardium and myocardium separate from a common precursor pool by crosstalk between bone morphogenetic protein- and fibroblast growth factor-signaling pathways. *Circ. Res.* **105**, 431–441.

Varadkar, P., Kraman, M., Despres, D., Ma, G., Lozier, J. and McCright, B. (2008). Notch2 is required for the proliferation of cardiac neural crest-derived smooth muscle cells. *Dev. Dyn.* **237**, 1144–1152.

Vermot, J., Niederreither, K., Garnier, J.-M., Chambon, P. and Dollé, P. (2003). Decreased embryonic retinoic acid synthesis results in a DiGeorge syndrome phenotype in newborn mice. *Proc. Natl. Acad. Sci. USA* **100**, 1763–1768.

Verzi, M. P., McCulley, D. J., De Val, S., Dodou, E. and Black, B. L. (2005). The right ventricle, outflow tract, and ventricular septum comprise a restricted expression domain within the secondary/anterior heart field. *Dev. Biol.* **287**, 134–145.

Vincentz, J. W., McWhirter, J. R., Murre, C., Baldini, A. and Furuta, Y. (2005). Fgf15 is required for proper morphogenesis of the mouse cardiac outflow tract. *Genesis* **41**, 192–201.

Vitelli, F., Huynh, T. and Baldini, A. (2009). Gain of function of tbx1 affects pharyngeal and heart development in the mouse. *Genesis* **47**, 188–195.

Vitelli, F., Zhang, Z., Huynh, T., Sobotka, A., Mupo, A. and Baldini, A. (2006). Fgf8 expression in the tbx1 domain causes skeletal abnormalities and modifies the aortic arch but not the outflow tract phenotype of tbx1 mutants. *Dev. Biol.* **295**, 559–570.

von Both, I., Silvestri, C., Erdemir, T., Lickert, H., Walls, J. R., Henkelman, R. M., Rossant, J., Harvey, R. P., Attisano, L. and Wrana, J. L. (2004). Foxh1 is essential for development of the anterior heart field. *Dev. Cell* **7**, 331–345.

Waldo, K. L., Hutson, M. R., Stadt, H. A., Zdanowicz, M., Zdanowicz, J. and Kirby, M. L. (2005a). Cardiac neural crest is necessary for normal addition of the myocardium to the arterial pole from the secondary heart field. *Dev. Biol.* **281**, 66–77.

Waldo, K. L., Hutson, M. R., Ward, C. C., Zdanowicz, M., Stadt, H. A., Kumiski, D., Abu-Issa, R. and Kirby, M. L. (2005b). Secondary heart field contributes myocardium and smooth muscle to the arterial pole of the developing heart. *Dev. Biol.* **281**, 78–90.

Waldo, K. L., Kumiski, D. H., Wallis, K. T., Stadt, H. A., Hutson, M. R., Platt, D. H. and Kirby, M. L. (2001). Conotruncal myocardium arises from a secondary heart field. *Development* **128**, 3179–3188.

Ward, C., Stadt, H., Hutson, M. and Kirby, M. L. (2005). Ablation of the secondary heart field leads to tetralogy of fallot and pulmonary atresia. *Dev. Biol.* **284**, 72–83.

Washington Smoak, I., Byrd, N. A., Abu-Issa, R., Goddeeris, M. M., Anderson, R., Morris, J., Yamamura, K., Klingensmith, J. and Meyers, E. N. (2005). Sonic hedgehog is required for cardiac outflow tract and neural crest cell development. *Dev. Biol.* **283**, 357–372.

Watanabe, Y., Miyagawa-Tomita, S., Vincent, S. D., Kelly, R. G., Moon, A. M. and Buckingham, M. E. (2010). Role of mesodermal FGF8 and FGF10 overlaps in the development of the arterial pole of the heart and pharyngeal arch arteries. *Circ. Res.* **106**, 495–503.

Waxman, J. S., Keegan, B. R., Roberts, R. W., Poss, K. D. and Yelon, D. (2008). Hoxb5b acts downstream of retinoic acid signaling in the forelimb field to restrict heart field potential in zebrafish. *Dev. Cell* **15**, 923–934.

Wei, Y. and Mikawa, T. (2000). Fate diversity of primitive streak cells during heart field formation in ovo. *Dev. Dyn.* **219**, 505–513.

Weintraub, H., Davis, R., Tapscott, S., Thayer, M., Krause, M., Benezra, R., Blackwell, T. K., Turner, D., Rupp, R. and Hollenberg, S. (1991). The myoD gene family: nodal point during specification of the muscle cell lineage. *Science* **251**, 761–766.

Wessels, A. and Pérez-Pomares, J. M. (2004). The epicardium and epicardially derived cells (EPDCs) as cardiac stem cells. *Anat. Rec. A Discov. Mol. Cell Evol. Biol.* **276**, 43–57.

Winter, E. M. and Gittenberger-de Groot, A. C. (2007). Epicardium-derived cells in cardiogenesis and cardiac regeneration. *Cell Mol. Life Sci.* **64**, 692–703.

Wu, S. M., Fujiwara, Y., Cibulsky, S. M., Clapham, D. E., Lien, C. L., Schultheiss, T. M. and Orkin, S. H. (2006). Developmental origin of a bipotential myocardial and smooth muscle cell precursor in the mammalian heart. *Cell* **127**, 1137–1150.

Xu, H., Morishima, M., Wylie, J. N., Schwartz, R. J., Bruneau, B. G., Lindsay, E. A. and Baldini, A. (2004). Tbx1 has a dual role in the morphogenesis of the cardiac outflow tract. *Development* **131**, 3217–3227.

Yamagishi, H., Maeda, J., Hu, T., McAnally, J., Conway, S. J., Kume, T., Meyers, E. N., Yamagishi, C. and Srivastava, D. (2003). Tbx1 is regulated by tissue-specific forkhead proteins through a common sonic hedgehog-responsive enhancer. *Genes Dev.* **17**, 269–281.

Yang, L., Cai, C.-L., Lin, L., Qyang, Y., Chung, C., Monteiro, R. M., Mummery, C. L., Fishman, G. I., Cogen, A. and Evans, S. (2006). Isl1cre reveals a common bmp pathway in heart and limb development. *Development* **133**, 1575–1585.

Yashiro, K., Shiratori, H. and Hamada, H. (2007). Haemodynamics determined by a genetic programme govern asymmetric development of the aortic arch. *Nature* **450**, 285–288.

Yuan, S. and Schoenwolf, G. C. (2000). Islet-1 marks the early heart rudiments and is asymmetrically expressed during early rotation of the foregut in the chick embryo. *Anat. Rec.* **260**, 204–207.

Zaffran, S., Kelly, R. G., Meilhac, S. M., Buckingham, M. E. and Brown, N. A. (2004). Right ventricular myocardium derives from the anterior heart field. *Circ. Res.* **95**, 261–268.

Zhang, Z., Cerrato, F., Xu, H., Vitelli, F., Morishima, M., Vincentz, J., Furuta, Y., Ma, L., Martin, J. F., Baldini, A. and Lindsay, E. (2005). Tbx1 expression in pharyngeal epithelia is necessary for pharyngeal arch artery development. *Development* **132**, 5307–5315.

Zhang, Z., Huynh, T. and Baldini, A. (2006). Mesodermal expression of tbx1 is necessary and sufficient for pharyngeal arch and cardiac outflow tract development. *Development* **133**, 3587–3595.

Zhang, J., Lin, Y., Zhang, Y., Lan, Y., Lin, C., Moon, A., Schwartz, R., Martin, J. and Wang, F. (2008). Frs2{alpha}-deficiency in cardiac progenitors disrupts a subset of FGF signals required for outflow tract morphogenesis. *Development* **135**, 3611–3622.

Zhang, Y., Singh, M., Degenhardt, K., Lu, M., Bennett, J., Yoshida, Y. and Epstein, J. (2009). Tie2cre-mediated inactivation of plexinD1 results in congenital heart, vascular and skeletal defects. *Dev. Biol.* **325**, 82–93.

Zhou, W., Lin, L., Majumdar, A., Li, X., Zhang, X., Liu, W., Etheridge, L., Shi, Y., Martin, J., Van de Ven, W., Kaartinen, V., Wynshaw-Boris, A., McMahon, A. P., Rosenfeld,

M. G. and Evans, S. M. (2007). Modulation of morphogenesis by noncanonical wnt signaling requires ATF/CREB family-mediated transcriptional activation of TGFbeta2. *Nat. Genet.* **39**, 1225–1234.

Zhou, B., Ma, Q., Rajagopal, S., Wu, S., Domian, I., Rivera-Feliciano, J., Jiang, D., von Gise, A., Ikeda, S., Chien, K. and Pu, W. (2008a). Epicardial progenitors contribute to the cardiomyocyte lineage in the developing heart. *Nature* **454**, 109–113.

Zhou, B., von Gise, A., Ma, Q., Rivera-Feliciano, J. and Pu, W. T. (2008b). Nkx2-5- and isl1-expressing cardiac progenitors contribute to proepicardium. *Biochem. Biophys. Res. Commun.* **375**, 450–453.

CHAPTER TWO

Vascular Development–Genetic Mechanisms and Links to Vascular Disease

John C. Chappell[*,†] *and* Victoria L. Bautch[*,†,‡]

Contents

1. Introduction 44
   1.1. Vessel assembly 46
   1.2. Angiogenic expansion of blood vessel networks 47
   1.3. Vessel remodeling and stabilization 48
2. VEGF in Vascular Development 49
   2.1. VEGF-A signaling pathway 49
   2.2. VEGF-A mutations 50
   2.3. VEGF-C/VEGFR3 signaling pathway 52
   2.4. VEGF-C/VEGFR3 mutations 52
3. BMP in Vascular Development 53
   3.1. BMP signaling pathway 53
   3.2. BMP–VEGF crosstalk 53
   3.3. BMP mutations 54
4. Notch/Delta/Jagged in Vascular Development 55
   4.1. Notch/Delta/Jagged signaling pathway 55
   4.2. Notch coordination of endothelial crosstalk and heterogeneity 56
   4.3. Notch mutations 58
   4.4. Delta mutations 59
   4.5. Jagged mutations 59
5. Perspectives 60
Acknowledgments 62
References 62

[*] Department of Biology, The University of North Carolina at Chapel Hill, Chapel Hill, North Carolina, USA
[†] McAllister Heart Institute, The University of North Carolina at Chapel Hill, Chapel Hill, North Carolina, USA
[‡] Lineberger Comprehensive Cancer Center, The University of North Carolina at Chapel Hill, Chapel Hill, North Carolina, USA

Current Topics in Developmental Biology, Volume 90
ISSN 0070-2153, DOI 10.1016/S0070-2153(10)90002-1

Abstract

Vertebrate development depends on the formation of intricate vascular networks at numerous sites and in precise patterns; these vascular networks supply oxygen and nutrients to the rapidly expanding tissues of the embryo. Embryonic blood vessels are composed of endothelial cells and pericytes that organize and expand into highly branched conduits. Proper development of the vasculature requires heterogeneity in the response of endothelial cells to angiogenic cues provided by other tissues and organs. The pathogenesis of vascular diseases results from genetic mutations in pathways that provide these cues and in signals that coordinate endothelial heterogeneity during blood vessel formation. Here we provide a brief overview of different aspects of blood vessel formation and then discuss three essential signaling pathways that help establish vessel networks and maintain endothelial phenotypic heterogeneity during vascular development: the vascular endothelial growth factor (VEGF), bone morphogenetic protein (BMP), and the Notch/Delta/Jagged pathways. The VEGF pathway is critical for the initiation and spatial coordination of angiogenic sprouting and endothelial proliferation, BMP signaling appears to act in a context-dependent manner to promote angiogenic expansion and remodeling, and the Notch pathway is a critical integrator of endothelial cell phenotypes and heterogeneity. We also discuss human genetic mutations that affect these pathways and the resulting pathological conditions.

1. Introduction

Blood vessels are essential to the development and viability of vertebrate embryos. The vascular system facilitates oxygen and nutrient delivery to cells and removal of metabolic waste from cells, processes that require transport as embryos grow beyond a size that allows for passive diffusion. Thus, blood vessels and the heart are the first organs to function during mammalian development, yet vessel networks continue to form and remodel dynamically even as they function. All stages of vascular development require complex interactions among genetic programs, molecular cues, and cellular behaviors, and these inputs are tightly regulated both spatially and temporally (Adams and Alitalo, 2007; Carmeliet, 2005; Jain, 2003; Risau, 1997). Mutations in numerous vascular genes lead to pathologies characterized by aberrant vessel formation or function (Fig. 2.1).

Insight into the mechanisms underlying vascular development has led to clinically relevant therapies for diseases with vascular components. For example, the prominence of the vascular endothelial growth factor (VEGF) pathway in development and in tumor angiogenesis has instigated the development of several antiangiogenesis drugs that specifically target this pathway (Ferrara, 2005; Heath and Bicknell, 2009; Jain, 2005). However, therapeutic approaches have been less effective than hoped, and we now

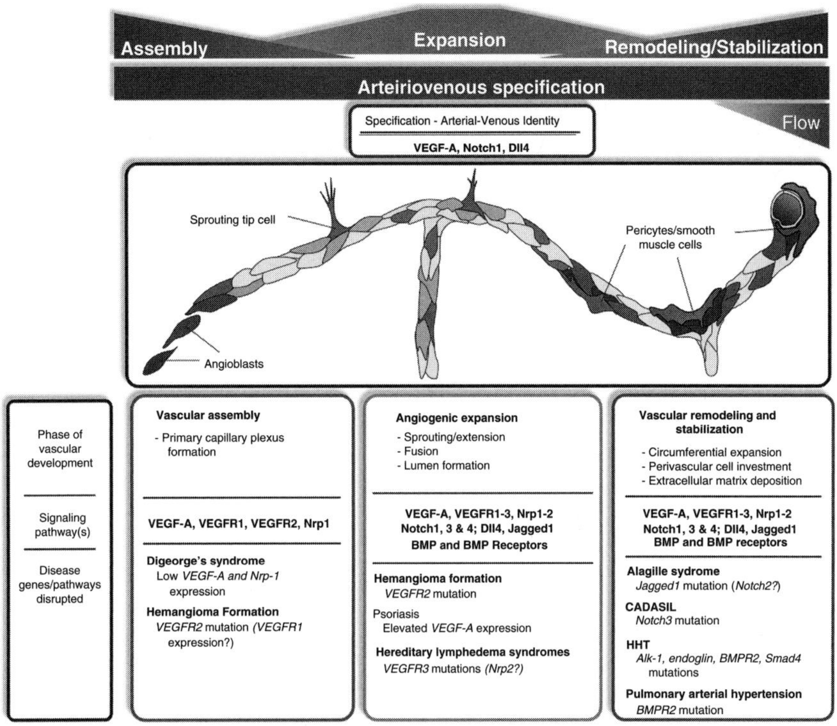

Figure 2.1 Overview of blood vessel formation. The morphological events constituting vessel assembly, expansion, and remodeling intensify in activity and then attenuate during phase transitions, while arteriovenous specification events likely occur throughout the vascular development program. The different stages of vascular development utilize overlapping molecular pathways for these events, and endothelial phenotypic heterogeneity (represented schematically by color variation, i.e., shades of red, brown, and violet) is an essential aspect of each stage. Genetic lesions in human genes that disrupt these pathways are associated with specific vascular diseases. (See Color Insert.)

realize that aspects of blood vessel formation need better elucidation. One such aspect is the concept that endothelial cells in developing vessel networks exhibit phenotypic heterogeneity. This heterogeneity is found at several levels in normal vessels and is essential for vascular development. For example, tip cells migrate to form new sprouts and stalk cells follow behind and proliferate, lateral base cells provide local guidance to emerging sprouts, and endothelial cells behind the sprouting front are said to form a phalanx and become quiescent (Bautch, 2009; Chappell *et al.*, 2009; Gerhardt *et al.*, 2003; Hellstrom *et al.*, 2007; Mazzone *et al.*, 2009). Perturbation of any of these relationships compromises vessel development and function. The endothelial cell heterogeneity that drives blood vessel formation will be highlighted throughout this review.

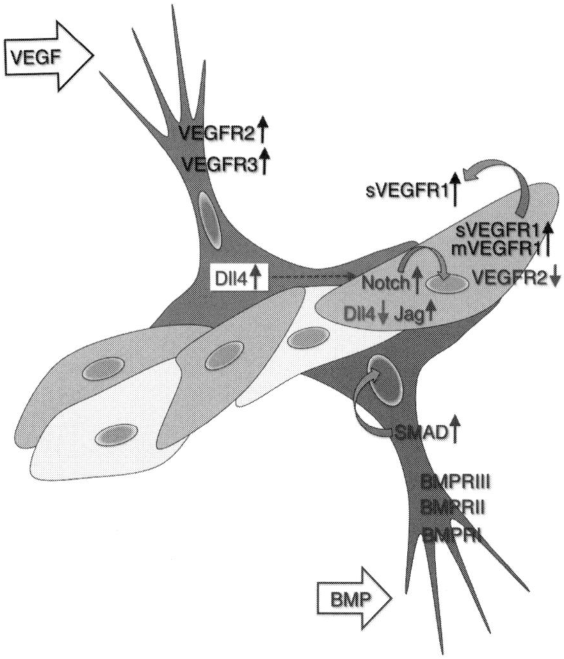

Figure 2.2 Molecular signaling and endothelial heterogeneity in blood vessel formation. As vessels undergo sprouting angiogenesis, tip cells (dark grey) upregulate VEGFR2 and VEGFR3 signaling in response to VEGF-A, and this leads to Dll4 upregulation. Dll4 binds Notch on a neighboring stalk/lateral base cell (medium grey) and activates Notch signaling, which leads to upregulation of VEGFR1, secretion of soluble VEGFR1, and downregulation of VEGFR2. Jagged antagonizes Dll4 in stalk cells. An alternative mode of sprouting (lower cell) involves activation of BMP receptors on some endothelial cells.

We first present a brief overview of vascular development and then focus on how three critical signaling pathways function and interact to promote endothelial heterogeneity: VEGF, bone morphogenetic protein (BMP), and Notch/Delta/Jagged (Fig. 2.2). These pathways were chosen from among many molecular inputs because they both contribute to and respond to endothelial heterogeneity in important ways. We also describe genetic mutations that disrupt these pathways and affect vascular processes in human development and disease.

1.1. Vessel assembly

Blood vessels are one of the first embryonic organ systems to develop and function during vertebrate development. Mesoderm cells give rise to endothelial precursor cells in several ways. Most angioblasts likely derive from progressive restriction of mesoderm to the endothelial lineage in response to signals such as IHH (Indian Hedgehog), FGF2 (fibroblast growth factor), BMP, and VEGF (Goldie *et al.*, 2008). Some endothelial cells may derive from hemangioblasts, bipotential cells that give rise to both hematopoietic and endothelial cells (Goldie *et al.*, 2008; Vogeli *et al.*, 2006), and some endothelial cells retain the ability to produce hematopoietic cells and are called hemogenic endothelium (Goldie *et al.*, 2008; Yoshimoto and Yoder, 2009).

During this initiation phase, called vasculogenesis, angioblasts aggregate into cords and differentiate into endothelial cells. For example, dorsal aorta formation in *Xenopus* and zebrafish models involves the recruitment and orientation of angioblasts at the midline by VEGF and other molecular signals (Cleaver *et al.*, 1997; Jin *et al.*, 2005). In other tissue beds, however, it remains unclear exactly how these vessel cords arise and form basic networks. Nevertheless, the rudimentary structures within these networks provide initiation points for the subsequent expansion of vessels. A hierarchy of vessels (i.e., arteries, capillaries, veins) is initially established by the activity of arteriovenous specification molecules during vasculogenesis (Swift and Weinstein, 2009). Sonic hedgehog expression from the zebrafish notochord induces somite VEGF production, which upregulates arterial specification molecules such as ephrinB2 through Notch signaling (Lawson *et al.*, 2001, 2002). Subsequent maintenance of arteriovenous identity requires flow-mediated remodeling, since in the chick and mouse yolk sac alterations in flow dynamics perturb vessel maturation (le Noble *et al.*, 2004; Lucitti *et al.*, 2007).

1.2. Angiogenic expansion of blood vessel networks

Primitive vessels must branch and establish new connections. This angiogenic expansion phase requires phenotypic heterogeneity among endothelial cells in their response to stimulatory cues (Fig. 2.2). Sprouting angiogenesis entails the selection of an individual endothelial cell for outward migration from a parent vessel (Gerhardt *et al.*, 2003; Phng and Gerhardt, 2009). This leading tip cell responds to a VEGF gradient by migrating up the gradient (Ruhrberg *et al.*, 2002). Endothelial cells behind the tip cell in the emerging sprout, termed *stalk cells*, do not migrate independently, but instead proliferate and eventually lumenize. The mechanism by which vessels acquire a patent lumen remains controversial and may actually be region specific (Iruela-Arispe and Davis, 2009). The lumen of the mouse dorsal aorta forms by organized changes in endothelial cell shape (Strilic *et al.*, 2009), while an alternative model for lumenization has been proposed in which intracellular and intercellular vacuoles form and fuse along connected endothelial cells (Blum *et al.*, 2008; Kamei *et al.*, 2006). Before expanding the vessel lumen, the stalk cells are initially lateral to the emerging tip cell. In these lateral base areas, endothelial cells provide soluble Flt-1/VEGFR1, which neutralizes VEGF to provide local guidance cues for the emerging sprout (Chappell *et al.*, 2009). The tip cell must eventually fuse with its target to establish a new branch, although the mechanism of endothelial cell fusion has yet to be elucidated. Observations from the developing mouse retina suggest that tip cell filopodia engage with those of a nearby tip cell to form a "bridge" and the foundation of a new vessel (Bentley *et al.*, 2009). This connection then develops a patent lumen so that blood can flow through the newly formed branch. During the transition from active sprouting to quiescence, endothelial cells adopt a "phalanx" phenotype that promotes vessel integrity and stabilizes the vasculature (Bautch, 2009; Mazzone *et al.*, 2009).

1.3. Vessel remodeling and stabilization

As the vasculature matures, vessel remodeling also contributes to network patterning. Changes in blood flow, metabolic demands, and growth factor secretion induce pruning of some vessels (Benjamin *et al.*, 1999). In contrast, other vessels become more stable through increased adhesion between endothelial cells and deposition of matrix and basement membrane (Stratman *et al.*, 2009). Hemodynamic stress and local molecular signals in the microenvironment regulate circumferential growth of vessels (Garcia-Cardena *et al.*, 2001; Masumura *et al.*, 2009; Skalak and Price, 1996; Zeng *et al.*, 2007). For example, skeletal muscle vessels exposed to elevated circumferential wall stress have increased diameters (Price and Skalak, 1996). Furthermore, mechanical and molecular factors also direct the recruitment and investment of mural cells such as pericytes and vascular smooth muscle cells (Armulik *et al.*, 2005; Gaengel *et al.*, 2009; Lindahl *et al.*, 1997). Platelet-derived growth factor (PDGF)-BB and transforming growth factor (TGF)-β promote vessel maturation by stimulating mural cell precursor migration and differentiation, respectively (Hirschi *et al.*, 1998). These stimuli work together with arteriovenous specification cues to facilitate proper development and enlargement of arteries and veins (Jones *et al.*, 2006).

Each phase of vascular development requires the integration of multiple signaling pathways to precisely coordinate cell behavior such that multicellular vessel networks form and expand. The following sections will describe how the VEGF, BMP, and Notch pathways provide this critical regulation for vascular development and morphogenesis, and how genetic mutations in components of these pathways contribute to human pathology.

2. VEGF in Vascular Development

The VEGF pathway is a vital regulator of vascular development and has been the subject of several comprehensive reviews (Ferrara *et al.*, 2003; Olsson *et al.*, 2006). We provide a brief characterization of this pathway, describe observations from genetic mutation experiments, and discuss human genetic disorders related to the pathway.

2.1. VEGF-A signaling pathway

VEGF-A is the predominant ligand for VEGF receptor 2 (VEGFR2, Flk-1 in mice), which positively signals for endothelial cell proliferation, migration, and survival (Shibuya and Claesson-Welsh, 2006). VEGF receptor 1 (VEGFR1, Flt-1 in mice), which has transmembrane and soluble isoforms due to alternative splicing, selectively binds VEGF-A, -B, and placental growth factor (Fischer *et al.*, 2008; Kendall and Thomas, 1993; Shibuya, 2006). Deletion of the intracellular signaling domain of Flt-1 is compatible with normal vascular development, indicating that signaling through Flt-1 is not required for its developmental role (Hiratsuka *et al.*, 1998). We showed that Flt-1 acts primarily as a ligand sink during vessel development, and it thus negatively modulates the amount of available VEGF-A that can bind and activate its receptor Flk-1 (Kappas *et al.*, 2008; Roberts *et al.*, 2004). This modulation negatively regulates endothelial proliferation but paradoxically positively regulates branching morphogenesis, and recent work from our laboratory illustrates a critical role for the soluble Flt-1 isoform in providing local sprout guidance cues necessary for proper vessel branching (Chappell *et al.*, 2009; Kearney *et al.*, 2002, 2004; Zeng *et al.*, 2007). The VEGF coreceptor Neuropilin (Nrp)-1 also provides regulation of the VEGF pathway (Larrivee *et al.*, 2009). Specifically, Nrp-1 enhances signaling by increasing Flk-1 affinity for VEGF-A (Whitaker *et al.*, 2001) and facilitating Flk-1 clustering (Soker *et al.*, 2002). This in turn promotes endothelial cell migration and guidance for proper vessel patterning (Gerhardt *et al.*, 2004; Jones *et al.*, 2008).

Spatial regulation of VEGF-A ligand availability is critical for coordinating its effects on vessel network formation. Alternative splicing of VEGF-A mRNA generates three primary isoforms (Tischer *et al.*, 1991), and the absence or presence of heparin-binding domains determines the affinity for the extracellular matrix and thus the spatial distribution of each isoform (Ruhrberg *et al.*, 2002; Stalmans *et al.*, 2002). Recent work suggests that matrix-bound and soluble VEGF-A isoforms provide distinct signaling cues to endothelial cells (Chen *et al.*, 2010). Endothelial cells themselves contribute to the spatial regulation of VEGF by responding differentially to stimulatory cues to express VEGF receptors in a spatially heterogeneous manner, and Flk-1 receptor activation is also heterogeneous in vessel networks (Jakobsson *et al.*, 2009; Kappas *et al.*, 2008; Sainson *et al.*, 2008). Moreover, heterogeneity of soluble Flt-1 expression from endothelial cells is thought to be required for local sprout guidance (Chappell *et al.*, 2009). Thus, proper VEGF signaling, and in turn blood vessel morphogenesis, requires the appropriate spatial distribution of VEGF ligand and receptors and heterogeneous activation of VEGF-mediated signaling.

2.2. VEGF-A mutations

2.2.1. In development

Genetic mutations in the mouse VEGF signaling pathway show critical roles for both ligands and receptors. Heterozygous disruption of the *Vegf-A* gene is lethal at approximately embryonic day (E) 9.5 due to abnormal blood island formation, perturbed angiogenesis, and disruption of vessel organization and lumen formation, suggesting that strong quantitative regulation of the VEGF pathway is important for development (Carmeliet *et al.*, 1996; Ferrara *et al.*, 1996). Vascular-specific deletion of *Vegf-A* leads to endothelial apoptosis and postnatal lethality despite intact paracrine VEGF (Lee *et al.*, 2007), indicating that vessel-derived VEGF is required for vascular maintenance. Genetic deletion of *flt-1* causes overproliferation of endothelial cells but reduced branching that results in embryonic lethality at E9.5 (Fong *et al.*, 1995; Kearney *et al.*, 2002, 2004). The vessel dysmorphogenesis observed in the *flt-1*$^{-/-}$ loss-of-function situation is consistent with the model described above in which an effective gain of function for VEGF signaling results from perturbed *flt-1* function (Kappas *et al.*, 2008; Roberts *et al.*, 2004). Lack of *flk-1 (VEGFR2)*, the primary transducer of VEGF signaling, disrupts both vasculogenesis and angiogenesis, and *flk-1*$^{-/-}$ mice die *in utero* (E8.5–9.5) from lack of blood island and primitive vessel network development (Shalaby *et al.*, 1995). *Nrp-1* knockout mice suffer from developmental defects in both the nervous system and the vascular system, including impaired brain, spinal cord, and yolk sac vascularization, and endothelial specific deletion of *Nrp-1* leads to dilated and poorly branched vessels (Gu *et al.*, 2003; Kawasaki *et al.*, 1999). Additionally, the yolk sacs of mice with mutations in both *Nrp-1* and *-2* are avascular, and these mice exhibit severely disrupted embryonic angiogenesis (Takashima *et al.*, 2002). These co-receptors thus make important contributions to the precise regulation of endothelial responses to VEGF.

2.2.2. In human disease

How genetic defects in VEGF signaling influence human development is unclear. Because VEGF signaling is tightly regulated both quantitatively and qualitatively in normal vessel development, gross disruptions in pathway components are likely to cause early fetal lethality in humans. Recent data, however, suggest that defective expression of VEGF pathway components contributes to several diseases. DiGeorge syndrome is a common genetic disorder (affecting 1 in 4000 infants) with multiple developmental abnormalities, including defects in blood vessel formation (Lindsay, 2001). Deletion of approximately 3 million base pairs in chromosome 22q11 (del22q11) results in haploinsufficiency of *tbx1*, but its variable expression suggests a role for additional modifiers. Stalmans and colleagues demonstrated interactions between Tbx1 and VEGF-A as well as an association between VEGF promoter haplotypes and cardiovascular birth defects in DiGeorge syndrome patients (Stalmans *et al.*, 2003). In addition, aberrant pharyngeal arch artery formation was exacerbated experimentally by VEGF knockdown in developing zebrafish with a concurrent tbx-1 knockdown, and regression of this artery was preceded by downregulation of Nrp-1, suggesting an increased risk for vascular complications in del22q11 patients with genetic lesions in VEGF pathway components (Stalmans *et al.*, 2003).

Genetic lesions in VEGF-A have also been associated with early-onset psoriasis, a chronic inflammatory disease with abnormal vessel expansion in the skin (Creamer *et al.*, 1997; Young *et al.*, 2004). Dermal microvessels within psoriatic plaques are tortuous, excessively dilated and permeable, and composed of hyperproliferative endothelial cells (Braverman and Sibley, 1982). Genetic analysis of psoriasis patients revealed the presence of single nucleotide polymorphisms in *Vegf-A* (Young *et al.*, 2004), and indeed elevated levels of VEGF-A protein are associated with this disease (Detmar *et al.*, 1994; Young *et al.*, 2004). Interestingly, the levels of circulating VEGFR1 were also higher in affected patients, which may represent hyperactivity of a feedback loop regulating VEGF-A availability or a compounding defect in VEGFR1 gene expression and activity.

Genes encoding the VEGF receptors have also been linked to vascular pathologies. In endothelial cells derived from infantile hemangiomas, a missense mutation in *Vegfr2* causes increased complex formation with tumor endothelial marker-8 (TEM-8) and β1-integrin (Boscolo and Bischoff, 2009; Jinnin *et al.*, 2008). The aberrant association of these molecules leads to decreased nuclear factor of activated T cells (NFAT) transcription of VEGFR1 and ultimately to focal regions of overgrown and disorganized cutaneous vessels. As stated previously, loss of VEGFR1 is essentially a gain of function for VEGF activity and results in elevated VEGF signaling and vessel dysmorphogenesis. Therefore, it follows that genetic mutations disrupting the tight regulation of VEGF signaling would result in vascular malformations. Nevertheless, more work remains to be done to uncover the exact mechanistic role for the VEGF signaling components in hemangioma formation and other disorders.

2.3. VEGF-C/VEGFR3 signaling pathway

Blood and lymphatic endothelial cells express VEGFR3 (Flt-4 in mice), a receptor for VEGF-C and -D (Saharinen *et al.*, 2004). During vascular development, higher *Vegfr3* expression in endothelial tip cells relative to neighboring stalk cells (Tammela *et al.*, 2008) indicates that this receptor also participates in the endothelial phenotypic heterogeneity that is necessary for proper network expansion. Expression of *Vegfr3* gradually becomes limited to the lymphatic endothelial cells, where it mediates VEGF-C signaling for lymphangiogenesis (Karpanen *et al.*, 2006b; Makinen *et al.*, 2001). Neuropilin-2, which interacts with VEGFR3, is also expressed by the lymphatic vessel endothelium and can bind VEGF-C (Karpanen *et al.*, 2006a; Xu *et al.*, 2010; Yuan *et al.*, 2002).

2.4. VEGF-C/VEGFR3 mutations

2.4.1. In development

Loss of *Vegf-C* results in impaired lymphatic vessel formation and abnormal fluid accumulation in various tissues (Karkkainen *et al.*, 2004). Cardiovascular failure occurs around E9.5 in *Vegfr3*$^{-/-}$ embryos, which also have vessel remodeling abnormalities and pericardial edema, highlighting a role for this receptor in vascular development in addition to its role in lymphangiogenesis (Dumont *et al.*, 1998). Genetic loss of *Nrp-2* impairs lymphatic endothelial cell proliferation as well as the development of capillaries and lymphatic microvessels (Yuan *et al.*, 2002). Given the importance of these VEGF pathway components in both angiogenesis and lymphangiogenesis, it is not surprising that developmental defects arise when their expression is disrupted.

2.4.2. In human disease

Genetic mutations in *Vegfr3* are linked to lymphedema in humans. Lymphedema arises from dilated lymphatic capillaries, which prevent adequate removal of lymph fluid from tissues (Ferrell, 2002). Congenital lymphedema in some families is associated with the *Vegfr3* locus on distal chromosome 5q (Cueni and Detmar, 2006). Moreover, missense mutations in the *Vegfr3* gene have been identified in patients with hereditary, early-onset lymphedema (Witte *et al.*, 2001). These genetic abnormalities impair the signaling activity of the VEGFR3 receptor (Alitalo and Carmeliet, 2002), thus leading to defective formation of lymphatic vessels and in turn excessive tissue fluid accumulation. Mutations in *Vegf-C* and *Nrp-2* have yet to be linked to human pathology, but establishing their contribution to lymphatic diseases could lead to improved therapies.

3. BMP in Vascular Development

BMPs are part of the TGF-β superfamily of signaling molecules, and the role of TGF-β family members in angiogenesis has been thoroughly reviewed (Holderfield and Hughes, 2008). The importance of BMP signaling in vascular development is increasingly apparent, although a global model to describe BMP effects on vessels is lacking (David *et al.*, 2009; Moreno-Miralles *et al.*, 2009). One effect of perturbing BMP signaling in vascular development is that distinctions between arteries and veins are often not maintained, leading to arteriovenous shunts, and an emerging theme is that BMP signaling has differential effects on arteries versus veins.

3.1. BMP signaling pathway

There are multiple components of BMP signaling that interact in complex ways during vascular development. Thus, it is quite likely that numerous unique combinations of BMP signaling components occur in different vessels, leading to context-dependent effects of BMP signaling on development and maintenance programs. In general, a Type II BMP receptor (i.e., BmprII or ActRII) and a Type I receptor (i.e., Alk 1/3/6) form heterodimers via ligand binding (i.e., BMP2, BMP4, BMP7) and also utilize a co-receptor (sometimes called a Type III receptor, i.e., endoglin) to initiate signaling. Signaling usually goes through transcription factors called SMADS (i.e., SMAD 1/5/8) that complex with a co-SMAD (SMAD 4) for translocation to the nucleus and modulation of transcription. However, BMP can also signal through MEK/ERK and p38 MAPK. BMP inhibitors such as noggin, chordin, gremlin, and BMP-binding endothelial cell precursor-derived regulator (BMPER) also modulate BMP signaling. In fact, the notochord in amniotes expresses BMP antagonists that promote the formation of an avascular zone at the embryonic midline (Bressan *et al.*, 2009; Reese *et al.*, 2004).

3.2. BMP–VEGF crosstalk

There is evidence that BMP stimulates VEGF expression and secretion in different cell types. Several studies showed that BMPs stimulate VEGF expression in osteoblast cells, and BMP-dependent enhanced secretion of VEGF was reported in retinal pigment epithelial cells (Deckers *et al.*, 2002; Kozawa *et al.*, 2001; Vogt *et al.*, 2006; Yeh and Lee, 1999). However, BMPs also induce endothelial migration, filopodia formation, and tube formation independent of VEGF activity, via activation of the transcription factor Id1 through SMADs or through ERK signaling (Valdimarsdottir *et al.*, 2002; Pi *et al.*, 2007). Moreover, there appear to be distinct requirements for VEGF vs. BMP signaling in blood vessel formation in zebrafish, as VEGF perturbations selectively affect the dorsal aorta and intersegmental vessels, while BMP perturbations selectively affect the axial vein and venous plexus immediately ventral to the aorta (D. M. Wiley, J. Hao, C. C. Hong, V. L. B. and S-W. Jin, submitted). These studies suggest that in some cases these pathways allow closely apposed vessels to expand networks in different directions.

3.3. BMP mutations

3.3.1. In development

Genetic evidence supports a role for BMP signaling in vascular development. Loss of the Type I receptor Alk 1 leads to lethality with angiogenesis defects in both mice and zebrafish (Oh *et al.*, 2000; Roman *et al.*, 2002; Urness *et al.*, 2000), while endothelial specific deletion of Alk 1 leads to arteriovenous malformations (Park *et al.*, 2006). Loss of the Type II receptor BMPRII is lethal early in development, and conditional loss in endothelial cells was associated with pulmonary arterial hypertension (PAH) later in life (Beppu *et al.*, 2004; Hong *et al.*, 2008). However, another group found that shRNA knockdown of the same receptor leads to vascular defects in multiple vascular beds, and signaling from isolated endothelial cells was attenuated, along with defects in smooth muscle investment (Liu *et al.*, 2007). Genetic deletion of co-receptors is lethal with effects on vessels. Loss of β-glycan leads to lack of coronary vessel development, while loss of endoglin results in both vascular and smooth muscle defects and arteriovenous malformations (Arthur *et al.*, 2000; Li *et al.*, 1999; Sorensen *et al.*, 2003). BMP ligands are proangiogenic by several criteria. BMP2, BMP4, and BMP6 stimulate endothelial cells in culture, and BMP-expressing beads stimulate angiogenesis in embryos (de Jesus Perez *et al.*, 2009; Nimmagadda *et al.*, 2005; Pi *et al.*, 2007; Ren *et al.*, 2007; Teichert-Kuliszewska *et al.*, 2006; Yang *et al.*, 2005).

3.3.2. In human disease

The strongest genetic evidence linking BMP to angiogenesis in humans is a set of mutations called hereditary hemorrhagic telangiectasia (HHT). These mutations are characterized by vessel malformations including dilated and fragile vessels and arteriovenous shunts (Abdalla and Letarte, 2006; Fernandez *et al.*, 2006). *Alk-1* and *endoglin* mutations primarily contribute to the development of HHT (Johnson *et al.*, 1996; McAllister *et al.*, 1994), while defective *BMPRII* and *Smad4* may contribute to disease progression and associated pulmonary artery disease (Abdalla *et al.*, 2004; Gallione *et al.*, 2004; Harrison *et al.*, 2003; Trembath *et al.*, 2001). PAH, a condition characterized by a decrease in the small, distal pulmonary arteries and excessive muscularity of remaining arteries (Rabinovitch, 2008), is associated with mutations in *BMPR2* (Deng *et al.*, 2000; Lane *et al.*, 2000; Machado *et al.*, 2006). The penetrance of *BMPR2* mutations is only about 20%, suggesting that additional genetic lesions in BMP signaling may contribute to the phenotype (Morrell, 2006).

There is still much to learn regarding the role of BMP signaling in vascular development. The number of pathway components, the role of BMP in early development, and potential redundancy have all contributed to the lack of a consensus model to date. In this regard, a recent study in zebrafish provides strong evidence for context-dependent requirements for BMP in vascular development, as described above (Wiley *et al.*, submitted). The elucidation of vascular BMP signaling requirements in the relatively simple fish embryo may inform hypotheses regarding the role of BMP signaling in vessel formation in higher vertebrates.

4. Notch/Delta/Jagged in Vascular Development

The Notch signaling pathway is essential for vascular development. Recent evidence implicates Notch signaling as a critical integration node for the establishment and/or maintenance of endothelial heterogeneity within developing vessels and as a pathway that intersects and perhaps integrates input from numerous other pathways critical to vascular development. There are several recent and comprehensive reviews, so this review will highlight experimental manipulations of Notch pathway genes and vascular pathologies associated with Notch genetic mutations (Hofmann and Iruela-Arispe, 2007; Phng and Gerhardt, 2009; Roca and Adams, 2007; Siekmann *et al.*, 2008).

4.1. Notch/Delta/Jagged signaling pathway

Briefly, the four Notch receptors (Notch1–4) are transmembrane proteins that engage with five different ligands – Delta-like (Dll)1, Dll3, and Dll4, and Jagged-1 and -2. All but Notch 2 are expressed in developing vessels. Several other regulators are essential for proper Notch signaling, including γ-secretase, which cleaves the Notch intracellular domain (NICD) following receptor binding of a ligand. Subsequently, the NICD enters the nucleus and forms a complex that triggers target gene activation (i.e., Hey-1, Hey-2, Hes, Nrarp) (Fischer *et al.*, 2004; Phng *et al.*, 2009). The Notch signaling system provides a means for cell–cell communication, because the Notch receptors and ligands interact at the interface between two neighboring cells (Bray, 2006). Thus, adjacent cells can establish distinctions between themselves and their neighbors, which is an important mechanism for establishing and maintaining phenotypic heterogeneity within a group of cells (Ghabrial and Krasnow, 2006).

The Notch pathway is implicated at several stages of blood vessel development, including endothelial cell differentiation and arteriovenous specification. In the developing zebrafish, Notch signaling in the nascent mesoderm acts as a cell-fate switch for the specification of endothelial and hematopoietic progenitor cells (Lee *et al.*, 2009). Elevated levels of Notch signaling in zebrafish embryos led to fewer endothelial cells, while reduced Notch activity increased their number at the expense of cells specified for the hematopoietic lineage. In addition, vascular smooth muscle cell differentiation, recruitment, and investment are perturbed by disruptions in the Notch pathway (Domenga *et al.*, 2004; High *et al.*, 2008; Kim *et al.*, 2008). Endothelial-specific deletion of *Jagged1* in mice, for instance, results in deficient smooth muscle cell differentiation, cardiovascular defects, and embryonic lethality (High *et al.*, 2008). The Notch pathway also plays a critical role in arteriovenous specification. Several Notch receptors and ligands are specifically expressed by arterial endothelial cells (Mailhos *et al.*, 2001; Shutter *et al.*, 2000; Villa *et al.*, 2001), suggesting the importance of the Notch pathway in establishing broader artery vs. vein specification within the developing vasculature (Lawson *et al.*, 2001; Siekmann and Lawson, 2007). Notch signaling is implicated in the transcriptional regulation of artery specification (Zhong *et al.*, 2000, 2001) via upregulation of artery-specific markers such as ephrinB2 (Grego-Bessa *et al.*, 2007; Iso *et al.*, 2006; Lawson *et al.*, 2001). The VEGF pathway intersects with Notch signaling to regulate arterial endothelial cell identification during development (Lawson *et al.*, 2002). The intersection of these pathways may also be important after the onset of blood flow to reinforce arterial identity and promote subsequent expansion of arteries (Masumura *et al.*, 2009). However, the exact molecular crosstalk between the pathways following exposure to blood flow remains unknown.

4.2. Notch coordination of endothelial crosstalk and heterogeneity

4.2.1. Notch–VEGF

It is becoming increasingly clear that endothelial cells have heterogeneous responses to angiogenic stimuli, and that this functional heterogeneity is important for proper blood vessel formation. These differences in signaling between adjacent endothelial cells are organized by the Notch pathway to establish the proper relationships between these cells, and their subsequent behaviors. During sprouting angiogenesis, for example, Notch receptor–

ligand interactions are proposed to establish phenotypic heterogeneity among vessel endothelial cells by critically regulating their responsiveness to VEGF stimulation. Local heterogeneity in VEGF responsiveness causes endothelial cells that experience relatively higher levels of VEGF signaling to increase Dll4 expression. Elevated Dll4 in turn further increases the cell's sensitivity to VEGF (via upregulation of VEGFR2 and VEGFR3), and this cell becomes the tip cell selected for outward migration from the parent vessel (Hellstrom *et al.*, 2007; Lobov *et al.*, 2007; Sainson *et al.*, 2005; Siekmann and Lawson, 2007; Suchting *et al.*, 2007; Tammela *et al.*, 2008). Notch receptors on neighboring stalk cells are engaged by the Dll4 ligands, and this activation of Notch signaling results in decreased VEGFR2 expression that is thought to inhibit sprouting from these cells (Suchting *et al.*, 2007). Evidence for interactions between the Notch pathway and VEGFR1 also exists (Funahashi *et al.*, 2010; Harrington *et al.*, 2008; J.C.C. and V.L.B., unpublished results; Suchting *et al.*, 2007; Taylor *et al.*, 2002). Thus, it is intriguing to speculate that VEGFR1 expression in lateral base cells is induced by Notch signaling to reduce VEGF ligand availability, preventing their outward migration and guiding the sprouting tip cell (Chappell *et al.*, 2009). Elevated expression of Jagged1 on stalk cells also contributes to endothelial phenotypic heterogeneity during angiogenesis (Benedito *et al.*, 2009). Stalk cell Jagged1 antagonizes Dll4 activity and thus reduces the induction of Notch signaling in the adjacent tip cell. The tip cell therefore maintains its responsiveness to VEGF stimulation and migrates outward to establish a new branch (Benedito *et al.*, 2009). Endothelial cell proliferation is also regulated by Notch (Hellstrom *et al.*, 2007; Leslie *et al.*, 2007; Lobov *et al.*, 2007; Siekmann and Lawson, 2007; Suchting *et al.*, 2007). However, it remains unclear how stalk cells, which experience elevated Notch signaling (i.e., suppression of proliferation) (Liu *et al.*, 2006; Noseda *et al.*, 2004), still undergo increased proliferation for vessel lengthening (Gerhardt *et al.*, 2003).

Although there is strong genetic evidence that Notch signaling and endothelial heterogeneity are important in vessel formation, the endothelial expression patterns of Notch pathway components do not fully follow the predictions of the models for establishing heterogeneity. This lack of concordance may reflect temporal oscillations in Notch signaling that are thought to be involved in establishing endothelial heterogeneity (Bentley *et al.*, 2009). Moreover, the mechanisms responsible for the initial endothelial heterogeneity that allows endothelial cells to respond differentially to comparable levels of angiogenic stimulation are still unknown.

4.2.2. Notch–BMP

In contrast to the extensive information regarding Notch–VEGF intersections, little is known about if and how Notch intersects with BMP signaling.

Both Notch and BMP stimulate expression of ER71, an Ets family transcription factor that stimulates formation of Flk-1-positive mesoderm, giving rise to both blood and vessel progenitors (Lee *et al.*, 2008). However, each pathway seems to independently regulate ER71 expression, since combined blockade is additive. There are a few examples of pathway intersections (Kluppel and Wrana, 2005). In myogenic cells, Smad1 activation downstream of BMP4 acts in both Notch-dependent and Notch-independent ways to activate Notch target genes such as Hey-1, and a physical interaction between Smad1 and the NICD was reported (Dahlqvist *et al.*, 2003). In endothelial cells, lack of cell contact (and Notch signaling) allowed BMP to activate Smad1 and upregulate Id1, which led to endothelial cell migration. Cell–cell contact activated Notch signaling and this synergized with BMP signals (via a Smad1–NICD complex) to activate Herp2, another Notch target that blocks endothelial migration (Itoh *et al.*, 2004).

4.3. Notch mutations

4.3.1. In development

Genetic manipulations of components in the Notch signaling system reveal the importance of this pathway for normal vascular development. However, because the Notch pathway is important in numerous embryonic tissues for developmental decisions, vascular-specific gene manipulations have been helpful for interpreting observations from the global genetic knockouts. Defects in vascular remodeling, and particularly of the large caliber vessels, arise from mutations in the *Notch1* gene (Krebs *et al.*, 2000). This same study showed that, while vascular development proceeds normally in *Notch4*$^{-/-}$ mice, *Notch1*$^{-/-}$ *Notch4*$^{-/-}$ double mutant embryos had more severe abnormalities in angiogenesis and in the formation of major vessels. Furthermore, endothelial-specific deletion of Notch1 recapitulates the vascular defects of the global Notch1 knockout mouse, suggesting an endothelial-intrinsic role for Notch1 in blood vessel formation (Limbourg *et al.*, 2005). Eye, heart, and kidney vessels in *Notch2* mutant mice exhibited dysmorphogenesis, but perinatal death was attributed to aberrant glomerular development in the kidney (McCright *et al.*, 2001). *Notch3*$^{-/-}$ mice are viable despite marked arterial defects resulting from impaired smooth muscle cell differentiation and maturation (Domenga *et al.*, 2004).

4.3.2. In human disease

Disrupted expression of Notch signaling components has been identified in congenital disorders that have distinct vascular pathologies. Cerebral autosomal dominant arteriopathy with subcortical infarcts and leukoencephalopathy, or CADASIL, results from inherited missense mutations in *Notch3* (Joutel *et al.*, 1996). Congruent with the arterial defects seen in *Notch3*$^{-/-}$ mice (Ruchoux *et al.*, 2003), CADASIL patients suffer from irregularities in dermal and cerebral arteries (Joutel and Tournier-Lasserve, 1998).

Degeneration of vascular smooth muscle cells and fibrotic accumulation around vessels contribute to the reduction in artery lumen diameter, resulting in migraines, dementia, and stroke (Chabriat *et al.*, 1995). Vascular abnormalities in experimental *Notch3* disruption are consistent with the clinically observed defects in CADASIL patients, but they do not fully explain the underlying pathogenesis. Thus, compounding genetic defects are likely to be involved, highlighting the importance of further investigations of the Notch pathway in human disease.

4.4. Delta mutations

4.4.1. In development

Dll1$^{+/-}$ mice survive to adulthood, and these mice show defects in postnatal arteriogenesis when challenged by an arterial occlusion (Limbourg *et al.*, 2007). By contrast, disruption of one allele of *Dll4* is lethal owing to irregularities in arterial branching, the specification of arterial endothelial cells, and vessel remodeling (Domenga *et al.*, 2004; Duarte *et al.*, 2004; Gale *et al.*, 2004). Analysis of heterozygous *Dll4*$^{+/-}$ embryos and early postnatal mice provided insight into some of these defects, which included increased angiogenic sprouting and excessive endothelial proliferation (Hellstrom *et al.*, 2007; Lobov *et al.*, 2007; Suchting *et al.*, 2007). Thus, lethality associated with *Dll4* haploinsufficiency demonstrates a strong dose-dependent regulation of vascular morphogenesis by Notch signaling, similar to that seen by VEGF signaling. Genetic disorders associated with *Dll4* mutations have yet to be identified, perhaps because misregulation of this gene is likely to be embryonic lethal owing to aberrant vascular development.

4.5. Jagged mutations

4.5.1. In development

Genetic loss of *Jagged1* does not compromise formation of the primary vascular networks, but remodeling of these vessels in the yolk sac and embryo is perturbed and leads to lethality at E10 (Xue *et al.*, 1999). Recent observations of mice with endothelial-specific manipulations of *Jagged1* suggest that deficient remodeling may arise from loss of both Jagged1-regulated angiogenesis and defective vascular smooth muscle cell differentiation (Benedito *et al.*, 2009; High *et al.*, 2008). As described previously, Jagged1 is proposed to antagonize Dll4 signaling and may activate intracellular signaling events within the ligand-presenting cell itself; and therefore, loss of *Jagged1* likely leads to aberrant Notch pathway activation in both the endothelial and smooth muscle cell compartments.

4.5.2. In human disease

Alagille syndrome (AGS) is a hereditary disorder arising from mutations in *Jagged1* (Li *et al.*, 1997; Oda *et al.*, 1997). Haploinsufficiency or truncations in *Jagged1* contribute to frequently observed abnormalities in blood vessel formation, including the narrowing of the aorta (coarctation) and other major arteries (stenosis) (Kamath *et al.*, 2004). *Notch2* mutations may also exacerbate the AGS phenotype. *Jagged1* heterozygous mice carrying a *Notch2* hypomorphic allele displayed several developmental defects consistent with AGS (McCright *et al.*, 2002), although significant vascular abnormalities were not reported. Nevertheless, mutations in *Notch2* have been found in AGS patients who lack *Jagged1* mutations (McDaniell *et al.*, 2006). Experimental evidence for the role of Jagged1 in blood vessel development coincides with the observed vascular deformities in AGS, and mutations in other Notch components such as *Notch2* have been implicated as modifiers of *Jagged1* perturbations. Further investigation will be necessary to determine which Notch signaling molecules contribute to and exacerbate the pathogenesis of this disorder.

5. Perspectives

The significant crosstalk among signaling pathways involved in vascular development is increasingly evident (Holderfield and Hughes, 2008; Jakobsson *et al.*, 2009). However, despite the defined vascular pathologies associated with the individual pathways described above, there is surprisingly little genetic evidence for pathway crosstalk in human pathogenesis. If signaling intersections are indeed important aspects of human disease, further understanding the signaling crosstalk among these pathways may enhance treatment for certain vascular defects and conditions.

It is likely that rigorous dissection of both the pathways and the crosstalk among the pathways in vascular development will benefit from computational modeling approaches that supplement the experimental data derived from model systems (Bentley *et al.*, 2008; Mac Gabhann and Popel, 2008; Peirce, 2008). For example, one important question is how these pathways initiate and maintain phenotypic heterogeneity in endothelial cells during blood vessel formation. Appropriate computational models will allow us to input different pathway relationships and simulate how these relationships affect endothelial heterogeneity and vessel morphogenesis.

Another important future goal in the field is to combine high-resolution imaging with signaling readouts to better understand how information conveyed by signaling pathways is organized both spatially and temporally. This review highlights the importance of spatial organization of information in developing vessel networks; yet we have little information regarding the dynamic temporal regulation of regional signaling that likely overlies the static images produced so far. The use of reporter genes that reflect pathway activity at the single cell level, along with fluorescence resonance energy transfer (FRET)-based tools to evaluate protein interactions stimulated by signaling pathways, will allow us to add temporal information to our understanding of vascular development. For example, a recent study followed PI3 kinase activation in individual migrating zebrafish neutrophils and used a photoactivatable Rac to show that Rac activation is sufficient to direct neutrophil migration *in vivo* (Yoo *et al.*, 2010). It will be exciting to apply these innovative approaches to outstanding questions in vascular development.

Finally, it will be informative to extend the concept of endothelial heterogeneity and pathway crosstalk beyond current models. For example, extrinsic angiogenic factors are often not set up in obvious gradients (Czirok *et al.*, 2008; Damert *et al.*, 2002; Kearney and Bautch, 2003; Keller, 2005), and so vessel-intrinsic heterogeneity may be important for proper vessel patterning in numerous scenarios. We recently showed that soluble Flt-1 released from cells adjacent to emerging sprouts modulates local VEGF distribution for proper sprout guidance away from the parent vessel (Chappell *et al.*, 2009). Notch signaling, which can be induced by VEGF signaling, regulates Flt-1 expression (Harrington *et al.*, 2008; Suchting *et al.*, 2007) and may represent a means for modulating numerous signals within the local microenvironment. Other pathways have negative regulators that may exhibit vessel-intrinsic phenotypes. For example, the BMP antagonist BMPER is expressed in some endothelial cells (Moser *et al.*, 2003) and may provide local cues to modulate BMP signaling during vessel development.

Endothelial heterogeneity is potentially relevant in maturing regions of the vasculature as well. Mazzone *et al.* recently demonstrated the importance of the endothelial "phalanx" phenotype in promoting vessel quiescence through increased cell adhesions and dampened response to VEGF (Mazzone *et al.*, 2009). Phenotypic heterogeneity likely exists in these regions of the endothelium such that recruited pericytes interact with more quiescent endothelial cells, forming localized contacts that enhance vessel stability (Armulik *et al.*, 2005; Gaengel *et al.*, 2009; Gerhardt and Betsholtz, 2003; Jain and Booth, 2003). Thus, the signaling pathways described in this review likely influence endothelial heterogeneity in both actively sprouting and remodeling regions of developing vessels.

In summary, extrinsic cues are likely complemented by endothelial cell signaling networks that establish an integrated and heterogeneous response to angiogenic stimuli. Some vascular pathologies likely arise from genetic mutations that perturb the normal establishment and maintenance of endothelial heterogeneity in vascular development and remodeling.

ACKNOWLEDGMENTS

We thank members of the Bautch lab for fruitful discussions. This work was supported by National Institutes of Health (NIH) grants HL43174 and HL86564 to V.L.B., and fellowship support from the NIH (T32CA9156 and F32HL95359) and the American Heart Association (0826082E) to J.C.C. The authors declare no conflict of interest.

REFERENCES

Abdalla, S. A., Gallione, C. J., Barst, R. J., Horn, E. M., Knowles, J. A., Marchuk, D. A., Letarte, M., and Morse, J. H. (2004). Primary pulmonary hypertension in families with hereditary haemorrhagic telangiectasia. *Eur. Respir. J.* **23**, 373–377.

Abdalla, S. A., and Letarte, M. (2006). Hereditary haemorrhagic telangiectasia: Current views on genetics and mechanisms of disease. *J. Med. Genet.* **43**, 97–110.

Adams, R. H., and Alitalo, K. (2007). Molecular regulation of angiogenesis and lymphangiogenesis. *Nat. Rev. Mol. Cell Biol.* **8**, 464–478.

Alitalo, K., and Carmeliet, P. (2002). Molecular mechanisms of lymphangiogenesis in health and disease. *Cancer Cell* **1**, 219–227.

Armulik, A., Abramsson, A., and Betsholtz, C. (2005). Endothelial/pericyte interactions. *Circ. Res.* **97**, 512–523.

Arthur, H. M., Ure, J., Smith, A. J., Renforth, G., Wilson, D. I., Torsney, E., Charlton, R., Parums, D. V., Jowett, T., Marchuk, D. A., Burn, J., and Diamond, A. G. (2000). Endoglin, an ancillary TGFbeta receptor, is required for extraembryonic angiogenesis and plays a key role in heart development. *Dev. Biol.* **217**, 42–53.

Bautch, V. L. (2009). Endothelial cells form a phalanx to block tumor metastasis. *Cell* **136**, 810–812.

Benedito, R., Roca, C., Sorensen, I., Adams, S., Gossler, A., Fruttiger, M., and Adams, R. H. (2009). The notch ligands Dll4 and Jagged1 have opposing effects on angiogenesis. *Cell* **137**, 1124–1135.

Benjamin, L. E., Golijanin, D., Itin, A., Pode, D., and Keshet, E. (1999). Selective ablation of immature blood vessels in established human tumors follows vascular endothelial growth factor withdrawal. *J. Clin. Invest.* **103**, 159–165.

Bentley, K., Gerhardt, H., and Bates, P. A. (2008). Agent-based simulation of notch-mediated tip cell selection in angiogenic sprout initialisation. *J. Theor. Biol.* **250**, 25–36.

Bentley, K., Mariggi, G., Gerhardt, H., and Bates, P. A. (2009). Tipping the balance: Robustness of tip cell selection, migration and fusion in angiogenesis. *PLoS Comput. Biol.* **5**, e1000549.

Beppu, H., Ichinose, F., Kawai, N., Jones, R. C., Yu, P. B., Zapol, W. M., Miyazono, K., Li, E., and Bloch, K. D. (2004). BMPR-II heterozygous mice have mild pulmonary hypertension and an impaired pulmonary vascular remodeling response to prolonged hypoxia. *Am. J. Physiol. Lung Cell Mol. Physiol.* **287**, L1241–L1247.

Blum, Y., Belting, H. G., Ellertsdottir, E., Herwig, L., Luders, F., and Affolter, M. (2008). Complex cell rearrangements during intersegmental vessel sprouting and vessel fusion in the zebrafish embryo. *Dev. Biol.* **316**, 312–322.

Boscolo, E., and Bischoff, J. (2009). Vasculogenesis in infantile hemangioma. *Angiogenesis* **12**, 197–207.

Braverman, I. M., and Sibley, J. (1982). Role of the microcirculation in the treatment and pathogenesis of psoriasis. *J. Invest. Dermatol.* **78**, 12–17.

Bray, S. J. (2006). Notch signalling: A simple pathway becomes complex. *Nat. Rev. Mol. Cell Biol.* **7**, 678–689.

Bressan, M., Davis, P., Timmer, J., Herzlinger, D., and Mikawa, T. (2009). Notochord-derived BMP antagonists inhibit endothelial cell generation and network formation. *Dev. Biol.* **326**, 101–111.

Carmeliet, P. (2005). Angiogenesis in life, disease and medicine. *Nature* **438**, 932–936.

Carmeliet, P., Ferreira, V., Breier, G., Pollefeyt, S., Kieckens, L., Gertsenstein, M., Fahrig, M., Vandenhoeck, A., Harpal, K., Eberhardt, C., Declercq, C., Pawling, J., *et al.* (1996). Abnormal blood vessel development and lethality in embryos lacking a single VEGF allele. *Nature* **380**, 435–439.

Chabriat, H., Vahedi, K., Iba-Zizen, M. T., Joutel, A., Nibbio, A., Nagy, T. G., Krebs, M. O., Julien, J., Dubois, B., Ducrocq, X., Levasseur, M., Mas, J. L., Dubois, B., Homeyer, P., and Lyon-Caen, O. (1995). Clinical spectrum of CADASIL: A study of 7 families. Cerebral autosomal dominant arteriopathy with subcortical infarcts and leukoencephalopathy. *Lancet* **346**, 934–939.

Chappell, J. C., Taylor, S. M., Ferrara, N., and Bautch, V. L. (2009). Local guidance of emerging vessel sprouts requires soluble flt-1. *Dev. Cell.* **17**, 377–386.

Chen, T. T., Luque, A., Lee, S., Anderson, S. M., Segura, T., and Iruela-Arispe, M. L. (2010). Anchorage of VEGF to the extracellular matrix conveys differential signaling responses to endothelial cells. *J. Cell Biol.* **188**, 595–609.

Cleaver, O., Tonissen, K. F., Saha, M. S., and Krieg, P. A. (1997). Neovascularization of the Xenopus embryo. *Dev. Dyn.* **210**, 66–77.

Creamer, D., Allen, M. H., Sousa, A., Poston, R., and Barker, J. N. (1997). Localization of endothelial proliferation and microvascular expansion in active plaque psoriasis. *Br. J. Dermatol.* **136**, 859–865.

Cueni, L. N., and Detmar, M. (2006). New insights into the molecular control of the lymphatic vascular system and its role in disease. *J. Invest. Dermatol.* **126**, 2167–2177.

Czirok, A., Zamir, E. A., Szabo, A., and Little, C. D. (2008). Multicellular sprouting during vasculogenesis. *Curr. Top. Dev. Biol.* **81**, 269–289.

Dahlqvist, C., Blokzijl, A., Chapman, G., Falk, A., Dannaeus, K., Ibanez, C. F., and Lendahl, U. (2003). Functional notch signaling is required for BMP4-induced inhibition of myogenic differentiation. *Development* **130**, 6089–6099.

Damert, A., Miquerol, L., Gertsenstein, M., Risau, W., and Nagy, A. (2002). Insufficient VEGFA activity in yolk sac endoderm compromises haematopoietic and endothelial differentiation. *Development* **129**, 1881–1892.

David, L., Feige, J. J., and Bailly, S. (2009). Emerging role of bone morphogenetic proteins in angiogenesis. *Cytokine Growth Factor Rev.* **20**, 203–212.

de Jesus Perez, V. A., Alastalo, T. P., Wu, J. C., Axelrod, J. D., Cooke, J. P., Amieva, M., and Rabinovitch, M. (2009). Bone morphogenetic protein 2 induces pulmonary angiogenesis via wnt-beta-catenin and wnt-RhoA-rac1 pathways. *J. Cell Biol.* **184**, 83–99.

Deckers, M. M., van Bezooijen, R. L., van der Horst, G., Hoogendam, J., van Der Bent, C., Papapoulos, S. E., and Lowik, C. W. (2002). Bone morphogenetic proteins stimulate angiogenesis through osteoblast-derived vascular endothelial growth factor A. *Endocrinology* **143**, 1545–1553.

Deng, Z., Morse, J. H., Slager, S. L., Cuervo, N., Moore, K. J., Venetos, G., Kalachikov, S., Cayanis, E., Fischer, S. G., Barst, R. J., Hodge, S. E., and Knowles, J. A. (2000). Familial primary pulmonary hypertension (gene PPH1) is caused by mutations in the bone morphogenetic protein receptor II gene. *Am. J. Hum. Genet.* **67**, 737–744.

Detmar, M., Brown, L. F., Claffey, K. P., Yeo, K. T., Kocher, O., Jackman, R. W., Berse, B., and Dvorak, H. F. (1994). Overexpression of vascular permeability factor/vascular endothelial growth factor and its receptors in psoriasis. *J. Exp. Med.* **180**, 1141–1146.

Domenga, V., Fardoux, P., Lacombe, P., Monet, M., Maciazek, J., Krebs, L. T., Klonjkowski, B., Berrou, E., Mericskay, M., Li, Z., Tournier-Lasserve, E., Gridley, T., and

Joutel, A. (2004). Notch3 is required for arterial identity and maturation of vascular smooth muscle cells. *Genes Dev.* **18**, 2730–2735.

Duarte, A., Hirashima, M., Benedito, R., Trindade, A., Diniz, P., Bekman, E., Costa, L., Henrique, D., and Rossant, J. (2004). Dosage-sensitive requirement for mouse dll4 in artery development. *Genes Dev.* **18**, 2474–2478.

Dumont, D. J., Jussila, L., Taipale, J., Lymboussaki, A., Mustonen, T., Pajusola, K., Breitman, M., and Alitalo, K. (1998). Cardiovascular failure in mouse embryos deficient in VEGF receptor-3. *Science* **282**, 946–949.

Fernandez, L. A., Sanz-Rodriguez, F., Blanco, F. J., Bernabeu, C., and Botella, L. M. (2006). Hereditary hemorrhagic telangiectasia, a vascular dysplasia affecting the TGF-beta signaling pathway. *Clin. Med. Res.* **4**, 66–78.

Ferrara, N. (2005). VEGF as a therapeutic target in cancer.*Oncology* **69**(Suppl. 3), 11–16.

Ferrara, N., Carver-Moore, K., Chen, H., Dowd, M., Lu, L., O'Shea, K. S., Powell-Braxton, L., Hillan, K. J., and Moore, M. W. (1996). Heterozygous embryonic lethality induced by targeted inactivation of the VEGF gene. *Nature* **380**, 439–442.

Ferrara, N., Gerber, H. P., and LeCouter, J. (2003). The biology of VEGF and its receptors. *Nat. Med.* **9**, 669–676.

Ferrell, R. E. (2002). Research perspectives in inherited lymphatic disease. *Ann. N. Y. Acad. Sci.* **979**, 39–51.

Fischer, C., Mazzone, M., Jonckx, B., and Carmeliet, P. (2008). FLT1 and its ligands VEGFB and PlGF: Drug targets for anti-angiogenic therapy? *Nat. Rev. Cancer* **8**, 942–956.

Fischer, A., Schumacher, N., Maier, M., Sendtner, M., and Gessler, M. (2004). The notch target genes hey1 and hey2 are required for embryonic vascular development. *Genes Dev.* **18**, 901–911.

Fong, G. H., Rossant, J., Gertsenstein, M., and Breitman, M. L. (1995). Role of the flt-1 receptor tyrosine kinase in regulating the assembly of vascular endothelium. *Nature* **376**, 66–70.

Funahashi, Y., Shawber, C. J., Vorontchikhina, M., Sharma, A., Outtz, H. H., and Kitajewski, J. (2010). Notch regulates the angiogenic response via induction of VEGFR-1. *J. Angiogenes Res.* **2**, 3.

Gaengel, K., Genove, G., Armulik, A., and Betsholtz, C. (2009). Endothelial-mural cell signaling in vascular development and angiogenesis. *Arterioscler. Thromb. Vasc. Biol.* **29**, 630–638.

Gale, N. W., Dominguez, M. G., Noguera, I., Pan, L., Hughes, V., Valenzuela, D. M., Murphy, A. J., Adams, N. C., Lin, H. C., Holash, J., Thurston, G., and Yancopoulos, G. D. (2004). Haploinsufficiency of delta-like 4 ligand results in embryonic lethality Due To major defects in arterial and vascular development. *Proc. Natl. Acad. Sci. USA* **101**, 15949–15954.

Gallione, C. J., Repetto, G. M., Legius, E., Rustgi, A. K., Schelley, S. L., Tejpar, S., Mitchell, G., Drouin, E., Westermann, C. J., and Marchuk, D. A. (2004). A combined syndrome of juvenile polyposis and hereditary haemorrhagic telangiectasia associated with mutations in MADH4 (SMAD4). *Lancet* **363**, 852–859.

Garcia-Cardena, G., Comander, J., Anderson, K. R., Blackman, B. R., and Gimbrone, M. A.Jr. (2001). Biomechanical activation of vascular endothelium as a determinant of its functional phenotype. *Proc. Natl. Acad. Sci. USA* **98**, 4478–4485.

Gerhardt, H., and Betsholtz, C. (2003). Endothelial-pericyte interactions in angiogenesis. *Cell Tissue Res.* **314**, 15–23.

Gerhardt, H., Golding, M., Fruttiger, M., Ruhrberg, C., Lundkvist, A., Abramsson, A., Jeltsch, M., Mitchell, C., Alitalo, K., Shima, D., and Betsholtz, C. (2003). VEGF guides angiogenic sprouting utilizing endothelial tip cell filopodia. *J. Cell Biol.* **161**, 1163–1177.

Gerhardt, H., Ruhrberg, C., Abramsson, A., Fujisawa, H., Shima, D., and Betsholtz, C. (2004). Neuropilin-1 is required for endothelial tip cell guidance in the developing central nervous system. *Dev. Dyn.* **231**, 503–509.

Ghabrial, A. S., and Krasnow, M. A. (2006). Social interactions among epithelial cells during tracheal branching morphogenesis. *Nature* **441**, 746–749.

Goldie, L. C., Nix, M. K., and Hirschi, K. K. (2008). Embryonic vasculogenesis and hematopoietic specification. *Organogenesis* **4**, 257–263.

Grego-Bessa, J., Luna-Zurita, L., del Monte, G., Bolos, V., Melgar, P., Arandilla, A., Garratt, A. N., Zang, H., Mukouyama, Y. S., Chen, H., Shou, W., Ballestar, E., Esteller, M., Rojas, A., Perez-Pomares, J. M., and de la Pompa, J. L. (2007). Notch signaling is essential for ventricular chamber development. *Dev. Cell.* **12**, 415–429.

Gu, C., Rodriguez, E. R., Reimert, D. V., Shu, T., Fritzsch, B., Richards, L. J., Kolodkin, A. L., and Ginty, D. D. (2003). Neuropilin-1 conveys semaphorin and VEGF signaling during neural and cardiovascular development. *Dev. Cell.* **5**, 45–57.

Harrington, L. S., Sainson, R. C., Williams, C. K., Taylor, J. M., Shi, W., Li, J. L., and Harris, A. L. (2008). Regulation of multiple angiogenic pathways by dll4 and notch in human umbilical vein endothelial cells. *Microvasc. Res.* **75**, 144–154.

Harrison, R. E., Flanagan, J. A., Sankelo, M., Abdalla, S. A., Rowell, J., Machado, R. D., Elliott, C. G., Robbins, I. M., Olschewski, H., McLaughlin, V., Gruenig, E., Kermeen, F., *et al.* (2003). Molecular and functional analysis identifies ALK-1 as the predominant cause of pulmonary hypertension related to hereditary haemorrhagic telangiectasia. *J. Med. Genet.* **40**, 865–871.

Heath, V. L., and Bicknell, R. (2009). Anticancer strategies involving the vasculature. *Nat. Rev. Clin. Oncol.* **6**, 395–404.

Hellstrom, M., Phng, L. K., Hofmann, J. J., Wallgard, E., Coultas, L., Lindblom, P., Alva, J., Nilsson, A. K., Karlsson, L., Gaiano, N., Yoon, K., Rossant, J., Iruela-Arispe, M. L., *et al.* (2007). Dll4 signalling through notch1 regulates formation of tip cells during angiogenesis. *Nature* **445**, 776–780.

High, F. A., Lu, M. M., Pear, W. S., Loomes, K. M., Kaestner, K. H., and Epstein, J. A. (2008). Endothelial expression of the notch ligand jagged1 is required for vascular smooth muscle development. *Proc. Natl. Acad. Sci. USA* **105**, 1955–1959.

Hiratsuka, S., Minowa, O., Kuno, J., Noda, T., and Shibuya, M. (1998). Flt-1 lacking the tyrosine kinase domain is sufficient for normal development and angiogenesis in mice. *Proc. Natl. Acad. Sci. USA* **95**, 9349–9354.

Hirschi, K. K., Rohovsky, S. A., and D'Amore, P. A. (1998). PDGF, TGF-beta, and heterotypic cell-cell interactions mediate endothelial cell-induced recruitment of 10T1/2 cells and their differentiation to a smooth muscle fate. *J. Cell Biol.* **141**, 805–814.

Hofmann, J. J., and Iruela-Arispe, M. L. (2007). Notch signaling in blood vessels: Who is talking to whom about what? *Circ. Res.* **100**, 1556–1568.

Holderfield, M. T., and Hughes, C. C. (2008). Crosstalk between vascular endothelial growth factor, notch, and transforming growth factor-beta in vascular morphogenesis. *Circ. Res.* **102**, 637–652.

Hong, K. H., Lee, Y. J., Lee, E., Park, S. O., Han, C., Beppu, H., Li, E., Raizada, M. K., Bloch, K. D., and Oh, S. P. (2008). Genetic ablation of the BMPR2 gene in pulmonary endothelium is sufficient to predispose to pulmonary arterial hypertension. *Circulation* **118**, 722–730.

Iruela-Arispe, M. L., and Davis, G. E. (2009). Cellular and molecular mechanisms of vascular lumen formation. *Dev. Cell.* **16**, 222–231.

Iso, T., Maeno, T., Oike, Y., Yamazaki, M., Doi, H., Arai, M., and Kurabayashi, M. (2006). Dll4-selective notch signaling induces ephrinB2 gene expression in endothelial cells. *Biochem. Biophys. Res. Commun.* **341**, 708–714.

Itoh, F., Itoh, S., Goumans, M. J., Valdimarsdottir, G., Iso, T., Dotto, G. P., Hamamori, Y., Kedes, L., Kato, M., and ten Dijke, P. (2004). Synergy and antagonism between notch and BMP receptor signaling pathways in endothelial cells. *EMBO J.* **23**, 541–551.

Jain, R. K. (2003). Molecular regulation of vessel maturation. *Nat. Med.* **9**, 685–693.

Jain, R. K. (2005). Normalization of tumor vasculature: An emerging concept in antiangiogenic therapy. *Science* **307**, 58–62.

Jain, R. K., and Booth, M. F. (2003). What brings pericytes to tumor vessels? *J. Clin. Invest.* **112**, 1134–1136.

Jakobsson, L., Bentley, K., and Gerhardt, H. (2009). VEGFRs and notch: A dynamic collaboration in vascular patterning. *Biochem. Soc. Trans.* **37**, 1233–1236.

Jin, S. W., Beis, D., Mitchell, T., Chen, J. N., and Stainier, D. Y. (2005). Cellular and molecular analyses of vascular tube and lumen formation in zebrafish. *Development* **132**, 5199–5209.

Jinnin, M., Medici, D., Park, L., Limaye, N., Liu, Y., Boscolo, E., Bischoff, J., Vikkula, M., Boye, E., and Olsen, B. R. (2008). Suppressed NFAT-dependent VEGFR1 expression and constitutive VEGFR2 signaling in infantile hemangioma. *Nat. Med.* **14**, 1236–1246.

Johnson, D. W., Berg, J. N., Baldwin, M. A., Gallione, C. J., Marondel, I., Yoon, S. J., Stenzel, T. T., Speer, M., Pericak-Vance, M. A., Diamond, A., Guttmacher, A. E., Jackson, C. E., *et al.* (1996). Mutations in the activin receptor-like kinase 1 gene in hereditary haemorrhagic telangiectasia type 2. *Nat. Genet.* **13**, 189–195.

Jones, E. A., le Noble, F., and Eichmann, A. (2006). What determines blood vessel structure? Genetic prespecification vs. hemodynamics. *Physiology (Bethesda)* **21**, 388–395.

Jones, E. A., Yuan, L., Breant, C., Watts, R. J., and Eichmann, A. (2008). Separating genetic and hemodynamic defects in neuropilin 1 knockout embryos. *Development* **135**, 2479–2488.

Joutel, A., Corpechot, C., Ducros, A., Vahedi, K., Chabriat, H., Mouton, P., Alamowitch, S., Domenga, V., Cecillion, M., Marechal, E., Maciazek, J., Vayssiere, C., Cruaud, C., *et al.* (1996). Notch3 mutations in CADASIL, a hereditary adult-onset condition causing stroke and dementia. *Nature* **383**, 707–710.

Joutel, A., and Tournier-Lasserve, E. (1998). Notch signalling pathway and human diseases. *Semin. Cell. Dev. Biol.* **9**, 619–625.

Kamath, B. M., Spinner, N. B., Emerick, K. M., Chudley, A. E., Booth, C., Piccoli, D. A., and Krantz, I. D. (2004). Vascular anomalies in Alagille syndrome: A significant cause of morbidity and mortality. *Circulation* **109**, 1354–1358.

Kamei, M., Saunders, W. B., Bayless, K. J., Dye, L., Davis, G. E., and Weinstein, B. M. (2006). Endothelial tubes assemble from intracellular vacuoles in vivo. *Nature* **442**, 453–456.

Kappas, N. C., Zeng, G., Chappell, J. C., Kearney, J. B., Hazarika, S., Kallianos, K. G., Patterson, C., Annex, B. H., and Bautch, V. L. (2008). The VEGF receptor Flt-1 spatially modulates Flk-1 signaling and blood vessel branching. *J. Cell Biol.* **181**, 847–858.

Karkkainen, M. J., Haiko, P., Sainio, K., Partanen, J., Taipale, J., Petrova, T. V., Jeltsch, M., Jackson, D. G., Talikka, M., Rauvala, H., Betsholtz, C., and Alitalo, K. (2004). Vascular endothelial growth factor C is required for sprouting of the first lymphatic vessels from embryonic veins. *Nat. Immunol.* **5**, 74–80.

Karpanen, T., Heckman, C. A., Keskitalo, S., Jeltsch, M., Ollila, H., Neufeld, G., Tamagnone, L., and Alitalo, K. (2006a). Functional interaction of VEGF-C and VEGF-D with neuropilin receptors. *FASEB J.* **20**, 1462–1472.

Karpanen, T., Wirzenius, M., Makinen, T., Veikkola, T., Haisma, H. J., Achen, M. G., Stacker, S. A., Pytowski, B., Yla-Herttuala, S., and Alitalo, K. (2006b). Lymphangiogenic growth factor responsiveness is modulated by postnatal lymphatic vessel maturation. *Am. J. Pathol.* **169**, 708–718.

Kawasaki, T., Kitsukawa, T., Bekku, Y., Matsuda, Y., Sanbo, M., Yagi, T., and Fujisawa, H. (1999). A requirement for neuropilin-1 in embryonic vessel formation. *Development* **126**, 4895–4902.

Kearney, J. B., Ambler, C. A., Monaco, K. A., Johnson, N., Rapoport, R. G., and Bautch, V. L. (2002). Vascular endothelial growth factor receptor Flt-1 negatively regulates developmental blood vessel formation by modulating endothelial cell division. *Blood* **99**, 2397–2407.

Kearney, J. B., and Bautch, V. L. (2003). In vitro differentiation of mouse ES cells: Hematopoietic and vascular development. *Methods Enzymol.* **365**, 83–98.

Kearney, J. B., Kappas, N. C., Ellerstrom, C., DiPaola, F. W., and Bautch, V. L. (2004). The VEGF receptor flt-1 (VEGFR-1) is a positive modulator of vascular sprout formation and branching morphogenesis. *Blood* **103**, 4527–4535.

Keller, G. (2005). Embryonic stem cell differentiation: Emergence of a new era in biology and medicine. *Genes Dev.* **19**, 1129–1155.

Kendall, R. L., and Thomas, K. A. (1993). Inhibition of vascular endothelial cell growth factor activity by an endogenously encoded soluble receptor. *Proc. Natl. Acad. Sci. USA* **90**, 10705–10709.

Kim, Y. H., Hu, H., Guevara-Gallardo, S., Lam, M. T., Fong, S. Y., and Wang, R. A. (2008). Artery and vein size is balanced by Notch and ephrin B2/EphB4 during angiogenesis. *Development* **135**, 3755–3764.

Kluppel, M., and Wrana, J. L. (2005). Turning it up a Notch: Cross-talk between TGF beta and Notch signaling. *BioEssays* **27**, 115–118.

Kozawa, O., Matsuno, H., and Uematsu, T. (2001). Involvement of p70 S6 kinase in bone morphogenetic protein signaling: Vascular endothelial growth factor synthesis by bone morphogenetic protein-4 in osteoblasts. *J. Cell. Biochem.* **81**, 430–436.

Krebs, L. T., Xue, Y., Norton, C. R., Shutter, J. R., Maguire, M., Sundberg, J. P., Gallahan, D., Closson, V., Kitajewski, J., Callahan, R., Smith, G. H., Stark, K. L., *et al.* (2000). Notch signaling is essential for vascular morphogenesis in mice. *Genes Dev.* **14**, 1343–1352.

Lane, K. B., Machado, R. D., Pauciulo, M. W., Thomson, J. R., Phillips, J. A., 3rd, Loyd, J. E., Nichols, W. C., and Trembath, R. C., (2000). Heterozygous germline mutations in BMPR2, encoding a TGF-beta receptor, cause familial primary pulmonary hypertension. *Nat. Genet.* **26**, 81–84.

Larrivee, B., Freitas, C., Suchting, S., Brunet, I., and Eichmann, A. (2009). Guidance of vascular development: Lessons from the nervous system. *Circ. Res.* **104**, 428–441.

Lawson, N. D., Scheer, N., Pham, V. N., Kim, C. H., Chitnis, A. B., Campos-Ortega, J. A., and Weinstein, B. M. (2001). Notch signaling is required for arterial-venous differentiation during embryonic vascular development. *Development* **128**, 3675–3683.

Lawson, N. D., Vogel, A. M., and Weinstein, B. M. (2002). Sonic hedgehog and vascular endothelial growth factor act upstream of the Notch pathway during arterial endothelial differentiation. *Dev. Cell.* **3**, 127–136.

le Noble, F., Moyon, D., Pardanaud, L., Yuan, L., Djonov, V., Matthijsen, R., Breant, C., Fleury, V., and Eichmann, A. (2004). Flow regulates arterial-venous differentiation in the chick embryo yolk sac. *Development* **131**, 361–375.

Lee, S., Chen, T. T., Barber, C. L., Jordan, M. C., Murdock, J., Desai, S., Ferrara, N., Nagy, A., Roos, K. P., and Iruela-Arispe, M. L. (2007). Autocrine VEGF signaling is required for vascular homeostasis. *Cell* **130**, 691–703.

Lee, D., Park, C., Lee, H., Lugus, J. J., Kim, S. H., Arentson, E., Chung, Y. S., Gomez, G., Kyba, M., Lin, S., Janknecht, R., Lim, D. S., and Choi, K. (2008). ER71 acts downstream of BMP, Notch, and Wnt signaling in blood and vessel progenitor specification. *Cell Stem Cell* **2**, 497–507.

Lee, C. Y., Vogeli, K. M., Kim, S. H., Chong, S. W., Jiang, Y. J., Stainier, D. Y., and Jin, S. W. (2009). Notch signaling functions as a cell-fate switch between the endothelial and hematopoietic lineages. *Curr. Biol.* **19**, 1616–1622.

Leslie, J. D., Ariza-McNaughton, L., Bermange, A. L., McAdow, R., Johnson, S. L., and Lewis, J. (2007). Endothelial signalling by the Notch ligand Delta-like 4 restricts angiogenesis. *Development* **134**, 839–844.

Li, L., Krantz, I. D., Deng, Y., Genin, A., Banta, A. B., Collins, C. C., Qi, M., Trask, B. J., Kuo, W. L., Cochran, J., Costa, T., Pierpont, M. E., Rand, E. B., Piccoli, D. A., *et al.* (1997). Alagille syndrome is caused by mutations in human Jagged1, which encodes a ligand for Notch1. *Nat. Genet.* **16**, 243–251.

Li, D. Y., Sorensen, L. K., Brooke, B. S., Urness, L. D., Davis, E. C., Taylor, D. G., Boak, B. B., and Wendel, D. P. (1999). Defective angiogenesis in mice lacking endoglin. *Science* **284**, 1534–1537.

Limbourg, A., Ploom, M., Elligsen, D., Sorensen, I., Ziegelhoeffer, T., Gossler, A., Drexler, H., and Limbourg, F. P. (2007). Notch ligand Delta-like 1 is essential for postnatal arteriogenesis. *Circ. Res.* **100**, 363–371.

Limbourg, F. P., Takeshita, K., Radtke, F., Bronson, R. T., Chin, M. T., and Liao, J. K. (2005). Essential role of endothelial Notch1 in angiogenesis. *Circulation* **111**, 1826–1832.

Lindahl, P., Johansson, B. R., Leveen, P., and Betsholtz, C. (1997). Pericyte loss and microaneurysm formation in PDGF-B-deficient mice. *Science* **277**, 242–245.

Lindsay, E. A. (2001). Chromosomal microdeletions: Dissecting del22q11 syndrome. *Nat. Rev. Genet.* **2**, 858–868.

Liu, D., Wang, J., Kinzel, B., Mueller, M., Mao, X., Valdez, R., Liu, Y., and Li, E. (2007). Dosage-dependent requirement of BMP type II receptor for maintenance of vascular integrity. *Blood* **110**, 1502–1510.

Liu, Z. J., Xiao, M., Balint, K., Soma, A., Pinnix, C. C., Capobianco, A. J., Velazquez, O. C., and Herlyn, M. (2006). Inhibition of endothelial cell proliferation by Notch1 signaling is mediated by repressing MAPK and PI3K/Akt pathways and requires MAML1. *FASEB J.* **20**, 1009–1011.

Lobov, I. B., Renard, R. A., Papadopoulos, N., Gale, N. W., Thurston, G., Yancopoulos, G. D., and Wiegand, S. J. (2007). Delta-like ligand 4 (Dll4) is induced by VEGF as a negative regulator of angiogenic sprouting. *Proc. Natl. Acad. Sci. USA* **104**, 3219–3224.

Lucitti, J. L., Jones, E. A., Huang, C., Chen, J., Fraser, S. E., and Dickinson, M. E. (2007). Vascular remodeling of the mouse yolk sac requires hemodynamic force. *Development* **134**, 3317–3326.

Mac Gabhann, F., and Popel, A. S. (2008). Systems biology of vascular endothelial growth factors. *Microcirculation* **15**, 715–738.

Machado, R. D., Aldred, M. A., James, V., Harrison, R. E., Patel, B., Schwalbe, E. C., Gruenig, E., Janssen, B., Koehler, R., Seeger, W., Eickelberg, O., Olschewski, H., Elliot, C. G., Glissmeyer, E., Carlquist, J., Kim, M., Torbicki, A., Fijalkowska, A., Szewczyk, G., Parma, J., Abramowicz, M. J., Galie, N., Morisaki, H., Kyotani, S., Nakanishi, N., Morisaki, T., Humbert, M., Simonneau, G., Sitbon, O., Soubrier, F., Coulet, F., Morrell, N. W., and Trembath, R. C. (2006). Mutations of the TGF-beta type II receptor BMPR2 in pulmonary arterial hypertension. *Hum. Mutat.* **27**, 121–132.

Mailhos, C., Modlich, U., Lewis, J., Harris, A., Bicknell, R., and Ish-Horowicz, D. (2001). Delta4, an endothelial specific notch ligand expressed at sites of physiological and tumor angiogenesis. *Differentiation* **69**, 135–144.

Makinen, T., Jussila, L., Veikkola, T., Karpanen, T., Kettunen, M. I., Pulkkanen, K. J., Kauppinen, R., Jackson, D. G., Kubo, H., Nishikawa, S., Yla-Herttuala, S., and Alitalo, K. (2001). Inhibition of lymphangiogenesis with resulting lymphedema in transgenic mice expressing soluble VEGF receptor-3. *Nat. Med.* **7**, 199–205.

Masumura, T., Yamamoto, K., Shimizu, N., Obi, S., and Ando, J. (2009). Shear stress increases expression of the arterial endothelial marker ephrinB2 in murine ES cells via the VEGF-Notch signaling pathways. *Arterioscler. Thromb. Vasc. Biol.* **29**, 2125–2131.

Mazzone, M., Dettori, D., de Oliveira, L., Loges, R., Schmidt, S., Jonckx, T., Tian, B., Lanahan, Y. M., Pollard, A. A., Ruiz, P., de Almodovar, C., De Smet, F., *et al.* (2009). Heterozygous deficiency of PHD2 restores tumor oxygenation and inhibits metastasis via endothelial normalization. *Cell* **136**, 839–851.

McAllister, K. A., Grogg, K. M., Johnson, D. W., Gallione, C. J., Baldwin, M. A., Jackson, C. E., Helmbold, E. A., Markel, D. S., McKinnon, W. C., Murrell, J. *et al.* (1994). Endoglin, a TGF-beta binding protein of endothelial cells, is the gene for hereditary haemorrhagic telangiectasia type 1. *Nat. Genet.* **8**, 345–351.

McCright, B., Gao, X., Shen, L., Lozier, J., Lan, Y., Maguire, M., Herzlinger, D., Weinmaster, G., Jiang, R., and Gridley, T. (2001). Defects in development of the kidney, heart and eye vasculature in mice homozygous for a hypomorphic Notch2 mutation. *Development* **128**, 491–502.

McCright, B., Lozier, J., and Gridley, T. (2002). A mouse model of Alagille syndrome: Notch2 as a genetic modifier of Jag1 haploinsufficiency. *Development* **129**, 1075–1082.

McDaniell, R., Warthen, D. M., Sanchez-Lara, P. A., Pai, A., Krantz, I. D., Piccoli, D. A., and Spinner, N. B. (2006). NOTCH2 mutations cause Alagille syndrome, a heterogeneous disorder of the notch signaling pathway. *Am. J. Hum. Genet.* **79**, 169–173.

Moreno-Miralles, I., Schisler, J. C., and Patterson, C. (2009). New insights into bone morphogenetic protein signaling: Focus on angiogenesis. *Curr. Opin. Hematol.* **16**, 195–201.

Morrell, N. W. (2006). Pulmonary hypertension due to BMPR2 mutation: A new paradigm for tissue remodeling? *Proc. Am. Thorac. Soc.* **3**, 680–686.

Moser, M., Binder, O., Wu, Y., Aitsebaomo, J., Ren, R., Bode, C., Bautch, V. L., Conlon, F. L., and Patterson, C. (2003). BMPER, a novel endothelial cell precursor-derived protein, antagonizes bone morphogenetic protein signaling and endothelial cell differentiation. *Mol. Cell. Biol.* **23**, 5664–5679.

Nimmagadda, S., Geetha Loganathan, P., Huang, R., Scaal, M., Schmidt, C., and Christ, B. (2005). BMP4 and noggin control embryonic blood vessel formation by antagonistic regulation of VEGFR-2 (Quek1) expression. *Dev. Biol.* **280**, 100–110.

Noseda, M., Chang, L., McLean, G., Grim, J. E., Clurman, B. E., Smith, L. L., and Karsan, A. (2004). Notch activation induces endothelial cell cycle arrest and participates in contact inhibition: Role of p21Cip1 repression. *Mol. Cell. Biol.* **24**, 8813–8822.

Oda, T., Elkahloun, A. G., Pike, B. L., Okajima, K., Krantz, I. D., Genin, A., Piccoli, D. A., Meltzer, P. S., Spinner, N. B., Collins, F. S., and Chandrasekharappa, S. C. (1997). Mutations in the human Jagged1 gene are responsible for Alagille syndrome. *Nat. Genet.* **16**, 235–242.

Oh, S. P., Seki, T., Goss, K. A., Imamura, T., Yi, Y., Donahoe, P. K., Li, L., Miyazono, K., ten Dijke, P., Kim, S., and Li, E. (2000). Activin receptor-like kinase 1 modulates transforming growth factor-beta 1 signaling in the regulation of angiogenesis. *Proc. Natl. Acad. Sci. USA* **97**, 2626–2631.

Olsson, A. K., Dimberg, A., Kreuger, J., and Claesson-Welsh, L. (2006). VEGF receptor signalling—in control of vascular function. *Nat. Rev. Mol. Cell Biol.* **7**, 359–371.

Park, C., Lavine, K., Mishina, Y., Deng, C. X., Ornitz, D. M., and Choi, K. (2006). Bone morphogenetic protein receptor 1A signaling is dispensable for hematopoietic development but essential for vessel and atrioventricular endocardial cushion formation. *Development* **133**, 3473–3484.

Peirce, S. M. (2008). Computational and mathematical modeling of angiogenesis. *Microcirculation* **15**, 739–751.

Phng, L. K., and Gerhardt, H. (2009). Angiogenesis: A team effort coordinated by notch. *Dev. Cell.* **16**, 196–208.

Phng, L. K., Potente, M., Leslie, J. D., Babbage, J., Nyqvist, D., Lobov, I., Ondr, J. K., Rao, S., Lang, R. A., Thurston, G., and Gerhardt, H. (2009). Nrarp coordinates endothelial Notch and Wnt signaling to control vessel density in angiogenesis. *Dev. Cell.* **16**, 70–82.

Pi, X., Ren, R., Kelley, R., Zhang, C., Moser, M., Bohil, A. B., Divito, M., Cheney, R. E., and Patterson, C. (2007). Sequential roles for myosin-X in BMP6-dependent filopodial extension, migration, and activation of BMP receptors. *J. Cell Biol.* **179**, 1569–1582.

Price, R. J., and Skalak, T. C. (1996). Chronic alpha 1-adrenergic blockade stimulates terminal and arcade arteriolar development. *Am. J. Physiol.* **271**, H752–H759.

Rabinovitch, M. (2008). Molecular pathogenesis of pulmonary arterial hypertension. *J. Clin. Invest.* **118**, 2372–2379.

Reese, D. E., Hall, C. E., and Mikawa, T. (2004). Negative regulation of midline vascular development by the notochord. *Dev. Cell.* **6**, 699–708.

Ren, R., Charles, P. C., Zhang, C., Wu, Y., Wang, H., and Patterson, C. (2007). Gene expression profiles identify a role for cyclooxygenase 2-dependent prostanoid generation in BMP6-induced angiogenic responses. *Blood* **109**, 2847–2853.

Risau, W. (1997). Mechanisms of angiogenesis. *Nature* **386**, 671–674.

Roberts, D. M., Kearney, J. B., Johnson, J. H., Rosenberg, M. P., Kumar, R., and Bautch, V. L. (2004). The vascular endothelial growth factor (VEGF) receptor Flt-1 (VEGFR-1) modulates Flk-1 (VEGFR-2) signaling during blood vessel formation. *Am. J. Pathol.* **164**, 1531–1535.

Roca, C., and Adams, R. H. (2007). Regulation of vascular morphogenesis by Notch signaling. *Genes Dev.* **21**, 2511–2524.

Roman, B. L., Pham, V. N., Lawson, N. D., Kulik, M., Childs, S., Lekven, A. C., Garrity, D. M., Moon, R. T., Fishman, M. C., Lechleider, R. J., and Weinstein, B. M. (2002). Disruption of acvrl1 increases endothelial cell number in zebrafish cranial vessels. *Development* **129**, 3009–3019.

Ruchoux, M. M., Domenga, V., Brulin, P., Maciazek, J., Limol, S., Tournier-Lasserve, E., and Joutel, A. (2003). Transgenic mice expressing mutant Notch3 develop vascular alterations characteristic of cerebral autosomal dominant arteriopathy with subcortical infarcts and leukoencephalopathy. *Am. J. Pathol.* **162**, 329–342.

Ruhrberg, C., Gerhardt, H., Golding, M., Watson, R., Ioannidou, S., Fujisawa, H., Betsholtz, C., and Shima, D. T. (2002). Spatially restricted patterning cues provided by heparin-binding VEGF-A control blood vessel branching morphogenesis. *Genes Dev.* **16**, 2684–2698.

Saharinen, P., Tammela, T., Karkkainen, M. J., and Alitalo, K. (2004). Lymphatic vasculature: Development, molecular regulation and role in tumor metastasis and inflammation. *Trends Immunol.* **25**, 387–395.

Sainson, R. C., Aoto, J., Nakatsu, M. N., Holderfield, M., Conn, E., Koller, E., and Hughes, C. C. (2005). Cell-autonomous notch signaling regulates endothelial cell branching and proliferation during vascular tubulogenesis. *FASEB J.* **19**, 1027–1029.

Sainson, R. C., Johnston, D. A., Chu, H. C., Holderfield, M. T., Nakatsu, M. N., Crampton, S. P., Davis, J., Conn, E., and Hughes, C. C. (2008). TNF primes endothelial cells for angiogenic sprouting by inducing a tip cell phenotype. *Blood* **111**, 4997–5007.

Shalaby, F., Rossant, J., Yamaguchi, T. P., Gertsenstein, M., Wu, X. F., Breitman, M. L., and Schuh, A. C. (1995). Failure of blood-island formation and vasculogenesis in Flk-1-deficient mice. *Nature* **376**, 62–66.

Shibuya, M. (2006). Vascular endothelial growth factor receptor-1 (VEGFR-1/Flt-1): A dual regulator for angiogenesis. *Angiogenesis* **9**, 225–230.

Shibuya, M., and Claesson-Welsh, L. (2006). Signal transduction by VEGF receptors in regulation of angiogenesis and lymphangiogenesis. *Exp. Cell Res.* **312**, 549–560.

Shutter, J. R., Scully, S., Fan, W., Richards, W. G., Kitajewski, J., Deblandre, G. A., Kintner, C. R., and Stark, K. L. (2000). Dll4, a novel Notch ligand expressed in arterial endothelium. *Genes Dev.* **14**, 1313–1318.

Siekmann, A. F., Covassin, L., and Lawson, N. D. (2008). Modulation of VEGF signalling output by the Notch pathway. *BioEssays* **30**, 303–313.

Siekmann, A. F., and Lawson, N. D. (2007). Notch signalling limits angiogenic cell behaviour in developing zebrafish arteries. *Nature* **445**, 781–784.

Skalak, T. C., and Price, R. J. (1996). The role of mechanical stresses in microvascular remodeling. *Microcirculation* **3**, 143–165.
Soker, S., Miao, H. Q., Nomi, M., Takashima, S., and Klagsbrun, M. (2002). VEGF165 mediates formation of complexes containing VEGFR-2 and neuropilin-1 that enhance VEGF165-receptor binding. *J. Cell. Biochem.* **85**, 357–368.
Sorensen, L. K., Brooke, B. S., Li, D. Y., and Urness, L. D. (2003). Loss of distinct arterial and venous boundaries in mice lacking endoglin, a vascular-specific TGFbeta coreceptor. *Dev. Biol.* **261**, 235–250.
Stalmans, I., Lambrechts, D., De Smet, F., Jansen, S., Wang, J., Maity, S., Kneer, P., von der Ohe, M., Swillen, A., Maes, C., Gewillig, M., Molin, D. G., *et al.* (2003). VEGF: A modifier of the del22q11 (DiGeorge) syndrome? *Nat. Med.* **9**, 173–182.
Stalmans, I., Ng, Y. S., Rohan, R., Fruttiger, M., Bouche, A., Yuce, A., Fujisawa, H., Hermans, B., Shani, M., Jansen, S., Hicklin, D., Anderson, D. J., *et al.* (2002). Arteriolar and venular patterning in retinas of mice selectively expressing VEGF isoforms. *J. Clin. Invest.* **109**, 327–336.
Stratman, A. N., Malotte, K. M., Mahan, R. D., Davis, M. J., and Davis, G. E. (2009). Pericyte recruitment during vasculogenic tube assembly stimulates endothelial basement membrane matrix formation. *Blood* **114**, 5091–5101.
Strilic, B., Kucera, T., Eglinger, J., Hughes, M. R., McNagny, K. M., Tsukita, S., Dejana, E., Ferrara, N., and Lammert, E. (2009). The molecular basis of vascular lumen formation in the developing mouse aorta. *Dev. Cell.* **17**, 505–515.
Suchting, S., Freitas, C., le Noble, F., Benedito, R., Breant, C., Duarte, A., and Eichmann, A. (2007). The Notch ligand Delta-like 4 negatively regulates endothelial tip cell formation and vessel branching. *Proc. Natl. Acad. Sci. USA* **104**, 3225–3230.
Swift, M. R., and Weinstein, B. M. (2009). Arterial-venous specification during development. *Circ. Res.* **104**, 576–588.
Takashima, S., Kitakaze, M., Asakura, M., Asanuma, H., Sanada, S., Tashiro, F., Niwa, H., Miyazaki Ji, J., Hirota, S., Kitamura, Y., Kitsukawa, T., Fujisawa, H., *et al.* (2002). Targeting of both mouse neuropilin-1 and neuropilin-2 genes severely impairs developmental yolk sac and embryonic angiogenesis. *Proc. Natl. Acad. Sci. USA* **99**, 3657–3662.
Tammela, T., Zarkada, G., Wallgard, E., Murtomaki, A., Suchting, S., Wirzenius, M., Waltari, M., Hellstrom, M., Schomber, T., Peltonen, R., Freitas, C., Duarte, A., *et al.* (2008). Blocking VEGFR-3 suppresses angiogenic sprouting and vascular network formation. *Nature* **454**, 656–660.
Taylor, K. L., Henderson, A. M., and Hughes, C. C. (2002). Notch activation during endothelial cell network formation in vitro targets the basic HLH transcription factor HESR-1 and downregulates VEGFR-2/KDR expression. *Microvasc. Res.* **64**, 372–383.
Teichert-Kuliszewska, K., Kutryk, M. J., Kuliszewski, M. A., Karoubi, G., Courtman, D. W., Zucco, L., Granton, J., and Stewart, D. J. (2006). Bone morphogenetic protein receptor-2 signaling promotes pulmonary arterial endothelial cell survival: Implications for loss-of-function mutations in the pathogenesis of pulmonary hypertension. *Circ. Res.* **98**, 209–217.
Tischer, E., Mitchell, R., Hartman, T., Silva, M., Gospodarowicz, D., Fiddes, J. C., and Abraham, J. A. (1991). The human gene for vascular endothelial growth factor. Multiple protein forms are encoded through alternative exon splicing. *J. Biol. Chem.* **266**, 11947–11954.
Trembath, R. C., Thomson, J. R., Machado, R. D., Morgan, N. V., Atkinson, C., Winship, I., Simonneau, G., Galie, N., Loyd, J. E., Humbert, M., Nichols, W. C., Morrell, N. W., *et al.* (2001). Clinical and molecular genetic features of pulmonary hypertension in patients with hereditary hemorrhagic telangiectasia. *N. Engl. J. Med.* **345**, 325–334.
Urness, L. D., Sorensen, L. K., and Li, D. Y. (2000). Arteriovenous malformations in mice lacking activin receptor-like kinase-1. *Nat. Genet.* **26**, 328–331.

Valdimarsdottir, G., Goumans, M. J., Rosendahl, A., Brugman, M., Itoh, S., Lebrin, F., Sideras, P., and ten Dijke, P. (2002). Stimulation of Id1 expression by bone morphogenetic protein is sufficient and necessary for bone morphogenetic protein-induced activation of endothelial cells. *Circulation* **106**, 2263–2270.

Villa, N., Walker, L., Lindsell, C. E., Gasson, J., Iruela-Arispe, M. L., and Weinmaster, G. (2001). Vascular expression of Notch pathway receptors and ligands is restricted to arterial vessels. *Mech. Dev.* **108**, 161–164.

Vogeli, K. M., Jin, S. W., Martin, G. R., and Stainier, D. Y. (2006). A common progenitor for haematopoietic and endothelial lineages in the zebrafish gastrula. *Nature* **443**, 337–339.

Vogt, R. R., Unda, R., Yeh, L. C., Vidro, E. K., Lee, J. C., and Tsin, A. T. (2006). Bone morphogenetic protein-4 enhances vascular endothelial growth factor secretion by human retinal pigment epithelial cells. *J. Cell. Biochem.* **98**, 1196–1202.

Whitaker, G. B., Limberg, B. J., and Rosenbaum, J. S. (2001). Vascular endothelial growth factor receptor-2 and neuropilin-1 form a receptor complex that is responsible for the differential signaling potency of VEGF(165) and VEGF(121). *J. Biol. Chem.* **276**, 25520–25531.

Witte, M. H., Bernas, M. J., Martin, C. P., and Witte, C. L. (2001). Lymphangiogenesis and lymphangiodysplasia: From molecular to clinical lymphology. *Microsc. Res. Tech.* **55**, 122–145.

Xu, Y., Yuan, L., Mak, J., Pardanaud, L., Caunt, M., Kasman, I., Larrivee, B., Del Toro, R., Suchting, S., Medvinsky, A., Silva, J., Yang, J., *et al.* (2010). Neuropilin-2 mediates VEGF-C-induced lymphatic sprouting together with VEGFR3. *J. Cell Biol.* **188**, 115–130.

Xue, Y., Gao, X., Lindsell, C. E., Norton, C. R., Chang, B., Hicks, C., Gendron-Maguire, M., Rand, E. B., Weinmaster, G., and Gridley, T. (1999). Embryonic lethality and vascular defects in mice lacking the Notch ligand Jagged1. *Hum. Mol. Genet.* **8**, 723–730.

Yang, X., Long, L., Southwood, M., Rudarakanchana, N., Upton, P. D., Jeffery, T. K., Atkinson, C., Chen, H., Trembath, R. C., and Morrell, N. W. (2005). Dysfunctional Smad signaling contributes to abnormal smooth muscle cell proliferation in familial pulmonary arterial hypertension. *Circ. Res.* **96**, 1053–1063.

Yeh, L. C., and Lee, J. C. (1999). Osteogenic protein-1 increases gene expression of vascular endothelial growth factor in primary cultures of fetal rat calvaria cells. *Mol. Cell. Endocrinol.* **153**, 113–124.

Yoo, S. K., Deng, Q., Cavnar, P. J., Wu, Y. I., Hahn, K. M., and Huttenlocher, A. (2010). Differential regulation of protrusion and polarity by PI3K during neutrophil motility in live zebrafish. *Dev. Cell.* **18**, 226–236.

Yoshimoto, M., and Yoder, M. C. (2009). Developmental biology: Birth of the blood cell. *Nature* **457**, 801–803.

Young, H. S., Summers, A. M., Bhushan, M., Brenchley, P. E., and Griffiths, C. E. (2004). Single-nucleotide polymorphisms of vascular endothelial growth factor in psoriasis of early onset. *J. Invest. Dermatol.* **122**, 209–215.

Yuan, L., Moyon, D., Pardanaud, L., Breant, C., Karkkainen, M. J., Alitalo, K., and Eichmann, A. (2002). Abnormal lymphatic vessel development in neuropilin 2 mutant mice. *Development* **129**, 4797–4806.

Zeng, G., Taylor, S. M., McColm, J. R., Kappas, N. C., Kearney, J. B., Williams, L. H., Hartnett, M. E., and Bautch, V. L. (2007). Orientation of endothelial cell division is regulated by VEGF signaling during blood vessel formation. *Blood* **109**, 1345–1352.

Zhong, T. P., Childs, S., Leu, J. P., and Fishman, M. C. (2001). Gridlock signalling pathway fashions the first embryonic artery. *Nature* **414**, 216–220.

Zhong, T. P., Rosenberg, M., Mohideen, M. A., Weinstein, B., and Fishman, M. C. (2000). Gridlock, an HLH gene required for assembly of the aorta in zebrafish. *Science* **287**, 1820–1824.

CHAPTER THREE

Lung Organogenesis

David Warburton,[*] Ahmed El-Hashash,[*] Gianni Carraro,[*] Caterina Tiozzo,[*] Frederic Sala,[*] Orquidea Rogers,[*] Stijn De Langhe,[†] Paul J. Kemp,[‡] Daniela Riccardi,[‡] John Torday,[§] Saverio Bellusci,[*,¶] Wei Shi,[*,‖] Sharon R Lubkin,[**] *and* Edwin Jesudason[††]

Contents

1. Introduction 74
2. Developmental Anatomy of the Lung 75
   2.1. The bauplan: key steps in lung morphogenesis 75
   2.2. The histological stages of lung development 75
   2.3. Focus on branching morphogenesis: simplifying the complexities 78
   2.4. The impact of abnormal lung development 80
3. Molecular Embryology of the Lung 81
   3.1. Process-driven molecular embryology of the lung 81
   3.2. Cataloguing the biochemical regulators of lung development 87
4. Mechanobiology of the Developing Lung 111
   4.1. Lessons on mechanobiology from human and *in vivo* studies 112
   4.2. The impact of hydraulic pressure on lung organogenesis 113
   4.3. The impact of embryonic airway peristalsis in lung organogenesis 114
   4.4. Lung stretch transduction and parathyroid hormone-related protein (PTHrP) 115
   4.5. Lung development and the $Ca2^{+}$-sensing receptor (CaSR) 116
5. Stem/Progenitor Cell Biology of the Lung 118
   5.1. Endogenous epithelial progenitor cells 118

[*] Developmental Biology and Regenerative Medicine Program, California Institute for Regenerative Medicine Training Program, The Saban Research Institute, Childrens Hospital Los Angeles, Los Angeles, California, USA
[†] National Jewish Health, Denver, Colorado, USA
[‡] School of Biosciences, Cardiff University, Cardiff, South Glamorgan, UK
[§] Los Angeles County Harbor-UCLA Medical Center, Los Angeles, California, USA
[¶] Excellence Cluster in Cardio-Pulmonary Systems, University of Giessen, ECCPS/Medical Clinic II Klinkstr, Giessen, Germany
[‖] Systems Biology and Disease Graduate Program, Keck School of Medicine and School of Dentistry, University of Southern California, Los Angeles, California, USA
[**] Department of Mathematics and Department of Biomedical Engineering, North Carolina State University, Raleigh, North Carolina, USA
[††] Alder Hey Children's Hospital & Division of Child Health, University of Liverpool, Liverpool, Merseyside, UK

Current Topics in Developmental Biology, Volume 90
ISSN 0070-2153, DOI 10.1016/S0070-2153(10)90003-3

5.2. Endogenous mesenchymal progenitors 122
5.3. Control of lung progenitor cell proliferation 123
5.4. Embryonic lung progenitors and proximal–distal patterning 124
5.5. Emergence of specific cell types during lung organogenesis 125
5.6. Stem and progenitor cells in the postnatal respiratory system 127
5.7. Potential strategies to protect lung progenitors 128
6. Postnatal and Adult Lung 128
6.1. The transition to air breathing 128
6.2. Lung aging and involution 130
7. Conclusions 130
Acknowledgments 131
References 131

Abstract

Developmental lung biology is a field that has the potential for significant human impact: lung disease at the extremes of age continues to cause major morbidity and mortality worldwide. Understanding how the lung develops holds the promise that investigators can use this knowledge to aid lung repair and regeneration. In the decade since the "molecular embryology" of the lung was first comprehensively reviewed, new challenges have emerged—and it is on these that we focus the current review. Firstly, there is a critical need to understand the progenitor cell biology of the lung in order to exploit the potential of stem cells for the treatment of lung disease. Secondly, the current familiar descriptions of lung morphogenesis governed by growth and transcription factors need to be elaborated upon with the reinclusion and reconsideration of other factors, such as mechanics, in lung growth. Thirdly, efforts to parse the finer detail of lung bud signaling may need to be combined with broader consideration of overarching mechanisms that may be therapeutically easier to target: in this arena, we advance the proposal that looking at the lung in general (and branching in particular) in terms of clocks may yield unexpected benefits.

1. Introduction

The concept that lung organogenesis is instructed by coordinated mesenchymal-to-epithelial crosstalk originates in the classical recombination experiments of Alescio and Cassini (1962), in which replacing tracheal mesenchyme with mesenchyme from the lung periphery induced ectopic branching of tracheal epithelium in murine embryonic lung organ culture. This idea was extended in an early review by Warburton and Olver (1997) to include the coordination of genetic, epigenetic, and environmental factors in lung development, injury, and repair. Thereafter, a molecular basis of lung morphogenesis was attempted by Warburton *et al.* (2000). Over the last decade, significant progress has been made in this field as reviewed by Cardoso and Lu (2006), Maeda *et al.* (2007), and others. Nevertheless, the ultimate goal remains as stated by Warburton and Olver (1997), "to devise new rational and gene therapeutic approaches to ameliorate lung injury and augment lung repair ... the ideal agent or agents would therefore mimic the instructive role of lung mesenchyme and would correctly induce the temporospatial pattern of lung-specific gene expression necessary to instruct lung regeneration." To this overall strategy, we can now add (i) the modulation of lung mechanobiology to favor appropriate lung regeneration and (ii) the stimulation of endogenous stem/progenitor cells or supply of exogenous ones for lung regeneration. Therefore, the current review draws together three important strands of information on lung organogenesis as of April 2010: (i) molecular embryology of the lung, (ii) mechanobiology of the developing lung, and (iii) pulmonary stem/progenitor cell biology. Applying advances in these complementary areas of research to lung regeneration and correction of lung diseases remains the therapeutic goal of this field. With the recent human transplanation of a stem/progenitor cell-derived tissue-engineered major airway (Macchiarini *et al.*, 2008), we can clearly see the potential of this field, while recognizing the many problems yet to be solved.

Before concentrating on the molecular biology, mechanobiology, and stem cell biology of the lung, a first step in regenerative strategies is to consider the developmental anatomy of the lung. From this, we can at least see what type of structures we need to generate.

2. Developmental Anatomy of the Lung

2.1. The bauplan: key steps in lung morphogenesis

A diagrammatic overview of lung morphogenesis is given in Fig. 3.1. Three lobes form on the right side and two lobes on the left side in human lung; in mice four lobes form on the right (cranial, medial, and caudal lobes, plus the accessory lobe) and one on the left. In contrast to humans, in the mouse, there are only 12 airway generations and alveolarization occurs entirely postnatally.

2.2. The histological stages of lung development

Histologically, lung development and maturation has been divided into four stages: pseudoglandular, canalicular, terminal saccular, and alveolar (Fig. 3.2).

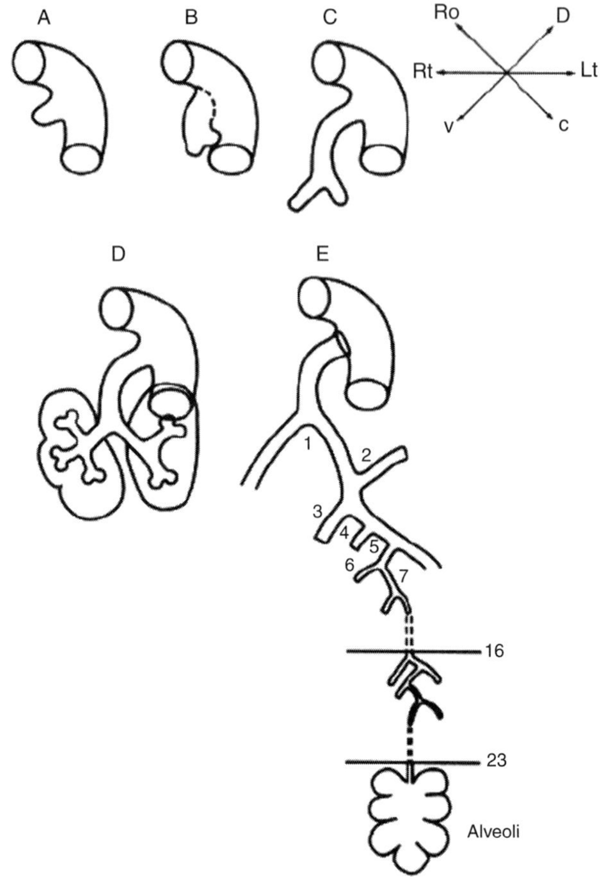

Figure 3.1 (A) The primitive lung anlage emerges as the laryngotracheal groove from the ventral surface of the primitive foregut at 5 weeks' gestation in the human. (B) The primitive trachea separates dorsoventrally from the primitive esophagus as the two primary bronchial branches arise from the lateral aspects of the laryngotracheal groove at 5 or 6 weeks' gestation in the human. (C) The embryonic larynx and trachea with the two primary bronchial branches are separated dorsoventrally from the embryonic esophagus at 6 weeks in the human. (D) The primitive lobar bronchi branching from the primary bronchi at 7 weeks in the human. (E) A schematic rendering of the airway at term in the human. The stereotypical first 16 airway generations are complete by 16 weeks in humans; between 16 and 24 weeks, further branching is nonstereotyped. Alveolarization begins about 20 weeks in humans and is complete by 7 years of age at the earliest. (After West, Burri, Warburton, and others).

The pseudoglandular stage (5–17 weeks of human pregnancy, E9.5–16.6 days in mouse embryo). During this, the earliest developmental stage, epithelial tubes lined with cuboidal epithelial cells undergo branching morphogenesis and resemble an exocrine gland (hence the nomenclature). However, this

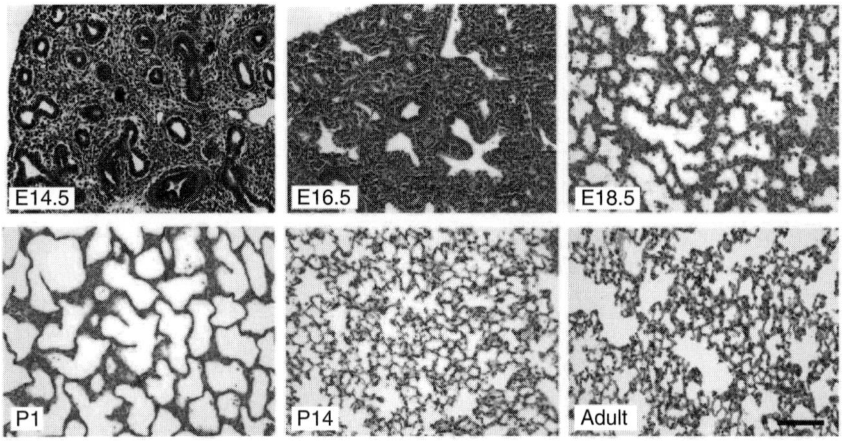

Figure 3.2 Histology of mouse lung at characteristic stages of development. Embryonic mouse lung develops from pseudoglandular stage (E14.5) to canalicular stage (E16.5) and further terminal sac stage (E18.5 and P1). Neonatal mouse lungs undergo alveolarization, resulting in the formation of many septa (P14). Finally, a mature honeycomb-like structure with alveoli surrounding alveolar ducts conferring normal respiratory structure and function is formed, as observed in the adult. Scale bar: 100 μm.

fluid-filled primitive respiratory tree structure is too immature to support efficient gas exchange.

The canalicular stage (16–25 weeks of human pregnancy, E16.6–17.4 days in mouse embryo). The cranial part of the lung develops faster than the caudal part, resulting in partial overlap between this stage and the previous stage. During the canalicular stage, the respiratory tree is further expanded in diameter and length, accompanied by vascularization and angiogenesis along the airway. A massive increase in the number of capillaries occurs. The terminal bronchioles are then divided into respiratory bronchioles and alveolar ducts, and the airway epithelial cells are differentiated into peripheral squamous cells and proximal cuboidal cells.

The terminal saccular stage (24 weeks to late fetal period in human, E17.4 to postnatal day 5 (P5) in mouse). There is substantial thinning of the interstitium during the terminal saccular stage. This results from apoptosis as well as ongoing differentiation of mesenchymal cells (Hashimoto *et al.*, 2002; Lu *et al.*, 2002). Additionally, at this stage, the alveolar epithelial cells (AECs) are more clearly differentiated into mature squamous type I pneumocytes and secretory rounded type II pneumocytes bearing lamellar bodies that contain surfactant. The capillaries also grow rapidly in the mesenchyme surrounding the saccules to form a complex network. In addition, the lymphatic network in lung tissue becomes well developed during this stage. The thick wall of these saccules, also called primary septae, comprises lining epithelial cells on both sides of a connective tissue core, within which there is a double parallel network of capillaries. Toward the end of this stage, the fetal lung can support air exchange in prematurely born human neonates. Maturation of surfactant synthesis and secretion is a key factor in determining whether the newborn lung can sustain gas exchange without collapsing.

The alveolar stage (late fetal period to childhood in human, P5–P30 in mouse). Alveolarization is the last step of lung development. The majority of the gas exchange surface is formed during this stage.

Genome-wide expression profiling has measured developing human lung transcriptomes in pregnancies terminated between 7 and 22 weeks post conception (Kho *et al.*, 2009). Within the 3,223 gene developing lung-characteristic subtranscriptome, transitions in gene expression correlated with some histological stages, as well as suggesting novel substages exist. For example, induction of surfactant gene expression identifies a "molecular transition" in the pseudoglandular phase.

Hence, the histological account of lung development is complimented by the molecular embryology that we consider in the next main section of the review.

2.3. Focus on branching morphogenesis: simplifying the complexities

Branching morphogenesis is a critical part of overall lung development and a crucial phenomenon in the development of several other organs. Understanding this key hurdle in lung regeneration strategies requires us to appreciate that despite the apparent and beautiful complexity of the lung, there are key simplicities that can help us in our task.

Fractal mathematics has revealed that relatively straightforward algorithms, when applied iteratively, could generate patterning of great complexity (Mandelbrot, 1982). Moreover, when the mathematical parameters were chosen appropriately, the approximation to living branched structures was striking. At first sight, intrapulmonary airway branching (distal to the primary bronchi) appears to become increasingly complicated as it proceeds distally and the number of individual branches increases into the millions. But, once the laryngotracheal complex and left–right laterality are established, distal airway branching is now thought to be driven by a relatively simple set of genetically encoded control routines. Echoing the fractal pattern formation achieved in mathematical models, these include the following: (i) a master branch generator routine, with three slave subroutines instructing a periodicity clock which times the appearance of subsequent branches, (ii) a rotational orientation subroutine which determines the orientation of the branches around the axis of the airway, and (iii) a branch tip division subroutine (Metzger *et al.*, 2008; Warburton, 2008). Thus, branching morphogenesis of the bronchi in early mouse embryo lung can be parsed anatomically into three simple geometric forms, termed domain branching, planar bifurcation, and orthogonal bifurcation (Fig. 3.3). These basic forms are repeated iteratively to form different arrangements of branches. The planar or bottlebrush array describes the sequential proximal to distal emergence of secondary branches along the lateral axis of the primary bronchial airway. The bottlebrush mechanism is then reoriented around the branch axis to form a second row of branches at right angles to

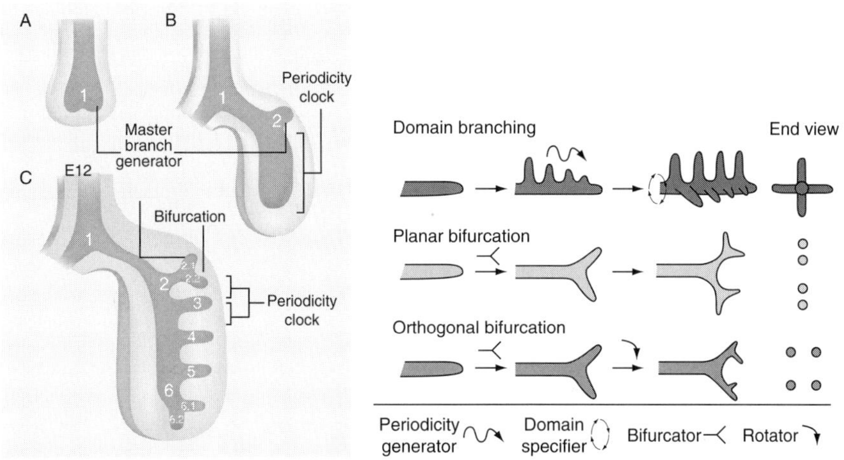

Figure 3.3 How the airways can form in a sequential manner by reiterating a few, relatively simple sets of genetic instructions. In the upper panel, which is drawn after Warburton (2008), a master branch generator, a periodicity clock, and a bifurcator program are shown as controlling the layout of the mainstem and lobar branches. At embryonic day (E) 10.5, (A) the primary bronchial branch (1) forms, followed by (B) the development of the left upper-lobe branch (2) by E11, and then (C) the first two segmental branches of the left upper-lobe branch (2.2 and 2.3) form and the subsequent formation of branches 3–6 occurs by E12. The master branch generator is active throughout these events, and the inferred sites of action of the periodicity clock and bifurcator subroutines are shown. Then, in the lower panel, following the views of Metzger *et al.*, (2008), a series of inferred genetic subroutines are shown, all driven by one master branch generator, shown as giving rise to domain or "bottle brush" branching along the lateral proximodistal axis of the main stem bronchi, which can then be rotated at right angles to give rise to a second rank of branches. Then, in subsequent rounds of branching, arising from the tips of the primary and secondary branches, it is shown how the same relatively simple periodicity generator, domain specifier, bifurcator, and rotator subroutines can give rise to apparently more complex patterns of peripheral branching to achieve an ever larger number of space filling terminal branches.

the first row. The terms *planar array* and *rosette array* describe the patterns formed by sequential bifurcation of the tips of secondary, tertiary, and subsequent buds at right angles to each other. Repetition of these simple branching modules, together with the hierarchical control and coupling of them, may therefore explain how the genome could possibly encode the highly complex yet stereotypic pattern of early bronchial branch formation, using a relatively simple toolbox of genetic modules. In a further illustration of how the mammalian lung uses simple routines and subroutines to construct itself, substantial homology has been identified between the genetic regulation of lung organogenesis and airway morphogenesis in *Drosophila* (Hacohen *et al.*, 1998; Tefft *et al.*, 1999). Despite the latter's relative simplicity, it is striking to note not only the genetic homology but also the similar epistatic signaling hierarchy into which these regulators are arranged in the fly.

Using real-time microscopic cinematography, individual airway tip branching can be parsed temporally into a branch extension phase, a branch tip arrest phase, and a tip-splitting budding phase, followed once again when the branch budding phase is completed by branch extension until the next round of budding follows once more. A clock mechanism mediated by FGF–FGFR–Sprouty signaling plays a key role in timing the rate of bud extension and hence the inter-branch distance (Unbekandt *et al.*, 2008; Warburton, 2008). Indeed a nested hierarchy of clock routines are likely to be present throughout lung development given the number of oscillating systems intrinsic to the lung (branching, airway peristalsis, calcium oscillations) or visited extrinsically upon it (fetal breathing, circadian rhythms). Branching morphogenesis is accompanied by contractile oscillations (airway peristalsis) that are themselves underpinned by periodic calcium waves (Featherstone *et al.*, 2006; Jesudason *et al.*, 2005). These oscillators appear to be coupled to lung growth, and their precise relation to the timing of branching remains to be determined. However, we postulate that clock routines underlying the linear process of somitogenesis are redeployed three-dimensionally for branching morphogenesis in the lung and other organs (Pourquie, 2003).

2.4. The impact of abnormal lung development

The airway is developed sequentially by early epithelial tube branching and later septation of terminal air sacs. Pulmonary vasculature develops within lung mesenchyme in close conjunction with epithelial morphogenesis. Airway and vascular smooth muscle also develop during early morphogenesis. Perturbation of these developmental processes results in abnormal lung structure, deficiency of gas exchange, and neonatal respiratory failure. Clinical examples of such disruption of normal lung growth include cystic adenomatoid malformation of the lung, bronchopulmonary dysplasia (BPD) (a sequel of premature human delivery), and hypoplasia of the lung (seen in congenital diaphragmatic hernia (CDH), a birth defect as common as cystic fibrosis). More subtle lung dysplasias that are not lethal neonatally may emerge later in life with asthmatic wheezing and perhaps predisposition to early onset of chronic obstructive pulmonary disease (COPD). One of the clearest examples of how early development can affect not only lung organogenesis but also long-term health is the ciliary dyskinesia encountered in Kartagener syndrome and primary ciliary dyskinesia (Storm van's Gravesande, 2005). Early in embryogenesis, failure of ciliary function leads to randomization of organ situs and hence a supranormal rate of dextrocardia. This is accompanied by randomization of lung asymmetry. Persisting ciliary dysfunction impairs mucociliary clearance in the sinuses as well as the airways and predisposes to chronic lung disease in later life. Crucially one can note that disruption of lung asymmetry does not itself lead to lung malformation: hence the lung "bauplan" is conserved despite the lungs' left–right asymmetry being the reverse of normal. This observation reiterates to us that the complexities of lung organogenesis may actually be broken down into nested routines and subroutines used to accomplish particular tasks in the overall process.

The implication for lung regeneration is that one need not understand the formation of every last alveolus, but rather that elucidating the iterative routines could suffice to promote pulmonary "self-assembly."

3. Molecular Embryology of the Lung

This section of the review serves as a comprehensive reference source. For those with no requirement for such detail, the reader is directed to the summary Fig. 3.4.

We will first use a step-wise "process-driven" description of lung growth followed by a catalogue of the biochemical factors involved: many such factors are involved at multiple stages and do not map neatly on to the "process-driven" account. The biochemical factors are considered as follows: growth and transcription factors in order of first appearance and then other participating factors such as extracellular matrix (ECM) and miRNA.

3.1. Process-driven molecular embryology of the lung

3.1.1. Induction of the early lung anlagen

Early lung induction is regulated by genes that act cooperatively to define the location of laryngotracheal groove and help specify the spatial axes of the developing organ.

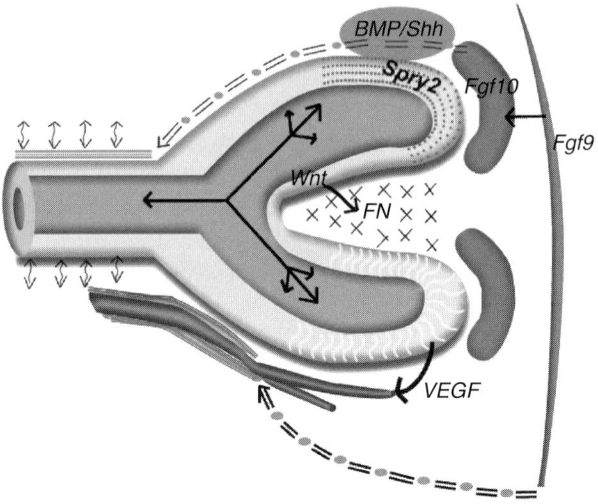

Figure 3.4 Schematic illustration of the variety of biochemical and biomechanical regulators of lung growth. The epithelium of the end-bud (yellow) encloses a fluid filled lumen (blue-green) in which oscillatory fluid flows are depicted (solid bi-directional arrows) that result from periodic peristaltic contractions (bi-directional curvilinear arrows) of airway smooth muscle (ASM: sample shown in brown running parallel and above the main epithelial trunk). The ASM derives from the FGF10$^+$ precursor pool (seen in purple on the right), and these ASM progenitors are seen becoming more proximal (double-lined "=" at the top of the epithelial outline) relative to the growing epithelium. Examples of key biochemical signaling are given: FGF9 from the mesothelium (far right) regulates the FGF10$^+$ mesenchyme (purple), which in turn interacts locally with epithelial Sprouty (SPRY2), BMP4, and Sonic Hedgehog (SHH) signaling (the latter two epithelial signals are shown in green and, due to space constraint, adjacent to the schematized epithelium). Epithelial Wnt signaling regulates fibronectin (FN) elaboration (shown as "xx") in the cleft between epithelial branches. Epithelial VEGF signals to developing vasculature shown at the base of the figure. These vessels attract vascular smooth muscle precursors from the mesothelium (shown as "=" at the base of the figure). (See Color Insert.)

Among the earliest endodermal signals essential for gut morphogenesis and gut tube closure are the GATA (zinc-finger proteins that recognize GATA DNA sequence) and hepatocyte nuclear factor (HNF/Fox) transcription factors. *Foxa2* is required for gut tube closure, while *Gata-6* is required for activation of the lung developmental program within the foregut endoderm. *Hnf-3/Foxa2β* is a survival factor for the endoderm; its expression is induced by Sonic hedgehog (*Shh*). Retinoids and their transcriptional factor receptors also play key roles in induction of early lung branching: retinoic acid (RA) deficiency and compound null mutation of retinoid receptors prevent induction of the laryngotracheal groove. Most recently, Wnt2/2b and β-catenin signaling have been shown to be necessary and sufficient to specify lung progenitors in the foregut (Goss *et al.*,

2009; Harris-Johnson *et al.*, 2009). Embryos lacking Wnt2/2b expression exhibit complete lung agenesis and do not express Nkx2.1, the earliest marker of the lung endoderm. This phenotype is recapitulated by an endoderm-restricted deletion of β-catenin. Conversely, conditional expression of an activated form of β-catenin leads to ectopic expansion of the Nkx2.1 expression domain into esophagus and stomach epithelium. Thus, gain or loss of trachea/lung progenitor identity is accompanied, respectively, by contraction or expansion of esophagus/stomach progenitor identity. Taken together, these findings suggest that Wnt2/2b signaling through the canonical Wnt pathway is required to specify lung endoderm progenitors within the foregut. Furthermore, ectopic lung bud formation can be induced in the esophagus by *Tbx4* misexpression activating *Fgf10* expression (Sakiyama *et al.*, 2003). In addition, left–right asymmetry is controlled by several genes, including *nodal, Lefty-1,2,* and *Pitx-2.* For example, single-lobed lungs are found bilaterally in *Lefty-1*$^{-/-}$ mice, and bilateral isomerism of the lung is found in *Pitx2*-null mutants.

3.1.2. Tracheoesophageal septation

The processes whereby trachea and esophagus form from primitive foregut is of clinical interest due to the common birth defect, tracheoesophageal fistula (TEF) (Fig. 3.5).

Usually encountered in conjunction with esophageal atresia (EA), the combined sequence is sometimes found together with other anomalies of heart, vertebrae, anorectum, and limbs. Genetic defects identified in patients with EA-TEF have recently been comprehensively reviewed (Felix *et al.*, 2009). Transgenic murine mutants with deletions in RA receptors or *Gli2/*

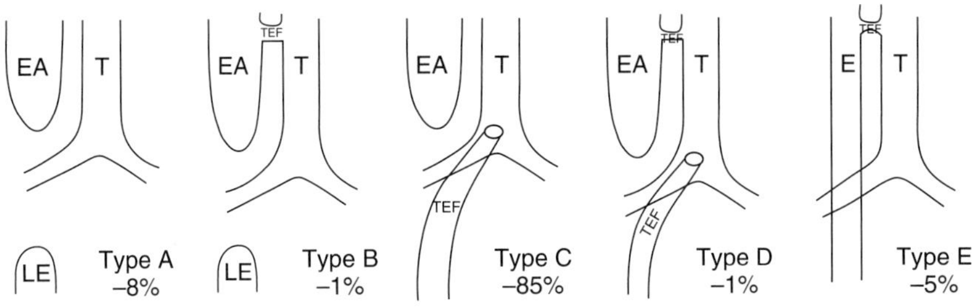

Figure 3.5 Subtypes of esophageal atresia (EA) and/or tracheo(T)–esophageal (E) fistula (TEF) with percentage frequency amongst EA-TEF cases. Type A: 'Pure' EA without TEF. Lower esophagus shown (LE); Type B: EA with TEF from proximal esophageal pouch but without any distal TEF; Type C: EA with distal TEF only (the most common variant); Type D: EA with both proximal TEF and distal TEF; Type E: TEF in the absence of EA. Laryngotracheal clefts (not shown here) are still rarer anomalies in which trachea and esophagus form a single lumen for a variable length. In severe variants, a combined tracheo-esophagus connects to the stomach whilst also giving rise to the main bronchi.

Gli3 feature a form of EA-TEF. Moreover, the transcriptomic changes associated with budding of the lung from the foregut have recently been enumerated. Alongside identifying the known regulators described above, further candidates will need experimental evaluation (Millien *et al.*, 2008). Illustrating that environmental factors may play a role, a EA-TEF phenotype can be generated by exposure of murine embryos to adriamycin (Diez-Pardo *et al.*, 1996). Interestingly, despite the major anomaly of foregut development, lung formation in EA-TEF patients is usually grossly normal. Their respiratory tract morbidity tends to derive from tracheomalacia and, more chronically, reactive airways disease. Whilst the latter is traditionally attributed to gastroesophageal reflux and pulmonary aspiration, it remains possible that some of this pulmonary morbidity stems from subtly abnormal early lung development.

3.1.3. Tracheal cartilage formation

Children with EA-TEF may also suffer from tracheal weakness (tracheomalacia) in which inadequate formation of the tracheal cartilages results in potentially life-threatening airway closure during expiration.

Dorsoventral patterning of the trachea during embryonic development is associated with formation of C-shaped cartilage rings ventrally and trachealis muscle dorsally. Ventral mesenchyme segregates into successive cartilaginous and noncartilaginous domains, providing a compromise between flexibility and rigidity.

Tracheomalacia describes weakness of the walls of the trachea and it may result in life-threatening episodes and/or recurrent hospitalizations for lower airway infections (Austin and Ali, 2003; Boogaard *et al.*, 2005; Carden *et al.*, 2005; McNamara and Crabbe, 2004). It can be an isolated idiopathic anomaly or associated with EA-TEF, primary defects of cartilage synthesis (e.g., dyschondroplasia), cartilage degeneration due to trauma (e.g., long-term tracheal intubation), or extrinsic compressive lesions such as vascular rings or tumors (Berdon, 2000). Tracheal stenosis is narrowing of the trachea: it can follow prolonged intubation or accompany a cartilaginous sleeve malformation of the trachea or may again be associated with extrinsic compressive lesions. Tracheal cartilaginous sleeve comprises fusion of the ventral cartilage rings. It is a rare malformation associated with craniosynostosis syndromes like Crouzon syndrome, Pfeiffer syndrome, Goldenhar syndrome or Apert syndrome (Lin *et al.*, 1995).

Tracheal maldevelopment can be modeled: nitrofen administration to pregnant dams results in tracheal malformations as well as CDH in offspring (Diez-Pardo *et al.*, 1996; Xia *et al.*, 1999). Although genetic control of the regulation of tracheal cartilage versus smooth muscle cell (SMC) formation remains unclear, relevant transgenic murine phenotypes have been observed. Miller *et al.* (2004) showed that partial *Shh* inactivation causes tracheobronchial cartilage abnormalities indicative of tracheomalacia. Park *et al.* (2009)

demonstrated *Shh* augments *Sox9* expression: Sox9 induces type II collagen (Col2a1) expression and promotes the chondrocyte lineage amongst mesenchymal cells. Bone morphogenic protein 4 (BMP4) also regulates Sox9 to induce chondroprogenitors amongst mesenchymal cells (Hatakeyama *et al.*, 2004). Impaired BMP signaling induces tracheal cartilage defects with EA-TEF (Que *et al.*, 2006). β-Catenin also interacts with Sox9 but to inhibit differentiation of tracheal chondroprogenitor cells (Akiyama *et al.*, 2004).

Recently, FGF18 and FGF10 have also been described to play important roles in tracheal cartilage ring formation (Elluru *et al.*, 2009; Tiozzo *et al.*, 2009; Whitsett *et al.*, 2002). Specific targeted inactivation of *Fgf18*, using the SP-C promoter driving Cre, induced malformation of the cartilage rings (Whitsett *et al.*, 2002). Overexpression of *Fgf18* also resulted in malformation of the cartilage rings, possibly via Sox9 upregulation (Elluru *et al.*, 2009). Tiozzo *et al.* (2009) reported that ectopic fibroblast growth factor receptor (FGFR)2b expression in tracheal mesenchyme renders this hyper-responsive to FGF10, resulting in cartilaginous sleeve formation reminiscent of the Apert syndrome tracheal phenotype (Fig. 3.6). This abnormal cartilage structure arises secondary to increased proliferation of cartilage progenitor cells within tracheal mesenchyme.

Despite incomplete understanding of such genetics, tissue engineered airway (formed using stem cells and cadaveric scaffold) has been successfully transplanted into adult and a pediatric patients to replace damaged bronchus and trachea, respectively (Macchiarini *et al.*, 2008)

3.1.4. Branching morphogenesis of airway and vasculature

A multiplicity of factors are required for normal airway branching morphogenesis. Much of the research in this area is focused on epithelial morphogenesis: this is discussed in detail in subsequent sections dealing with individual signaling pathways. A key insight has been that epithelial morphogenesis proceeds interdependently with vascular development. Indeed

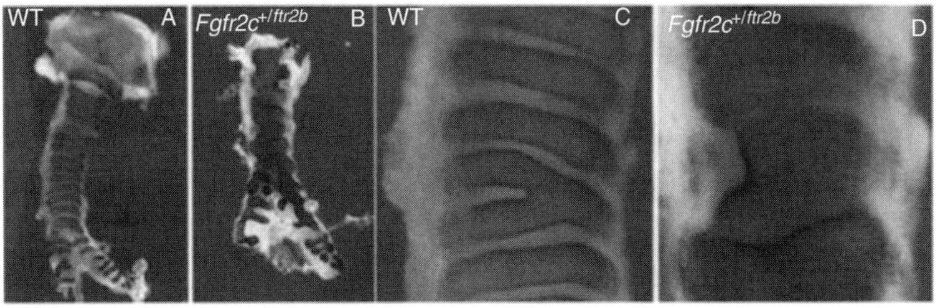

Figure 3.6 Excessive mesenchymal FGF signaling leads to overgrowth of tracheal rings. Wild-type and mutant tracheas are stained with Alcian blue. (A) Wild-type trachea at P0 exhibiting regular cartilage rings separated by noncartilaginous mesenchyme; (B) *Fgfr2c*$^{+/Fgfr2b}$ trachea at P0 showing excessive growth of the cartilage with absence of noncartilaginous mesenchyme; (C, D) high magnification of A and B, respectively. (See Color Insert.)

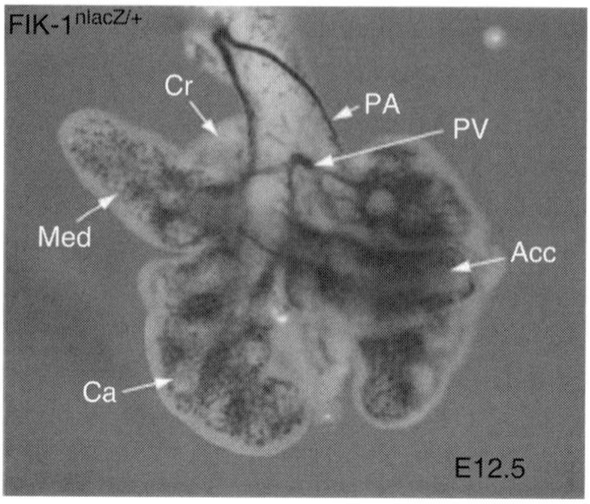

Figure 3.7 Vascular endothelial development in E12.5 mouse lung is shown in whole mount as a blue signal resulting from transgenic expression of *Flk1*-β-galactosidase (Flk-1$^{nLacZ/+}$): pulmonary artery (PA); pulmonary vein (PV); cranial lobe (Cr); medial lobe (Med); caudal lobe (Ca); accessory lobe (Acc). (See Color Insert.)

tight coupling of endothelial with epithelial development is required for efficient gas transport and fluid clearance at birth. Vascular endothelial growth factor (VEGF) signaling from the epithelium to the developing endothelium is essential for the primitive hemangioblasts to develop into mature capillary networks. Likewise, the endothelium probably signals back to coordinate epithelial morphogenesis. There is a stereotyped anatomical relationship between the developing pulmonary capillaires, arteries, and veins (Fig. 3.7). The arteries run along the superior surface of the developing lobules, while the veins run along the interior surface.

3.1.5. Alveolar septum formation

Septation of terminal sacs generates alveoli and involves interacting mesenchymal myofibroblasts, epithelial cells, and endothelial cells. Myofibroblasts are smooth muscle precursors with fibroblast morphology that migrate within nascent septa and deposit elastin (particularly at tips) as the first step of secondary septa development (Bostrom *et al.*, 1996; Lindahl *et al.*, 1997). Alveolar myofibroblast differentiaion requires Lunatic fringe and Notch signaling, and their elastin deposition is PGFa- and FGFR3/4-regulated (Xu *et al.*, 2010). Septal thinning and maturation of the alveolar capillary network are also needed. Interstitial thinning proceeds with expansion of septal epithelial, vascular, and airspace compartments but also myofibroblast apoptosis (Awonusonu *et al.*, 1999; Schittny *et al.*, 1998) that affects lipid-filled interstitial fibroblasts (LFIF) rather than non-LFIF (NLFIF). This apoptosis is associated with downregulation of insulin-like growth factor I receptor (*Igf-IR*) mRNA and cell surface protein expression (Srinivasan *et al.*, 2002).

Finally, the new alveolar septum differentiates into a functional respiratory membrane that consists of type I AECs (AECI), basement membrane, and capillary endothelial cells. The respiratory membrane provides a short distance for diffusion thereby facilitating gas exchange. It is estimated that about 50 million alveoli are present in neonatal lung. However, by age 7–8 years, when alveolarization is largely complete, the number of alveolar units in the lung has increased six-fold to about 300 million. Meanwhile the adut alveolar capillary bed is capable of accommodating the entire adult cardiac output of 5 L/min, rising five-fold to 25 L/min during maximal exercise.

Alveolarization can be adversely affected by premature delivery, hyperoxia, postnatal steroid exposure, and prolonged mechanical ventilation, even with room air (Mokres *et al.*, 2010). Thus, prematurity and hyperoxia plus pressure plus time are well-recognized risk factors for the hypoalveolarization characteristic of BPD in human premature infants. It is postulated that a cascade of events including endotoxin exposure, inflammation, and expression and activation of excessive amounts of transforming growth factor (TGF)-β ligand inhibit aleolarization in BPD and therefore portend an adverse outcome of this disease (reviewed in Shi *et al.*, 2009).

3.2. Cataloguing the biochemical regulators of lung development

Having considered the process of building the lung, we next turn to catalogue the factors required for lung growth and maturation. Transgenic mouse technology has allowed us to evaluate roles in organogenesis by, for example, overexpressing or knocking out a specific genes (Costa *et al.*, 2001). Some murine pulmonary phenotypes resulting from a loss or gain of gene function are listed in Table 3.1, modified from Cardoso and Lu (2006).

3.2.1. Transcription factors

At least four groups of transcription factors, forkhead box, Nkx homeodomain, RA receptors, and *Gli* family play important roles in lung development.

Forkhead box transcription factor family

Members of the forkhead box family transcription factors, such as Foxa1, Foxa2, HFH8, and HFH4, share homology in the winged-helix DNA-binding domain and regulate pulmonary cellular proliferation and differentiation.

HNF-3α (Foxa1) and HNF-3β (Foxa2) share 93% homology in amino acid sequences and were first identified as factors in hepatocyte differentiation (Qian and Costa, 1995). However, *Hnf-3β* is expressed in developing lung, with higher levels in proximal epithelial cells and lower levels in distal type II epithelial cells (Zhou *et al.*, 1996b). Overexpression of *Hnf-3β* under

Table 3.1 Examples of mutations in mouse giving a reported lung and/or tracheal phenotype

Gene symbol	Gene name	Expression pattern	Phenotype	Reference
Signaling molecule				
Egfr	Epidermal growth factor receptor	Epithelium and mesenchyme	Impaired branching and deficient alveolization	Miettinen *et al.* (1997)
Fgf18	Fibroblast growth factor 18	Mesenchyme	Deficient alveolization	Usui *et al.* (2004)
Fgf9	Fibroblast growth factor 9	Epithelium and pleura	Impaired branching, reduced mesenchyme	Colvin *et al.* (2001)
Grem1	Gremlin 1	Epithelium and mesenchyme	Deficient alveolization	Michos *et al.* (2004)
Hip1	Huntingtin-interacting protein 1	Mesenchyme	Impaired branching	Chuang *et al.* (2003)
Shh	Sonic hedgehog	Epithelium	Impaired branching, tracheoesophageal fistula	Litingtung *et al.* (1998)
Tgfb3	Transforming growth factor, β3	Epithelium and pleura	Impaired branching	Kaartinen *et al.* (1995)
Wnt7b	Wingless-related MMTV integration site 7B	Epithelium	Vascular defect, reduced mesenchyme	Shu *et al.* (2002)
Catnnb1	β-Catenin	Epithelium	Impaired branching, proximal/distal specification	Mucenski *et al.* (2003)

Ltbp4	Latent transforming growth factor β binding protein 4	Not reported	Pulmonary emphysema	Sterner-Kock *et al.* (2002)
Wnt5a	Wingless-related MMTV integration site 5A	Mesenchyme and epithelium	Increased branching, tracheal defect	Li *et al.* (2002)
Fgf10	Fibroblast growth factor 10	Mesenchyme	Lung agenesis	Sekine *et al.* (1999)
Fgfr2b	Fibroblast growth factor receptor 2b	Epithelium	Lung agenesis	De Moerlooze *et al.* (2000)
Fgf8	Fibroblast growth factor 8	Not reported	Right pulmonary isomerism	Fischer *et al.* (2002)
TAB1	TGF-B activated kinase-1 binding protein-1	Epithelium	Lung dysmorphogenesis	Komatsu *et al.* (2002)
Acvr2b	Activin receptor IIB	Not reported	Right pulmonary isomerism	Oh and Li (1997)
Nodal	Nodal	Not reported	Right pulmonary isomerism	Lowe *et al.* (2001)
Lefty1	Left right determination factor 1	Not reported	Left pulmonary isomerism	Meno *et al.* (1998)
Traf4	Tnf receptor associated factor 4	Not reported	Tracheal defect	Shiels *et al.* (2000)
Fgfr3/Fgfr4	Fibroblast growth factor receptor 3/4	Epithelium and mesenchyme	Defective elastin production, alveolarization defect	Weinstein *et al.* (1998)

(*Continued*)

Table 3.1 (*Continued*)

Gene symbol	Gene name	Expression pattern	Phenotype	Reference
Nog	Noggin	Mesenchyme	Lobation defect	Weaver *et al.* (2003)
Pitx-2	Paired-like homeodomain transcription factor 2	Not reported	Bilateral isomerism	Kitamura *et al.* (1999)
Dermo1	Twist homolog 2	Mesenchyme	Impaired branching	De Langhe *et al.* (2008)
BMP4	Bone morphogenic protein 4	Epithelium and mesenchyme	Abnormal lung morphogenesis with cystic terminal sacs	Bellusci *et al.* (1996)
Igf1r	Insulin-like growth factor 1 receptor	Not reported	Impaired development	Liu *et al.* (1993).
Notch2/3	Notch gene homolog 2/3	Epithelium	Defective myofibroblast differentiation, alveolarization defect	Xu *et al.* (2009)
PDGFa	Platelet derived growth factor a	Epithelium	Defective myofibroblast elastin production, alveolarization defect	Bostrom *et al.* (2002)
Timp3	Tissue inhibitor of metalloproteinase 3	Mesenchyme	Reduced number of bronchioles and attenuated alveogenesis	Gill *et al.* (2003)

Transcription factor				
Cebpa	CCAAT/enhancer binding protein (C/EBP), α	Epithelium	Hyperproliferation of type II cells	Sugahara *et al.* (2001)
Foxa1/Foxa2	Forkhead box A1/A2	Epithelium	Impaired branching, reduced smooth muscle	Wan *et al.* (2005)
Foxf1a	Forkhead box F1a	Mesenchyme	Impaired branching, lobation defect	Lim *et al.* (2002)
Hoxa5	Homeobox A5	Mesenchyme	Impaired branching, tracheal defect	Aubin *et al.* (1997)
Klf2	Kruppel-like factor 2 (lung)	Not reported	Impaired sacculation	Wani *et al.* (1999)
Mycn	Neuroblastoma myc-related oncogene 1	Epithelium	Impaired branching	Moens *et al.* (1992)
Trp63	Transformation-related protein 63	Epithelium	Tracheobronchial defect	Daniely *et al.* (2004)
Titf1	Thyroid transcription factor 1	Epithelium	Loss of distal lung fate, impaired branching, tracheoesophageal fistula	Kimura *et al.* (1996)
Nfib	Nuclear factor I/B	Epithelium and mesenchyme	Sacculation defect	Steele-Perkins *et al.* (2005)
Sox11	SRY-box-containing gene 11	Epithelium	Hypoplastic lung	Sock *et al.* (2004)
Tcf21	Transcription factor 21 (Pod1)	Mesenchyme	Impaired branching	Quaggin *et al.* (1999)
Rarb/Rara	Retinoic acid receptor α/β	Epithelium and mesenchyme	Left lung agenesis and right lung hypoplasia	Mendelsohn *et al.* (1994)

(Continued)

Table 3.1 *(Continued)*

Gene symbol	Gene name	Expression pattern	Phenotype	Reference
Pitx2	Paired-like homeodomain transcription factor 2	Mesenchyme	Right pulmonary isomerism	Lin *et al.* (1999)
Foxj1	Forkhead box J1	Epithelium	Left–right asymmetry, loss of ciliated cells	Brody *et al.* (2000)
Gata6	GATA-binding protein 6	Epithelium	Impaired sacculation	Yang *et al.* (2002)
Gli2/Gli3	GLI-Kruppel family member GLI2/GLI3	Mesenchyme	Lung agenesis	Motoyama *et al.* (1998)
Ascl1	Achaete-scute complex homolog-like 1	Neuroendocrine cells	Loss of neuroendocrine cells	Ito *et al.* (2000)
Erm	Ets variant gene 5	Epithelium	Impaired type I cell formation	Liu and Hogan (2002), Liu *et al.* (2003)
Wnt2/2b	Wingless-related MMTV integration site 2/2b	Mesenchyme	Complete lung agenesis	Goss *et al.* (2009) Harris-Johson *et al.* (2009)

Alk3	Aurora-like kinase	Epithelium	Retardation of lung branching, reduced cell proliferation and differentiation	Sun *et al.* (2008)
Others				
Eln	Elastin	Mesenchyme	Deficient alveolization	Wendel *et al.* (2000)
Lmnb1	Lamin B1	Epithelium and mesenchyme	Deficient alveolization	Vergnes *et al.* (2004)
Lama5	Laminin α5	Epithelium and pleura	Defective lobation	Nguyen *et al.* (2002)
Pcaf	p300/CBP-associated factor	Epithelium and mesenchyme	Defective proximal and distal epithelial cell differentiation	Shikama *et al.* (2003)
Adam17	A disintegrin and metallopeptidase domain 17	Epithelium	Impaired epithelial differentiation, impaired branching	Zhao *et al.* (2001), Peschon *et al.* (1998)
Crh	Corticotropin-releasing hormone	Epithelium	Defective epithelial and mesenchymal maturation	Muglia *et al.* (1999)
Pthlh	Parathyroid hormone-like peptide	Epithelium	Deficient alveolization	Rubin *et al.* (2004)
Itga3	Integrin α3	Epithelium	Impaired branching	Kreidberg *et al.* (1996)
Cutl1	Cut-like 1	Epithelium	Impaired epithelial differentiation	Ellis *et al.* (2001)
RXRa	Retinoic X receptor alpha	Epithelium and mesenchyme	Decrease in alveolar surface area and alveolar number	McGowan *et al.* (2000)

(*Continued*)

Table 3.1 (*Continued*)

Gene symbol	Gene name	Expression pattern	Phenotype	Reference
Tmem16a	Transmembrane protein 16a	Epithelium	Abnormal tracheal cartilages resulting in tracheomegaly	Rock *et al.* (2008)
TACE	Tumor necrosis factor-α converting enzyme	Not reported	Failure to form saccular structures	Zhao *et al.* (2001)
PDGFa	Platelet derived growth factor a	Epithelium	Defective myofibroblast elastin production, alveolarization defect	Bostrom *et al.* (2002)
Na/K ATPase	Sodium/ Potassium ATPase	Epithelium	Failure to absorb fetal lung liquid, causes significant respiratory distress and neonatal lethality	Hummel *et al.*
Lfng	Lunatic Fringe	Epithelium	Impaired myofibroblast differentiation and alveogenesis	Xu *et al.* (2009)

control of epithelial specific *SP-C* promoter inhibits lung branching morphogenesis and vasculogenesis *in vivo* (Zhou *et al.*, 1997). HNF-3α and HNF-3β also regulate expression of CCSP and surfactant proteins in bronchiolar and type II epithelial cells (Bingle *et al.*, 1995; Bohinski *et al.*, 1994; He *et al.*, 2000). *Hnf-3β* is inducible by interferon, and regulates in turn the expression of the *Nkx* homeodomain transcription factor *Nkx2.1* (also termed *Ttf-1* and *CebpI*), which in turn regulates transcription of the surfactant protein genes in peripheral lung epithelium (Ikeda *et al.*, 1996; Samadani *et al.*, 1995).

HFH8 is restricted to splanchnic mesoderm contacting embryonic gut and presumptive lung at E9.5 suggesting that *Hfh-8* may participate in lung induction. HFH-8 expression continues in lateral mesoderm-derived tissue during development. By E18.5, *Hfh-8* expression is restricted to distal lung mesenchyme and bronchial muscle (Peterson *et al.*, 1997). The level of *Hfh-8* expression is important for normal development: alveolar hemorrhage is observed in *Hfh8*(+/−) mice, while *Hfh8*(−/−) mice die *in utero*. Reduced *Hfh-8* expression in *Hfh-8*$^{+/-}$ mutants is accompanied by decreased expression of VEGF and its receptor 2 (Flk-1), bone morphogenetic protein 4 (BMP-4), and the transcription factors of the Brachyury T-Box family (Tbx2–Tbx5) and Lung Kruppel-like factor (Kalinichenko *et al.*, 2001). HFH8 regulates mesebchymal *Pdgf* receptor (Bostrom *et al.*, 1996; Shinbrot *et al.*, 1994; Souza *et al.*, 1996). HFH8 binding sites are also found in the promoter region of genes, such as *Bmp4, Hgf,* and *Hoxa5*, that are critical regulators of lung morphogenesis (Ohmichi *et al.*, 1998; Weaver *et al.*, 1999).

Hfh4 (*Foxj1*) regulates ciliated epithelial cell differentiation. It is expressed in E15.5 airway epithelium just before ciliated cells appear (Hackett *et al.*, 1995) and *Hfh4*$^{-/-}$-null mutant mice feature defective ciliogenesis in airway epithelial cells and randomized left–right asymmetry (mimicking human Kartagener syndrome). The latter can result in perinatal lethality, but in low penetrance it gives rise to situs inversus, sinusitis, bronchiectasis, and sterility, all caused by defects in ciliary beat (Brody *et al.*, 2000; Chen *et al.*, 1998). Interestingly, HFH4 and other proximal lung markers such as CCSP are upregulated by BMP antagonist Noggin in mesenchyme-free airway epithelial culture (Hyatt *et al.*, 2002).

Foxp1, Foxp2, and *Foxp4* are highly expressed in mouse lung and gut. *Foxp1* and *Foxp4* are expressed in both proximal and distal airway epithelium while Foxp2 is expressed primarily in distal epithelium. Foxp1 protein expression is also observed in the mesenchyme and vascular endothelial cells of the lung (Lu *et al.*, 2002).

Nkx and Hox homeodomain transcription factors

One of the most prominent homeodomain transcription factors in lung development is NKX2.1, also called TTF-1 (thyroid-specific transcription factor) or CEBP-1. *Nkx2.1* is expressed in epithelial cells derived from foregut endoderm in lungs, thyroid, and pituitary, as well as restricted regions of fetal brain (Guazzi *et al.*, 1990; Lazzaro *et al.*, 1991). Hence, human Nkx2.1 mutants may feature benign hereditary chorea, congenital hypothyroidism, and neonatal respiratory distress at term (sometimes retrieved by the transactivating activity of Pax8) (Carre *et al.*, 2009). *Nkx2.1*$^{-/-}$ mice exhibit impaired tracheoesophageal separation and early arrest of lung development featuring two main bronchi but no distal branches (Kimura *et al.*, 1996; Minoo *et al.*, 1999). In developing mouse airway epithelium, *Nkx2.1* is initially expressed proximally and distally becoming restricted at later stages to distal AECs (Zhou *et al.*, 1996b). Overexpression of *Nkx2.1* causes dose-dependent morphological alterations in postnatal lung: modest overexpression raises type II pneumocyte proliferation and SP-B levels; greater overexpression disrupts alveolar septation with emphysema due to alveolar hypoplasia. The highest overexpression of *Nkx2.1* in transgenic mice causes severe pulmonary inflammation, fibrosis, and respiratory failure, associated with eosinophil infiltration and increased eotaxin and IL-6 expression (Wert *et al.*, 2002). Nkx2.1 signaling is critical for surfactant protein, *T1a*, and *CC10* gene expression (Boggaram, 2003; Bruno *et al.*, 1995; Guazzi *et al.*, 1990; Ramirez *et al.*, 1997; Whitsett and Glasser, 1998; Yan *et al.*, 1995; Zhang *et al.*, 1997). *Nkx2.1*-deficient pulmonary epithelial cells fail to express nonciliated marker genes, including differentiated *Sp-B*, *Sp-C*, and *CC10. Bmp4* expression in these cells is also reduced. In addition to modulating expression of other lung-related genes, it is clear that NKX2.1 phosphorylation plays a crucial role in its signaling: mice with point mutation of seven serine phosphorylation sites of NKX2.1 died immediately following birth with malformation of acinar tubules, pulmonary hypoplasia, and reduced expression of surfactant proteins, CC10/secretoglobulin 1A, and *Vegf* (DeFelice *et al.*, 2003). Whilst regulating expression of numerous genes, *Nkx2.1* expression can itself be activated by transcription factors HNF-3β (Ikeda *et al.*, 1996) and GATA-6 (Shaw-White *et al.*, 1999) during lung morphogenesis.

Hox family transcription factors

Hox transcription factors are expressed with proximodistal polarity in developing lung: *Hoxa5, Hoxb2*, and *Hoxb5* are restricted to distal lung mesenchyme, whilst *Hoxb3* and *Hoxb4* are expressed in proximal and distal mesenchyme (Aubin *et al.*, 1997; Bogue *et al.*, 1996; Volpe *et al.*, 1997). Illustrating their functional role, *Hoxa5*$^{-/-}$-null mutant mice have tracheal defects and occlusions, impaired lung branching morphogenesis, diminished surfactant protein expression, and alveolar wall thickening (Aubin *et al.*, 1997).

GLI family zinc-finger transcription factors

GLI 1, 2, 3 are zinc-finger transcription factors and activated by SHH. All are mesodermally expressed, particularly in the distal lung (Grindley *et al.*, 1997). Combined *Gli2*$^{-/-}$ and *Gli3*$^{-/-}$ mutant mice feature lung agenesis. *Gli3*$^{-/-}$ mice are viable but have small dysmorphic lungs (Grindley *et al.*, 1997). *Gli2* regulates normal lung asymmetry: *Gli2*$^{-/-}$ mice have a fused right and left lung (a small single lobe with defective primary branching in the right lung) and hypoplastic trachea and esophagus that are nevertheless distinct and retain normal proximal–distal differentiation (Motoyama *et al.*, 1998).

3.2.2. Peptide growth factors

Embryonic lung mesenchymal and epithelial cells communicate through autocrine and paracrine factors, as demonstrated by effects of added growth factors on cultured embryonic lung growth (Jaskoll *et al.*, 1988; Warburton *et al.*, 1992).

FGF family

FGF family members are found throughout the vertebrates and invertebrates. Their functions in respiratory organogenesis are conserved from *Drosophila* to mammals (Glazer and Shilo, 1991; Sutherland *et al.*, 1996). Based on protein sequence homology, FGFs have been divided into 23 subgroups. Similarly, their cognate transmembrane protein tyrosine kinase receptors (FGFRs) are classified into four types, contributing to the specificity of FGF ligand binding (Ornitz and Itoh, 2001). Heparan sulfate proteoglycan, an ECM glycoprotein, has been reported to be essential for FGF ligand–receptor binding and activation (Izvolsky *et al.*, 2003a,b; Lin *et al.*, 1999). FGFs play critical roles in cell proliferation, migration, and differentiation during development. Early inhibition of murine FGFR signaling shows it is required for early lung branching morphogenesis. Later FGFR inhibition in E14.5 lung decreases prenatal airway tubule formation and is associated with severe emphysema at maturity. At E16.5, FGFR inhibition causes mild focal emphysema. Murine mutants lacking FGFR3 and FGFR4 fail to undergo normal alveolarization, with poorly organized myofibroblasts and excessive amounts of poorly organized elastin. However, inhibition of FGFR signaling after birth did not appear to alter postnatal alveolarization (Hokuto *et al.*, 2003).

FGF10 is one of the most-studied family members during lung development. *Fgf10*-null mice lack distal lung despite formation of larynx and trachea (Min *et al.*, 1998). *Fgf10* is expressed focally in E11–12 mouse peripheral lung mesenchyme and signals through adjacent distal epithelial FGFR2IIIb (whose loss also disrupts lung development) (De Moerlooze *et al.*, 2000). These sites of expression change dynamically, compatible with the idea that FGF10 appears at sites of bud formation (Bellusci *et al.*, 1997b).

FGF10 has a chemotactic effect on nearby epithelium in culture: epithelial tips will proliferate and migrate toward FGF10 in mesenchyme or on beads (Park *et al.*, 1998; Weaver *et al.*, 2000). FGF10 controls epithelial differentiation, inducing *Sp-C* expression and downregulating *Bmp4* expression (Hyatt *et al.*, 2002). FGF10 dosage and signal transduction level is critical: mice with 20% of normal FGF10 expression (due to an enhancer trap bearing LacZ inserted 100Kb upstream in the FGF10 promoter) feature lung hypoplasia (Ramasamy *et al.*, 2007); similarly, downstream signaling inhibition by misexpression of Sprouty2 under control of the *Sftpc* promoter induces lung hypoplasia (Mailleux *et al.*, 2001). Several key regulatory molecules such as SHH, BMPs, and TGF-βs crosstalk with FGF10 during embryonic lung morphogenesis: their interactions will be discussed later.

FGF7 (KGF) is found in developing lung mesenchyme at late stages (Post *et al.*, 1996). In early cultured mouse embryonic lung, addition of FGF7 promotes epithelial growth and formation of cyst-like structures with extensive cell proliferation. FGF7 can also contribute to distal airway epithelial cell differentiation (Cardoso *et al.*, 1997; Deterding *et al.*, 1996). *Erm* and *Pea3* are ETS domain transcription factors known to be downstream of FGF signaling. FGF7 can induce *Erm/Pea3* expression more effectively than FGF10. *Erm* is transcribed exclusively in the epithelium, while *Pea3* is expressed in both epithelium and mesenchyme. When examined at E18.5, transgenic expression of a repressor form of Erm specifically in the embryonic lung epithelium shows that the distal epithelium of *Sp-C-Erm* transgenic lungs is composed predominantly of immature type II cells, while no mature type I cells are observed. By contrast, the differentiation of proximal epithelial cells, including ciliated cells and Clara cells, appears to be unaffected (Liu and Hogan, 2002; Liu *et al.*, 2003). FGF7 does not seem to protect against hyperoxic inhibition of normal postnatal alveoli formation and early pulmonary fibrosis, but FGF7 consistently had a significant protective/preventive effect against the development of pulmonary hypertension during hyperoxia (Frank, 2003). However, *Fgf7*$^{-/-}$ mutant mice have no gross lung abnormalities (Guo *et al.*, 1996), suggesting a FGF7 redundancy during lung development.

FGF9, which signals through FGFR2IIIc, also regulates branching morphogenesis. In E10.5 lung, *Fgf9* is expressed in visceral pleura outlining the lung bud and in bronchial epithelium, while *Fgfr2IIIc* is predominantly expressed in lung mesenchyme. At E12.5 and E14.5, *Fgf9* expression persists in visceral pleura but is no longer detected in epithelium (Colvin *et al.*, 1999). *Fgf9*-null mice exhibit reduced mesenchyme and decreased airway branching but show significant distal airspace formation and pneumocyte differentiation. The reduction in the amount of mesenchyme in *Fgf9*$^{-/-}$ lungs limits expression of mesenchymal *Fgf10* (Colvin *et al.*, 2001). By contrast, addition of recombinant FGF9 protein inhibits the differentiation response of the mesenchyme to N-SHH, but does not affect proliferation (Weaver *et al.*, 2003).

The signaling cascade activated by FGF10 and FGF9 involves FGFR2b and 2c, respectively, as well as Shp2, Raf, MAP ERK kinase (MEK), and extracellular-regulated kinases (ERK) 1 and 2 as signal transducers. MEK inhibition has been shown to reduce lung branching and epithelial cell proliferation, but increase mesenchyme cell apoptosis in fetal lung explants (Papadakis *et al.*, 1997). FGF signaling is regulated at several levels. One of the key inducible negative regulators is the Sprouty family. There are four sprouty (Spry) genes in mouse (*mSpry1–4*) and human (*hSpry1–4*). Murine *Spry2* is expressed in the distal tip of embryonic lung epithelial branches, but is downregulated between the sites of new bud formation. Murine *Spry4* is predominantly expressed in the distal mesenchyme of the embryonic lung (Mailleux *et al.*, 2001), and may play roles in branching morphogenesis. Sprouties (SPRY1, 2, 4) act as suppressors of Ras–MAP kinase signaling (Hacohen *et al.*, 1998; Kramer *et al.*, 1999; Reich *et al.*, 1999). Overexpression of *mSpry2* or *mSpry4* can inhibit lung branching morphogenesis through reducing epithelium cell proliferation (Hadari *et al.*, 1998; Perl *et al.*, 2003; Tefft *et al.*, 2002). SPRED-1 and SPRED-2 are two sprouty related proteins, which contain Enabled/VAsodilator-Stimulated Phosphoprotein (VASP) Homology-1 (EVH-1) domains. *Spreds* are predominantly expressed in mesenchymal cells. Expression of *Spreds* is especially strong in the peripheral mesenchyme and epithelium of new bud formation. After birth, *Spreds* expression decreases, while the expression of *Sprouties* expression remains high. Both *Sprouties* and *spreds* play important roles in mesenchyme–epithelium interaction during lung development (Hashimoto *et al.*, 2002).

TGF-β/BMP family

The TGF-β superfamily comprises numerous structurally related polypeptide growth factors including TGF-β, BMP, and activin subfamilies. TGF-β ligands bind to cognate cell surface receptors, and activate Smad proteins, which translocate to the nucleus and modulate target gene expression (Massague, 1998; Shi and Massague, 2003).

TGF-β subfamily

The TGF-β ligand subfamily comprises three isoforms, TGF-β1, 2, and 3. TGF-β1 is expressed in early embryonic lung mesenchyme, particularly underlying distal epithelial branch points; TGF-β2 is localized mainly in distal epithelium; TGF-β3 is mainly expressed in proximal mesenchyme and mesothelium (Bragg *et al.*, 2001; Millan *et al.*, 1991; Pelton *et al.*, 1991a,b; Schmidt *et al.*, 1991). Each TGF-β isoform has nonredundant roles revealed by isoform-specific knockouts. Mice lacking TGF-β1 develop apparently normally, but die within 2 months of life from aggressive pulmonary or gut inflammation, as a result of failure to negatively modulate the immune system (McLennan *et al.*, 2000). TGF-β2$^{-/-}$ mutation results in embryonic lethality around E14.5 in mice featuring complex cardiac anomalies and lung dysplasia amongst others (Bartram *et al.*, 2001). TGF-β3$^{-/-}$ mutant mice display cleft palate, retarded lung development, and neonatal lethality with difficulty swallowing and breathing (Kaartinen *et al.*, 1995; Shi *et al.*, 1999). Furthermore, blockade of TGF-β signaling by null mutation of TGF-β activated kinase-1 binding protein-1 (TAB1) results in lethal cardiovascular and lung dysmorphogenesis (Komatsu *et al.*, 2002).

As with the FGFs, the timing and dosage of TGF-β signaling are critical during lung development. Optimal physiological levels of TGF-β-Smad3 signaling appear essential for secondary alveolar septa formation: abrogation of TGF-β type II receptor in lung epithelial cells reduces alveolar septation and allows emergence of AECI (Chen *et al.*, 2008). However, TGF-β1 overexpression in early mouse embryonic lung epithelium inhibits branching morphogenesis (Zhao *et al.*, 1999), whereas misexpression of *Sp-C* promoter-controlled TGF-β1 in embryonic lung epithelium arrests embryonic lung growth and epithelial cell differentiation whilst inhibiting pulmonary vasculogenesis (Zhou *et al.*, 1996a, 2001). Suggesting a crucial role for optimal TGF-β1 levels in human lung maturation, excessive activated TGF-β1 has been reported in tracheal aspirates of human premature infants who develop more severe BPD (Lecart *et al.*, 2000; Toti *et al.*, 1997). Furthermore, misexpression of TGF-β1 in neonatal rat lung using recombinant adenoviral vectors resulted in neonatal alveolar hypoplasia and interstitial fibrosis; this histological picture closely phenocopies human BPD (Gauldie *et al.*, 2003). By contrast, misexpression of TGF-β1 in adult rats results in chronic progressive interstitial pulmonary fibrosis with increased proliferation and matrix secretion by the mesenchyme (Sime *et al.*, 1997; Zhao *et al.*, 2002). In addition, TGF-β1 may be centrally involved in pulmonary fibrotic responses to bleomycin, or endotoxin and infection (Bonniaud *et al.*, 2005). Blockade of TGF-β signaling via Smad3-null mutation strongly attenuates bleomycin-induced pulmonary fibrosis (Zhao *et al.*, 2002).

The activity of TGF-β signaling is multiply regulated: β6-integrin, Latent transforming growth factor-beta binding proteins (LTBPs), and thrombospondin regulate TGF-β release, whilst β-glycan, endoglin, or decorin modulate TGF-β receptor binding affinity. As expected, mutation of the above genes causes some phenotypes similar to those of TGF-β mutants: loss-of-function mutation in human and mouse endoglin (whose protein binds TGF-β and Alk1, its type I receptor) causes hereditary hemorrhagic telangiectasia (Li *et al.*, 1999; Massague, 2000; McAllister *et al.*, 1994; Urness *et al.*, 2000). Null mutation of LTBP-3 or LTBP-4 causes profound defects in elastin fiber structure and lung alveolarization similar to Smad3 knockout mouse lung (Sterner-Kock *et al.*, 2002; Colarossi *et al.*, 2005; Chen *et al.*, 2005).

In addition, TGF-β signaling blockade has distinct impacts on lung branching morphogenesis and alveolarization depending on whether epithelial or mesenchymal cells are targeted. Mesenchymal TGF-β signaling blockade driven by Dermo1 retards branching after mid-gestation; by contrast, epithelial TGF-β signaling abrogation lacks prenatal impact and only disrupts postnatal lung alveolarization (Chen *et al.*, 2008). Meanwhile, human TGF-β pathway mutations in, for example, TGF-β type II receptor (Alk5) or TGF-β binding proteins such as fibrillin underlie dysplastic matrix elastin defects that predispose to aortic dissection or sudden alveolar rupture in Marfan's syndrome patients (Kaartinen and Warburton, 2003). Thus, TGF-β signaling has to be regulated "just right" with both deficiency and excess deleterious to normal alveolarization (a concept we term the Goldilocks hypothesis).

BMP subfamily

BMPs, with more than 20 ligand family members, regulate many processes, including lung development (Hogan, 1996). Expression of *Bmp3, 4, 5,* and 7 is detected in embryonic lung (Bellusci *et al.*, 1996; King *et al.*, 1994; Takahashi and Ikeda, 1996). BMP4 plays a central role in normal lung development (Hogan, 1996). Addition of BMP4 to whole embryonic lung explants stimulates branching (Bragg *et al.*, 2001; Shi *et al.*, 2001). However, BMP4 inhibited FGF10-induced growth of isolated E11.5 mouse lung endoderm cultured in Matrigel (Weaver *et al.*, 2000). Transgenic overexpression of BMP4 in distal fetal lung endoderm, driven by a 3.7 kb human surfactant protein C (SP-C) promoter, causes abnormal morphogenesis with cystic terminal air sacs (Bellusci *et al.*, 1996). Conversely, *Sftp-C* promoter-driven overexpression of BMP antagonists Noggin or Gremlin severely reduces distal epithelial cell phenotype whilst increasing proximal cell types (Lu *et al.*, 2001; Weaver *et al.*, 1999). Interestingly, blockade of endogenous BMP4 in embryonic mouse lung epithelial cells using a conditional gene knockout approach results in abnormal lung development with similar dilated terminal sacs as seen in BMP4 transgenic mouse lung (Eblaghie *et al.*, 2006). This suggests optimal BMP4 levels are essential for normal lung development. As extracellular growth factors, BMPs bind heteromeric complexes of BMP serine/threonine kinase type I and type II receptors to activate intracellular signal pathway (Massague, 1998; Shi and Massague, 2003). Three cognate BMP type I receptors (Alk2, Alk3, and Alk6) have been identified. Among them, Alk3 is expressed predominantly in distal airway epithelial cells during mouse lung development. Alk3 abrogation in mouse lung epithelia either from early lung organogenesis or from late gestation resulted in similar neonatal respiratory distress phenotypes, accompanied with collapsed lungs (Sun *et al.*, 2008). Early induction of Alk3 knockout in lung epithelial cells causes retardation of early lung branching morphogenesis and reduces cell proliferation and differentiation. But late gestation induction of Alk3 knockout also causes significant epithelial apoptosis accompanied by lack of surfactant secretion (Sun *et al.*, 2008). Furthermore, canonical Wnt signaling was perturbed, possibly through reduced WIF-1 expression in Alk3 knockout lungs (Sun *et al.*, 2008).

Therefore, deficiency of appropriate BMP signaling in lung epithelial cells results in prenatal lung malformation, neonatal atelectasis, and respiratory failure.

In addition, BMP signaling is also important in lung vasculogenesis and angiogenesis. Mutations of BMP type II receptor (BMPRII) and change in expression of BMP antagonist Gremlin are associated with primary pulmonary hypertension (PPH) (Lane *et al.*, 2000; Costello *et al.*, 2008). Moreover, upregulation of Gremlin is also associated with pulmonary fibrosis and the severity of the fibrotic pathology (Koli *et al.*, 2006; Myllarniemi *et al.*, 2008)

Sonic hedgehog (Shh) pathway

Sonic hedgehog is a vertebrate homolog of *Hedgehog* (Hh) that patterns the segment, leg, wing, eye, and brain in *Drosophila.* Hh binds to patched (Ptc), a transmembrane protein, and releases its inhibitory effect on downstream smoothened (Smo), which is a G protein-coupled transmembrane spanning receptor. This leads to the activation of cubitus interruptus (Ci), a 155-kDa transcription factor that is cleaved to form a 75-kDa transcription inhibitor in cytosol. Elements of the *Drosophila* Hh signaling pathway and their general functions in the pathway are highly conserved in vertebrates, albeit with increased levels of complexity. Gli1, 2, and 3 are the three vertebrate Ci gene orthologs (van Tuyl and Post, 2000).

The SHH signal transduction pathway plays important roles in mesenchyme–epithelium interaction. In developing mouse lung, *Shh* is detected in the tracheal diverticulum, the esophagus, and later in the trachea and lung endoderm. *Shh* is expressed at low levels throughout the epithelium, whilst at higher level in the growing distal buds (Bellusci *et al.*, 1997a; Urase *et al.*, 1996). Null mutation of *Shh* produces profound lung hypoplasia and failed trachea–esophageal septation. Mesenchymal *Ptc, Gli1*, and *Gli3* expression are all downregulated in *Shh* knockout lung. Nevertheless, proximodistal differentiation of airway epithelium is preserved (Litingtung *et al.*, 1998; Pepicelli *et al.*, 1998). Also, *Fgf10* expression is dysregulated in *Shh*-null mutant lung compared to the precisely restricted expression seen normally. Lung-specific *Shh* overexpression results in severe alveolar hypoplasia and significant increase in interstitial tissue caused by increased epithelial and mesenchymal proliferation (Bellusci *et al.*, 1997a). Defective hedgehog signaling may lead to EA and TEF (Spilde *et al.*, 2003).

The membrane-bound Hedgehog interacting protein 1 (HIP1) directly binds mammalian Hedgehog (HH) proteins and attenuates their signaling (Chuang and McMahon, 1999). *Hip1* is transcriptionally activated in response to HH signaling, overlapping the expression domains of *Ptc1* (Chuang and McMahon, 1999; Goodrich *et al.*, 1996). Targeted disruption of *Hip1* results in upregulated Hedgehog signaling and lethal neonatal respiratory failure: left–right asymmetry persists but initial branching from the two primary buds is absent; *Fgf10* expression is slightly downregulated at the tips of the primary buds in *Hip1*$^{-/-}$ lungs at E10.5 but completely absent from the mesenchyme where secondary branching normally initiates (Goodrich *et al.*, 1996). Attenuated PTC1 activity in a *Hip1*$^{-/-}$ mutant lungs leads to an accelerated lethality. *Hip1* and *Ptch1* have redundant roles in lung branching control (Goodrich *et al.*, 1996). Both of them can attenuate SHH signal in lung development and pancreas development (Goodrich *et al.*, 1996; Kawahira *et al.*, 2003).

Wnt/β-catenin pathway

Wnt signals are transduced via seven transmembrane Wnt receptors encoded by *Frizzled* (*Fzd*) genes to activate the β-catenin T Cell transcription Factor (TCF) pathway, the c-Jun N-terminal kinases (JNK) pathway, or the intracellular Ca2^{+}-releasing pathway. The Wnt/β-catenin pathway plays a critical role in many developmental and tumorigenesis processes. Following Wnt binding to the receptor, β-catenin is dephosphorylated and translocates to the nucleus to activate downstream gene expression (Wodarz and Nusse, 1998).

TOPGAL and BATGAL reporter transgenes have been used to analyze patterns of β-catenin stabilization in developing lung. Within the respiratory precursor region, the TOPGAL reporter is expressed in the undivided proximal endodermal tube and then the lung buds as early as E9.5 (Okubo and Hogan, 2004). This pattern is maintained as the trachea and esophagus separate and the lung buds grow out between E10 and E11.5 (Dean *et al.*, 2005; De Langhe *et al.*, 2005; Okubo and Hogan, 2004; Shu *et al.*, 2005). Between E12.5 and E18.5, analysis of TOPGAL and BATGAL transgene activity suggests a dynamic pattern of TCF/β-catenin-dependent gene expression. Reporter gene activity is found in the tracheal epithelium and cartilaginous condensations at E12.5 but is restricted to the bronchial mesenchyme at E13.5 (De Langhe *et al.*, 2005; Shu *et al.*, 2005). The distal lung epithelium expresses both reporters by E9.5. The pattern of TCF/β-catenin-dependent gene activity in the distal lung at later time points is somewhat variable and dependent on the reporter transgene analyzed. In general, transgene activity clears from the central airways between E13.5 and postnatal day 14 (Okubo and Hogan, 2004; Shu *et al.*, 2005). At E14.5, expression in the distal tip epithelium is either extinguished (TOPGAL) (De Langhe *et al.*, 2005) or restricted to a subset of early alveolar type 2 cells (BATGAL) (Shu *et al.*, 2005). In the adult lung, the TOPGAL transgene is highly expressed in the distal trachea and in clusters of airway secretory and ciliated cells but rarely in the alveolar region (De Langhe, unpublished data).

β-catenin deletion in proximal airway epithelium during development resulted in no obvious alteration to lung structure (Mucenski *et al.*, 2003). By contrast, embryonic deletion of *β-catenin* in the distal lung epithelium resulted in profound perturbation of normal epithelial, mesenchymal, and vascular development. The latter mice feature proximalization of lung epithelium with decreased expression of alveolar type 2 cell marker Sftpc, vascular endothelial marker PECAM, and α-smooth muscle actin; upper airway epithelial markers (Scgb1a1, FoxJ1, and β-tubulin) were unaltered.

Stabilization of β-catenin in proximal epithelium using the *Catnb*^{floxedExon3} allele raised epithelial β-catenin levels, resulting in squamous, cuboidal, and goblet cell dysplasia in intrapulmonary conducting airways and the appearance of alveolar type 2-like cells in the bronchioles (Mucenski *et al.*, 2005). Epithelial levels of Scgb1a1 immunopositive cells were low whilst SPC expression increased, indicating an increase in Scgb1a1/Sftpc double-positive cells. Similar expansion of Scgb1a1/Sftpc double-positive bronchioalveolar stem cells (BASCs) in response to increased canonical Wnt signaling has been shown in the lung epithelium upon *Gata6* loss (Zhang *et al.*, 2008). These authors also showed that canonical Wnt signaling is activated within the niche containing BASCs during lung epithelial regeneration, while forced Wnt activation greatly increases BASC numbers.

Li *et al.*, (2009) stabilized β-catenin in the entire developing lung epithelium using Nkx2.1-cre and Catnb[+/lox(ex3)] mice: in trachea and main bronchi, polyp-like structures formed featuring intracellular β-catenin accumulation suggesting blocked differentiation of spatially-appropriate airway epithelial cell types, Clara cells, ciliated cells, and basal cells (BCs), while activating UCHL1, a marker for pulmonary neuroendocrine cells.

Alternatively, the method of using a *Spc* promoter-regulated *Lef1-dN89β-catenin* to stabilize β-catenin from about E10.5 was employed by Okubo and Hogan (2004) to generate mice with widened primary bronchial tubes opening directly into saccules (lined with simple cuboidal or columnar epithelium), decreased progenitor differentiation into secretory and ciliated cells, and absence of alveolar type 2 and type 1 cells. Thus, constitutive β-catenin signaling in developing foregut endoderm partially inhibited branching morphogenesis and blocked expression of lung-specific differentiation genes.

Using a hypomorphic *Fgf10* allele, Ramasamy *et al.* (2007) showed that FGF10 signaling via FGFR2b controls the proliferation of the pulmonary epithelial progenitors in part by autoregulation of β-catenin signaling in the epithelium. This correlation of a reduction in epithelial FGF signaling and epithelial TOPGAL activity has also been demonstrated in lungs of a mouse Apert disease model (De Langhe *et al.*, 2006). Intriguingly, the regulation of epithelial β-catenin signaling by FGF10 and concomitant upregulation of *Fgfr2b* receptor expression result in potentiating this signaling cascade locally, thus maintaining the distal epithelial progenitor state.

By contrast, the lack of significant activity of well-established Wnt reporters in mesenchyme (including TOPGAL and BATGAL mice) does not support an important role for mesenchymal Wnt signaling during organogenesis. However, expression of several mesenchymal Wnt receptors in the lung has been reported (De Langhe *et al.*, 2005). Furthermore, Wnt5a overexpression either directly or indirectly regulates mesenchymal *Fgf10* expression (Li *et al.*,

2005), while Wnt7b acts on lung vascular SMCs via Frizzled 1 and LRP5 (Wang *et al.*, 2005). Besides Lef1/TCF-mediated β-catenin signaling, β-catenin also acts via PITX family transcription factors (Kioussi *et al.*, 2002), which are abundantly expressed in developing mesenchyme (Kitamura *et al.*, 1999). Using *Dermo1*$^{Cre/+}$-mediated conditional inactivation (CKO) of *β-catenin*, De Langhe *et al.* (2008) showed *Dermo1-cre/β-catenin* CKO embryos have multiple defects reminiscent of double knockout of *Pitx1* and *Pitx2* genes (Marcil *et al.*, 2003). Combining fate analysis and global gene expression studies, mesenchymal β-catenin signaling was shown to have dual, lineage-dependant functions: it regulates formation and amplification but not differentiation of *Fgf10*-expressing parabronchial smooth muscle progenitors (in part via regulation of *Fgfr2c* expression) but is required for normal endothelial cell differentiation (De Langhe *et al.*, 2008). Cohen *et al.* (2009) confirmed the role of Wnt in parabronchial smooth muscle development and showed Wnt pathway upregulation in experimental asthma.

Epidermal growth factor (EGF) family growth factors

EGF, TGF-α, and amphiregulin are EGF receptor (EGFR) ligands. Loss- or gain-of-function experiments in mouse, rat, or other animal models prove that EGF ligands positively modulate early mouse embryonic lung branching morphogenesis and cytodifferentiation through EGFR (Schuger *et al.*, 1996a; Seth *et al.*, 1993; Warburton *et al.*, 1992). EGF is expressed in mature AECs and regulates type 2 cell proliferation via autocrine mechanism in culture and *in vivo* (Raaberg *et al.*, 1992). However, epithelial TGF-α overexpression under *Sp-C* promoter control induces postnatal lung fibrosis (Korfhagen *et al.*, 1994). TGF-α overexpression caused severe pulmonary vascular disease mediated via EGFR in distal epithelium, but reductions in VEGF may also contribute (Le Cras *et al.*, 2003).

EGFR is a tyrosine kinase receptor whose deletion (*Egfr*$^{-/-}$) causes abnormal branching, poor alveolarization, and aberrant matrix metalloprotease protein (MMP) expression (Kheradmand *et al.*, 2002). EGFR phosphorylation in response to stretch induces, at least in part, fetal epithelial cell differentiation via ERK pathway activation. Specific EGFR or ERK pathway blockade reduces stretch-inducible *Sp-C* mRNA expression. Thus, EGFR may represent a mechanical signal sensor during lung development (Sanchez-Esteban *et al.*, 2003).

Tumor necrosis factor-α (TNFα)-converting enzyme (TACE) is a transmembrane metalloprotease disintegrin that functions as a membrane sheddase to release the ectodomain portions of many transmembrane proteins, including the precursors of TNFα and several other cytokines, as well as the receptors for TNFα, and neuregulin (ErbB4) (Shi *et al.*, 2003). Neonatal TACE-deficient mice had visible respiratory distress and their lungs failed to form normal saccular structures. Mouse embryonic lung explant cultures show that TGF-α and EGF can rescue the inhibition of TACE activity (Zhao *et al.*, 2001).

Platelet-derived growth factors (PDGFs)

PDGF-A and PDGF-B form homodimers (AA or BB) or heterodimers (AB). Two PDGF receptor types, α and β, occur in embryonic mouse lung and are differentially regulated in fetal rat lung epithelium and fibroblasts (Buch *et al.*, 1994). PDGF-A regulates DNA synthesis and branching in cultured embryonic murine lung epithelium (Souza *et al.*, 1995b). *Pdgf-a*$^{-/-}$ or *Pdgf-α*$^{-/-}$ mutants die perinatally with loss of alveolar myofibroblasts and SMCs and reduced parenchymal elastin fiber deposition; furthermore, failed alveolar septation is associated with emphysema (Bostrom *et al.*, 1996, 2002). Antisense oligodeoxynucleotide abrogation of PDGF-B reduces the epithelial component of embryonic mouse lung explants but not branch number (Souza *et al.*, 1994). PDGF-B and its receptor are crucial for vascular development during the alveolar phase (Lindahl *et al.*, 1997). PDGF-C and D also dimerize and bind PDGF α or β receptor (LaRochelle *et al.*, 2001; Li *et al.*, 2000). PDGF-C mRNA expression shows a significant increase in bleomycin-induced lung fibrosis (Zhuo *et al.*, 2003).

Insulin-like growth factors (IGFs)

IGFs and their receptors are expressed in rodent and human fetal lung (Batchelor *et al.*, 1995; Lallemand *et al.*, 1995; Maitre *et al.*, 1995; Retsch-Bogart *et al.*, 1996; Schuller *et al.*, 1995). Null mutant mice for the cognate type 1 IGF receptor (*Igf1r*) gene die at birth with respiratory failure and growth deficiency (45% of normal birth weight). Dwarfism is further exacerbated (70% of size reduction) in either *Igf1* or *Igf2* double null mutants or in *Igf1r* and *Igf2* double null mutants. There does not appear to be a gross defect in branching morphogenesis *per se*; the lungs merely appear hypoplastic (Liu *et al.*, 1993). IGF signaling may play a role in facilitating other peptide growth factor pathways during lung morphogenesis, e.g., IGF1R signaling is required for both mitogenic and transforming activities of EGFR (Coppola *et al.*, 1994). *Igf1*-deficient mice have reduced airspaces, which are exacerbated in mice with additional leukemia inhibitory factor (*Lif*) deletion (featuring abnormal epithelium and decreased Sp3, *Nkx2.1*, and *Sp-B* expression) (Pichel *et al.*, 2003). IGF1 is also trophic for fetal lung endothelial cells: in human fetal lung explants, IGF-IR inactivation results in endothelial cell loss, attenuates time-dependent increase in budding of distal airway, and increases mesenchymal cell apoptosis (Han *et al.*, 2003).

Vascular endothelial growth factor (VEGF) isoforms and cognate receptors

Effective pulmonary gas exchange requires alignment between alveoli and pulmonary capillaries. VEGFs regulate pulmonary vascular development (reviewed extensively in Pauling and Vu, 2004 and in Warburton *et al.*,

2000, 2003; Shi *et al.*, 2007) and signal through cognate receptors Flk-1 (fetal liver kinase-1, VEGFR2) and Flt-1 (fetal liver tyrosinase-1, VEGFR1) (Larrivee and Karsan, 2000). VEGF is regulated by hypoxia-inducible transcription factor-2α (Compernolle *et al.*, 2002) and expression is controlled transcriptionally by hypoxia inducible factor-1 (HIF1) (see Dunwoodie, 2009 for an excellent review). Isoforms VEGF 120, 164, and 188 are expressed in E12.5 murine lung epithelial and mesenchymal cells and regulate endothelial proliferation and microvascular structure (Greenberg *et al.*, 2002; Ng *et al.*, 2001). As epithelial branching progresses, *Vegf-A* expression becomes restricted to distal lung (Healy *et al.*, 2000), which may be partly due to the high affinity of VEGF-A for matrix components concentrated around branching tips (Acosta *et al.*, 2001; Ng *et al.*, 2001). Epithelial and vascular branching morphogenesis of cultured mouse embryonic lung features epithelial, mesenchymal, and endothelial crosstalk mediated in part by VEGF-A signaling via Flk-1 (Del Moral *et al.*, 2006a). Inhibited VEGF signaling disrupts pulmonary endothelial survival and reduces postnatal alveolarization (Kasahara *et al.*, 2000). Neonatal lungs treated with antibodies to Flt-1 are small with simplified alveoli (Gerber *et al.*, 1999). VEGR3$^{-/-}$ mice have lymphatic hypoplasia and lethally delayed removal of lung liquid at birth. Excessive VEGF signaling also disrupts vascular and epithelial lungmorphogenesis: *Vegf* misexpression under SP-C promotercontrol yields decreased acinar tubules and mesenchyme (Zeng *et al.*, 1998).

ROBO/SLIT

Roundabout (ROBO) is a receptor involved in repellent signaling and controlling axonal extension and with its ligand SLIT regulates non-neuronal cell migration (Wu *et al.*, 2001). Suggesting interaction during perinatal lung development, *Slit-2* is expressed in saccular mesenchyme whilst *Robo* is expressed in adjacent apical epithelium (Anselmo *et al.*, 2003). A *Robo* knockout mouse loses alvelolar septation with thickened mesenchyme (Xian *et al.*, 2001).

3.2.3. Other biochemical factors

Small noncoding microRNAs (miRNAs) and lung development

A class of noncoding RNAs called microRNAs (miRNAs) has been recognized for their abundance and conservation between species (Ambros, 2001; Lau *et al.*, 2001; Lagos-Quintana *et al.*, 2001). An intranuclear primary transcript (pri-miRNA) is cleaved by Drosha RNase III endonuclease to liberate the pre-miRNA, a 60–70 nt stem loop intermediate (Lee *et al.*, 2002). Pre-miRNA is transported to the cytoplasm by Exportin-5 (Yi *et al.*, 2003) where Dicer, another RNase III endonuclease, generates the mature miRNA. Mature miRNAs are incorporated as single-stranded RNAs into a ribonucleoprotein complex, the RNA-induced silencing complex (RISC) that downregulates gene expression by mRNA cleavage or translational repression. Mice lacking Dicer die before gastrulation indicating a fundamental developmental role of miRNAs. In lung, conditional epithelial Dicer inactivation arrests branching and disrupts Fgf10 expression (Harris *et al.*, 2006).

Our studies (Carraro *et al.*, 2009) and others' (Lu *et al.*, 2007) highlighted specific miRNAs' importance in lung epithelial cell development. Ubiquitous developing epithelial overexpression of the miR-17–92 cluster augments multipotent Sox9-expressing epithelial cell number at the expense of delayed epithelial differentiation (Lu *et al.*, 2007). In addition, miR-17 and its paralogs miR-20a and miR-106b target Stat3 and Mapk14 and thereby regulate E-Cadherin expression by modulating FGF10–FGFR2b downstream signaling. Thus, mir17 paralogs control a key pathway modulating the FGF10–FGFR2b–Sprouty-driven bud morphogenesis periodicity clock (Carraro *et al.*, 2009; Warburton *et al.*, 2008).

Extracellular matrix and lung development

Differentially expressed protein components of extracellular basement membrane, laminins (LNs), entactin/nidogen, type IV collagen, perlecan, SPARC, and fibromodulin mediate cell–cell and cell–ECM interaction during lung morphogenesis. ECM components provide tissue support, may modulate cell proliferation and differentiation (Lwebuga-Mukasa, 1991), and serve as barrier and reservoir for growth factors. Preventing epithelial interaction with basement membrane disrupts lung development (Hilfer, 1996; Minoo and King, 1994).

LNs are glycoproteins involved in cell adhesion, migration, proliferation, and differentiation during tissue development and remodeling. LNs are composed of three chains, one central (α) and two lateral (β and γ), linked by disulfide bonds to form a cross-shaped molecule (Burgeson *et al.*, 1994). To date five α, three β, and three γ chain isoforms have been identified, which suggests their combination can lead to approximately 30 LN variants (Bernier *et al.*, 1995; Ehrig *et al.*, 1990; Galliano *et al.*, 1995; Iivanainen *et al.*, 1995a,b, 1997, 1999; Koch *et al.*, 1999; Pierce *et al.*, 1998; Vuolteenaho *et al.*, 1994). The α1 chain is principally localized in basement membrane at the epithelial–mesenchymal interface with a predilection for specific zones. LN α1 chain also surrounds some mesenchymal cells. A domain in the cross-region of the α1 chain influences lung epithelial cell proliferation (Schuger *et al.*, 1992). The α4 chain in LN8 and LN9 variants is highly expressed developing murine lung and heart (Frieser *et al.*, 1997; Iivanainen *et al.*, 1995a,b, 1997). LN α4 chain is localized around vessels in fetal lung and may assist organization of lung mesenchyme (Miner *et al.*, 1997; Pierce *et al.*, 1998). The α5 chain in LN10 and LN11 is abundantly expressed during lung morphogenesis (Miner *et al.*, 1995, 1998; Pierce *et al.*,

1998): mutated LN α5 chains impair lobe septation and bronchiolar branching in mutant murine lung.

With roles in cell adhesion, β1 and γ1 are constantly expressed during fetal lung development (Durham and Snyder, 1995): globular domains near their N-termini help regulate cell polarization (Schuger *et al.*, 1995, 1996b). LN β2 chain isoform localizes to the basement membrane of prealveolar ducts, airways, SMCs of airways, and arterial blood vessels, as well as type II pneumocytes.

Nidogen (150 kDa) in basement membrane binds γ1 and γ3 chains helping link LN to collagen IV (Dziadek, 1995; Koch *et al.*, 1999; Reinhardt *et al.*, 1993). Nidogen is mesenchymally synthesized and appears to help basement membrane organization during lung morphogenesis (Senior *et al.*, 1996). Blocking Nidogen's interaction with LN disrupts lung development (Dziadek, 1995; Ekblom *et al.*, 1994; Senior *et al.*, 1996). Nidogen to degradation by matrix metalloproteinases (MMPs) may facilitate remodeling via basement membrane degradation (Mayer *et al.*, 1993).

Proteoglycans comprise a protein core with sulfated carbohydrate side chains. They form flexible structures and function as a reservoir for growth factors, water, and ions. Inhibiting proteoglycan sulfation disrupts branching of E13 mouse lung explants and inhibits epithelial migration toward lung mesenchyme or FGF10-soaked beads (Shannon *et al.*, 2003). Perlecan is a predominant basement membrane proteoglycan, composed of an approximately 450 kDa core protein with three heparan sulfate chains. Pulmonary perlecan synthesis rises sharply with increased fetal SMC proliferation (Belknap *et al.*, 1999).

Fibronectin (FN) is essential for clefting during initial epithelial branching in salivary gland and is accumulated at sites of epithelial constriction and indentation in lung and kidney (Sakai *et al.*, 2003), supporting roles for FN in lung branching morphogenesis (Roman, 1997). Indeed, FN is a Wnt target required in the control mechanism of lung bud tip splitting (De Langhe *et al.*, 2005). Treating lung rudiments with anti-FN antibody or siRNA inhibited branching morphogenesis, while FN supplementation promoted branching (Sakai *et al.*, 2003). Located in epithelium and mesenchyme, the EIIIA segment is one of the alternatively spliced FN segments, modulating FN's cell proliferative effect: expression decreases from pseudoglandular to saccular stages, increases into the alveolar stage, and parallels changes in Proliferating Cell Nuclear Antigen (PCNA)-positive distal pulmonary cell number (Kikuchi *et al.*, 2003).

MMPs are ECM-degrading enzymes that are inhibited by tissue inhibitors of metalloproteinases (TIMPs). MMPs may alter cell fate and behavior by ECM modulation and modulate signaling of bioactive molecules by their cleavage, by their release from bound stores, or by altering activity of their inhibitors (Vu and Werb, 2000). *Timp3*-null mutant mice have attenuated airway branching and alveologenesis (Gill *et al.*, 2003) and develop progressive emphysema-like changes (Leco *et al.*, 2001). MMP2, MMP9, and their tissue inhibitors (TIMP1 and TIMP2) are influenced by early postnatal dexamethasone exposure: *Timp2* expression falls and *Mmp9* expression increases. Such changes may contribute to steroids' effects on neonatal lung structure (Valencia *et al.*, 2003). MT1-MMP, a potent MMP2 activator, is a major downstream EGFR target. *Egfr*$^{-/-}$ mice had low *MT1-Mmp* expression with 10-fold reduction in active MMP2. Abnormal lung alveolarization in *Mmp2*$^{-/-}$ mice is similar to but less severe than that in neonatal *Egfr*$^{-/-}$ mice (Kheradmand *et al.*, 2002).

Retinoic acid signaling

Retinoids (all-trans, 9-cis, and 13-cis) are fundamental for development and homeostasis of numerous systems including lung, and there is a well-described mammalian RA synthesis and degradation system (Chambon, 1996). Retinaldehyde dehydrogenase-2 (RALDH-2) plays a prominent role in generating RA during organogenesis (Niederreither *et al.*, 1997, 1999; Ulven *et al.*, 2000). RA signaling is mediated by its nuclear receptors of the steroid hormone receptor superfamily: RAR (α, β, γ) and retinoid RXR (α, β, γ) (Chambon, 1996). RAR/RXR heterodimers also transduce RA signaling *in vivo* (Kastner *et al.*, 1997). Within E13.5 lung, *Rar-β* isoform transcripts are localized to proximal airway epithelium and adjacent mesenchyme, whereas *Rarα1, Rarα2,* and *Rarγ2* isoforms are ubiquitously expressed (Chazaud *et al.*, 2003).

RA signaling is required for lung bud initiation. Acute vitamin A deprivation in pregnant rats at the onset of lung development results in blind-ending tracheae and lung agenesis in some embryos, which is similar to the case of *Fgf10*$^{-/-}$ mutant mice (Dickman *et al.*, 1997; Sekine *et al.*, 1999). Disruption of RA signaling in *Rarα/β2* knockout mice causes left lung agenesis and right lung hypoplasia (Mendelsohn *et al.*, 1994). Lung branching morphogenesis is characterized by dramatic downregulation of RA signaling. Preventing this by treating embryonic lung explants with high RA concentrations (10^{-6}–10^{-5} M) disrupts distal budding with formation of proximal-like immature airways (Cardoso *et al.*, 1995; Malpel *et al.*, 2000). Continued RA activation by overexpression of constitutively activated *Rara* chimeric receptors meant lungs did not form saccules or identifiable type I cells: raised epithelial *Sp-C, Nkx2.1,* and *Gata6* levels (but not *Sp-A* or *Sp-B*) at birth suggested differentiation was arrested early in these lungs. Downregulation of RA signaling is required to allow differentiation to form mature type I and II cells (Wongtrakool *et al.*, 2003). RA inhibits expression and alters *Fgf10* and *Bmp4* distribution (Cardoso *et al.*, 1995; Malpel *et al.*, 2000). Pan-RAR antagonism alters *Tgf-β3, Hnf-3β,* and *Cftr* expression in proximal tubules and *Bmp4, Fgf10,* and *Shh* expression in distal buds (Chazaud *et al.*, 2003).

In early E11–12.5 murine lung, *Raldh-2* is concentrated in trachea (mesenchyme) and proximal lung (mesothelium) at sites of limited branching: this lack of overlap with *Fgf10* suggests RA may restrict *Fgf10* expression thereby defining the proximal–distal lung axis. However, during postnatal lung development, RA increases alveolar number, partially rescuing dexamethasone's effects. In adult rats, RA reverses features of elastase-induced emphysema (Maden and Hind, 2003; Massaro and Massaro, 1996, 2000). *RAR-γ*-null mice have defective alveolar septation consistent with abnormal elastin deposition (Chytil, 1996). Additional deletion of a retinoid X receptor (*RXRα*) allele decreases alveolar surface area and number (McGowan *et al.*, 2000). RA is a vitamin A metabolite. Vitamin A deficiency injures lung and impairs rat type II pneumocyte function (McGowan *et al.*, 2000). This combined evidence suggests RA may have a complex role in alveolar development.

4. Mechanobiology of the Developing Lung

Mechanical stimuli to lung development have been long appreciated. Recent advances in molecular and stem cell biology allow these fields to be integrated with modern mechanobiology (Ingber, 2003). For example, ECM polymers, in addition to binding and presenting growth factors, provide resistance to deformation, fluid flow, and diffusion, and transmit force over surprisingly long distances. Adhesion molecules regulate cell motility and tissue structure, and these in turn interact through forces and deformations. Cytoskeletal components generate and transmit forces and conversely provide resistance to deformation. If we do not understand the relevant physics, we miss major factors in development, physiology, and regulation.

In an inflated rubber glove, raising internal pressure *decreases* curvature with *simplification* of form. By contrast, in lung increased internal pressure is associated with *increased* curvature and more complex morphology. Why this difference? Part of the answer is that the rubber glove is a thin elastic shell whilst embryonic tissue better resembles a viscoelastic fluid (Forgacs *et al.*, 1998; Jakab *et al.*, 2008). Over the timescale of growth, the elastic component of tissue viscoelasticity may be neglected; the tissue therefore behaves mechanically as a fluid, such that a simple mechanobiological model (Lubkin and Murray, 1995) predicts pulmonary pressure–morphology relationships (Unbekandt *et al.*, 2008).

Mechanics also influence differentiation: *in vitro* mesenchymal stem cells differentiate toward neurons at 1 kPa, muscle at 10 kPa, and cartilage at 30 kPa (Engler *et al.*, 2006). Hypothesizing that lung seeks to equilibrate tangential epithelial stress, a mechanobiological model of pseudoglandular lung (Lubkin and Murray, 1995) treated the epithelium as a viscous fluid with surface tension (Foty *et al.*, 1994) to predict that branch size will be inversely related to pressure difference between external medium (and native mesenchyme) and lumen. Indeed embryonic lung epithelium appears to regulate tangential stress by modulating cytoskeletal tension via the Rho–ROCK system (Moore *et al.*, 2005).

4.1. Lessons on mechanobiology from human and *in vivo* studies

Human birth defects and *in utero* experiments have demonstrated lung development is subject to mechanics. For example, CDH (Smith *et al.*, 2005) comprises a diaphragmatic defect, intrathoracic herniation of abdominal viscera and lung hypoplasia: affected newborns retain a high mortality rate due to inadequate lung growth. Traditionally, lung hypoplasia was attributed to lung compression by herniated abdominal viscera. Indeed lung growth is impaired when fetal CDH is created surgically (Starrett and de Lorimier, 1975). Similarly, human fetuses with renal agenesis or profound renal failure exhibit Potter's syndrome, in which an underfilled amniotic cavity is thought to cause lung hypoplasia due to excessive lung fluid loss and/or fetal thorax compression. Certainly, bilateral fetal nephrectomy impairs ovine lung growth (Wilson *et al.*, 1993). Alternatively, lung hypoplasia may result from developmental insults to the lung that precede or coincide with the origins of CDH and renal agenesis, respectively. For example, in the nitrofen-induced CDH model, early lung malformation precedes CDH (Jesudason *et al.*, 2000). Similarly, lung hypoplasia emerges before fetal urine output normally contributes to amniotic fluid in a transgenic murine model of renal dysgenesis (Smith *et al.*, 2006). Synthesizing these positions argues for an early developmental insult to the lung that is then compounded by unfavorable mechanical influences (Keijzer *et al.*, 2000). In addition to extrinsic forces acting on fetal lung, a distending pressure is generated by lung liquid production. Draining this fluid by fetal tracheostomy is associated with lung hypoplasia (Fewell *et al.*, 1983). Likewise, retention of this fluid in congenital laryngeal atresia is associated with lung overgrowth and distension (Harding and Hooper, 1996). This led to development of fetal tracheal occlusion to rescue hypoplastic lung growth in human CDH (Harrison *et al.*, 2003; Hedrick *et al.*, 1994). The normal fetal larynx appears to open only during diaphragmatic contraction (fetal breathing movements: FBMs), which restricts lung liquid efflux (Fewell and Johnson, 1983). Hence, failure of FBM in CDH may also contribute to lung hypoplasia. Experimental FBM abolition by phrenic nerve section is associated with lung hypoplasia (Miller *et al.*, 1993). Loss of skeletal muscle formation also causes lung hypoplasia: thinned diaphragms in MyoD$^{-/-}$ mice cannot support FBM and the lungs are hypoplastic with reduced cell proliferation at E18.5 (Inanlou and Kablar, 2003).

Neonatally, mechanical ventilation conspires with factors such as inflammation to generate BPD in premature newborns (Warburton *et al.*, 2001). Mechanical factors appear influential beyond this period: compensatory lung growth follows lung resection (Thurlbeck, 1983) comprising lung distension and parenchymal growth. This postpneumonectomy effect suggests the lung responds to altered mechanics and that the organism to reduced alveolar surface area. At a smaller scale, airway smooth muscle (ASM) hypertrophy and hyperreactivity in asthma are associated with air trapping and acute lung distension; however, with time, this is associated with airway remodeling and chronic lung hyperexpansion. ASM-led airway occlusions in asthma may therefore have analogous effects to fetal tracheal occlusion (which distends and remodels prenatal lung) (Jesudason, 2007). Moreover, transient endogenous ASM-led airway occlusions occur in fetal lung (known as airway peristalsis), and this contractility may be an important regulator of lung growth (discussed below) (Jesudason, 2006a). With this in mind, we next focus on three areas of interest in lung mechanobiology: (i) lung liquid, (ii) airway contractility, and (iii) calcium signaling in this secretory, contractile environment.

4.2. The impact of hydraulic pressure on lung organogenesis

Prenatal lung liquid is neither plasma ultrafiltrate nor "inhaled" amniotic fluid (Adamson *et al.*, 1969). Lung liquid is produced throughout prenatal lung development by incompletely understood mechanisms that involve active Cl^- transport from blood/interstitium into lumen (Olver and Strang, 1974). Intracellular Cl^- accumulation is energized by the basolateral Na^+/K^+-ATPase (Bland and Boyd, 1986) and accomplished via Na^+-linked cellular Cl^- uptake through the $Na^+/K^+/2Cl^-$ co-transporter (Thom and Perks, 1990); indeed Cl^- secretion rate depends on NKCC1 expression (Gillie *et al.*, 2001). Movement of accumulated Cl^- down its concentration gradient via apical Cl^- channels results in accompanying Na^+ and water flux to generate fetal lung fluid (see Olver *et al.*, 2004 for comprehensive review). Whilst active Cl^- and fluid secretion are crucial to lung growth (Alcorn *et al.*, 1977), they may not contribute to branching per se (Souza *et al.*, 1995a).

The identity of the apical Cl^- channel remains unclear. Numerous channels are demonstrated in fetal alveolar type II cells, including a G protein-regulated maxiCl channel (Kemp *et al.*, 1994), cystic fibrosis transmembrane conductance regulator (CFTR) (McCray *et al.*, 1993), at least one member of the Chloride Channel (CLC) channel family (Blaisdell *et al.*, 2004; Murray *et al.*, 1995), and a Ca^{2+}-activated Cl^- channel, TMEM16a (Rock *et al.*, 2008). $CFTR^{-/-}$ mice have normal prenatal lungs (Wallace *et al.*, 2008), suggesting CFTR plays no role in producing lung liquid or there is functional redundancy. Although a definitive link between CLC channels and lung liquid production remains to be established *in vivo*, there is evidence that CLC-2 contributes to fluid secretion and cyst expansion *in vitro* (Blaisdell *et al.*, 2004). Interestingly, murine TMEM16a$^{-/-}$ mutants die of respiratory failure at an interval after birth with characteristic tracheomegaly and disruption of trachealis formation (Rock *et al.*, 2008).

The rate of liquid production and the laryngeal valve function help determine hydraulic pressure in the lung. Obstructing the prenatal trachea increases intraluminal pressure two- to three-fold and airway branching three-fold; the rate of bud extension increases about two-fold whilst inter-bud distance is halved. These effects depend on FGF10–FGFR2b–Sprouty signaling (Unbekandt *et al.*, 2008). Several studies have used tracheal obstruction to try to improve lung growth in human CDH (Harrison *et al.*, 2003; Jani *et al.*, 2005). However, clinical evidence of benefit of this potentially hazardous intervention remains limited. An alternative being explored is to exploit spontaneous airway occlusions that may be important for lung growth and perhaps avoid invasive fetal interventions (Jesudason, 2009).

4.3. The impact of embryonic airway peristalsis in lung organogenesis

Early mammalian airway exhibits spontaneous transient airway occlusions due to airway peristalsis. This is mediated by spontaneous ASM contractions that occur in birds and humans and which increase in frequency from embryonic stages to birth (Schittny *et al.*, 2000). Peristaltic contractions and airway occlusions direct waves of fluid toward the lung's tips. This results in rhythmic stretch and relaxation of growing buds (Fig. 3.8). Hence airway peristalsis and occlusions are well placed to regulate both pressure and stretch in the tips of developing lung (Jesudason, 2009). These ASM waves emanate from pacemaker areas in proximal airway before transmission distally (Jesudason *et al.*, 2005). This pacemaker-driven airway contractility may even be important postnatally in asthma (Jesudason *et al.*, 2006b). Thus, putative pulmonary pacemakers might be targeted for ablation by bronchial thermoplasty for asthma (Jesudason, 2009). Studying frequency of peristalsis in embryonic lung culture revealed that it is amenable to acceleration by cholingergic agents as well as growth factors (FGF10). These accelerated rates accompany enhanced *in vitro* lung growth. Similarly, *in vitro* inhibition of peristalsis is associated with reduced lung growth (Jesudason *et al.*, 2005). This apparent coupling raised interest in mechanisms linking morphogenesis and peristalsis-led airway occlusions. In particular, $Ca2^{+}$-imaging studies revealed that prenatal lung features spontaneous regenerative intercellular ASM calcium waves that propagate along main airways immediately prior to the wave of peristaltic contractility (Featherstone *et al.*, 2005). Using pharmacological inhibitors, we showed that ASM calcium waves depend on extra- and intracellular calcium as well

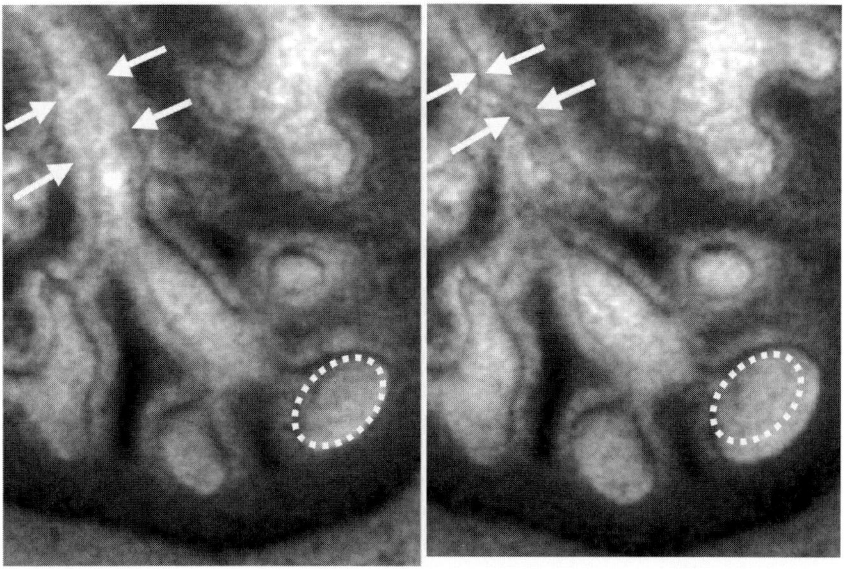

Figure 3.8 Peristalsis occludes airway proximally whilst distending it distally. Sequential photomicrographs of cultured embryonic lung: in both panels the arrows outline proximal airway (open in the top picture but occluded by contraction in the lower one). The peristaltic airway occlusion's distal effect is shown in a terminal bud, which rapidly rhythmically increases in size (the same ovoid outline is applied to both pictures). This distension and relaxation recurs with each wave.

as gap junction integrity. Moreover, these calcium waves are abnormal in experimental lung hypoplasia (Featherstone *et al.*, 2006). Thus, if peristaltic airway contractions do regulate lung growth, it means that underlying calcium oscillations govern lung development.

4.4. Lung stretch transduction and parathyroid hormone-related protein (PTHrP)

Airway peristalsis is coupled to lung growth, responsible for phasic lung stretch and underpinned by calcium oscillations. Transduction of such mechanical activity involves key modulators and sensors of serum $Ca2^{+}$. For example, stretching alveolar type II cells produces parathyroid hormone-related protein (PTHrP) (Torday *et al.*, 2002), which binds in paracrine fashion to its cognate receptor on adjacent adepithelial fibroblasts (Torday and Rehan, 2002); the latter are induced to differentiate into a lipofibroblast lineage (Schultz *et al.*, 2002). The lipofibroblasts, discovered by Vaccaro and Brody (1978), protect against oxidant injury (Torday *et al.*, 2001) and produce leptin, which stimulates synthesis of surfactant phospholipids and proteins, by ligating its receptor on alveolar type II cells (Torday *et al.*, 2002a,b). Alveolar stretch releases PTHrP, which dilates alveolar capillaries (Gao and Raj, 2005) and coordinates surfactant production with alveolar perfusion. The Torday group showed that fluid distension permits cultured lung to differentiate like lung *in vivo*, agreeing with the concept discussed elsewhere that rhythmic mechanical distension serves as a clock or "zeitgeber" mechanism for organogenesis (Gross *et al.*, 1978). Conversely, inhibiting lung fluid in explants inhibits the putative "zeitgeber," while PTHrP treatment rescues the temporal molecular development of the lung. Fluid distension also stimulates endodermal Shh expression, which in turn stimulates the mesodermal Wnt pathway. Wnt increases endodermal PTHrP expression, which feeds back negatively on the Wnt pathway by stimulating protein kinase A. Wnt/β-catenin signaling has recently been linked to both PTHrP and the CFTR (Cohen *et al.*, 2008).

4.5. Lung development and the Ca2$^+$-sensing receptor (CaSR)

Stretch-induced PTHrP regulates Ca2$^+$ and lung growth, and stretch-inducing airway peristalsis is underpinned by Ca2$^+$ signaling. However, the developing fetus is actually hypercalcemic compared to the adult, with free extracellular Ca^{2+} concentration ($[Ca^{2+}]_o$) of around 1.7 mM, while maternal $[Ca^{2+}]_o$ is 1.1.-1.3 mM. Studies in mouse lung explant culture demonstrated fetal $[Ca^{2+}]_o$ suppresses branching and cellular proliferation while enhancing Cl^--dependent fluid secretion (Finney *et al.*, 2008). Since effects of fetal $[Ca^{2+}]_o$ are mimicked by pharmacological activation of Ca^{2+}-sensing receptor (CaSR), the G protein-coupled extracellular CaSR and fetal Ca^{2+} via CaSR may balance branching morphogenesis and lumen distension. CaSR expression has been documented in mouse (Finney *et al.*, 2008) and humans (Finney, Wilkinson, Kemp, and Riccardi, unpublished observations), throughout the pseudoglandular stage. At the start of this phase, CaSR is restricted to the epithelium where its expression is maintained for at least 48 h in the organ culture. From E14.5 in mouse (and week 9 in humans) mesenchymal CaSR expression appears. CaSR expression wanes during the canalicular phase and is absent from adult mouse, rat, and bovine lung (Brown *et al.*, 1993; Finney *et al.*, 2008; Riccardi *et al.*, 1995). CaSR-dependent regulation of branching and fluid secretion occurs via phospholipase C-dependent rises in intracellular Ca^{2+} concentration and phosphoinositide 3 (PI_3) kinase activation (Fig. 3.9). Whether CaSR-dependent rises in Ca^{2+} affects lung development through peristaltic waves (discussed above) and/or controls growth by altering epithelial integrity via activation of PI_3 kinase-dependent β-catenin signaling remains unclear. CaSR activation could also affect clock mechanisms of airway branching by interfering with FGF–FGFR–Sprouty signaling events: lungs cultured in 1.7 mM Ca^{2+} delay both bud extension and tip-splitting,

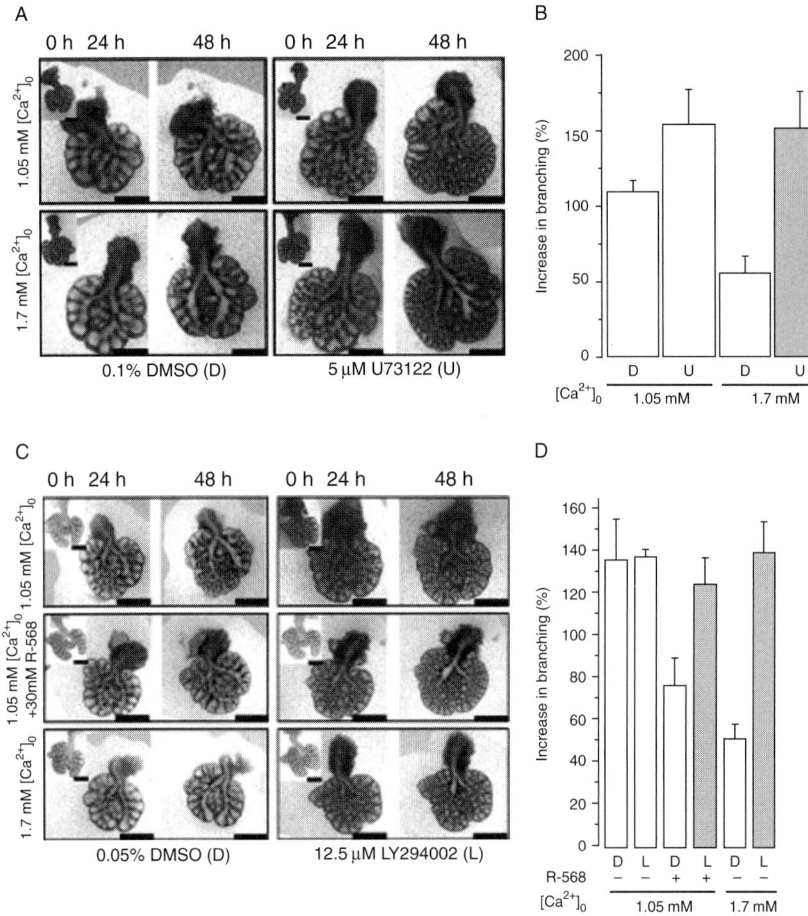

Figure 3.9 CaSR-evoked inhibition of branching and its dependence on Pulmonary lymphangitiic carcinomatosis (PLC) and phosphoinositide 3 (PI_3) kinase signaling. Effect on branching of 1.05 mM (A, upper panels) or 1.7 mM (A, lower panels) $Ca^{2+}{}_o$ in the absence (0.1% DMSO vehicle control; A left panels) or presence (A, right panels) of 5 μM of the PLC inhibitor, U73122. Quantification of branching at 48 h in the four conditions is shown in (B). Inhibition of PLC rescues suppression of branching evoked by 1.7 mM $Ca^{2+}{}_o$. Bars = 700 μm. Effect on branching of 1.05 mM $Ca^{2+}{}_o$ (C, upper panels), 1.05 mM $Ca^{2+}{}_o$ plus 30 nM R-568 (C, middle panels), and 1.7 mM $Ca^{2+}{}_o$ (C, lower panels) in the absence (0.05% DMSO vehicle control) or presence of 12.5 μM of the PI3K inhibitor, LY294002. Bars = 750 μm. Quantification of branching at 48 h in the six conditions is shown in (D). The calcimimetic R-568 mimics the suppressive effect of high $Ca^{2+}{}_o$ on branching, further implicating CaSR in the process. Note PI_3 kinase inhibition rescues suppression of branching, whether evoked by high $Ca^{2+}{}_o$ or calcimimetic. Adapted from Finney *et al.* (2008).

when compared to lungs cultured in lower, adult levels of $[Ca^{2+}]_o$ (1.2 mM) (Finney *et al.*, 2008). Finally, patients with inactivating CaSR mutations have increased risk of interstitial lung disease (Auwerx *et al.*, 1985a,b). Given that CaSR expression appears confined to the pseudoglandular phase, we need to test if this phenotype originates prenatally.

5. Stem/Progenitor Cell Biology of the Lung

A stem cell describes a self-renewing, primitive, undifferentiated, multipotent source of multiple cell lineages. Such cells are critical for development and growth; pools of adult stem cells are hypothetical sources for tissue regeneration and repair as well as cancers. In contrast to embryonic stem cells and tumor cells, adult stem cells reduce telomere length with age (Warburton *et al.*, 2008).

In the lung, there is limited knowledge about existence of self-renewing cells, whether such cells conform to classic or nonclassic stem cell hierarchies and whether a single stem/progenitor cell suffices to generate the more than 40 distinct cell types required in mature lung. At least five putative epithelial stem/progenitor cell niches are reported in adult mouse airway (Liu and Engelhardt, 2008), as well as endothelial stem cells in the pulmonary vasculature and ASM stem cells. Additionally, circulating stem/progenitors may lodge in the lung. Unlike skin and gastrointestinal tract, postnatal lung turns over slowly, which hampers putative progenitor identification. Specific markers and clonality assays also limit attempts at isolation. Herein we review new information regarding the development and function of lung stem/progenitor cells in organogenesis.

5.1. Endogenous epithelial progenitor cells

Failing regeneration and repair with age has been suggested to be due to stem cell failure. In pseudoglandular lung, tips of the branching tubules appear to contain undifferentiated multipotent epithelial progenitors. In adult lung, putative endogenous epithelial progenitors have been located in the basal layer of the upper airways, within or near pulmonary neuroendocrine cell rests as well as at the bronchoalveolar junction and in the alveolar epithelium (Engelhardt, 2001; Giangreco *et al.*, 2002, 2004; Kim *et al.*, 2005; Rawlins and Hogan 2006; Reddy *et al.*, 2004; Reynolds *et al.*, 2000, 2004). The distal-most epithelial cells were shown to be a multipotent progenitor cell population during branching morphogenesis of the lung (Rawlins *et al.*, 2009a). A recent model suggests, as the lung branches, descendents of distal tip progenitors are left to differentiate in the stalks, whereas self-renewing progenitors remain in the epithelial tips: distal epithelial cells have a unique gene expressionpattern, including high levels of *Sox 9, Id2* (inhibitor of differentiation 2), *N-myc*, and *Etv5/ERM* (ets variant gene 5). In addition, they are exposed to high levels of FGF, Wnt, Shh, and BMP signaling (Bellusci *et al.*, 1996; Liu *et al.*, 2002; Shu *et al.*, 2005). Many of these pathways including *sox9, etv5*, and FGFR signaling are associated with stem/progenitor cells in other endodermally derived organs (Seymour *et al.*, 2007; Zhou *et al.*, 2007). Distal epithelial cells also have different cell cycle kinetics compared with the rest of the epithelium; a higher proportion of them incorporate the thymidine analog bromodeoxyuridine (BrdU) during a short pulse (Okubo *et al.*, 2005).

5.1.1. Tracheal and bronchial epithelial progenitors

Candidate endogenous stem or progenitor cells have been identified in trachea and lung, using lung injury/repair models. For example, CC10/ *Scgb1a1*$^+$ Clara cells self-renew and proliferate after tracheal injury, but seem not the main source for tracheal epithelial regeneration (Rawlins *et al.*, 2009b). However, subsets of Keratin-14 (K-14)-expressing BCs in the trachea (Hong *et al.*, 2004a) can restore differentiated epithelium after injury and are distinct from bronchial BCs (Hong *et al.*, 2004b). Lineage tracing in the adult mouse lung and trachea showed K-14-positive cells act as progenitors and ciliated cells cannot (Hong, *et al.*, 2004a; Rawlins, *et al.*, 2007). Human *SpC* and rat *CC10* promoters have also been used for lineage tracing in lung (Perl *et al.*, 2002, 2005a).

Since pseudostratified epithelium of mouse trachea and human airways contains a BC population expressing cytokeratin 5 (*Krt5*), a recent study using *Krt5-CreER*^{T2} transgenic mouse line for lineage tracing showed BCs of mouse trachea function as progenitors during postnatal growth, in adult homeostasis and also in epithelial repair of experimentally-induced SO_2 damage (Rock *et al.*, 2009). A clonality assay also found BCs of mouse and human airways self-renew and differentiate into mucus and ciliated lineages in the absence of stroma or columnar epithelial cells.

There is also a rare mixed population of pluripotent cells in lower respiratory tract characterized as a Hoechst dye effluxing side population (SP) cells. They express molecular markers of airway and mesenchymal origin (Giangreco *et al.*, 2004). CD45$^-$ SP cells isolated from human tracheobronchial epithelium have proliferative potential. Increased numbers of these cells in asthmatic airways suggest that dysregulation of pluripotent cells may play a role in this chronic disorder (Hackett *et al.*, 2008). In developing lung, some SP cells, which are CD45$^+$ and CD45$^-$ have endothelial progenitor cell (EPC) potential in response to hyperoxia (Irwin *et al.*, 2007).

The submucosal gland ducts in proximal airway are likewise suspected to contain stem cells (Liu and Engelhardt, 2008). However, relatively little is known about glandular stem/progenitor cells and their niche(s). Studies have suggested regenerating tracheal epithelium after naphthalene injury arises from cells migrating from gland ducts (Borthwick *et al.*, 2001).

Rawlins *et al.* (2009b) used lineage tracing to circumvent obstacles that hampered earlier studies of lung stem/progenitor cells. Using the restricted expression of *CC10/Scgb1a1*, they generated a "knocking" transgenic mouse with a tamoxifen (TM) inducible *Cre-recombinase* (*ScgB1a1-CreER*™) that lineage-tags Clara cells of the airway. By varying dose and timing of TM administration, they discovered that epithelial reconstitution in the bronchioles involves Clara cell self-renewal and differentiation into ciliated cells. These data argue that bronchiolar *ScgB1a1*-expressing cells, largely mature Clara cells, are a self-renewing progenitor pool. The observation that lineage tags are chased into ciliated cells over time is consistent with early findings of Evans and colleagues (1978) that Clara cells are progenitors for ciliated cell renewal. On the other hand, Rawlins *et al.* (2009b) showed that lineage tags introduced into *ScgB1a1*-expressing cells of tracheobronchial airways were depleted within the *ScgB1a1*-expressing population over time. Collectively, these data suggest that *ScgB1a1*-expressing cells of proximal airways behave like transit amplifying (TA) cells, like those of intestinal epithelium, whereas *ScgB1a1* cells of bronchiolar airways behave like self-renewing progenitors present in the interfollicular epidermis (reviewed by Chen *et al.*, 2009).

A different approach by Giangreco and colleagues (2009) to investigate long-term behavior of airway progenitors in normal and injured airways showed in concordance with Rawlins *et al.* (2009b) that during homeostasis an abundant progenitor cell pool maintains the airway epithelium (rather than rare tissue stem cells). However, clonal patches of labeled cells emanate from tissue-specific stem cells located at airway branch points or bronchioalveolar duct junctions, after Clara cell depletion resulting from naphthalene exposure. In this naphthalene injury case, repairing bronchiolar airways more closely resemble the renewing epidermis after wounding, wherein stem cells are recruited from the hair follicle bulge to replace the depleted BC pool of the interfollicular epidermis (Zemke *et al.*, 2009).

Varying dose and timing of TM administration, Rawlins *et al.* (2009a) discovered that reconstitution of bronchiolar epithelium involves Clara cell self-renewal and differentiation into ciliated cells and that Clara cells contribute to tracheal repair. Using lineage tracing, this study showed that a special population of BASCs which coexpress *CC10* and *SP-C*, which have been proposed to contribute to both bronchioles and alveoli, has no apparent function during postnatal growth, adult homeostasis, or alveolar repair. Thus, they propose that trachea, bronchioles, and alveoli are maintained by distinct progenitor populations (Rawlins *et al.*, 2009a). Currently, the significance that some *Scgb1a1*$^{+}$ bronchiolar Clara cells express *SftpC* and some alveolar type 2 cells express *Scgb1a1* is not understood. There is accumulating evidence the (*Scgb1a1*$^{+}$, *SftpC*$^{+}$) coexpressing cell population increases in number in murine lung cancer models (Ventura *et al.*, 2007; Yang *et al.*, 2008). However, it is unclear if this is due to preferential proliferation of preexisting (*Scgb1a1*$^{+}$, *SftpC*$^{+}$) cells or oncogenic upregulation of *SftpC* or *Scgb1a1*.

In a recent study (Tompkins *et al.*, 2009), selective *Sox2* deletion in Clara cells with *Scgb1a1-Cre* showed that Clara cell Sox2 is required for differentiation and/or maintenance of ciliated, Clara, and goblet cells in bronchiolar epithelium after birth and caused progressive loss of ciliated, Clara, and goblet cells and an inability to produce goblet cells in response to allergen. The findings indicate Clara cells can serve as common progenitors of ciliated, Clara, and goblet cells in a process requiring *Sox2*.

5.1.2. Alveolar epithelial progenitors

Epithelial progenitors of the alveoli have yet to be identified. An interesting model is that the alveolar progenitors are located in distal epithelial tips during the canalicular stage. However, there is no published evidence to support, or refute, this hypothesis. Presumably, because of its vast area a large number of AECs must function as a "ready reserve" to repair damaged alveolar surface. For instance, the expression of telomerase, a stem/progenitor cell marker, after acute oxygen injury is widely upregulated in AECs during recovery (Driscoll *et al.*, 2000). This suggests that either AECs contain a progenitor cell subpopulation or that the majority of AECs undergo reactivation progenitor-like states after injury (Driscoll *et al.*, 2000). In addition, without telomerase, resistance to injury and repopulation of damaged alveoli are compromised, indicating this pathway is likely critical for alveolar progenitor cell activity (Driscoll *et al.*, personal communication). Moreover, Kim and colleagues (2005) described BASCs, which possess stem cell characteristics, are resistant to naphthaline injury and proliferate after airway or alveolar injury. Such BASCs reside near bronchioalveolar junctions and coexpress both alveolar (SP-C) and airway (CC10) epithelial cell markers, as well as coexpressing Sca-1. They are capable of self-renewal and differentiation into Clara cells and alveolar cells, and are also multipotent in clonal assays. Moreover, studies by Hong and coworkers (2001) identified variant Clara cells as endogenous lung stem cells, which infrequently proliferate during steady state but are held responsible for repopulating distal airway epithelium after injury. Variant Clara cells express Clara cell secretory protein, but survive naphthalene injury. As the lung continues to grow postnatally, Clara cells both self-renew and act as progenitors for ciliated cells, based on kinetics of cell labeling after a pulse of tritiated [3H]-thymidine (McDowell *et al.*, 1985; Plopper, *et al.*, 1992). This is supported by recent lineage labeling (Perl *et al.*, 2005a). Whether all Clara cells have this capacity requires investigation. Moreover, type II cells proliferate and give rise to type I cells after adult alveolar injury, and this probably also occurs during postnatal growth (Evans *et al.*, 1975). Several putative endogenous alveolar stem cell populations thus provide targets for directed regenerative therapies. Taking acute oxygen injury as an example, AECs undergo DNA and other forms of damage such as mitochondrial failure, glutathione depletion, and apoptosis (Buckley *et al.*, 1998; Lee *et al.*, 2006; Roper *et al.*, 2004).

5.2. Endogenous mesenchymal progenitors

Several studies have shown that signals from lung mesenchyme are essential for branching morphogenesis. Mesothelium-derived FGF9 activates and controls FGF10 signaling from peripheral mesenchyme via FGFR2b, SHP2, Grb2, Sos, Ras, and Sprouty in epithelium (Del Moral *et al.*, 2006b; Bellusci *et al.*, 1997b; Tefft *et al.*, 2002, 2005).

5.2.1. Smooth muscle progenitors

Distal *Fgf10*-expressing mesenchymal cells serve as progenitors for peripheral ASM (De Langhe *et al.*, 2006; Mailleux *et al.*, 2005; Ramasamy *et al.*, 2007). *Fgf10-lacz* lineage tracing reveals ASM progenitors begin as *Fgf10*-expressing cells that, as the airway elongates, become distributed along peripheral airway. Transdifferentiation to express alpha–smooth muscle actin occurs under the control of SHH and BMP4, which are expressed proximal to the airway tip. Thus, increase in population size and localization of peripheral ASM progenitors occur early in development. Another population of ASM progenitors arise in proximal mesenchyme and advance peripherally (Shan *et al.*, 2008).

5.2.2. Vascular progenitors

Lung microcirculation is rich in progenitors, but our understanding of these is limited. Mesothelium overlying the lung contains progenitors that give rise to pulmonary vascular (but not airway) SMCs during embryonic development (Que *et al.* 2008). Endothelial progenitors arise from endogenous vascular wall or from circulating progenitors. Similar to lung epithelial cells, heterogenous pulmonary endothelial cells may require a site-specific niche (Clark *et al.*, 2008); alternatively, putative resident endothelial progenitors may constitute a universal pool of progenitors that lack segmental specification (Blaisdell *et al.*, 2009).

Distal airspace and vascular growth are coordinated so injury can affect both (Jakkula *et al.*, 2000). Balasubramaniam *et al.* (2007) examined endothelial progenitors in BPD to show that hyperoxia disrupts alveolar and vascular growth, limiting surface area for gas exchange. In the lung, nitric oxide, VEGF, and erythropoietin contribute to mobilization and homing of EPCs. Several related developmental changes occur after hyperoxia in neonatal mice: expression of endothelial nitric oxide synthase, VEGF, and erythropoietin receptor and the number of EPCs in the blood, bone marrow, and lung were all reduced (Balasubramaniam *et al.*, 2007).

Primitive capillaries surround the laryngotracheal groove as the lung buds from foregut and can be visualized by β-galactosidase expression under control of Flk1 promoter. This promoter is active and the earliest known marker of hemangioblasts. Under stimulation of epithelial VEGF, these hemangioblasts differentiate into a capillary network that surrounds bronchial, lobar, and segmental airways (Del Moral *et al.*, 2006a; Ramasamy *et al.*, 2007). Organization of this plexus appears essential for correct branching and perfusion. Thus, mesothelial–mesenchymal–epithelial–endothelial crosstalk matches epithelial and vascular progenitor function and will likely be essential for lung regeneration to succeed. Further studies are needed to define phenotypes of the pulmonary endothelial cell but also SMCs within the vasculature (Stevens *et al.*, 2008).

5.3. Control of lung progenitor cell proliferation

Embryonic progenitors undergo symmetric and asymmetric divisions. To distinguish these, one can look at differences in spindle orientation or differential inheritance of cytoplasmic or membrane-bound proteins such as cell fate determinant Numb and atypical protein kinase C (PKC) (Huttner and Kosodo, 2005; Morrison and Kimble, 2006; Wang *et al.*, 2009; El-Hashash and Warburton, unpublished data). Cells divide asymmetrically in response to extrinsic or intrinsic fate determinants: extrinsically, daughter cells placed in different microenvironments adopt different fates; intrinsically, cytoplasmic cell fate determinants (e.g., Numb) are asymmetrically localized within a cell and segregate differentially into daughters that adopt different fates (reviewed by Yamashita, 2009). Comparing progenitor numbers in mutant and sibling control lungs, we infer that certain molecules promote progenitor self-renewal or differentiation (Rawlins, 2008).

Several transcription factors and signaling molecules control lung growth and therefore probably affect progenitor cell proliferation. Thyroid transcription factor 1 (Ttf-1/Nkx2.1) expression marks lung lineage commitment in the early embryo and is critical for distal lung progenitor development (Kimura *et al.*, 1999). *Ttf1*$^{-/-}$-null mice have insufficiently differentiated lungs for survival (Kimura, *et al.*, 1996). HMG box transcription factor, *Sox9* is intensively expressed in distal epithelial progenitors from E11.5 to E16.5 (Liu and Hogan, 2002). However, lung-specific conditional deletion has no effect on progenitor cell behavior (Perl *et al.*, 2005b). *Sox9* may therefore act redundantly with other, as yet unknown, regulators: N-myc is also essential for maintaining a distal population of undifferentiated, proliferating progenitor cells, and may promote their self-renewal (Okubo *et al.*, 2005).

In addition, several forkhead/winged helix (fox) family transcription factors have mutant knockout phenotypes and may promote lung epithelial progenitor proliferation. For example, conditional deletion of both *foxa1* and *foxa2* genes in lung results in small lungs with decreased cell division rates (Wan *et al.*, 2005). A similar phenotype was reported after conditional deletion of both *foxp1* and *foxp2*, which are enriched in the distal epithelial progenitors. In *foxp22/2; foxp11/2* double mutants, the lungs are smaller than normal, with inhibited proliferation, but normal proximal–distal patterning (Shu *et al.*, 2007). This suggests an essential role of fox transcription factors in the maintenance of the progenitor cell population and their self-renewing divisions.

Similarly, five key signaling molecules regulate many processes in embryonic development: Wnt, Notch, Hedgehog, FGF, and TGF-β family. Embryos lacking Wnt2/2b exhibit lung agenesis and do not express Nkx2.1, the earliest marker of lung endoderm. Endoderm-restricted deletion of β-catenin replicates this, suggesting canonical Wnt2/2b signaling is required to specify lung endoderm progenitors in the foregut (Goss *et al.*, 2009, Harris-Johnson, 2009).FGF signaling plays an essential role in specification of distal lung lineages (De Langhe *et al.*, 2008; Ramasamy *et al.*, 2007) and others (Serls *et al.*, 2005). FGF10 is expressed by lung mesenchyme and is a chemotaxin during morphogenesis. FGF10 overexpression maintains epithelial progenitor cell proliferation and leads to goblet cell metaplasia (Nyeng *et al.*, 2008). In addition, FGF10 coordinates alveolar SMC formation and vascular development (Ramasamy *et al.*, 2007). RA signaling is also essential for expansion of lung progenitors and formation of primary lung buds, by affecting *Fgf10* expression through TGF-β signaling (Chen *et al.*, 2007). Similarly, Shh in distal epithelium controls proliferation and branching and is believed to promote progenitor proliferation (Pepicelli *et al.*, 1998). Autocrine Bmp signaling is likewise important for proliferation of the distal epithelial progenitor cell compartment. Wnt5a is also highly expressed around distal epithelial tips. *Wnt5a*$^{-/-}$ lungs have increased cell proliferation and an additional airway branch (Li *et al.*, 2002), but it is unknown if this phenotype relates to defective progenitors. The details of how these signaling pathways regulate distal epithelial progenitor cells remain to be determined.

5.4. Embryonic lung progenitors and proximal–distal patterning

Recent studies suggest that Wnt and Bmp signaling controls proximal–distal lung patterning, but there is currently no evidence to confirm that this is mediated through progenitors. Shu *et al.* (2005) demonstrated that proximal–distal lung patterning depends on Wnt/β-catenin signaling and is mediated, in part, through regulation of N-myc, Bmp-4, and FGF signaling. Potentiation of β-catenin signaling in proximal airway results in arrested differentiation of immature bronchiolar stem cells, but β-catenin is unnecessary for adult bronchiolar stem cell maintenance (Zemke *et al.*, 2009). Fortunately, reporters of Wnt pathway activity are highly active in distal lung epithelial cells. Recent studies suggested that Wnt signaling regulates proximal–distal patterning and progenitor proliferation independently, and that Wnt promotes distal airway fate at the expense of the proximal. (Mucenski *et al.*, 2003; Shu *et al.*, 2005). Shu and coworkers overexpressed Dickkopf-1 to inhibit Wnt pathway activity throughout developing epithelium: this expands proximal (conducting) airways at the expense of the distal, without effects on total levels of cell proliferation (Shu *et al.*, 2005). Similarly, Mucenski *et al.* (2003) showed that lung-specific deletion of β-catenin abrogates distal epithelial differentiation. Notch signaling favors progenitor identity at the expense of differentiated phenotypes in different organs (Jadhav *et al.*, 2006; Mizutani *et al.*, 2007) and is also required for lung epithelial progenitors. Notch1 is highly expressed in distal epithelial progenitors during the pseudoglandular stage (Post *et al.*, 2000). Notch controls cell fates in developing airways (Tsao *et al.*, 2009), and arrests normal differentiation of distal lung progenitors before they initiate an alveolar program (Guseh *et al.*, 2009). Notch misexpression in the distal lung prevented the differentiation of alveolar cell types (Guseh *et al.*, 2009); expression of a constitutively active form of Notch3 throughout the developing lung epithelium prevents cell differentiation (Dang *et al.*, 2003).

Furthermore, BMP signaling is also required for lung epithelium development, probably by promoting distal and repressing proximal cell fate. Inactivation of Bmp signaling by overexpression of a dominant-negative BMP receptor, or BMP antagonists Gremlin or Noggin, results in proximalization of lung epithelium (Weaver *et al.*, 1999; Lu *et al.*, 2001). Thus, reduction of BMP or Wnt signaling causes lung proximalization phenotypes (Eblaghie *et al.*, 2006; Li *et al.*, 2002).

5.5. Emergence of specific cell types during lung organogenesis

At least 40 differentiated cell types emerge during lung organogenesis. Early trachea and esophagus are both lined with ciliated epithelium; following their septation, esophageal epithelium becomes squamous, while tracheal epithelium retains cilia. Primitive airway epithelium expresses several marker proteins including cGRP, Clara cell protein, and SP-A: its differentiation starts around E16 in mouse with emergence of pulmonary neurendocrine (PNE) cell rests, surrounded shortly after by Clara cells. In the periphery, AEC2 differentiation in E18 mouse is denoted by glycogen granules' disappearance and emergence of surfactant-containing lamellar bodies with increased SP-C expression.

In mature lung, epithelial lineages are arranged proximodistally along the airways. Cartilage lies outside the submucosa and decreases in amount as bronchial caliber decreases; it is absent from bronchioles. The two major epithelial cell types in proximal bronchi are pseudostratified ciliated columnar cells and mucous (goblet) cells. Both arise from BCs, but ciliated cells predominate. Goblet cells begin to mature around 13 weeks' gestation in humans (mature ciliated columnar cells are already present), express mucin markers (MUC5B, 5A, 5C), and release mucus granules into the airway which reduces drying and, through ciliary-driven cephalad mucus flow, cleanses the airway. In cystic fibrosis, mutation of the cystic fibrosis transmembrane conductance regulator (*Cftr*) gene disrupts expression of the encoded transmembrane Na^+ ion transporter protein leading to thick mucus that overwhelms ciliary clearance and increases susceptibility to infection. In chronic airway injury, goblet cell hyperplasia may follow repair or experimental epithelial IL-9 exposure; the latter increases epithelial lysozyme and mucus production (Vermeer *et al.*, 2003). IL-4, IL-13, and allergens enhance TGF-α release, which is a ligand for the EGFR that also stimulates goblet cell differentiation (Lordan *et al.*, 2002).

There are three types of cells in bronchial submucosal glands. Myoepithelial cells surround the gland, while mucous cells (pale cytoplasm) and serous cells (basophilic cytoplasm) produce mucins. These secreted mucins mix with lysozyme and IgA on airway surface.

Kulchitsky cells are also found next to bronchial glands, but their function is unclear. It is believed they are pulmonary neuroendocrine cells (PNECs) producing peptides such as serotonin and calcitonin. Their cytoplasmic extensions usually reach the airway lumen. Kulchitsky cells expressing gastrin-releasing peptide (GRP), calcitonin gene-related peptide (CGRP), and chromogranin may be related to small cell carcinoma and carcinoid tumors. However, PNEC differentiate earlier by 10 weeks' human gestation and are the first fully differentiated murine airway epithelial cells.

Clara cells reside in distal bronchiolar airway epithelium (normally lacking mucous cells) and produce mucus-poor, watery secretion. They emerge during the 19th week in humans and appear to assist with clearance, detoxification, and surface tension reduction in small airways. Clara cell-specific protein (CC10, CCSP, or uteroglobin) and cytochrome P450 reductase CC10 can be used as Clara cell markers. Whilst normal mice feature few mucin-positive cells in the airway, mucus metaplasia is associated with numerous Clara cell-derived mucous cells with excess mucin production or reduced secretion (Evans *et al.*, 2004).

Most of the alveolar surface is covered by flat type I epithelial cells that are believed to be terminal differentiated and express several markers, such as T1a and aquaporin 5. T1a is developmentally regulated and encodes an apical membrane protein of unknown function. Absence of T1a protein blocks type I cell differentiation. Homozygous T1a null mice die at birth of respiratory failure with lungs that will not inflate normally (Ramirez *et al.*, 2003). Aquaporin 5 is a water-channel in type I epithelial cells. Recently, a knock in of Cre-ERT2 into the *Aqp5* locus has been reported (Borok, personal communication). These mice will be useful, not only to target gene deletion in type I cells but also for lineage tracing of the type I cells under development, injury, and repair.

Whilst type I epithelial cells cover more than 95% of the alveolar surface area, they account for only 40% of total AECs: the other 60% are rounded type II pneumocytes. These plump or cuboidal cells can regenerate and replace type I cells post injury and have finely stippled cytoplasm and surface microvilli. They manufacture surfactant phospholipids and proteins that modulate alveolar surface tension, such that despite varied size, alveoli remain open and end-expiratory atelectasis is reduced. Surfactant protein C (*SftpC*) is a commonly used type II cell marker. The four surfactant proteins, SP-A, B, C, and D play critical roles: SP-A and SP-D participate in airway host defense; SP-B and SP-C contribute to surfactant's surface tension reduction (Whitsett *et al.*, 2002).

Macrophages, although a small percentage of alveolar cells, are a major sentinel of host defense and derived primarily from blood monocytes; once in the lung, their turnover is extremely slow.

5.6. Stem and progenitor cells in the postnatal respiratory system

Stem and progenitor cells presumably help repair damaged lung, but identification of such cells remains problematic. The large surface area, numerous branches, and folded topography suggest lung harbors several stem or progenitor cell types. In trachea and bronchi, certain BCs and mucous-gland duct cells are believed to be stem/progenitor cells. Clara-like cells and type II pneumocytes are also thought to function as stem/progenitor cells in bronchioles and alveoli, respectively. Another population of stem/progenitor cells apparently lies at the bronchoalveolar duct junction (Kim *et al.*, 2005). They have been proposed to function as bipotential precursors of both the SP-C- and CC10-expressing cell lineages.

It has recently been reported that bone marrow-derived mesenchymal stem cells can differentiate into airway epithelial cells and alveolar type I pneumocytes, particularly post-injury. By contrast, *in vitro* cell culture indicates Syrian hamster fetal lung epithelial M3E3/C3 cells differentiate into Clara cells and type II pneumocytes under different culture conditions. Whether CCSP-expressing cells with pre-Clara cell phenotypes are stem cells for the entire respiratory tract remains to be determined. In addition, the concept of a pluripotent stem cell for the whole lung needs to be further investigated due to the great differences between identified stem cell or progenitor candidates in proximal bronchi and distal alveoli. Recently, we discovered lung contains cells with stem or progenitor characteristics that can be FACS sorted from adult rat and mouse lung and which are relatively apoptosis-resistant and perhaps responsible for post-injury alveolar repopulation. Another such population sorted as "side cells" on FACS may repopulate several tissues including bone marrow.

ASM derives from at least two progenitor populations: one comes from the lung periphery and arises from *Fgf10*-expressing cells in subepithelial mesenchyme. By virtue of FGF10 expression, these cells first help mediate epithelial branching; however, as the airway elongates into peripheral mesenchyme, these progenitors remain to lie along the more proximal stalk of the distal bud. Here they differentiate into ASM, most probably under paracrine induction of BMP4 and Sonic hedgehog from the adjacent epithelium (DeLanghe *et al.*, 2006). The second ASM progenitor population arises around the proximal large airways (Shan *et al.*, 2008) and appears to meet with the distally-derived counterparts beyond major lobar and segmental branches. It is speculated that whilst size of such ASM progenitor populations are determined during embryonic airway branching, they may nevertheless determine susceptibility to later BPD and asthma. Moreover, maternal smoking may dysregulate ASM progenitors and their progeny via the cholinergic-agonist, nicotine.

5.7. Potential strategies to protect lung progenitors

Both FGF7 and inosine treatment ameliorate DNA damage in AECs, as well as enhancing mitochondrial protection and the ability of AEC to migrate and repair in an *in vitro* scratch assay (Buckley *et al.*, 1997). FGF7 has also been evaluated by others *in vivo* as a treatment to enhance resistance to alveolar injury in animal models (Plantier *et al.*, 2007; Ray *et al.*, 2003). Also, FGF10 has a protective effect against lung injury and fibrosis (Gupte *et al.*, 2009). We have also shown that inosine has protective properties against oxygen injury, including glutathione repletion, mitochondrial protection, decreased apoptosis, and increased VEGF expression (Buckley *et al.*, 2005). Thus, it appears that protection or enhancement of alveolar progenitor cell function may be a viable therapeutic option that could possibly be evaluated in clinical trials of lung progenitor cell protection using small molecules such as inosine or FGF7 or FGF10.

6. Postnatal and Adult Lung

6.1. The transition to air breathing

Maturation of the surfactant system is one of two key steps to prepare fetal lung for air breathing. During the last 8 weeks of human gestation, fetal lung glycogen is converted into surfactant phospolipids, the most important of which is disaturated phosphatidylcholine (DSPC). This maturation is under the control of, and can be stimulated by, corticosteroids since it is blocked in mice with null mutations of glucocorticoid receptors or corticotrophin-releasing hormone. Human mutations have been found, such as surfactant protein B, that adversely affect stability of surfactant and hence the ability to maintain lung inflation.

The transition to air breathing occurs rapidly in mature neonatal lung. Immediately following severance of the umbilical circulation, a spike in catecholamine levels switches off chloride secretion and stimulates sodium/potassium ATPase (Brown *et al.*, 1983; Olver and Strang, 1974; Olver *et al.*, 1986). This replaces tracheal fluid production with its rapid absorption into lung interestiitum (and thence to lymphatic and capillary circulations). Null mutation of Na/K ATPase in mice leads to failure to absorb fetal lung

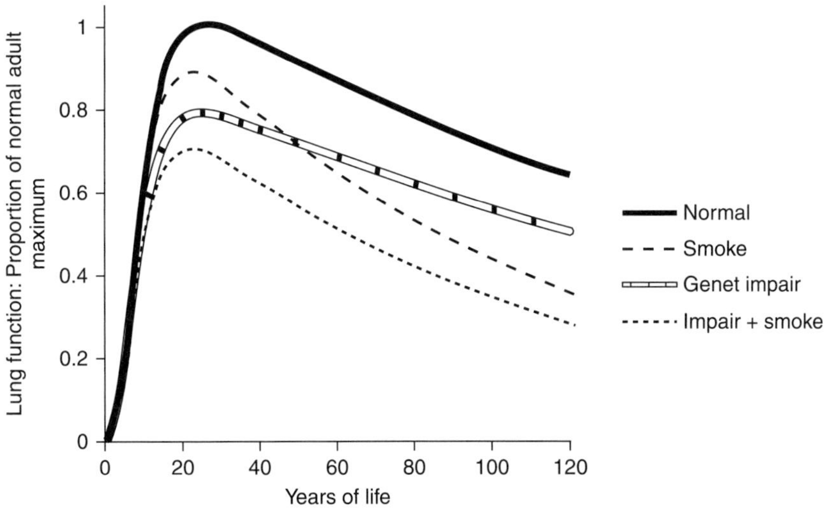

Figure 3.10 Smoking and genetics synergize to degrade lung function with age (modified after Fletcher and Peto, 1977; Shi, W. and Warburton, D. 2010). Wild-type lungs "grown in room air" achieve greatest capacity and remain healthy despite age-related degradation. Smoking exacerbates degradation of lung function even in the healthy. Genetic alterations decrease the potential to develop maximal lung capacity compared to wild type and smoking exacerbates lung degradation still further. A 0.5% loss of lung function per year is assumed for normal lungs and double that for smokers. Lungs with a genetic defect were assumed to have 95% of the growth rate of a normal lung. Growth rate is considered as a process of doublings, but the rate of doublings declines from maturity exponentially with age. This is expressed in a simple differential equation for modeling lung function $L(t)$: rate of change of lung function = rate of increase of lung function – rate of decline of lung function; $dL/dt = r \star \exp(-\mathrm{mat} \star t) \star L - a \star L$, where mat is the maturation coefficient, r is the doubling coefficient, and a is the loss-of-function coefficient.

liquid, which causes significant respiratory distress and even neonatal lethality (Hummler *et al.*, 1996). In humans delayed lung liquid absorption manifests as transient tachypnea of the newborn.

6.2. Lung aging and involution

From middle age in normal humans, an inexorable decline in lung function supervenes (illustrated by FEV1). By 120 years, FEV1 resembles that end-stage COPD in a younger person; hence, degenerating lung function appears currently to limit human life expectancy (Fletcher and Peto, 1977; Shi, W. and Warburton, D. 2010). Whilst some genetic mutations and/or environmental exposures fundamentally disrupt lung development and result in pre- or perinatal death, less critical leions may only be manifest as lung disease in infancy, childhood, or beyond. For example, minor genetic changes such as DNA polymorphisms may have very subtle impacts on lung organogenesis with apparently normal neonatal phenotype. Nevertheless, such lungs may have abnormal responses to subsequent environmental injury (e.g., cigarette smoke or vehicular pollution) that degrade lung anatomy and physiology faster than normal and predispose to, for example, COPD (Figure 3.10). Therefore, by understanding, protecting, and re-entraining developmental processes, amelioration or reversal of lung degeneration may permit enhanced duration and quality of life.

7. Conclusions

Appreciating that distal lung mesenchyme could trigger epithelial airway development has stimulated the search for controls of lung development. Given the mortality and morbidity of lung disease at all stages of life, lung regeneration is a global therapeutic priority. To achieve such goals, clinicians and scientists need to decipher how the lung is formed. Whilst this understanding began with histological analyses, advances in biology have allowed the "molecular embryology" of the lung to be elucidated. In parallel with this progress, lessons from human lung maldevelopment illustrate the importance of mechanical forces to normal lung growth. Such forces encompass both extrinsic factors (thoracic size, FBMs) and intrinsic ones (lung fluid, airway peristalsis, endogenous airway occlusions). Attempting to weave these diverse influences to facilitate regenerative lung growth appears a daunting task. Nevertheless, there are reasons for optimism: first, following Alan Turing's insight, complex (lung) morphogenesis may arise via simple iterative biochemical signaling; secondly, Benoit Mandelbrot illustrated that simple mathematics can be applied to generate apparently complex form; thirdly, D'Arcy Thompson made clear that the set of genetically possible forms are vastly constrained by fundamental physical constraints; fourth, despite huge uncertainties about the regulation of lung development, regenerative medicine has already allowed transplantation of autologous tissue-engineered airway to aid patients. Hence, despite the structural complexity of the lung, its organogenesis is governed by simpler routines more readily susceptible to discovery and therapeutic exploitation. In pursuing the latter, we may similarly be reassured that physical constraints limit the possible structures we may engineer. Finally, despite all that we do not know, clinically important aspects of pulmonary regeneration can already be achieved. The challenge for the future will be the generation of more complex and vascularized structures that can ultimately support and/or replace impaired lung function.

ACKNOWLEDGMENTS

We apologize to those colleagues whose important work in this field we have failed to cite.

Funding sources: National Heart, Lung and Blood Institute, National Institutes of Health, USA, National Science Foundation, USA, California Institute for Regenerative Medicine, Medical Research Council UK, Biotechnology and Biological Sciences Research Council, UK, Foreign and Commonwealth Office UK/USA stem cell collaboration grant, American Heart Association, American Lung Association, and Pasadena Guild of Childrens Hospital Los Angeles.

Editorial assistance: Zoe Ly and Theresa Webster.

Facilitation of US/UK collaborations on this review: UK Science & Innovation Network, British Consulate-General Los Angeles.

REFERENCES

Acosta, J. M., Thebaud, B., Castillo, C., Mailleux, A., Tefft, D., Wuenschell, C., Anderson, K. D., Bourbon, J., Thiery, J. P., Bellusci, S., and Warburton, D. (2001). Novel mechanisms in murine nitrofen-induced pulmonary hypoplasia: FGF-10 rescue in culture. *Am. J. Physiol. Lung Cell Mol. Physiol.* **281**, L250–257.

Adamson, T. M., Boyd, R. D., Platt, H. S., and Strang, L. B. (1969). Composition of alveolar liquid in the foetal lamb. *J. Physiol.* **204**, 159–168.

Akiyama, H., Lyons, J. P., Mori-Akiyama, Y., Yang, X., Zhang, R., Zhang, Z., Deng, J. M., Taketo, M. M., Nakamura, T., Behringer, R. R., McCrea, P. D., and de Crombrugghe, B. (2004). Interactions between sox9 and beta-catenin control chondrocyte differentiation. *Genes Dev.* **18**, 1072–1087.

Alcorn, D., Adamson, T. M., Lambert, T. F., Maloney, J. E., Ritchie, B. C., and Robinson, P. M. (1977). Morphological effects of chronic tracheal ligation and drainage in the fetal lamb lung. *J. Anat.* **123**, 649–660.

Ambros, V. (2001). microRNAs: tiny regulators with great potential. *Cell* **107**, 823–826.

Anselmo, M. A., Dalvin, S., Prodhan, P., Komatsuzaki, K., Aidlen, J. T., Schnitzer, J. J., Wu, J. Y., and Bernard, K. T. (2003). Slit and robo: Expression patterns in lung development. *Gene Expr. Patterns* **3**, 13–19.

Aubin, J., Lemieux, M., Tremblay, M., Berard, J., and Jeannotte, L. (1997). Early postnatal lethality in hoxa-5 mutant mice is attributable to respiratory tract defects. *Dev. Biol.* **192**, 432–445.

Austin, J., and Ali, T. (2003). Tracheomalacia and bronchomalacia in children: Pathophysiology, assessment, treatment and anaesthesia management. *Pediatr. Anaesth.* **13**, 3–11.

Auwerx, J., Demedts, M., Bouillon, R., and Desmet, J. (1985a). Coexistence of hypocalciuric hypercalcaemia and interstitial lung disease in a family: A cross-sectional study. *Eur. J. Clin. Invest.* **15**(**1**), 6–14.

Auwerx, J., Boogaerts, M., Ceuppens, J. L., and Demedts, M. (1985b). Defective host defence mechanisms in a family with hypocalciuric hypercalcaemia and coexisting interstitial lung disease. *Clin. Exp. Immunol.* **62**, 57–64.

Awonusonu, F., Srinivasan, S., Strange, J., Al Jumaily, W., and Bruce, M. C. (1999). Developmental shift in the relative percentages of lung fibroblast subsets: Role of apoptosis postseptation. *Am. J. Physiol.* **277**, L848–L859.

Balasubramaniam, V., Mervis, C. F., Maxey, A. M., Markham, N. E., and Abman, S. H. (2007). Hyperoxia reduces bone marrow, circulating, and lung endothelial progenitor cells in the developing lung: Implications for the pathogenesis of bronchopulmonary dysplasia. *Am. J. Physiol. Lung Cell Mol. Physiol.* **292**, L1073–L1084.

Bartram, U., Molin, D. G., Wisse, L. J., Mohamad, A., Sanford, L. P., Doetschman, T., Speer, C. P., Poelmann, R. E., and Gittenberger-de Groot, A. C. (2001). Double-outlet right ventricle and overriding tricuspid valve reflect disturbances of looping, myocardialization, endocardial cushion differentiation, and apoptosis in TGF-beta(2)-knockout mice. *Circulation* **103**, 2745–2752.

Batchelor, D. C., Hutchins, A. M., Klempt, M., and Skinner, S. J. (1995). Developmental changes in the expression patterns of IGFs, type 1 IGF receptor and IGF-binding proteins-2 and -4 in perinatal rat lung. *J. Muscoskel. Pain* **15**, 105–115.

Belknap, J. K., Weiser-Evans, M. C., Grieshaber, S. S., Majack, R. A., and Stenmark, K. R. (1999). Relationship between perlecan and tropoelastin gene expression and cell replication in the developing rat pulmonary vasculature. *Am. J. Respir. Cell Mol. Biol.* **20**, 24–34.

Bellusci, S., Furuta, Y., Rush, M. G., Henderson, R., Winnier, G., and Hogan, B. L. (1997a). Involvement of Sonic hedgehog (Shh) in mouse embryonic lung growth and morphogenesis. *Development* **124**, 53–63.

Bellusci, S., Grindley, J., Emoto, H., Itoh, N., and Hogan, B. L. (1997b). Fibroblast growth factor 10 (FGF10) and branching morphogenesis in the embryonic mouse lung. *Development* **124**, 4867–4878.

Bellusci, S., Henderson, R., Winnier, G., Oikawa, T., and Hogan, B. L. (1996). Evidence from normal expression and targeted misexpression that bone morphogenetic protein (bmp-4) plays a role in mouse embryonic lung morphogenesis. *Development* **122**, 1693–1702.

Berdon, W. E. (2000). Rings, slings, and other things: Vascular compression of the infant trachea updated from the midcentury to the millennium—the legacy of Robert E. Gross, M.D., and Edward B. D. Neuhauser, M.D. *Radiology* **216**, 624–632.

Bernier, S. M., Utani, A., Sugiyama, S., Doi, T., Polistina, C., and Yamada, Y. (1995). Cloning and expression of laminin alpha 2 chain (M-chain) in the mouse. *Matrix Biol.* **14**, 447–455.

Bingle, C. D., Hackett, B. P., Moxley, M., Longmore, W., and Gitlin, J. D. (1995). Role of hepatocyte nuclear factor-3 alpha and hepatocyte nuclear factor-3 beta in Clara cell secretory protein gene expression in the bronchiolar epithelium. *Biochem. J.* **308**(**Pt 1**), 197–202.

Blaisdell, C. J., Gail, D. B., and Nabel, E. G. (2009). National heart, lung, and blood institute perspective: Lung progenitor and stem cells–gaps in knowledge and future opportunities. *Stem Cells* **27**(**9**), 2263–2270.

Blaisdell, C. J., Morales, M. M., Andrade, A. C., Bamford, P., Wasicko, M., and Welling, P. (2004). Inhibition of CLC-2 chloride channel expression interrupts expansion of fetal lung cysts. *Am. J. Physiol. Lung Cell Mol. Physiol.* **286**(**2**), L420–L426.

Bland, R. D., and Boyd, C. A. (1986). Cation transport in lung epithelial cells derived from fetal, newborn, and adult rabbits. *J. Appl. Physiol.* **61**(**2**), 507–515.

Boggaram, V. (2003). Regulation of lung surfactant protein gene expression. *Front. Biosci.* **8**, d751–d764.

Bogue, C. W., Lou, L. J., Vasavada, H., Wilson, C. M., and Jacobs, H. C. (1996). Expression of Hoxb genes in the developing mouse foregut and lung. *Am. J. Respir. Cell Mol. Biol.* **15**, 163–171.

Bohinski, R. J., Di Lauro, R., and Whitsett, J. A. (1994). The lung-specific surfactant protein B gene promoter is a target for thyroid transcription factor 1 and hepatocyte nuclear factor 3, indicating common factors for organ-specific gene expression along the foregut axis. *Mol. Cell. Biol.* **14**, 5671–5681.

Bonniaud, P., Margetts, P. J., Ask, K., Flanders, K., Gauldie, J., and Kolb, M. (2005). TGF-beta and Smad3 signaling link inflammation to chronic fibrogenesis. *J. Immunol.* **175**, 5390–5395.

Boogaard, R., Huijsmans, S. H., Pijnenburg, M. W., Tiddens, H. A., de Jongste, J. C., and Merkus, P. J. (2005). Tracheomalacia and bronchomalacia in children: Incidence and patient characteristics. *Chest* **128**, 3391–3397.

Borthwick, D. W., Shahbazian, M., Krantz, Q. T., Dorin, J. R., and Randell, S. H. (2001). Evidence for stem cell niches in the tracheal epithelium. *Am. J. Respir. Cell Mol. Biol.* **24**, 662–670.

Bostrom, H., Gritli-Linde, A., and Betsholtz, C. (2002). PDGF-A/PDGF alpha-receptor signaling is required for lung growth and the formation of alveoli but not for early lung branching morphogenesis. *Dev. Dyn.* **223**, 155–162.

Bostrom, H., Willetts, K., Pekny, M., Leveen, P., Lindahl, P., Hedstrand, H., Pekna, M., Hellstrom, M., Gebre-Medhin, S., Schalling, M., Nilsson, M., Kurland, S., Tornell, J., Heath, J. K., and Betsholtz, C. (1996). PDGF-A signaling is a critical event in lung alveolar myofibroblast development and alveogenesis. *Cell* **85**, 863–873.

Bragg, A. D., Moses, H. L., and Serra, R. (2001). Signaling to the epithelium is not sufficient to mediate all of the effects of transforming growth factor beta and bone morphogenetic protein 4 on murine embryonic lung development. *Mech. Dev.* **109**, 13–26.

Brody, S. L., Yan, X. H., Wuerffel, M. K., Song, S. K., and Shapiro, S. D. (2000). Ciliogenesis and left-right axis defects in forkhead factor HFH-4-null mice. *Am. J. Respir. Cell Mol. Biol.* **23**, 45–51.

Brown, E. M., Gamba, G., Riccardi, D., Lombardi, M., Butters, R., Kifor, O., Sun, A., Hediger, M. A., Lytton, J., and Hebert, S. C. (1993). Cloning and characterization of an extracellular Ca(2+)-sensing receptor from bovine parathyroid. *Nature* **366**(**6455**), 575–580.

Brown, L., Erdmann, E., and Thomas, R. (1983). Digitalis structure-activit relationship analyses. Conclusions from indirect binding studies with cardiac (Na+ +K+)-ATPase. *Biochem. Pharmacol.* **32**, 2767–2774.

Bruno, M. D., Bohinski, R. J., Huelsman, K. M., Whitsett, J. A., and Korfhagen, T. R. (1995). Lung cell-specific expression of the murine surfactant protein A (SP-A) gene is mediated by interactions between the SP-A promoter and thyroid transcription factor-1. *J. Biol. Chem.* **270**, 6531–6536.

Buch, S., Jassal, D., Cannigia, I., Edelson, J., Han, R., Liu, J., Tanswell, K., and Post, M. (1994). Ontogeny and regulation of platelet-derived growth factor gene expression in distal fetal rat lung epithelial cells. *Am. J. Respir. Cell Mol. Biol.* **11**, 251–261.

Buckley, S., Barsky, L., Weinberg, K., and Warburton, D. (2005). In vivo inosine protects alveolar epithelial type 2 cells against hyperoxia-induced DNA damage through MAP kinase signaling. *Am. J. Physiol. Lung Cell Mol. Physiol.* **288**, L569–L575.

Buckley, S., Driscol, B., Anderson, K. D., and Warburton, D. (1998). Apoptosis and DNA damage in AEC2 cultured from hyperoxic rats. *Am. J. Physio.* **274(5 Pt 1)**, L714–L720.

Buckley, S., Driscoll, B., Anderson, K. D., and Warburton, D. (1997). Cell cycle in alveolar epithelial type II cells: Integration of matrigel and KGF. *Am. J. Physiol.* **273**, L572–L580.

Burgeson, R. E., Chiquet, M., Deutzmann, R., Ekblom, P., Engel, J., Kleinman, H., Martin, G. R., Meneguzzi, G., Paulsson, M., Sanes, J., Timpl, R., Tryggvason, K., Yamada, Y., Yurchenco, P. D. (1994). A new nomenclature for the laminins. *Matrix Biol.* **14**, 209–211.

Butt, S. J., Sousa, V. H., Fuccillo, M. V., Hjerling-Leffler, J., Miyoshi, G., Kimura, S., and Fishell, G. (2008). The requirement of nkx2-1 in the temporal specification of cortical interneuron subtypes. *Neuron* **59**, 722–732.

Carden, K. A., Boiselle, P. M., Waltz, D. A., and Ernst, A. (2005). Tracheomalacia and tracheobronchomalacia in children and adults: An in-depth review. *Chest* **127**, 984–1005.

Cardoso, W. V., and Lu, J. (2006). Regulation of early lung morphogenesis: Auestions, facts and controversies. *Development* **133(9)**, 1611–1624.

Cardoso, W. V., Itoh, A., Nogawa, H., Mason, I., and Brody, J. S. (1997). FGF-1 and FGF-7 induce distinct patterns of growth and differentiation in embryonic lung epithelium. *Dev. Dyn.* **208**, 398–405.

Cardoso, W. V., Williams, M. C., Mitsialis, S. A., Joyce-Brady, M., Rishi, A. K., and Brody, J. S. (1995). Retinoic acid induces changes in the pattern of airway branching and alters epithelial cell differentiation in the developing lung in vitro. *Am. J. Respir. Cell Mol. Biol.* **12**, 464–476.

Carraro, G., El-Hashash, A., Guidolin, D., Tiozzo, C., Turcatel, G., Young, B. M., De Langhe, S. P., Bellusci, S., Shi, W., Parnigotto, P. P., and Warburton, D. (2009). MiR-17 family of microRNAs controls FGF10-mediated embryonic lung epithelial branching morphogenesis through MAPK14 and STAT3 regulation of E-cadherin distribution. *Dev. Biol.* **333(2)**, 238–250.

Carre, A., Szinnai, G., Castanet, M., Sura-Trueba, S., Tron, E., Broutin-L'Hermite, I., Bara, P., Goizet, C., Lacombe, D., Moutard, M. L., Raybaud, C., Raynaud-Ravni, C., Romana, S., Ythier, H., Leger, J., and Polka, M., (2009). Five new TTF1/NKX2.1 mutations in brain-lung-thyroid syndrome: rescue by PAX8 synergism in one case. *Hum Mol Genet.* **18**, 2266–2276.

Chambon, P. (1996). A decade of molecular biology of retinoic acid receptors. *FASEB J.* **10**, 940–954.

Chazaud, C., Dolle, P., Rossant, J., and Mollard, R. (2003). Retinoic acid signaling regulates murine bronchial tubule formation. *Mech. Dev.* **120**, 691–700.

Chen, F., Desai, T. J., Qian, J., Niederreither, K., Lu, J., and Cardoso, W. V. (2007). Inhibition of TGF b signaling by endogenous retinoic acid is essential for primary lung bud induction. *Development* **134**, 2969–2979.

Chen, H., Sun, J., Buckley, S., Chen, C., Warburton, D., Wang, X. F., and Shi, W. (2005). Abnormal mouse lung alveolarization caused by Smad3 deficiency is a developmental antecedent of centrilobular emphysema. *Am. J. Physiol. Lung Cell Mol. Physiol.* **288**, L683–L691.

Chen, J. Lecuona, E., Briva, A., Welch, L. C., and Sznajder, J. I. (2008). Carbonic anhydrase II and alveolar fluid reabsorption during hypercapnia. (2008). *Am. J. Repir. Cell Mol. Biol.* **38**, 32–37.

Chen, J., Knowles, H. J., Hebert, J. L., and Hackett, B. P. (1998). Mutation of the mouse hepatocyte nuclear factor/forkhead homologue 4 gene results in an absence of cilia and random left-right asymmetry. *J. Clin. Invest.* **102**, 1077–1082.

Chen, M., Przyboroski, M., and Berthaume, F. (2009). Stem cells for skin tissue engineering and would healing. *Crit. Rev. Biomed. Eng.* **37**, 399–421.

Chuang, P. T., Kawcak, T., and McMahon, A. P. (2003). Feedback control of mammalian hedgehog signaling by the hedgehog-binding protein, hip1, modulates fgf signaling during branching morphogenesis of the lung. *Genes Dev.* **17**, 342–347.

Chuang, P. T., and McMahon, A. P. (1999). Vertebrate hedgehog signalling modulated by induction of a hedgehog-binding protein. *Nature* **397**, 617–621.

Chytil, F. (1996). Retinoids in lung development. *FASEB J.* **10**, 986–992.

Clark, J., Alvarez, D. F., Alexeyev, M., King, J. A., Huang, L., Yoder, M. C., and Stevens, T. (2008). Regulatory role for nucleosomeassembly protein-1 in the proliferative and vasculogenic phenotypeof pulmonary endothelium. *Am. J. Physiol. Lung Cell Mol. Physiol.* **294**, L431–L439.

Cohen, E. D., Ihida-Stansbury, K., Lu, M. M., Panettieri, R. A., Jones, P. L., and Morrisey, E. E. (2009). Wnt signaling regulates smooth muscle precursor development in the mouse lung via a tenascin C/PDGFR pathway. *J. Clin. Invest.* **119**(**9**), 2538–2549. doi: 38079 [pii] 10.1172/JCI38079

Cohen, J. C., Larson, J. E., Killeen, E., Love, D., and Takemaru, K. (2008). CFTR and Wnt/beta-catenin signaling in lung development. *BMC Dev. Biol.* **6**(**8**), 70.

Colarossi, C., Chen, Y., Obata, H., Jurukovski, V., Fontana, L., Dabovic, B., and Rifkin, D. B. (2005). Lung alveolar septation defects in Ltbp-3-null mice. *Am. J. Pathol.* **167**, 419–428.

Colvin, J. S., White, A. C., Pratt, S. J., and Ornitz, D. M. (2001). Lung hypoplasia and neonatal death in fgf9-null mice identify this gene as an essential regulator of lung mesenchyme. *Development* **128**, 2095–2106.

Colvin, J. S., Feldman, B., Nadeau, J. H., Goldfarb, M., and Ornitz, D. M. (1999). Genomic organization and embryonic expression of the mouse fibroblast growth factor 9 gene. *Dev. Dyn.* **216**, 72–88.

Compernolle, V., Brusselmans, K., Acker, T., Hoet, P., Tjwa, M., Beck, H., Plaisance, S., Dor, Y., Keshet, E., Lupu, F., Nemery, B., Dewerchin, M., Van Veldhoven, P., Plate, K., Moons, L., Collen, D., and Carmeliet, P. (2002). Loss of HIF-2alpha and inhibition of VEGF impair fetal lung maturation, whereas treatment with VEGF prevents fatal respiratory distress in premature mice. *Nat. Med.* **8**, 702–710.

Coppola, D., Ferber, A., Miura, M., Sell, C., D'Ambrosio, C., Rubin, R., and Baserga, R. (1994). A functional insulin-like growth factor I receptor is required for the mitogenic and transforming activities of the epidermal growth factor receptor. *Mol. Cell. Biol.* **14**, 4588–4595.

Costa, R. H., Kalinichenko, V. V., and Lim, L. (2001). Transcription factors in mouse lung development and function. *Am. J. Physiol. Lung Cell Mol. Physiol.* **280**, L823–L838.

Costello, C. M., Howell, K., Cahill, E., McBryan, J., Konigshoff, M., Eickelberg, O., Gaine, S., Martin, F., and McLoughlin, P. (2008). Lung-selective gene responses to alveolar hypoxia: potential role for the bone morphogenetic antagonist gremlin in pulmonary hypertension. *Am. J. Physiol. Lung Cell Mol. Physiol.* **295**, L272–L284.

Dang, T. P., Eichenberger, S., Gonzalez, A., Olson, S., and Carbone, D. P. (2003). Constitutive activation of notch3 inhibits terminal epithelial differentiation in lungs of transgenic mice. *Oncogene* **22**, 1988–1997.

Daniely, Y., Liao, G., Dixon, D., Linnoila, R. I., Lori, A., Randell, S. H., Oren, M., and Jetten, A. M. (2004). Critical role of p63 in the development of a normal esophageal and tracheobronchial epithelium. *Am. J. Physiol. Cell Physiol.* **287**, C171–C181.

De Felice, M., Silberschmidt, D., DiLauro, R., Xu, Y., Wert, S. E., Weaver, T. E., Bachurski, C. J., Clark, J. C., and Whitsett, J. A. (2003). TTF-1 phosphorylation is required for peripheral lung morphogenesis, perinatal survival, and tissue-specific gene expression. *J. Biol. Chem.* **278**, 35574–35583.

De Langhe, S. P., Carraro, G., Tefft, D., Li, C., Xu, X., Chai, Y., Minoo, P., Hajihosseini, M. K., Drouin, J., Kaartinen, V., and Bellusci, S. (2008). Formation and differentiation of multiple mesenchymal lineages during lung development is regulated by beta-catenin signaling. *PLoS ONE* **3**(**1**), e1516. doi: 10.1371/journal.pone.0001516

De Langhe, S. P., Carraro, G., Warburton, D., Hajihosseini, M. K., and Bellusci, S. (2006). Levels of mesenchymal FGFR2 signaling modulate smooth muscle progenitor cell commitment in the lung. *Dev. Biol.* **299**(**1**), 52–62.

De Langhe, S. P., Sala, F. G., Del Moral, P. M., Fairbanks, T. J., Yamada, K. M., Warburton, D., Burns, R. C., and Bellusci, S. (2005). Dickkopf-1 (DKK1) reveals that fibronectin is a major target of wnt signaling in branching morphogenesis of the mouse embryonic lung. *Dev. Biol.* **277**(**2**), 316.

De Moerlooze, L., Spencer-Dene, B., Revest, J., Hajihosseini, M., Rosewell, I., and Dickson, C. (2000). An important role for the IIIb isoform of fibroblast growth factor receptor 2 (FGFR2) in mesenchymal-epithelial signalling during mouse organogenesis. *Development* **127**, 483–492.

Del Moral, P. M., Sala, F. G., Tefft, D., Shi, W., Keshet, E., Bellusci, S., and Warburton, D. (2006a). VEGF-A signaling through flk-1 is a critical facilitator of early embryonic lung epithelial to endothelial crosstalk and branching morphogenesis. *Dev. Biol.* **290**, 177–188.

Del Moral, P. M., De Langhe, S. P., Sala, F. G., Veltmaat, J. M., Tefft, D., Wang, K., Warburton, D., and Bellusci, S. (2006b). Differential role of FGF9 on epithelium and mesenchyme in mouse embryonic lung. *Dev. Biol.* **293**, 77–89.

Dean, C. H., Miller, L. A., Smith, A. N., Dufort, D., Lang, R. A., and Niswander, L. A. (2005). Canonical wnt signaling negatively regulates branching morphogenesis of the lung and lacrimal gland. *Dev. Biol.* **286**(**1**), 270–286. doi: S0012-1606(05)00517-8 [pii] 10.1016/j.ydbio.2005.07.034

Deterding, R. R., Jacoby, C. R., and Shannon, J. M. (1996). Acidic fibroblast growth factor and keratinocyte growth factor stimulate fetal rat pulmonary epithelial growth. *Am. J. Physiol.* **271**, L495–L505.

Dickman, E. D., Thaller, C., and Smith, S. M. (1997). Temporally-regulated retinoic acid depletion produces specific neural crest, ocular and nervous system defects. *Development* **124**, 3111–3121.

Diez-Pardo, J. A., Qi, B. Q., Navarro, C., and Tovar, J. A. (1996). A new rodent experimental model of esophageal atresia and tracheoesophageal fistula: Preliminary report. *J. Pediatr. Surg.* **31**(**4**), 498–502.

Driscoll, B., Buckley, S., Bui, K. C., Anderson, K. D., and Warburton, D. (2000). Telomerase in alveolar epithelial development and repair. *Am. J. Physiol. Lung Cell Mol. Physiol.* **279**, L1191–L1198.

Dunwoodie, S. L. (2009). The role of hypoxia in development of the Mammalian embryo. *Dev. Cell.* **17**, 755–773.

Durham, P. L., and Snyder, J. M. (1995). Characterization of alpha 1, beta 1, and gamma 1 laminin subunits during rabbit fetal lung development. *Dev. Dyn.* **203**, 408–421.

Dziadek, M. (1995). Role of laminin-nidogen complexes in basement membrane formation during embryonic development. *Experientia* **51**, 901–913.

Eblaghie, M. C., Reedy, M., Oliver, T., Mishina, Y., and Hogan, B. L. (2006). Evidence that autocrine signaling through bmpr1a regulates the proliferation, survival and morphogenetic behavior of distal lung epithelial cells. *Dev. Biol.* **291**, 67–82.

Ehrig, K., Leivo, I., Argraves, W. S., Ruoslahti, E., and Engvall, E. (1990). Merosin, a tissue-specific basement membrane protein, is a laminin-like protein. *Proc. Natl. Acad. Sci. USA* **87**, 3264–3268.

Ekblom, P., Ekblom, M., Fecker, L., Klein, G., Zhang, H. Y., Kadoya, Y., Chu, M. L., Mayer, U., and Timpl, R. (1994). Role of mesenchymal nidogen for epithelial morphogenesis in vitro. *Development* **120**, 2003–2014.

Ellis, T., Gambardella, L., Horcher, M., Tschanz, S., Capol, J., Bertram, P., Jochum, W., Barrandon, Y., and Busslinger, M. (2001). The transcriptional repressor CDP (cutl1) is essential for epithelial cell differentiation of the lung and the hair follicle. *Genes Dev.* **15**, 2307–2319.

Elluru, R. G., Thompson, F., and Reece, A. (2009). Fibroblast growth factor 18 gives growth and directional cues to airway cartilage. *Laryngoscope* **119**, 1153–1165.

Engler, A. J., Sen, S., Sweeney, H. L., and Discher, D. E. (2006). Matrix elasticity directs stem cell lineage specification. *Cell* **126**, 677–689.

Engelhardt, J. F. (2001). Stem cell niches in the mouse airway. *Am. J. Respir. Cell Mol. Biol.* **24**, 649–652.

Esposito, G., Fogolari, F., Damante, G., Formisano, S., Tell, G., Leonardi, A., Di Lauro, R., and Viglino, P. (1996). Analysis of the solution structure of the homeodomain of rat thyroid transcription factor 1 by 1h-NMR spectroscopy and restrained molecular mechanics. *Eur. J. Biochem.* **241**, 101–113.

Evans, C. M., Williams, O. W., Tuvim, M. J., Ngam, R., Mixides, G. P., Blackburn, M. R., DeMayo, F. J., Burns, A. R., Smith, C., Reynolds, S. D., Stripp, B. R., and Dickey, B. F. (2004). Mucin is produced by Clara cells in the proximal airways of antigen-challenged mice. *Am. J. Respir. Cell Mol. Biol.* **31**, 382–394.

Evans, M. J., Cabral, L. J., Stephens, R. J., and Freeman, G. (1975). Transformation of alveolar type 2 cells to type 1 cells following exposure to NO_2. *Exp. Mol. Pathol.* **22**, 142–150.

Featherstone, N. C., Connell, M. G., Fernig, D. G., Wray, S., Burdyga, T. V., Losty, P. D. and Jesudason, E. C. (2006). Airway smooth muscle dysfunction precedes teratogenic congenital diaphragmatic hernia and May contribute to hypoplastic lung morphogenesis. *Am. J. Respir. Cell Mol. Biol.* **35**(**5**), 571–578.

Featherstone, N. C., Jesudason, E. C., Connell, M. G., Fernig, D. G., Wray, S., Losty, P. D. and Burdyga, T. V. (2005). Spontaneous propagating calcium waves underpin airway peristalsis in embryonic rat lung. *Am. J. Respir. Cell Mol. Biol.* **33**(**2**), 153–160.

Felix, J. F., de Jong, E. M., Torfs, C. P., de Klein, A., Rottier, R. J., and Tibboel, D. (2009). Genetic and environemental factors in the etiology of esophageal atresia and/or tracheoesophageal fistula: an overview of the current concepts. Birth Defects Re. *A Clin. Mol. Teratol.* **85**, 747–754.

Fewell, J. E., and Johnson, P., (1983). Upper airway dynamics during breathing and during apnoea in fetal lambs. *J Physiol.* **339**, 495–505.

Fewell, J. E., Hislop, A. A., Kitterman, J. A., and Johnson, P. (2003). Effect of tracheostomy on lung development in fetal lambs. *J. Appl. Physiol.* **55**, 1103–1108.

Finney, B. A., del Moral, P. M., Wilkinson, W. J., Cayzac, S., Cole, M., Warburton, D., Kemp, P. J., and Riccardi, D. (2008). Regulation of mouse lung development by the extracellular calcium-sensing receptor, CaR. *J. Physiol.* **15**(**586 Pt 24**), 6007–6019.

Fischer, A., Viebahn, C., and Blum, M. (2002). FGF8 acts as a right determinant during establishment of the left-right axis in the rabbit. *Curr. Biol.* **12**, 1807–1816.

Fletcher, C., and Peto, R. (1977). The natural history of chronic airflow obstruction. *Br. Med. J.* **1**, 1645–1648.

Forgacs, G., Foty, R. A., Shafrir, Y., and Steinberg, M. S. (1998). Viscoelastic properties of living embryonic tissues: A quantitative study. *Biophys. J.* **74**(**5**), 2227–2234.

Foty, R. A., Forgacs, G., Pfleger, C. M., and Steinberg, M. S. (1994). Liquid properties of embryonic tissues: Measurement of interfacial tensions. *Phys. Rev. Lett.* **72**, 2298–2301.

Frank, L. (2003). Protective effect of keratinocyte growth factor against lung abnormalities associated with hyperoxia in prematurely born rats. *Biol. Neonate* **83**, 263–272.

Frieser, M., Nockel, H., Pausch, F., Roder, C., Hahn, A., Deutzmann, R., and Sorokin, L. M. (1997). Cloning of the mouse laminin alpha 4 cDNA. Expression in a subset of endothelium. *Eur. J. Biochem.* **246**, 727–735.

Galliano, M. F., Aberdam, D., Aguzzi, A., Ortonne, J. P., and Meneguzzi, G. (1995). Cloning and complete primary structure of the mouse laminin alpha 3 chain. Distinct expression pattern of the laminin alpha 3a and alpha 3b chain isoforms. *J. Biol. Chem.* **270**, 21820–21826.

Gao, Y., and Raj, J. U. (2005). Parathyroid hormone-related protein-mediated responses in pulmonary arteries and veins of newborn lambs. *Am. J. Physiol. Lung Cell Mol. Physiol.* **289**(**1**), L60–L66.

Gauldie, J. Galt, T. l., Bonniaud, P., Robbins, C., Kelly, M., and Warburton, D. (2003). Transfer of the active form of tranforming growth factor-beta 1 gene to newborn rat lung induces changes consistent with bronchopulmonary dysplasia. *Am. J. Pathol.* **163**, 2575–2584.

Gerber, H. P., Hillan, K. J., Ryan, A. M., Kowalski, J., Keller, G. A., Rangell, L., Wright, B. D., Radtke, F., Aguet, M., and Ferrara, N. (1999). VEGF is required for growth and survival in neonatal mice. *Development* **126**, 1149–1159.

Giangreco, A., Arwert, E. N., Rosewell, I. R., Snyder, J., Watt, F. M., and Stripp, B. R. (2009). Stem cells are dispensable for lung homeostasis but restore airways after injury. *Proc. Natl. Acad Sci. USA.* **106**, 9286–9291.

Giangreco, A., Shen, H., Reynolds, S. D., and Stripp, B. R. (2004). Molecular phenotype of airway side population cells. *Am. J. Physiol. Lung Cell Mol. Physiol.* **286**, L624–L630.

Giangreco, A., Reynolds, S. D., and Strip, B. R. (2002). Terminal bronchiales harbor a unique airway stem cell population that localized to the bronchoalveolar duct junction. *Am. J. Pathol.* **161**, 173–182.

Gill, S. E., Pape, M. C., Khokha, R., Watson, A. J., and Leco, K. J. (2003). A null mutation for tissue inhibitor of metalloproteinases-3 (timp-3) impairs murine bronchiole branching morphogenesis. *Dev. Biol.* **261**, 313–323.

Gillie, D. J., Pace, A. J., Coakley, R. J., Koller, B. H., and Barker, P. M. (2001). Liquid and ion transport by fetal airway and lung epithelia of mice deficient in sodium-potassium-2-chloride transporter. *Am. J. Respir. Cell Mol. Biol.* **25**(**1**), 14–20.

Glazer, L., and Shilo, B. Z. (1991). The drosophila FGF-R homolog is expressed in the embryonic tracheal system and appears to be required for directed tracheal cell extension. *Genes Dev.* **5**, 697–705.

Goodrich, L. V., Johnson, R. L., Milenkovic, L., McMahon, J. A., and Scott, M. P. (1996). Conservation of the hedgehog/patched signaling pathway from flies to mice: Induction of a mouse patched gene by hedgehog. *Genes Dev.* **10**, 301–312.

Goss, A. M., Tian, Y., Tsukiyama, T., Cohen, E. D., Zhou, D., Lu, M. M., Yamaguchi, T. P., and Morrisey, E. E. (2009). Wnt2/2b and beta-catenin signaling are necessary and sufficient to specify lung progenitors in the foregut. *Dev. Cell* **17**(**2**), 290–298.

Greenberg, J. M., Thompson, F. Y., Brooks, S. K., Shannon, J. M., McCormick-Shannon, K., Cameron, J. E., Mallory, B. P., and Akeson, A. L. (2002). Mesenchymal expression of vascular endothelial growth factors D and A defines vascular pattering in developing lung. *Dev. Dyn.* **224**, 144–153.

Grindley, J. C., Bellusci, S., Perkins, D., and Hogan, B. L. (1997). Evidence for the involvement of the Gli gene family in embryonic mouse lung development. *Dev. Biol.* **188**, 337–348.

Gross, I., Smith, G. J., Maniscalco, W. M., Czajka, M. R., Wilson, C. M., and Rooney, S. A. (1978). An organ culture model for study of biochemical development of fetal rat lung. *J. Appl. Physiol.* **45**(**3**), 355–362.

Guazzi, S., Price, M., De Felice, M., Damante, G., Mattei, M. G., and Di Lauro, R. (1990). Thyroid nuclear factor 1 (TTF-1) contains a homeodomain and displays a novel DNA binding specificity. *EMBO J.* **9**, 3631–3639.

Gupte, V. V., Ramasamy, S. K., Reddy, R., Lee, J., Weinreb, P. H., Violette, S. M., Guenther, A., Warburton, D., Driscoll, B., Minoo, P., and Bellusci, S. (2009). Overexpression of fibroblast growth factor-10 during both inflammatory and fibrotic phases attenuates bleomycin-induced pulmonary fibrosis in mice. *Am. J. Repir. Crit. Care Med.* **180**, 424–436.

Guo, L., Degenstein, L., and Fuchs, E. (1996). Keratinocyte growth factor is required for hair development but not for wound healing. *Genes Dev.* **10**, 165–175.

Guseh, J. S., Bores, S. A., Stanger, B. Z., Zhou, Q., Anderson, W. J., Melton, D. A., and Rajagopal, J. (2009). Notch signaling promotes airway mucous metaplasia and inhibits alveolar development. *Development* **136**(**10**), 1751–1759.

Hackett, T. L., Shaheen, F., Johnson, A., Wadsworth, S., Pechkovsky, D. V., Jacoby, D. B., Kicic, A., Stick, S. M., and Knight, D. A. (2008). Characterization of side population cells from human airway epithelium. *Stem Cells* **26**, 2576–2585.

Hackett, B. P., Brody, S. L., Liang, M., Zeitz, I. D., Bruns, L. A., and Gitlin, J. D. (1995). Primary structure of hepatocyte nuclear factor/forkhead homologue 4 and characterization of gene expression in the developing respiratory and reproductive epithelium. *Proc. Natl. Acad. Sci. USA* **92**, 4249–4253.

Hacohen, N., Kramer, S., Sutherland, D., Hiromi, Y., and Krasnow, M. A. (1998). Sprouty encodes a novel antagonist of FGF signaling that patterns apical branching of the drosophila airways. *Cell* **92**, 253–263.

Hadari, Y. R., Kouhara, H., Lax, I., and Schlessinger, J. (1998). Binding of shp2 tyrosine phosphatase to FRS2 is essential for fibroblast growth factor-induced PC12 cell differentiation. *Mol. Cell. Biol.* **18**, 3966–3973.

Han, R. N., Post, M., Tanswell, A. K., and Lye, S. J. (2003). Insulin-like growth factor-I receptor-mediated vasculogenesis/angiogenesis in human lung development. *Am. J. Respir. Cell Mol. Biol.* **28**, 159–169.

Harding, R. and Hooper, S. B. (1996). Regulation of lung expansion and lung growth before birth. *J. Appl. Physiol.* **81**(**1**), 209–224.

Harris, K. S., Zhang, Z., McManus, M. T., Harfe, B. D., and Sun, X. Dicer function is essential for lung epithelium morphogenesis. *Proc. Natl. Acad. Sci. USA.* **103**, 2208–2213.

Harris-Johnson, K. S., Domyan, E. T., Vezina, C. M., and Sun, X. (2009). Beta-catenin promotes respiratory progenitor identify in mouse foregut. *Pro. Natl. Acad. Sci. USA* **106** (**38**), 16287–16292.

Harrison, M. R., Keller, R. L., Hawgood, S. B., Kitterman, J. A., Sandberg, P. L., Farmer, D. L., Lee, H., Filly, R. A., Farrell, J. A., and Albanese, C. T. (2003). A randomized trial of fetal endoscopic tracheal occlusion for severe fetal congenital diaphragmatic hernia. *N. Engl. J. Med.* **349**(**20**), 1916–1924.

Hashimoto, S., Nakano, H., Singh, G., and Katyal, S. (2002). Expression of spred and sprouty in developing rat lung. *Mech. Dev.* **119**(**Suppl 1**), S303–S309.

Hatakeyama, Y., Tuan, R. S., and Shum, L. (2004). Distinct functions of BMP4 and GDF5 in the regulation of chondrogenesis. *J. Cell. Biochem.* **91**, 1204–1217.

He, Y., Crouch, E. C., Rust, K., Spaite, E., and Brody, S. L. (2000). Proximal promoter of the surfactant protein D gene: Regulatory roles of AP-1, forkhead box, and GT box binding proteins. *J. Biol. Chem.* **275**, 31051–31060.

Healy, A. M., Morgenthau, L., Zhu, X., Farber, H. W., and Cardoso, W. V. (2000). VEGF is deposited in the subepithelial matrix at the leading edge of branching airways and stimulates neovascularization in the murine embryonic lung. *Dev. Dyn.* **219**, 341–352.

Hedrick, M. H., Estes, J. M., Sullivan, K. M., Bealer, J. F., Kitterman, J. A., Flake, A. W., Adzick, N. S., and Harrison, M. R. (1994). Plug the lung until it grows (PLUG): A new method to treat congenital diaphragmatic hernia in utero. *J. Pediatr. Surg.* **29**(**5**), 612–617.

Hilfer, S. R. (1996). Morphogenesis of the lung: Control of embryonic and fetal branching. *Annu. Rev. Physiol.* **58**, 93–113.

Hogan, B. L. (1996). Bone morphogenetic proteins in development. *Curr. Opin. Genet. Dev.* **6**, 432–438.

Hokuto, I., Perl, A. K., and Whitsett, J. A. (2003). Prenatal, but not postnatal, inhibition of fibroblast growth factor receptor signaling causes emphysema. *J. Biol. Chem.* **278**, 415–421.

Hong, K. U., Reynolds, S. D., Giangreco, A., Hurley, C. M., and Stripp, B. R. (2001). Clara cell secretory protein- expressing cells of the airway neuroepithelial body microenvironment include a label-retaining subset and are critical for epithelial renewal after progenitor cell depletion. *Am. J. Respir. Cell Mol. Biol.* **24**, 671–681.

Hong, K. U., Reynolds, S. D., Watkins, S., Fuchs, E., and Stripp, B. R. (2004a). In vivo differentiation potential of tracheal basal cells: Evidence for multipotent and unipotent subpopulations. *Am. J. Physiol. Lung Cell Mol. Physiol.* **286**, L643–L649.

Hong, K. U., Reynolds, S. D., Watkins, S., Fuchs, E., and Stripp, B. R. (2004b). Basal cells are a multipotent progenitor capable of renewing the bronchial epithelium. *Am. J. Pathol.* **164**, 577–588.

Hummler, E., Barker, P., Gatzy, J., Beermann, F., Verdumo, C., Schmidt, A., Boucher, R., and Rossier, B. C. (1996). Early death due to defective neonatal lung liquid clearance in alpha ENaC-deficient mice. *Nat. Genet.* **12(3)**, 325–328.

Huttner, W. B., and Kosodo, Y. (2005). Symmetric versus asymmetric cell division during neurogenesis in the developing vertebrate central nervous system. *Curr. Opin. Cell Biol.* **17(6)**, 648–657.

Hyatt, B. A., Shangguan, X., and Shannon, J. M. (2002). BMP4 modulates fibroblast growth factor-mediated induction of proximal and distal lung differentiation in mouse embryonic tracheal epithelium in mesenchyme-free culture. *Dev. Dyn.* **225**, 153–165.

Iivanainen, A., Morita, T., and Tryggvason, K. (1999). Molecular cloning and tissue-specific expression of a novel murine laminin gamma3 chain. *J. Biol. Chem.* **274**, 14107–14111.

Iivanainen, A., Kortesmaa, J., Sahlberg, C., Morita, T., Bergmann, U., Thesleff, I., and Tryggvason, K. (1997). Primary structure, developmental expression, and immunolocalization of the murine laminin alpha4 chain. *J. Biol. Chem.* **272**, 27862–27868.

Iivanainen, A., Sainio, K., Sariola, H., and Tryggvason, K. (1995a). Primary structure and expression of a novel human laminin alpha 4 chain. *FEBS Lett.* **365**, 183–188.

Iivanainen, A., Vuolteenaho, R., Sainio, K., Eddy, R., Shows, T. B., Sariola, H., and Tryggvason, K. (1995b). The human laminin beta 2 chain (S-laminin): Structure, expression in fetal tissues and chromosomal assignment of the LAMB2 gene. *Matrix Biol.* **14**, 489–497.

Ikeda, K., Shaw-White, J. R., Wert, S. E., and Whitsett, J. A. (1996). Hepatocyte nuclear factor 3 activates transcription of thyroid transcription factor 1 in respiratory epithelial cells. *Mol. Cell. Biol.* **16**, 3626–3636.

Inanlou, M. R. and Kablar, B. (2003). Abnormal development of the diaphragm in mdx: MyoD-/-(9th) embryos leads to pulmonary hypoplasia. *Int. J. Dev. Biol.* **47**, 363–371.

Ingber, D. (2003). Mechanobiology and diseases of mechanotransduction. *Ann. Med.* **35(8)**, 564–577.

Irwin, D., Helm, K., Campbell, N., Imamura, M., Faggan, K., Harral, J., Carr, M., Young, K. A., Klemm, D., Gebb, S., Dempsey, E. C., West, J., and Majka, S. (2007). Neonatal lung side population cells demonstrate endothelial potential and are altered in response to hyperoxia-induced lung simplification. *Am. J. Physiol. Lung Cell Mol. Physiol.* **293**, L941–L951.

Ito, T., Udaka, N., Yazawa, T., Okudela, K., Hayashi, H., Sudo, T., Guillemot, F., Kageyama, R., and Kitamura, H. (2000). Basic helix-loop-helix transcriptionfactors regulate the neuroendocrine differentiation of fetal mouse pulmonary epithelium. *Development* **127**, 3913–3921.

Izvolsky, K. I., Shoykhet, D., Yang, Y., Yu, Q., Nugent, M. A., and Cardoso, W. V. (2003a). Heparan sulfate-FGF10 interactions during lung morphogenesis. *Dev. Biol.* **258**, 185–200.

Izvolsky, K. I., Zhong, L., Wei, L., Yu, Q., Nugent, M. A., and Cardoso, W. V. (2003b). Heparan sulfates expressed in the distal lung are required for fgf10 binding to the epithelium and for airway branching. *Am. J. Physiol. Lung Cell Mol. Physiol.* **285**, L838–L846.

Jadhav, A. P., Cho, S. H., and Cepko, C. L. (2006). Notch activity permits retinal cells to progress through multiple progenitor states and acquire a stem cell property. *Proc. Natl. Acad. Sci. USA* **103**, 18998–19003.

Jakab, K., Damon, B., Marga, F., Doaga, O., Mironov, V., Kosztin, I., Markwald, R., and Forgacs, G. (2008). Relating cell and tissue mechanics: Implications and applications. *Dev. Dyn.* **237**, 2438–2449.

Jakkula, M., Le Cras, T. D., Gebb, S., Hirth, K. P., Tuder, R. M., Voelkel, N. F., and Abman, S. H. (2000). Inhibition of angiogenesis decreases alveolarization in the developing rat lung. *Am. J. Physiol. Lung Cell Mol. Physiol.* **279**, L600–L607.

Jani, J., Gratacos, E., Greenough, A., Piero, J. L., Benachi, A., Harrison, M., Nicolaides, J., and Deprest, J. (2005). Percutaneous fetal endoscopic tracheal occlusion (FETO) for severe left-sided congenital diaphragmatic hernia. *Clin. Obstet. Gynecol.* **48**(**4**), 910–922.

Jaskoll, T. F., Don-Wheeler, G., Johnson, R., and Slavkin, H. C. (1988). Embryonic mouse lung morphogenesis and type II cytodifferentiation in serumless, chemically defined medium using prolonged in vitro cultures. *Cell Differ.* **24**, 105–117.

Jesudason, E. C. (2009). Airway smooth muscle: An architect of the lung? *Thorax* **64**(**6**), 541–545.

Jesudason, E. C. (2007). Exploiting mechanical stimuli to rescue growth of the hypoplastic lung. *Pediatr. Surg. Int.* **23**, 827–836.

Jesudason, E. C., Smith, N. P., Connell, M. G., Spiller, D. G., White, M. R., Fernig, D. G., and Losty, P. D. (2006a). Peristalsis of airway smooth muscle is developmentally regulated and uncoupled from hypoplastic lung growth. *Am. J. Physiol. Lung Cell Mol. Physiol.* **291**(**4**), L559–L565.

Jesudason, E. C. (2006b). Small lungs and suspect smooth muscle: Congenital diaphragmatic hernia and the smooth muscle hypothesis. *J. Pediatr. Surg.* **41**(**2**), 431–435.

Jesudason, E. C., Smith, N. P., Connell, M. G., Spiller, D. G., White, M. R., Fernig, D. G., and Losty, P. D. (2005). Developing rat lung has a sided pacemaker region for morphogenesis-related airway peristalsis. *Am. J. Respir. Cell Mol. Biol.* **32**(**2**), 118–127.

Jesudason, E. C., Connell, M. G., Fernig, D. G., Lloyd, D. A., and Losty, P. D. (2000). Early lung malformations in congenital diaphragmatic hernia. *J. Pediatr. Surg.* **35**(**1**), 124–127.

Kaartinen V., Voncken, J. W., Shuler, C., Warburton, D., Bu, D., Heisterkamp, N., and Groffen, J. (1995). Abnormal lung development and cleft palate in mice lacking TGF-beta 3 indicates defects of epithelial-mesenchymal interaction. *Nat. Genet.* **11**, 415–421.

Kaartinen, V., and Warburton, D. (2003). Fibrillin controls TGF-beta activation. *Nat. Genet.* **33**, 331–332.

Kalinichenko, V. V., Lim, L., Stolz, D. B., Shin, B., Rausa, F. M., Clark, J., Whitsett, J. A., Watkins, S. C., and Costa, R. H. (2001). Defects in pulmonary vasculature and perinatal lung hemorrhage in mice heterozygous null for the Forkhead Box f1 transcription factor. *Dev. Biol.* **235**, 489–506.

Kasahara, Y., Tuder, R. M., Taraseviciene-Stewart, L., Le Cras, T. D., Abman, S., Hirth, P. K., Waltenberger, J., and Voelkel, N. F. (2000). Inhibition of VEGF receptors causes lung cell apoptosis and emphysema. *J. Clin. Invest.* **106**, 1311–1319.

Kastner, P., Mark, M., Ghyselinck, N., Krezel, W., Dupe, V., Grondona, J. M., and Chambon, P. (1997). Genetic evidence that the retinoid signal is transduced by heterodimeric RXR/RAR functional units during mouse development. *Development* **124**, 313–326.

Kawahira, H., Ma, N. H., Tzanakakis, E. S., McMahon, A. P., Chuang, P. T., and Hebrok, M. (2003). Combined activities of hedgehog signaling inhibitors regulate pancreas development. *Development* **130**, 4871–4879.

Keijzer, R., Liu, J., Deimling, J., Tibboel, D., and Post, M. (2000). Dual-hit hypothesis explains pulmonary hypoplasia in the nitrofen model of congenital diaphragmatic hernia. *Am. J. Pathol.* **156(4)**, 1299–1306.

Kemp, P. J., Roberts, G. C., and Boyd, C. A. (1994). Identification and properties of pathways for K+ transport in guinea-pig and rat alveolar epithelial type II cells. *J Physiol.* **476**, 79–88.

Kheradmand, F., Rishi, K., and Werb, Z. (2002). Signaling through the EGF receptor controls lung morphogenesis in part by regulating MT1-MMP-mediated activation of gelatinase A/MMP2. *J. Cell. Sci.* **115**, 839–848.

Kho, A. T., Bhattacharya, S., Mecham, B. H., Hong, J., Kohane, I. S., and Mariani, T. J. (2009) Expression profiles of the mouse lung identify a molecular signature of time-to-birth. *Am. J. Repir. Cell Mol. Biol.* **40**, 47–57.

Kikuchi, W., Arai, H., Ishida, A., Takahashi, Y., and Takada, G. (2003). Distal pulmonary cell proliferation is associated with the expression of eiiia+ fibronectin in thedeveloping rat lung. *Exp. Lung Res.* **29**, 135–147.

Kim, C. F., Jackson, E. L., Woolfenden, A. E., Lawrence, S., Babar, I., Vogel, S., Crowley, D., Bronson, R. T., and Jacks, T. (2005). Identification of bronchioalveolar stem cells in normal lung and lung cancer. *Cell* **121**, 823–835.

Kimura, S., Ward, J. M., and Minoo, P. (1999). Thyroid-specific enhancer-binding protein/ thyroid transcription factor 1 is not required for the initial specificationof the thyroid and lung primordia. *Biochimie* **81**, 321–327.

Kimura, S., Hara, Y., Pineau, T., Fernandez-Salguero, P., Fox, C. H., Ward, J. M., and Gonzalez, F. J. (1996). The T/ebp null mouse: Thyroid-specific enhancer-binding protein is essential for the organogenesis of the thyroid, lung, ventral forebrain, and pituitary. *Genes Dev.* **10**, 60–69.

King, J. A., Marker, P. C., Seung, K. J., and Kingsley, D. M. (1994). BMP5 and the molecular, skeletal, and soft-tissue alterations in short ear mice. *Dev. Biol.* **166**, 112–122.

Kioussi, C., Briata, P., Baek, S. H., Rose, D. W., Hamblet, N. S., Herman, T., Ohgi, K. A., Lin, C., Gleiberman, A., Wang, J., Brault, V., Ruiz-Lozano, P., Nguyen, H. D., Kemler, R., Glass, C. K., Wynshaw-Boris, A., and Rosenfeld, M. G. (2002). Identification of a Wnt/Dvl/beta-Catenin –> Pitx2 pathway mediating cell-type-specific proliferation during development. *Cell* **111(5)**, 673–685.

Kitamura, K., Miura, H., Miyagawa-Tomita, S., Yanazawa, M., Katoh-Fukui, Y., Suzuki, R., Ohuchi, H., Suerhiro, A., Motegi, Y., Nakahara, Y., Kondo, S., and Yokoyama, M. (1999). Mouse pitx2 deficiency leads to anomalies of the ventral body wall, heart, extra- and periocular mesoderm and right pulmonary isomerism. *Development* **126(24)**, 5749–5758.

Koch, M., Olson, P. F., Albus, A., Jin, W., Hunter, D. D., Brunken, W. J., Burgeson, R. E., and Champliaud, M. F. (1999). Characterization and expression of the laminin gamma3 chain: A novel, non-basement membrane-associated, laminin chain. *J. Cell Biol.* **145**, 605–618.

Koli, K., Myllarniemi, M., Vuorinen, K., Salmenkivi, K., Ryynanen, M. J., Kinnula, V. L., and Keski-Oja, J. (2006). Bone morphogenetic protein-4 inhibitor gremlin is over-expressed in idiopathic pulmonary fibrosis. *Am. J. Pathol.* **169**, 61–71.

Komatsu, Y., Shibuya, H., Takeda, N., Ninomiya-Tsuji, J., Yasui, T., Miyado, K., Sekimoto, T., Ueno, N., Matsumoto, K., and Yamada, G. (2002). Targeted disruption of the tab1 gene causes embryonic lethality and defects in cardiovascular and lung morphogenesis. *Mech. Dev.* **119**, 239–249.

Korfhagen, T. R., Swantz, R. J., Wert, S. E., McCarty, J. M., Kerlakian, C. B., Glasser, S. W., and Whitsett, J. A. (1994). Respiratory epithelial cell expression of human transforming growth factor-alpha induces lung fibrosis in transgenic mice. *J. Clin. Invest.* **93**, 1691–1699.

Kramer, S., Okabe, M., Hacohen, N., Krasnow, M. A., and Hiromi, Y. (1999). Sprouty: A common antagonist of FGF and EGF signaling pathways in drosophila. *Development* **126**, 2515–2525.

Kreidberg, J. A., Donovan, M. J., Goldstein, S. L., Rennke, H., Shepherd, K., Jones, R. C., and Jaenisch, R. (1996). Alpha 3 beta 1 integrin has a crucial rolein kidney and lung organogenesis. *Development* **122**, 3537–3547.

Lagos-Quintana, M., Rauhut, R., Lendeckel, W., and Tuschl, T. (2001). Idenfication of novel genes coding for small expressed RNAs. *Science* **294**, 853–858.

Lallemand, A. V., Ruocco, S. M., Joly, P. M., and Gaillard, D. A. (1995). In vivo localization of the insulin-like growth factors I and II (IGF I and IGF II) gene expression during human lung development. *Int. J. Dev. Biol.* **39**, 529–537.

Lane, K. B., Machado, R. D., Pauciulo, M. W., Thomson, J. R., Phillips, J.A.I.I.I., Loyd, J. E., Nichols, W. C., and Trembath, R. C. (2000). Heterozygous germline mutations in BMPR2, encoding a TGF-beta receptor, cause familial primary pulmonary hypertension. The International PPH Consortium. *Nat. Genet.* **26**, 81–84.

Larrivee, B. and Karsan, A. (2000). Signaling pathways induced by vascular endothelial growth factor. *Int. J. Mol. Med.* **5**, 447–456.

LaRochelle, W. J., Jeffers, M., McDonald, W. F., Chillakuru, R. A., Giese, N. A., Lokker, N. A., Sullivan, C., Boldog, F. L., Yang, M., Vernet, C., Burgess, C. E., Fernandes, E., Deegler, L. L., Rittman, B., Shimkets, J., Shimkets, R. A., Rothberg, J. M., and Lichenstein, H. S. (2001). PDGF-D, a new protease-activated growth factor. *Nat. Cell. Biol.* **3**, 517–521.

Lau, N. C., Lim, L. P., Weinstein, E. G., and Bartel, D. P. (2001). An abundant class of tiny RNAs with probably regulatory role in *Caenorhabditis elegans*. *Science* **294**, 858–862.

Lazzaro, D., Price, M., De Felice, M., and Di Lauro, R. (1991). The transcription factor TTF-1 is expressed at the onset of thyroid and lung morphogenesis and in restricted regions of the foetal brain. *Development* **113**, 1093–1104.

Le Cras, T. D., Hardie, W. D., Fagan, K., Whitsett, J. A., and Korfhagen, T. R. (2003). Disrupted pulmonary vascular development and pulmonary hypertension in transgenic mice overexpressing transforming growth factor-α. *Am. J. Physiol. Lung Cell Mol. Physiol.* **258**, L1046–L1054.

Lecart, C., Cayabyab, R., Buckley, S., Morrison, J., Kwong, K. Y., Warburton, D., Ramanathan, R., Jones, C. A., and Minoo, P. (2000). Bioactive transforming growth factor-beta in the lungs of extremely low birthweight neonates predicts the need for home oxygen supplementation. *Biol. Neonate* **77**, 217–223.

Leco, K. J., Waterhouse, P., Sanchez, O. H., Gowing, K. L., Poole, A. R., Wakeham, A., Mak, T. W., and Khokha, R. (2001). Spontaneous air space enlargement in the lungs of mice lacking tissue inhibitor of metalloproteinases-3 (TIMP-3). *J. Clin. Invest.* **108**, 817–829.

Lee, J., Reddy, R., Barsky, L., Weinberg, K., and Driscoll, B. (2006). Contribution of proliferation and DNA damage repair to alveolar epithelial type 2 cell recovery from hyperoxia. *Am. J. Physiol. Lung Cell Mol. Physiol.* **290**, L685–L694.

Lee, Y., Jeon, K., Lee, J. T., Kim, S., and Kim, V. N. (2002). MicroRNA maturation: stepwise processing and subcellular localization. *EMBO J.* **21**, 4663–4670.

Li, C., Li, A., Li, M., Xing, Y., Chen, H., Hu, L., Tiozzo, C., Anderson, S., Taketo, M. M., and Minoo, P. (2009). Stabilized beta-catenin in lung epithelial cells changes cell fate and leads to tracheal and bronchial polyposis. *Dev. Biol.* **334**, 97–108.

Li, C., Hu, L., Xiao, J., Chen, H., Li, J. T., Bellusci, S., DeLanghe, S., and Minoo, P. (2005). Wnt5a regulates Shh and Fgf10 signaling during lung development. *Dev. Biol.* **287**(**1**), 86–97.

Li, C., Xiao, J., Hormi, K., Borok, Z., and Minoo, P. (2002). Wnt5a participates in distal lung morphogenesis. *Dev. Biol.* **248**, 68–81.

Li, D. Y., Sorensen, L. K., Brooke, B. S., Urness, L. D., Davis, E. C., Taylor, D. G., Boak, B. B., and Wendel, D. P. (1999). Defective angiogenesis in mice lacking endoglin. *Science* **284**, 1534–1537.

Li, X., Ponten, A., Aase, K., Karlsson, L., Abramsson, A., Uutela, M., Backstrom, G., Hellstrom, M., Bostrom, H., Li, H., Soriano, P., Betsholtz, C., Heldin, C. H., Alitalo, K., Ostman, A., and Eriksson, U. (2000). PDGF-C is a new protease-activated ligand for the PDGF alpha-receptor. *Nat. Cell. Biol.* **2**(**5**), 302–309.

Lim, L., Kalinichenko, V. V., Whitsett, J. A., and Costa, R. H. (2002). Fusion of lung lobes and vessels in mouse embryos heterozygous for the forkhead box f1 targeted allele. *Am. J. Physiol. Lung Cell Mol. Physiol.* **282**, L1012–L1022.

Lin, C. R., Kioussi, C., O'Connell, S., Briata, P., Szeto, D., Liu, F., Izpisua-Belmonte, J. C., and Rosenfeld, M. G. (1999). Pitx2 regulates lung asymmetry, cardiac positioning and pituitary and tooth morphogenesis. *Nature* **401**, 279–282.

Lin, S. Y., Chen, J. C., Hotaling, A. J., and Holinger, L. D. (1995). Congenital tracheal cartilaginous sleeve. *Laryngoscope* **105**, 1213–1219.

Lin, X., Buff, E. M., Perrimon, N., and Michelson, A. M. (1999). Heparan sulfate proteoglycans are essential for FGF receptor signaling during Drosophila embryonic development. *Development* **126**, 3715–3723.

Lindahl, P., Karlsson, L., Hellstrom, M., Gebre-Medhin, S., Willetts, K., Heath, J. K., and Betsholtz, C. (1997). Alveogenesis failure in PDGF-A-deficient mice is coupled to lack of distal spreading of alveolar smooth muscle cell progenitors during lung development. *Development* **124**, 3943–3953.

Litingtung, Y., Lei, L., Westphal, H., and Chiang, C. (1998). Sonic hedgehog is essential to foregut development. *Nat. Genet.* **20**, 58–61.

Liu, C., Glasser, S. W., Wan, H., and Whitsett, J. A. (2002). GATA-6 and thyroid transcription factor-1 directly interact and regulate surfactant protein-C gene expression. *J. Biol. Chem.* **277**(**4519–4525**), 22.

Liu, J. P., Baker, J., Perkins, A. S., Robertson, E. J., and Efstratiadis, A. (1993). Mice carrying null mutations of the genes encoding insulin-like growth factor I (Igf-1) and type 1 IGF receptor (Igf1r). *Cell* **75**, 59–72.

Liu, X. and Engelhardt, J. F. (2008). The glandular stem/progenitor cell niche in airway development and repair. *Proc. Am. Thorac. Soc.* **5**, 682–688.

Liu, Y. and Hogan, B. L. (2002). Differential gene expression in the distal tip endoderm of the embryonic mouse lung. *Gene Expr. Patterns* **2**, 229–233.

Liu, Y., Jiang, H., Crawford, H. C., and Hogan, B. L. (2003). Role for ETS domain transcription factors Pea3/Erm in mouse lung development. *Dev. Biol.* **261**, 10–24.

Lordan, J. L., Bucchieri, F., Richter, A., Konstantinidis, A., Holloway, J. W., Thornber, M., Puddicombe, S. M., Buchanan, D., Wilson, S. J., Djukanovic, R., Holgate, S. T., and Davies, D. E. (2002). Cooperative effects of Th2 cytokines and allergen on normal and asthmatic bronchial epithelial cells. *J. Immunol.* **169**, 407–414.

Lowe, L. A., Yamada, S., and Kuehn, M. R. (2001). Genetic dissection of nodalfunction in patterning the mouse embryo. *Development* **128**, 1831–1843.

Lu, M. M., Li, S., Yang, H., and Morrisey, E. E. (2002). Foxp4: A novel member of the Foxp subfamily of winged-helix genes co-expressed with Foxp1 and Foxp2 in pulmonary and gut tissues. *Mech. Dev.* **119**(**Suppl 1**), S197–S202.

Lu, M. M., Yang, H., Zhang, L., Shu, W., Blair, D. G., and Morrisey, E. E. (2001). The bone morphogenic protein antagonist gremlin regulates proximal-distal patterning of the lung. *Dev Dyn.* **222**, 667–680.

Lu, Y., Thomson, J. M., Wong, H. Y., Hammond, S. M., and Hogan, B. L. (2007). Transgenic over-expression of the microrna mir-17–92 cluster promotes proliferation and inhibits differentiation of lung epithelial progenitor cells. *Dev. Biol.* **310**, 442–453.

Lubkin, S. R., and Murray, J. D. (1995). A mechanism for early branching in lung morphogenesis. *J. Med. Entomol.* **34(1)**, 77–94.

Lwebuga-Mukasa, J. S. (1991). Matrix-driven pneumocyte differentiation. *Am. Rev. Respir. Dis.* **144(2)**, 452–457.

Macchiarini, P., Jungebluth, P., Go, T., Asnaghi, M. A., Rees, L. E., Cogan, T. A., Dodson, A., Martorell, J. Bellini, S., Parnigotto, P. P., Dickinson, S. C., Hollander, A. P., Mantero, S., Conconi, M. T., and Birchall, M. A., (2008) Clinical transplantation of a tissue-engineered airway. Lancet **372**, 2023–2030.

Maden, M., and Hind, M. (2003). Retinoic acid, a regeneration-inducing molecule. *Dev. Dyn.* **226**, 237–244.

Maeda, Y., Dave, V., and Whitsett, J. A. (2007). Transcriptional control of lung morphogenesis. *Physiol. Rev.* **87**, 219–244.

Mailleux, A. A., Kelly, R., Veltmaat, J. M., De Langhe, S. P., Zaffran, S., Thiery, J. P., and Bellusci, S. (2005). Fgf10 expression identifies parabronchial smooth muscle cell progenitors and is required for their entry into the smooth muscle cell lineage. *Development* **132**, 2157–2166.

Mailleux, A. A., Tefft, D., Ndiaye, D., Itoh, N., Thiery, J. P., Warburton, D., and Bellusci, S. (2001). Evidence that SPROUTY2 functions as an inhibitor of mouse embryonic lung growth and morphogenesis. *Mech. Dev.* **102**, 81–94.

Maitre, B., Clement, A., Williams, M. C., and Brody, J. S. (1995). Expression of insulin-like growth factor receptors 1 and 2 in the developing lung and their relation to epithelial cell differentiation. *Am. J. Respir. Cell Mol. Biol.* **13**, 262–270.

Malpel, S., Mendelsohn, C., and Cardoso, W. V. (2000). Regulation of retinoic acid signaling during lung morphogenesis. *Development* **127**, 3057–3067.

Mandelbrot, B. (1982). *The Fractal Geometry* of Nature. Freeman, San Francisco.

Maquet, E., Costagliola, S., Parma, J., Christophe-Hobertus, C., Oligny, L. L., Fournet, J. C., Robitaille, Y., Vuissoz, J. M., Payot, A., Laberge, S., Vassart, G., Van Vliet, G., and Deladoey, J. (2009). Lethal respiratory failure and mild primary hypothyroidism in a term girl with a de novo heterozygous mutation in the TITF1/NKX2.1 gene. *J. Clin. Endocrinol. Metab.* **94**, 197–203.

Marcil, A., Dumontier, E., Chamberland, M., Camper, S. A., and Drouin, J. (2003). Pitx1 and Pitx2 are required for development of hindlimb buds. *Development* **130(1)**, 45–55.

Massague, J. (2000). How cells read TGF-beta signals. *Nat. Rev. Mol. Cell Biol.* **1**, 169–178.

Massague, J. (1998). TGF-beta signal transduction. *Annu. Rev. Biochem.* **67**, 753–791.

Massaro, G. D., and Massaro, D. (2000). Retinoic acid treatment partially rescues failed septation in rats and in mice. *Am. J. Physiol. Lung Cell Mol. Physiol.* **278**, L955–L960.

Massaro, G. D., and Massaro, D. (1996). Postnatal treatment with retinoic acid increases the number of pulmonary alveoli in rats. *Am. J. Physiol.* **270**, L305–L310.

Mayer, U., Mann, K., Timpl, R., and Murphy, G. (1993). Sites of nidogen cleavage by proteases involved in tissue homeostasis and remodelling. *Eur. J. Biochem.* **217**, 877–884.

McAllister, K. A., Grogg, K. M., Johnson, D. W., Gallione, C. J., Baldwin, M. A., Jackson, C. E., Helmbold, E. A., Markel, D. S., McKinnon, W. C., and Murrell, J. (1994). Endoglin, a TGF-beta binding protein of endothelial cells, is the gene for hereditary haemorrhagic telangiectasia type 1. *Nat. Genet.* **8**, 345–351.

McCray, P. B.Jr., Bettencourt, J. D., Bastacky, J., Denning, G. M., and Welsh, M. J. (1993). Expression of CFTR and a cAMP-stimulated chloride secretory current in cultured human fetal alveolar epithelial cells. *Am. J. Respir. Cell Mol. Biol.* **9(6)**, 578–585.

McDowell, E. M., Newkirk, C., and Coleman, B. (1985). Development of hamster tracheal epithelium: II. Cell proliferation in the fetus. *Anat. Rec.* **213**, 448–456.

McGowan, S., Jackson, S. K., Jenkins-Moore, M., Dai, H. H., Chambon, P., and Snyder, J. M. (2000). Mice bearing deletions of retinoic acid receptors demonstrate reduced lung elastin and alveolar numbers. *Am. J. Respir. Cell Mol. Biol.* **23**, 162–167.

McLennan, I. S., Poussart, Y., and Koishi, K. (2000). Development of skeletal muscles in transforming growth factor-beta 1 (TGF-beta1) null-mutant mice. *Dev. Dyn.* **217**, 250–256.

McNamara, V. M., and Crabbe, D. C. (2004). Tracheomalacia. *Paediatr. Respir. Rev.* **5**, 147–154.

Mendelsohn, C., Lohnes, D., Decimo, D., Lufkin, T., LeMeur, M., Chambon, P., and Mark, M. (1994). Function of the retinoic acid receptors (RARs) during development (II). Multiple abnormalities at various stages of organogenesis in RAR double mutants. *Development* **120**, 2749–2771.

Meno, C., Shimono, A., Saijoh, Y., Yashiro, K., Mochida, K., Ohishi, S., Noji, S., Kondoh, H., and Hamada, H. (1998). Lefty-1 is required for left-rightdetermination as a regulator of lefty-2 and nodal. *Cell* **94**, 287–297.

Metzger, R. J., Klein, O. D., Martin, G. R., and Krasnow, M. A. (2008). The branching programme of mouse lung development. *Nature* **453(7196)**, 745–750.

Michos, O., Panman, L., Vintersten, K., Beier, K., Zeller, R., and Zuniga, A. (2004). Gremlin-mediated BMP antagonism induces the epithelial-mesenchymalfeedback signaling controlling metanephric. *Development* **131**, 3401–3410.

Miettinen, P. J., Warburton, D., Bu, D., Zhao, J. S., Berger, J. E., Minoo, P., Koivisto, T., Allen, L., Dobbs, L., Werb, Z., and Derynck, R. (1997). Impaired lung branching morphogenesis in the absence of functional EGF receptor. *Dev. Biol.* **186**, 224–236.

Miettinen, P. J., Berger, J. E., Meneses, J., Phung, Y., Pedersen, R. A., Werb, Z., and Derynck, R. (1995). Epithelial immaturity and multiorgan failure in mice lacking epidermal growth factor receptor. *Nature* **376**, 337–341.

Millan, F. A., Denhez, F., Kondaiah, P., and Akhurst, R. J. (1991). Embryonic gene expression patterns of TGF beta 1, beta 2 and beta 3 suggest different developmental functions in vivo. *Development* **111**, 131–143.

Miller, A. A., Hooper, S. B., and Harding, R. (1993). Role of fetal breathing movements in control of fetal lung distension. *J. Appl. Physiol.* **75(6)**, 2711–2717.

Miller, L. A., Wert, S. E., Clark, J. C., Xu, Y., Perl, A. K., and Whitsett, J. A. (2004). Role of Sonic hedgehog in patterning of tracheal-bronchial cartilage and the peripheral lung. *Dev. Dyn.* **231**, 57–71.

Millien, G., Beane, J., Lenburg, M., Tsao, P. N., Lu, J. N., Spira, A., and Ramirez, M. I. (2008). Characterization of the mid-foregut transcriptome identifies genes regulated during lung bud induction. *Gene Expr. Patterns* **8(2)**, 124–139. doi: 10.1016/j.modgep.2007.09.003

Min, H., Danilenko, D. M., Scully, S. A., Bolon, B., Ring, B. D., Tarpley, J. E., DeRose, M., and Simonet, W. S. (1998). Fgf-10 is required for both limb and lung development and exhibits striking functional similarity to Drosophila branchless. *Genes Dev.* **12**, 3156–3161.

Miner, J. H., Cunningham, J., and Sanes, J. R. (1998). Roles for laminin in embryogenesis: Exencephaly, syndactyly, and placentopathy in mice lacking the laminin alpha5 chain. *J. Cell Biol.* **143**, 1713–1723.

Miner, J. H., Patton, B. L., Lentz, S. I., Gilbert, D. J., Snider, W. D., Jenkins, N. A., Copeland, N. G., and Sanes, J. R. (1997). The laminin alpha chains: Expression, developmental transitions, and chromosomal locations of alpha1–5, identification of heterotrimeric laminins 8–11, and cloning of a novel alpha3 isoform. *J. Cell Biol.* **137**, 685–701.

Miner, J. H., Lewis, R. M., and Sanes, J. R. (1995). Molecular cloning of a novel laminin chain, alpha 5, and widespread expression in adult mouse tissues. *J. Biol. Chem.* **270**, 28523–28526.

Minoo, P., Su, G., Drum, H., Bringas, P., and Kimura, S. (1999). Defects in tracheoesophageal and lung morphogenesis in Nkx2.1(-/-) mouse embryos. *Dev. Biol.* **209**, 60–71.

Minoo, P., and King, R. J. (1994). Epithelial-mesenchymal interactions in lung development. *Annu. Rev. Physiol.* **56**, 13–45.

Mizutani, K., Yoon, K., Dang, L., Tokunaga, A., and Gaiano, N. (2007). Differential notch signalling distinguishes neural stem cells from intermediate progenitors. *Nature* **449**, 351–355.

Moens, C. B., Auerbach, A. B., Conlon, R. A., Joyner, A. L., and Rossant, J. (1992). A targeted mutation reveals a role for N-myc in branchingmorphogenesis in the embryonic mouse lung. *Genes Dev.* **6**, 691–704.

Moessinger, A. C., Harding, R., Adamson, T. M., Singh, M., and Kiu, G. T. (1990). Role of lung fluid volume in growth and maturation of the fetal sheep lung. *J. Clin. Invest.* **86**, 1270–1277.

Mokres, L. M., Parai, K., Hilgendorff, A., Ertsey, R., Alvira, C. M., Rabinovitch, M., and Bland, R. D. (2010). Prolonged mechanical ventilation with air induced apoptosis and causes failure of alveolar septation and angiogenesis in lungs of newborn mice. *Am. J. Physiol. Lung Cell Mol. Physiol.* **298**, L23–L35.

Moore, K. A., Polte, T., Huang, S., Shi, B., Alsberg, E., Sunday, M. E., and Ingber, D. E. (2005). Control of basement membrane remodeling and epithelial branching morphogenesis in embryonic lung by Rho and cytoskeletal tension. *Dev. Dyn.* **232**(**2**), 268–281.

Morris, D. G., Huang, X., Kaminski, N., Wang, Y., Shapiro, S. D., Dolganov, G., Glick, A., and Sheppard, D. (2003). Loss of integrin alpha(v)beta6-mediated TGF-beta activation causes Mmp12-dependent emphysema. *Nature* **422**, 169–173.

Morrison, S. J., and Kimble, J. (2006). Asymmetric and symmetric stem-cell divisions in development and cancer. *Nature* **441**, 1068–1074.

Motoyama, J., Liu, J., Mo, R., Ding, Q., Post, M., and Hui, C. C. (1998). Essential function of Gli2 and Gli3 in the formation of lung, trachea and oesophagus. *Nat. Genet.* **20**, 54–57.

Mucenski, M. L., Nation, J. M., Thitoff, A. R., Besnard, V., Xu, Y., Wert, S. E., Harada, N., Taketo, M. M., Stahlman, M. T., and Whitsett, J. A. (2005). Beta-catenin regulates differentiation of respiratory epithelial cells in vivo. *Am. J. Physiol. Lung Cell Mol. Physiol.* **289**(**6**), L971–L979.

Mucenski, M. L., Wert, S. E., Nation, J. M., Loudy, D. E., Huelsken, J., Birchmeier, W., Morrisey, E. E., and Whitsett, J. A. (2003). Beta-catenin is required for specification of proximal/distal cell fate during lung morphogenesis. *J. Biol. Chem.* **278**(**41**), 40231–40238.

Muglia, L. J., Bae, D. S., Brown, T. T., Vogt, S. K., Alvarez, J. G., Sunday, M. E., and Majzoub, J. A. (1999). Proliferation and differentiation defects during lung development in corticotropin-releasing hormone-deficient mice. *Am. J. Respir. Cell Mol. Biol.* **20**, 181–188.

Murray, C. B., Morales, M. M., Flotte, T. R., McGrath-Morrow, S. A., Guggino, W. B., and Zeitlin, P. L. (1995). CIC-2: A developmentally dependent chloride channel expressed in the fetal lung and downregulated after birth. *Am. J. Respir. Cell Mol. Biol.* **12**(**6**), 597–604.

Myllarniemi, M., Lindholm, P., Ryynanen, M. J., Kliment, C. R., Salmenkivi, K., Keski-Oja, J., Kinnula, V. L., Oury, T. D., and Koli, K. (2008). Gremlin-medicated decrease in bone morphogenetic protein signaling promotes pulmonary fibrosis. *Am. J. Respir. Crit. Care Med.* **177**, 321–329.

Ng, Y. S., Rohan, R., Sunday, M. E., Demello, D. E., and D'Amore, P. A. (2001). Differential expression of VEGF isoforms in mouse during development and in the adult. *Dev. Dyn.* **220**, 112–121.

Nguyen, N. M., Miner, J. H., Pierce, R. A., and Senior, R. M. (2002). Laminin alpha 5 is required for lobar septation and visceral pleural basement membrane formation in the developing mouse lung. *Dev. Biol.* **246**, 231–244.

Niederreither, K., Subbarayan, V., Dolle, P., and Chambon, P. (1999). Embryonic retinoic acid synthesis is essential for early mouse post-implantation development. *Nat. Genet.* **21**, 444–448.

Niederreither, K., McCaffery, P., Drager, U. C., Chambon, P., and Dolle, P. (1997). Restricted expression and retinoic acid-induced downregulation of the retinaldehyde dehydrogenase type 2 (RALDH-2) gene during mouse development. *Mech. Dev.* **62**, 67–78.

Nyeng, P., Norgaard, G. A., Kobberup, S., and Jensen, J. (2008). Fgf10 maintains distal lung bud epithelium and excessive signaling leads to progenitor state arrest, distalization, and goblet cell metaplasia. *BMC Dev. Biol.* **8**, 2.

Oh, S. P. and Li, E. (1997). The signaling pathway mediated by the type IIB activinreceptor controls axial patterning and lateral asymmetry in the mouse. *Genes Dev.* **11**, 1812–1826.

Ohmichi, H., Koshimizu, U., Matsumoto, K., and Nakamura, T. (1998). Hepatocyte growth factor (HGF) acts as a mesenchyme-derived morphogenic factor during fetal lung development. *Development* **125**, 1315–1324.

Okubo, T. Knoepfler, P. S., Eisenman, R. N., and Hogan, B. L. (2005). Nmyc plays an essential role during lung development as a dosage-sensitive regulator of progenitor cell proliferation and differentiation. *Development* **132**(**6**), 1363–1374. doi: dev.01678 [pii] 10.1242/dev.01678

Okubo, T. and Hogan, B. L. (2004). Hyperactive Wnt signaling changes the developmental potential of embryonic lung endoderm. *J. Biol.* **3**(**3**), 11.

Olver, R. E., Walters, D. V., and M, W. S. (2004). Developmental regulation of lung liquid transport. *Annu. Rev. Physiol.* **66**, 77–101.

Olver, R. E., Ramsden, C. A., Strang, L. B., and Walters, D. V. (1986). The role of amiloride-blockable sodium transport in adrenaline-induced lung liquid reabsorption in the fetal lamb. *J. Physiol.* **376**, 321–340.

Olver, R. E. and Strang, L. B. (1974). Ion fluxes across the pulmonary epithelium and the secretion of lung liquid in the foetal lamb. *J. Physiol.* **241**(**2**), 327–357.

Ornitz, D. M. and Itoh, N. (2001). Fibroblast growth factors. *Genome. Biol.* **2**, REVIEWS3005.

Papadakis, K., Luks, F. I., De Paepe, M. E., Piasecki, G. J., and Wesselhoeft, C. W. Jr. (1997). Fetal lung growth after tracheal ligation is not solely a pressure phenomenon. *J. Pediatr. Surg.* **32**, 347–351.

Park, J., Zhang, J. J., Moro, A., Kushida, M., Wegner, M., and Kim, P. C. (2009). Regulation of Sox9 by Sonic Hedgehog (Shh) is essential for patterning and formation of tracheal cartilage. *Dev. Dyn.* doi: dvdy.22192 [pii] 10.1002/dvdy.22192.

Park, W. Y., Miranda, B., Lebeche, D., Hashimoto, G., and Cardoso, W. V. (1998). FGF-10 is a chemotactic factor for distal epithelial buds during lung development. *Dev. Biol.* **201**, 125–134.

Pauling, M. H., and Vu, T. H. (2004). Mechanisms and regulation of lung vascular development. *Curr. Top. Dev. Biol.* **64**, 73-99.

Pelton, R. W., Johnson, M. D., Perkett, E. A., Gold, L. I., and Moses, H. L. (1991a). Expression of transforming growth factor-beta 1, -beta 2, and -beta 3 mRNA and protein in the murine lung. *Am. J. Respir. Cell Mol. Biol.* **5**, 522–530.

Pelton, R. W., Saxena, B., Jones, M., Moses, H. L., and Gold, L. I. (1991b). Immunohistochemical localization of TGF beta 1, TGF beta 2, and TGF beta 3 in the mouse embryo: Expression patterns suggest multiple roles during embryonic development. *J. Cell Biol.* **115**, 1091–1105.

Pepicelli, C. V., Lewis, P. M., and McMahon, A. P. (1998). Sonic hedgehog regulates branching morphogenesis in the mammalian lung. *Curr. Biol.* **8**, 1083–1086.

Perl, A. K., Wert, S. E., Loudy, D. E., Shan, Z., Blair, P. A., and Whitsett, J. A. (2005a). Conditional recombination reveals distinct subsets of epithelial cells in trachea, bronchi, and alveoli. *Am. J. Respir. Cell Mol. Biol.* **33**, 455–462.

Perl, A. K., Kist, R., Shan, Z., Scherer, G., and Whitsett, J. A. (2005b). Normal lung development and function after sox9 inactivation in the respiratory epithelium. *Genesis* **41**, 23–32.

Perl, A. K., Hokuto, I., Impagnatiello, M. A., Christofori, G., and Whitsett, J. A. (2003). Temporal effects of Sprouty on lung morphogenesis. *Dev. Biol.* **258**, 154–168.

Perl, A. K., Wert, S. E., Nagy, A., Lobe, C. G., and Whitsett, J. A. (2002). Early restriction of peripheral and proximal cell lineages during formation of the lung. *Proc. Natl. Acad. Sci. USA* **99**, 10482–10487.

Peschon, J. J., Slack, J. L., Reddy, P., Stocking, K. L., Sunnarborg, S. W., Lee, D. C., Russell, W. E., Castner, B. J., Johnson, R. S., Fitzner, J. N., Boyce, R. W., Nelson, N., Kozlosky, C. J., Wolfson, M. F., Rauch, C. T., Cerretti, D. P., Paxton, R. J., March, C. J., and Black, R. A. (1998). An essential role for ectodomain shedding in mammalian development. *Science* **282**, 1281–1284.

Peterson, R. S., Lim, L., Ye, H., Zhou, H., Overdier, D. G., and Costa, R. H. (1997). The winged helix transcriptional activator HFH-8 is expressed in the mesoderm of the primitive streak stage of mouse embryos and its cellular derivatives. *Mech. Dev.* **69**, 53–69.

Pichel, J. G., Fernandez-Moreno, C., Vicario-Abejon, C., Testillano, P. S., Patterson, P. H., and de Pablo, F. (2003). Developmental cooperation of leukemia inhibitory factor and insulin-like growth factor I in mice is tissue-specific and essential for lung maturation involving the transcription factors Sp3 and TTF-1. *Mech. Dev.* **120**, 349–361.

Pierce, R. A., Griffin, G. L., Mudd, M. S., Moxley, M. A., Longmore, W. J., Sanes, J. R., Miner, J. H., and Senior, R. M. (1998). Expression of laminin alpha3, alpha4, and alpha5 chains by alveolar epithelial cells and fibroblasts. *Am. J. Respir. Cell Mol. Biol.* **19**, 237–244.

Plantier, L., Marchand-Adam, S., Antico, V. G., Boyer, L., De Coster, C., Marchal, J., Bachoual, R., Mailleux, A., Boczkowski, J., and Crestani, B. (2007). Keratinocyte growth factor protects against elastase-induced pulmonary emphysema in mice. *Am. J. Physiol. Lung Cell Mol. Physiol.* **293**, L1230–L1239.

Plopper, C. G., Nishio, S. J., Alley, J. L., Kass, P., and Hyde, D. M. (1992). The role of the nonciliated bronchiolar epithelial (Clara) cell as the progenitor cell during bronchiolar epithelial differentiation in the perinatal rabbit lung. *Am. J. Respir. Cell Mol. Biol.* **7**, 606–613.

Post, L. C., Ternet, M., and Hogan, B. L. (2000). Notch/delta expression in the developing mouse lung. *Mech. Dev.* **98**, 95–98.

Post, M., Souza, P., Liu, J., Tseu, I., Wang, J., Kuliszewski, M., and Tanswell, A. K. (1996). Keratinocyte growth factor and its receptor are involved in regulating early lung branching. *Development* **122**, 3107–3115.

Pourquie, O. (2003). The segmentation clock: Converting embryonic time into spatial pattern. *Science* **301**, 328–330.

Qian, X., and Costa, R. H. (1995). Analysis of hepatocyte nuclear factor-3 beta protein domains required for transcriptional activation and nuclear targeting. *Nucleic Acids Res.* **23**, 1184–1191.

Quaggin, S. E., Schwartz, L., Cui, S., Igarashi, P., Deimling, J., Post, M., and Rossant, J. (1999). The basic-helix-loop-helix protein pod1 is critically important for kidney and lung organogenesis. *Development* **126**, 5771–5783.

Que, J., Wilm, B., Hasegawa, H., Wang, F., Bader, D., and Hogan, B. L. (2008). Mesothelium contributes to vascular smooth muscle and mesenchyme during lung development. *Proc. Natl. Acad. Sci. USA* **105**, 16626–16630.

Que, J., Choi, M., Ziel, J. W., Klingensmith, J., and Hogan, B. L. (2006). Morphogenesis of the trachea and esophagus: Current players and new roles for noggin and Bmps. *Differentiation* **74**, 422–437.

Raaberg, L., Nexo, E., Buckley, S., Luo, W., Snead, M. L., and Warburton, D. (1992). Epidermal growth factor transcription, translation, and signal transduction by rat type II pneumocytes in culture. *Am. J. Respir. Cell Mol. Biol.* **6**, 44–49.

Ramasamy, S. K., Mailleux, A. A., Gupte, V. V., Mata, F., Sala, F. G., Veltmaat, J. M., Del Moral, P. M., DeLanghe, S., Parsa, S., and Kelly, L. K. (2007). Fgf10 dosage is critical for the amplification of epithelial cell progenitors and for the formation of multiple mesenchymal lineages during lung development. *Dev. Biol.* **307**(**2**), 237–247.

Ramirez, M. I., Millien, G., Hinds, A., Cao, Y., Seldin, D. C., and Williams, M. C. (2003). T1alpha, a lung type I cell differentiation gene, is required for normal lung cell proliferation and alveolus formation at birth. *Dev. Biol.* **256**, 61–72.

Ramirez, M. I., Rishi, A. K., Cao, Y. X., and Williams, M. C. (1997). TGT3, thyroid transcription factor I, and Sp1 elements regulate transcriptional activity of the 1.3-kilobase pair promoter of T1alpha, a lung alveolar type I cell gene. *J. Biol. Chem.* **272**, 26285–26294.

Rawlins, E. L. (2008). Lung Epithelial Progenitor Cells Lessons from development. *Proc. Am. Thorac. Soc.* **5**, 675–681.

Rawlins, E. L., Clark, C. P., Xue, Y., and Hogan, B. L. (2009a). The Id2+ distal tip lung epithelium contains individual multipotent embryonic progenitor cells. *Development* **136**(**22**), 3741–3745.

Rawlins, E. L., Okubo, T., Xue, Y., Brass, D. M., Auten, R. L., Hasegawa, H., Wang, F., and Hogan, B. L. (2009b). The role of Scgb1a1+ Clara cellsin the long-term maintenance and repair of lung airway, but not alveolar epithelium. *Cell Stem Cell* **4**(**6**), 525–534.

Rawlins, E. L., Ostrowski, L. E., Randell, S. H., and Hogan, B. L. (2007). Lung development and repair: Contribution of the ciliated lineage. *Proc. Natl. Acad. Sci. USA* **104**, 410–417.

Rawlins, E. L., and Hogan, B. L. (2006). Epithelial stem cells of the lung: Privileged few or opportunities for many?. *Development* **133**, 2455–2465.

Ray, P., Devaux, Y., Stolz, D. B., Yarlagadda, M., Watkins, S. C., Lu, Y., Chen, L., Yang, X. F., and Ray, A. (2003). Inducible expression of keratinocyte growth factor (KGF) in mice inhibits lung epithelial cell death induced by hyperoxia. *Proc. Natl. Acad. Sci. USA* **100**, 6098–6103.

Reddy, R., Buckley, S., Doerken, M., Barsky, L., Weinberg, K., Anderson, K. D., Warburton, D., and Driscoll, B. (2004). Isolation of a putative progenitor subpopulation of alveolar epithelial type 2 cells. *Am. J. Physiol. Lung Cell Mol. Physiol.* **286**, L658–L667.

Reich, A., Sapir, A., and Shilo, B. (1999). Sprouty is a general inhibitor of receptor tyrosine kinase signaling. *Development* **126**, 4139–4147.

Reinhardt, D., Mann, K., Nischt, R., Fox, J. W., Chu, M. L., Krieg, T., and Timpl, R. (1993). Mapping of nidogen binding sites for collagen type IV, heparan sulfate proteoglycan, and zinc. *J. Biol. Chem.* **268**, 10881–10887.

Retsch-Bogart, G. Z., Moats-Staats, B. M., Howard, K., D'Ercole, A. J., and Stiles, A. D. (1996). Cellular localization of messenger RNAs for insulin-like growth factors (IGFs), their receptors and binding proteins during fetal rat lung development. *Am. J. Respir. Cell Mol. Biol.* **14**, 61–69.

Reynolds, S. D., Giangreco, A., Hong, K. U., McGrath, K. E., Ortiz, L. A., and Stripp, B. R. (2004). Airway injury in lung disease pathophysiology: Selective depletion of airway stem and progenitor cell pools potentiates lung inflammation and alveolar dysfunction. *Am. J. Physiol. Lung Cell Mol. Physiol.* **287**, L1256–L1265.

Reynolds, S. D., Giangreco, A., Power, J. H., and Stripp, B. R. (2000). Neuroepithelial bodies of pulmonary airways serve as a reservoir of progenitor cells capable of epithelial regeneration. *Am. J. Pathol.* **156**, 269–278.

Riccardi, D., Park, J., Lee, W. S., Gamba, G., Brown, E. M., and Hebert, S. C. (1995). Cloning and functional expression of a rat kidney extracellular calcium/polyvalent cation-sensing receptor. *Proc. Natl. Acad. Sci. USA* **92**(**1**), 131–135.

Rock, J. R., Onaitis, M. W., Rawlins, E. L., Lu, Y., Clark, C. P., Xue, Y., Randell, S. H., and Hogan, B. L. (2009). Basal cells as stem cells of the mouse trachea and human airway epithelium. *Proc. Natl. Acad. Sci. USA* **106**(**31**), 12771–12775.

Rock, J. R., Futtner, C. R., and Harfe, B. D. (2008). The transmembrane protein TMEM16A is required for normal development of the murine trachea. *Dev. Biol.* **321** (**1**), 141–149. doi: 10.1016/j.ydbio.2008.06.009 321(1), 141–149

Roman, J. (1997). Fibronectin and fibronectin receptors in lung development. *Exp. Lung Res.* **23**, 147–159.

Roper, J. M., Mazzatti, D. J., Watkins, R. H., Maniscalco, W. M., Keng, P. C., and O'Reilly, M. A. (2004). In vivo exposure to hyperoxia induces DNA damage in a population of alveolar type II epithelial cells. *Am. J. Physiol. Lung Cell Mol. Physiol.* **286**, L1045–L1054.

Rubin, L. P., Kovacs, C. S., De Paepe, M. E., Tsai, S. W., Torday, J. S., and Kronenberg, H. M. (2004). Arrested pulmonary alveolar cytodifferentiation and defective surfactant synthesis in mice missing the gene for parathyroid hormone-related protein. *Dev. Dyn.* **230**, 278–289.

Sakai, T., Larsen, M., and Yamada, K. M. (2003). Fibronectin requirement in branching morphogenesis. *Nature* **423**, 876–881.

Sakiyama, J., Yamagishi, A., and Kuroiwa, A. (2003). Tbx4-Fgf10 system controls lung bud formation during chicken embryonic development. *Development* **130**, 1225–1234.

Samadani, U., Porcella, A., Pani, L., Johnson, P. F., Burch, J. B., Pine, R., and Costa, R. H. (1995). Cytokine regulation of the liver transcription factor hepatocyte nuclear factor-3 beta is mediated by the C/EBP family and interferon regulatory factor 1. *Cell Growth Differ.* **6**, 879–890.

Sanchez-Esteban, J., Wang, Y., Gruppuso, P. A., and Rubin, L. P. (2003). Mechanical stretch induces fetal type II cell differentiation via an EGFR-ERK signaling pathway. *Am. J. Respir. Cell Mol. Biol.* **30**(**1**), 76–83.

Sanford, L. P., Ormsby, I., Gittenberger-de Groot, A. C., Sariola, H., Friedman, R., Boivin, G. P., Cardell, E. L., and Doetschman, T. (1997). TGFbeta2 knockout mice have multiple developmental defects that are non-overlapping with other TGFbeta knockout phenotypes. *Development* **124**, 2659–2670.

Schittny, J. C., Miseroccandi, G., and Sparrow, M. P. (2000). Spontaneous peristaltic airway contractions propel lung liquid through the bronchial tree of intact and fetal lung explants. *Am. J. Respir. Cell Mol. Biol.* **23**(**1**), 11–18.

Schittny, J. C., Djonov, V., Fine, A., and Burri, P. H. (1998). Programmed cell death contributes to postnatal lung development. *Am. J. Respir. Cell Mol. Biol.* **18**, 786–793.

Schmid, P., Cox, D., Bilbe, G., Maier, R., and McMaster, G. K. (1991). Differential expression of TGF beta 1, beta 2 and beta 3 genes during mouse embryogenesis. *Development* **111**, 117–130.

Schuger, L., Johnson, G. R., Gilbride, K., Plowman, G. D., and Mandel, R. (1996a). Amphiregulin in lung branching morphogenesis: Interaction with heparan sulfate proteoglycan modulates cell proliferation. *Development* **122**, 1759–1767.

Schuger, L., Skubitz, A. P., Gilbride, K., Mandel, R., and He, L. (1996b). Laminin and heparan sulfate proteoglycan mediate epithelial cell polarization in organotypic cultures of embryonic lung cells: Evidence implicating involvement of the inner globular region of laminin beta 1 chain and the heparan sulfate groups of heparan sulfate proteoglycan. *Dev. Biol.* **179**, 264–273.

Schuger, L., Skubitz, A. P., de las, M. A., and Gilbride, K. (1995). Two separate domains of laminin promote lung organogenesis by different mechanisms of action. *Dev. Biol.* **169**, 520–532.

Schuger, L., Varani, J., Killen, P. D., Skubitz, A. P., and Gilbride, K. (1992). Laminin expression in the mouse lung increases with development and stimulates spontaneous organotypic rearrangement of mixed lung cells. *Dev. Dyn.* **195**, 43–54.

Schuller, A. G., van Neck, J. W., Beukenholdt, R. W., Zwarthoff, E. C., and Drop, S. L. (1995). IGF, type I IGF receptor and IGF-binding protein mRNA expression in the developing mouse lung. *J. Muscoskel. Pain* **14**, 349–355.

Schultz, C. J., Torres, E., Londos, C., and Torday, J. S. (2002). Role of adipocyte differentiation-related protein in surfactant phospholipid synthesis by type II cells. *Am. J. Physiol. Lung Cell Mol. Physiol.* **283**, L288–L296.

Sekine, K., Ohuchi, H., Fujiwara, M., Yamasaki, M., Yoshizawa, T., Sato, T., Yagishita, N., Matsui, D., Koga, Y., Itoh, N., and Kato, S. (1999). Fgf10 is essential for limb and lung formation. *Nat. Genet.* **21**, 138–141.

Senior, R. M., Griffin, G. L., Mudd, M. S., Moxley, M. A., Longmore, W. J., and Pierce, R. A. (1996). Entactin expression by rat lung and rat alveolar epithelial cells. *Am. J. Resp. Cell Mol. Biol.* **14**(**3**), 239–247.

Serls, A. E., Doherty, S., Parvatiyar, P., Wells, J. M., and Deutsch, G. H. (2005). Different thresholds of fibroblast growth factors pattern the ventral foregut into liver and lung. *Development* **132**(**1**), 35–47.

Seth, R., Shum, L., Wu, F., Wuenschell, C., Hall, F. L., Slavkin, H. C., and Warburton, D. (1993). Role of epidermal growth factor expression in early mouse embryo lung branching morphogenesis in culture: Antisense oligodeoxynucleotide inhibitory strategy. *Dev. Biol.* **158**(**2**), 555–559.

Seymour, P. A., Freude, K. K., Tran, M. N., Mayes, E. E., Jensen, J., Kist, R., Scherer, G., and Sander, M. (2007). Sox9 is required for maintenance of the pancreatic progenitor cell pool. *Proc. Natl. Acad. Sci. USA* **104**(**6**), 1865–1870.

Shan, L., Subramaniam, M., Emanuel, R. L., Degan, S., Johnston, P., Tefft, D., Warburton, D., and Sunday, M. E. (2008). Centrifugal migration of mesenchymal cells in embryonic lung. *Dev. Dyn.* **237**, 750–757.

Shannon, J. M., McCormick-Shannon, K., Burhans, M. S., Shangguan, X., Srivastava, K., and Hyatt, B. A. (2003). Chondroitin sulfate proteoglycans are required for lung growth and morphogenesis in vitro. *Am. J. Physiol. Lung Cell Mol. Physiol.* **285**(**6**), L1323–L1336.

Shaw-White, J. R., Bruno, M. D., and Whitsett, J. A. (1999). GATA-6 activates transcription of thyroid transcription factor-1. *J. Biol. Chem.* 274, 2658–2664.

Shi, W., and Warburton, D. (2010). Is COPD in adulthood really so far removed from early development? *Eur. Respir. J.* **35**, 12–13.

Shi, W., Xu, J., and Warburton, D. (2009). Development, repair, and fibrosis: what is common and why it matters. *Respirology* **14**, 656–665.

Shi, W., Bellusci, S., and Warburton, D. (2007). Lung development and adult lung diseases. *Chest* **2**, 651–656.

Shi, W., Chen, H., Sun, J., Buckley, S., Zhao, J., Anderson, K. D., Williams, R. G., and Warburton, D. (2003). TACE is required for fetal murine cardiac development and modeling. *Dev. Biol.* **261**, 371–380.

Shi, W., Zhao, J., Anderson, K. D., and Warburton, D. (2001). Gremlin negatively modulates BMP-4 induction of embryonic mouse lung branching morphogenesis. *Am. J. Physiol. Lung Cell Mol. Physiol.* **280**, L1030–L1039.

Shi, W., Heisterkamp, N., Groffen, J., Zhao, J., Warburton, D., and Kaartinen, V. (1999). TGF-beta3-null mutation does not abrogate fetal lung maturation in vivo by glucocorticoids. *Am. J. Physiol.* **277**(**6 Pt 1**), L1205–L1213.

Shi, Y., and Massague, J. (2003). Mechanisms of TGF-beta signaling from cell membrane to the nucleus. *Cell* **113**, 685–700.

Shiels, H., Li, X., Schumacker, P. T., Maltepe, E., Padrid, P. A., Sperling, A., Thompson, C. B., and Lindsten, T. (2000). TRAF4 deficiency leads to tracheal malformation with resulting alterations in air flow to the lungs. *Am. J. Pathol.* **157**, 679–688.

Shikama, N., Lutz, W., Kretzschmar, R., Sauter, N., Roth, J. F., Marino, S., Wittwer, J., Scheidweiler, A., and Eckner, R. (2003). Essential function of p300 acetyltransferase activity in heart, lung and small intestine formation. *EMBO J.* **22**, 5175–5185.

Shinbrot, E., Peters, K. G., and Williams, L. T. (1994). Expression of the platelet-derived growth factor beta receptor during organogenesis and tissue differentiation in the mouse embryo. *Dev. Dyn.* **199**, 169–175.

Shu, W., Lu, M. M., Zhang, Y., Tucker, P. W., Zhou, D., and Morrisey, E. E. (2007). Foxp2 and foxp1 cooperatively regulate lung and esophagus development. *Development* **134**, 1991–2000.

Shu, W., Guttentag, S., Wang, Z., Andl, T., Ballard, P., Lu, M. M., Piccolo, S., Birchmeier, W., Whitsett, J. A., Millar, S. E., and Morrisey, E. E. (2005). Wnt/beta-catenin signaling acts upstream of N-myc, BMP4, and FGF signaling to regulate proximal-distal patterning in the lung. *Dev. Biol.* **283**(**1**), 226–239.

Shu, W., Jiang, Y. Q., Lu, M. M., and Morrisey, E. E. (2002). Wnt7b regulates mesenchymal proliferation and vascular development in the lung. *Development* **129**, 4831–4842.

Sime, P. J., Xing, Z., Graham, F. L., Csaky, K. G., and Gauldie, J. (1997). Adenovector-mediated gene transfer of active transforming growth factor-beta1 induces prolonged severe fibrosis in rat lung. *J. Clin. Invest.* **100**, 768–776.

Smith, N. P., Jesudason, E. C., Featherstone, N. C., Corbett, H. J., and Losty, P. D. (2005). Recent advances in congenital diaphragmatic hernia. *Arch. Dis. Child.* **90**(**4**), 426–428.

Smith, N. P., Losty, P. D., Connell, M. G., Mayer, U., and Jesudason, E. C. (2006). Abnormal lung development precedes oligohydramnios in a transgenic murine model of renal dysgenesis. *J. Urol.* **175**(**2**), 783–786.

Sock, E., Rettig, S. D., Enderich, J., Bosl, M. R., Tamm, E. R., and Wegner, M. (2004). Gene targeting reveals a widespread role for the high-mobility-group transcription factor Sox11 in tissue remodeling. *Mol. Cell. Biol.* **24**, 6635–6644.

Souza, P., Tanswell, A. K., and Post, M. (1996). Different roles for PDGF-alpha and -beta receptors in embryonic lung development. *Am. J. Respir. Cell Mol. Biol.* **15**, 551–562.

Souza, P., O'Brodovich, H., and Post, M. (1995a). Lung fluid restriction affects growth but not airway branching of embryonic rat lung. *Int. J. Dev. Biol.* **39**(**4**), 629–637.

Souza, P., Kuliszewski, M., Wang, J., Tseu, I., Tanswell, A. K., and Post, M. (1995b). PDGF-AA and its receptor influence early lung branching via an epithelial-mesenchymal interaction. *Development* **121**, 2559–2567.

Souza, P., Sedlackova, L., Kuliszewski, M., Wang, J., Liu, J., Tseu, I., Liu, M., Tanswell, A. K., and Post, M. (1994). Antisense oligodeoxynucleotides targeting PDGF-B mRNA inhibit cell proliferation during embryonic rat lung development. *Development* **120**, 2163–2173.

Spilde, T. L., Bhatia, A. M., Mehta, S., Ostlie, D. J., Hembree, M. J., Preuett, B. L., Prasadan, K., Li, Z., Snyder, C. L., and Gittes, G. K. (2003). Defective sonic hedgehog signaling in esophageal atresia with tracheoesophageal fistula. *Surgery.* **134**, 345–350.

Srinivasan, S., Strange, J., Awonusonu, F., and Bruce, M. C. (2002). Insulin-like growth factor I receptor is downregulated after alveolarization in an apoptotic fibroblast subset. *Am. J. Physiol. Lung Cell Mol. Physiol.* **282**, L457–L467.

Starrett, R. W. and de Lorimier, A. A. (1975). Congenital diaphragmatic hernia in lambs: Hemodynamic and ventilatory changes with breathing. *J. Pediatr. Surg.* **10(5)**, 575–582.

Steele-Perkins, G., Plachez, C., Butz, K. G., Yang, G., Bachurski, C. J., Kinsman, S. L., Litwack, E. D., Richards, L. J., and Gronostajski, R. M. (2005). The transcription factor gene Nfib is essential for both lung maturation and brain development. *Mol. Cell. Biol.* **25**, 685–698.

Sterner-Kock, A., Thorey, I. S., Koli, K., Wempe, F., Otte, J., Bangsow, T., Kuhlmeier, K., Kirchner, T., Jin, S., Keski-Oja, J., and von Melchner, H. (2002). Disruption of the gene encoding the latent transforming growth factor-beta binding protein 4 (LTBP-4) causes abnormal lung development, cardiomyopathy, and colorectal cancer. *Genes Dev.* **16**, 2264–2273.

Stevens, T., Phan, S., Frid, M. G., Alvarez, D., Herzog, E., and Stenmark, K. R. (2008). Lung vascular cell heterogeneity: Endothelium, smooth muscle, and fibroblasts. *Proc. Am. Thorac. Soc.* **5**, 783–791.

Storm van's Gravesande, K., and Omran, H. (2005). Primary ciliary dykskinesia: clinical presentation, diagnosis and genetics. *Ann. Med.* **37**, 439–449.

Sugahara, K., Iyama, K. I., Kimura, T., Sano, K., Darlington, G. J., Akiba, T., and Takiguchi, M. (2001). Mice lacking CCAAt/enhancer-binding protein-alphashow hyperproliferation of alveolar type II cells and increased surfactant protein mRNAs. *Cell Tissue Res.* **306**, 57–63.

Sun, J., Chen, H., Chen, C., Whitsett, J. A., Mishina, Y., Bringas, P.Jr., Ma, J. C., Warburton, D., and Shi, W. (2008). Prenatal lung epithelial cell-specific abrogation of Alk3-bone morphogenetic protein signaling causes neonatal respiratory distress by disrupting distal airway formation. *Am. J. Pathol.* **172(3)**, 571–582.

Sutherland, D., Samakovlis, C., and Krasnow, M. A. (1996). Branchless encodes a Drosophila FGF homolog that controls tracheal cell migration and the pattern of branching. *Cell* **87**, 1091–1101.

Takahashi, H., and Ikeda, T. (1996). Transcripts for two members of the transforming growth factor-beta superfamily BMP-3 and BMP-7 are expressed in developing rat embryos. *Dev. Dyn.* **207**, 439–449.

Tefft, D., De Langhe, S. P., Del Moral, P. M., Sala, F., Shi, W., Bellusci, S., and Warburton, D. (2005). A novel function for the protein tyrosine phosphatase Shp2 during lung branching morphogenesis. *Dev. Biol.* **282**, 422–431.

Tefft, D., Lee, M., Smith, S., Crowe, D. L., Bellusci, S., and Warburton, D. (2002). mSprouty2 inhibits FGF10-activated MAP kinase by differentially binding to upstream target proteins. *Am. J. Physiol. Lung Cell Mol. Physiol.* **283**, L700–L706.

Tefft, J. D., Lee, M. Smith, S., Leinwand, M., Zhao, J. Bringas, P. Jr., Crowe, D. L., and Warburton, D. (1999) Conserved function of mSpry-2, a murine homolog of Drosophila sprouty, which negatively modulates respiratory organogenesis. *Curr. Biol.* **9**, 219–222.

Thom, J. and Perks, A. M. (1990). The effects of furosemide and bumetanide on lung liquid production by in vitro lungs from fetal guinea pigs. *Can. J. Physiol. Pharmacol.* **68(8)**, 1131–1135.

Thurlbeck, W. (1983). Postpneumonectomy compensatory lung growth. *Am. Rev. Respir. Dis.* **128(6)**, 965–967.

Tiozzo, C., De Langhe, S., Carraro, G., Alam, D. A., Nagy, A., Wigfall, C., Hajihosseini, M. K., Warburton, D., Minoo, P., and Bellusci, S. (2009). Fibroblast growth factor 10 plays a causative role in the tracheal cartilage defects in a mouse model of apert syndrome. *Pediatr. Res.* **66**, 386–390.

Tompkins, D. H., Besnard, V., Lange, A. W., Wert, S. E., Keiser, A. R., Smith, A. N., Lang, R., and Whitsett, J. A. (2009). Sox2 is required for maintenance and differentiation of bronchiolar Clara, Ciliated, and Goblet cells. *PLoS One* **4**(**12**), e1932. doi: 10.1371/journal.pone.0008248

Torday, J. S., Sun, H., Wang, L., Torres, E., Sunday, M. E., and Rubin, L. P. (2002). Leptin mediates the parathyroid hormone-related protein paracrine stimulation of fetal lung maturation. *Am. J. Physiol. Lung Cell Mol. Physiol.* **282**(**3**), L405–L410.

Torday, J. S., and Rehan, V. K. (2002). Stretch-stimulated surfactant synthesis is coordinated by the paracrine actions of PTHrP and leptin. *Am. J. Physiol. Lung Cell Mol. Physiol.* **283**, L130–L135.

Torday, J. S., Torday, D. P., Gutnick, J., Qin, J., and Rehan, V. (2001). Biologic role of fetal lung fibroblast triglycerides as antioxidants. *Pediatr. Res.* **49**(**6**), 843–849.

Toti, P., Buonocore, G., Tanganelli, P., Catella, A. M., Palmeri, M. L., Vatti, R., and Seemayer, T. A. (1997). Bronchopulmonary dysplasia of the premature baby: An immunohistochemical study. *Pediatr. Pulmonol.* **24**, 22–28.

Tsao, P. N., Vasconcelos, M., Izvolsky, K. I., Qian, J., Lu, J., and Cardoso, W. V. (2009). Notch signaling controls the balance of ciliated and secretory cell fates in developing airways. *Development* **136**(**13**), 2297–2307.

Unbekandt, M., del Moral, P. M., Sala, F. G., Bellusci, S., Warburton, D., and Fleury, V. (2008). Tracheal occlusion increases the rate of epithelial branching of embryonic mouse lunch via the FGF10-FGFR2b-Sprouty2 pathway. *Mech. Dev.* **120**(**3–4**), 314–324.

Ulven, S. M., Gundersen, T. E., Weedon, M. S., Landaas, V. O., Sakhi, A. K., Fromm, S. H., Geronimo, B. A., Moskaug, J. O., and Blomhoff, R. (2000). Identification of endogenous retinoids, enzymes, binding proteins, and receptors during early postimplantation development in mouse: Important role of retinal dehydrogenase type 2 in synthesis of all-trans-retinoic acid. *Dev. Biol.* **220**, 379–391.

Urase, K., Mukasa, T., Igarashi, H., Ishii, Y., Yasugi, S., Momoi, M. Y., and Momoi, T. (1996). Spatial expression of Sonic hedgehog in the lung epithelium during branching morphogenesis. *Biochem. Biophys. Res. Commun.* **225**, 161–166.

Urness, L. D., Sorensen, L. K., and Li, D. Y. (2000). Arteriovenous malformations in mice lacking activin receptor-like kinase-1. *Nat. Genet.* **26**, 328–331.

Usui, H., Shibayama, M., Ohbayashi, N., Konishi, M., Takada, S., and Itoh, N. (2004). Fgf18 is required for embryonic lung alveolar development. *Biochem. Biophys. Res. Commun.* **322**, 887–892.

Vaccaro, C., and Brody, J. S. (1978). Ultrastructure of developing alveoli. I. The role of the interstitial fibroblast. *Anat. Rec.* **192**(**4**), 467–479.

van Tuyl, M., and Post, M. (2000). From fruitflies to mammals: Mechanisms of signalling via the Sonic hedgehog pathway in lung development. *Respir. Res.* **1**, 30–35.

Valencia, A. M., Beharry, K. D., Ang, J. G., Devarajan, K., Van Woerkom, R., Abrantes, M., Nishihara, K., Chang, E., Waltzman, J., and Modanlou, H. D. (2003). Early postnatal dexamethasone influences matrix metalloproteinase-2 and -9, and heir tissue inhibitors in the developing rat lung. *Pediatr. Pulmonol.* **35**, 456–462.

Ventura, J. J., Tenbaum, S., Perdiguero, E., Hth, M., Guerra, C., Barbacid, M., Pasparakis, M., and Nebreda, A. R. (2007). p38alpha MAP kinase is essential in lung stem and progenitor cell proliferation and differentiation. *Nat. Genet.* **39**, 750–758.

Vergnes, L., Peterfy, M., Bergo, M. O., Young, S. G., and Reue, K. (2004). Lamin B1 is required for mouse development and nuclear integrity. *Proc. Natl. Acad. Sci. USA* **101**, 10428–10433.

Vermeer, P. D., Harson, R., Einwalter, L. A., Moninger, T., and Zabner, J. (2003). Interleukin-9 induces goblet cell hyperplasia during repair of human airway epithelia. *Am. J. Respir. Cell Mol. Biol.* **28**, 286–295.

Volpe, M. V., Martin, A., Vosatka, R. J., Mazzoni, C. L., and Nielsen, H. C. (1997). Hoxb-5 expression in the developing mouse lung suggests a role in branching morphogenesis and epithelial cell fate. *Histochem. Cell Biol.* **108**, 495–504.

Vu, T. H. and Werb, Z. (2000). Matrix metalloproteinases: Effectors of development and normal physiology. *Genes Dev.* **14**, 2123–2133.

Vuolteenaho, R., Nissinen, M., Sainio, K., Byers, M., Eddy, R., Hirvonen, H., Shows, T. B., Sariola, H., Engvall, E., and Tryggvason, K. (1994). Human laminin M chain (merosin): Complete primary structure, chromosomal assignment, and expression of the M and A chain in human fetal tissues. *J. Cell Biol.* **124**, 381–394.

Wallace, H., Connell, M., Losty, P., Jesudason, E., and Southern, K. W. (2008). Embryonic lung growth is normal in a cftr-knockout mouse model. *Exp. Lung Res.* **34**(**10**), 717–727.

Wan, H., Dingle, S., Xu, Y., Besnard, V., Kaestner, K. H., Ang, S. L., Wert, S., Stahlman, M. T. and Whitsett, J. A. (2005). Compensatory roles of Foxa1 and Foxa2 during lung morphogenesis. *J. Biol. Chem.* **280**, 13809–13816.

Wang, C., Chang, K. C., Somers, G., Virshup, D., Ang, B. T., Tang, C., Yu, F., and Wang, H. (2009). Protein phosphatase 2A regulates self-renewal of Drosophila neural stem cells. *Development* **136**(**13**), 2287–2296.

Wang, Z., Shu, W., Lu, M. M., and Morrisey, E. E. (2005). Wnt7b activates canonical signaling in epithelial and vascular smooth muscle cells through interactions with Fzd1, Fzd10, and LRP5. *Mol. Cell. Biol.* **25**(**12**), 5022–5030.

Wani, M. A., Wert, S. E., and Lingrel, J. B. (1999). Lung Kruppel-like factor, a zinc finger transcription factor, is essential for normal lung development. *J. Biol. Chem.* **274**, 21180–21185.

Warburton, D., Perin, L., Defilippo, R., Bellusci, S., Shi, W., and Driscoll, B. (2008). Stem/progenitor cells in lung development, injury repair, and regeneration. *Proc. Am. Thorac. Soc.* **5**(**6**), 703–706.

Warburton, D., Schwarz, M., Tefft, D., Flores-Delgado, G., Anderson, K. D., and Cardoso, W. V. (2000). The molecular basis of lung morphogenesis. *Mech. Dev.* **92**(**1**), 55–81.

Warburton, D., Tefft, D., Mailleux, A., Bellusci, S., Thiery, J. P., Zhao, J., Buckley, S., Shi, W., and Driscoll, B. (2001). Do lung remodeling, repair, and regeneration recapitulate respiratory ontogeny?. *Am. J. Respir. Crit. Care Med.* **164**(**10 Pt 2**), S59–S62.

Warburton, D., and Olver, B. E. (1997). Coordination of genetic, epigenetic, and environmental factors in lung development, injury, and repair. *Chest* **111**(**6S**), 119S–122S.

Warburton, D., Seth, R., Shum, L., Horcher, P. G., Hall, F. L., Werb, Z., and Slavkin, H. C. (1992). Epigenetic role of epidermal growth factor expression and signalling in embryonic mouse lung morphogenesis. *Dev. Biol.* **149**, 123–133.

Weaver, M., Batts, L., and Hogan, B. L. (2003). Tissue interactions pattern the mesenchyme of the embryonic mouse lung. *Dev. Biol.* **258**, 169–184.

Weaver, M., Dunn, N. R., and Hogan, B. L. (2000). Bmp4 and Fgf10 play opposing roles during lung bud morphogenesis. *Development* **127**, 2695–2704.

Weaver, M., Yingling, J. M., Dunn, N. R., Bellusci, S., and Hogan, B. L. (1999). Bmp signaling regulates proximal-distal differentiation of endoderm in mouse lung development. *Development* **126**, 4005–4015.

Weinstein, M., Xu, X., Ohyama, K., and Deng, C. X. (1998). FGFR-3 and FGFR-4 function cooperatively to direct alveogenesis in the murine lung. *Development* **125**, 3615–3623.

Wendel, D. P., Taylor, D. G., Albertine, K. H., Keating, M. T., and Li, D. Y. (2000). Impaired distal airway development in mice lacking elastin. *Am. J. Respir. Cell Mol. Biol.* **23**, 320–326.

Wert, S. E., Dey, C. R., Blair, P. A., Kimura, S., and Whitsett, J. A. (2002). Increased expression of thyroid transcription factor-1 (TTF-1) in respiratory epithelial cells inhibits alveolarization and causes pulmonary inflammation. *Dev. Biol.* **242**, 75–87.

Whitsett, J. A., Clark, J. C., Picard, L., Tichelaar, J. W., Wert, S. E., Itoh, N., Perl, A. K., and Stahlman, M. T. (2002). Fibroblast growth factor 18 influences proximal programming during lung morphogenesis. *J. Biol. Chem.* **277**, 22743–22749.

Whitsett, J. A., and Glasser, S. W. (1998). Regulation of surfactant protein gene transcription. *Biochem. Biophys. Acta* **1408**, 303–311.

Wilson, J., DiFiore, J., and Peters, C. (1993). Experimental fetal tracheal ligation prevents the pulmonary hypoplasia associated with fetal nephrectomy: Possible application for congenital diaphragmatic hernia. *J. Pediatr. Surg.* **28**(**11**), 1433–1439.

Wodarz, A., and Nusse, R. (1998). Mechanisms of Wnt signaling in development. *Annu. Rev. Cell Dev. Biol.* **14**, 59–88.

Wongtrakool, C., Malpel, S., Gorenstein, J., Sedita, J., Ramirez, M. I., Underhill, T. M., and Cardoso, W. V. (2003). Downregulation of retinoic acid receptor alpha signaling is required for sacculation and type 1 cell formation in the developing lung. *J. Biol. Chem.* **278**(**47**), 46911–46918.

Wu, J. Y., Feng, L., Park, H. T., Havlioglu, N., Wen, L., Tang, H., Bacon, K. B., Jiang, Z., Zhang, X., and Rao, Y. (2001). The neuronal repellent Slit inhibits leukocyte chemotaxis induced by chemotactic factors. *Nature* **410**, 948–952.

Xia, H., Migliazza, L., Diez-Pardo, J. A., and Tovar, J. A. (1999). The tracheobronchial tree is abnormal in experimental congenital diaphragmatic hernia. *Pediatr. Surg. Int.* **15**(**3–4**), 184–187.

Xian, J., Clark, K. J., Fordham, R., Pannell, R., Rabbitts, T. H., and Rabbitts, P. H. (2001). Inadequate lung development and bronchial hyperplasia in mice with a targeted deletion in the Dutt1/Robo1 gene. *Proc. Natl. Acad. Sci. USA* **98**, 15062–15066.

Xu, K., Nieuwenhuis, E., Cohen, B. L., Wang, W., Canty, A. J., Danska, J. S., Coultas, L., Rossant, J., Wu, M. Y., Piscione, T. D., Nagy, A., Gossler, A., Hicks, G. G., Hui, C. C., Henkelman, R. M., Yu, L. X., Sled, J. G., Gridley, T., and Egan, S. E. (2010). Lunatic Fringe-mediated Notch signaling is required for lung alveogenesis. *Am. J. Physiol. Lung Cell Mol. Physiol.* **298**, L45–L56.

Yamashita, Y. (2009). Asymmetric stem cell division and pathology: insights from Drosophila stem cell systems. *J. Pathol.* **217**, 181–185.

Yan, C., Sever, Z., and Whitsett, J. A. (1995). Upstream enhancer activity in the human surfactant protein B gene is mediated by thyroid transcription factor 1. *J. Biol. Chem.* **270**, 24852–24857.

Yang, Y., Iwanaga, K., Raso, M. G., Wislez, M., Hanna, A. E., Wieder, E. D., Molidrem, J. J., Wistuba, I. I., Powis, G., Demayo, F. J., Kim C. F., and Kurie J. M. (2008). Phosphatidylinositol 3-kinase mediates bronchioalveolar stem cell expansion in mouse models of oncogenic K-ras-induced lung cancer. *PLoS One.* **3**, e2220.

Yang, H., Lu, M. M., Zhang, L., Whitsett, J. A., and Morrisey, E. E. (2002). GATA6 regulates differentiation of distal lung epithelium. *Development* **129**, 2233–2246.

Yi, R., Qin, Y., Macara, I. G., and Cullen, B. R. (2003). Exportin-5 mediates the nuclear export of pre-microRNAs and short hairpin RNAs. *Genes. Dev.* **17**, 3011–3016.

Zemke, A. C., Teisanu, R. M., Giangreco, A., Drake, J. A., Brockway, B. L., Reynolds, S. D., and Stripp, B. R. (2009). b-Catenin is not necessary or maintenance or repair of the bronchiolar epithelium. *Am. J. Respir. Cell Mol. Biol.* **41**(**5**), 535–543.

Zeng, X., Gray, M., Stahlman, M. T., and Whitsett, J. A. (2001). TGF-beta1 perturbs vascular development and inhibits epithelial differentiation in fetal lung in vivo. *Dev. Dyn.* **221**, 289–301.

Zeng, X., Wert, S. E., Federici, R., Peters, K. G., and Whitsett, J. A. (1998). VEGF enhances pulmonary vasculogenesis and disrupts lung morphogenesis in vivo. *Dev. Dyn.* **211**, 215–227.

Zhang, L., Whitsett, J. A., and Stripp, B. R. (1997). Regulation of Clara cell secretory protein gene transcription by thyroid transcription factor-1. *Biochim. Biophys. Acta* **1350**, 359–367.

Zhang, Y., Goss, A. M., Cohen, E. D., Kadzik, R., Lepore, J. J., Muthukumaraswamy, K., Yang, J., DeMayo, F. J., Whitsett, J. A., Parmacek, M. S., and Morrisey, E. E. (2008). A Gata6-Wnt pathway required for epithelial stem cell development and airway regeneration. *Nat. Genet.* **40(7)**, 862–870. doi: ng.157 [pii] 10.1038/ng.157

Zhao, J., Shi, W., Wang, Y. L., Chen, H., Bringas, P.Jr., Datto, M. B., Frederick, J. P., Wang, X. F., and Warburton, D. (2002). Smad3 deficiency attenuates bleomycin-induced pulmonary fibrosis in mice. *Am. J. Physiol. Lung Cell Mol. Physiol.* **282**, L585–L593.

Zhao, J., Chen, H., Peschon, J. J., Shi, W., Zhang, Y., Frank, S. J., and Warburton, D. (2001). Pulmonary hypoplasia in mice lacking tumor necrosis factor-alpha converting enzyme indicates an indispensable role for cell surface protein shedding during embryonic lung branching morphogenesis. *Dev. Biol.* **232**, 204–218.

Zhao, J., Sime, P. J., Bringas, P.Jr., Tefft, J. D., Buckley, S., Bu, D., Gauldie, J., and Warburton, D. (1999). Spatial-specific TGF-beta1 adenoviral expression determines morphogenetic phenotypes in embryonic mouse lung. *Eur. J. Cell Biol.* **78**, 715–725.

Zhou, L., Dey, C. R., Wert, S. E., Yan, C., Costa, R. H., and Whitsett, J. A. (1997). Hepatocyte nuclear factor-3beta limits cellular diversity in the developing respiratory epithelium and alters lung morphogenesis in vivo. *Dev. Dyn.* **210**, 305–314.

Zhou, L., Dey, C. R., Wert, S. E., and Whitsett, J. A. (1996a). Arrested lung morphogenesis in transgenic mice bearing an SP-C-TGF-beta 1 chimeric gene. *Dev. Biol.* **175**, 227–238.

Zhou, L., Lim, L., Costa, R. H., and Whitsett, J. A. (1996b). Thyroid transcription factor-1, hepatocyte nuclear factor-3beta, surfactant protein B, C, and Clara cell secretory protein in developing mouse lung. *J. Histochem. Cytochem.* **44**, 1183–1193.

Zhou, Q., Law, A. C., Rajagopal, J., Anderson, W. J., Gray, P. A., and Melton, D. A. (2007). A multipotent progenitor domain guides pancreatic organogenesis. *Dev. Cell* **13**, 103–114.

Zhuo, Y., Zhang, J., Laboy, M., and Lasky, J. A. (2003). Modulation of PDGF-C and PDGF-D expression during bleomycin-induced lung fibrosis. *Am. J. Physiol. Lung Cell Mol. Physiol.* **286(1)**, L182–L188. doi:10.1152/ajplung.00083.2003

CHAPTER FOUR

Transcriptional Networks and Signaling Pathways that Govern Vertebrate Intestinal Development

Joan K. Heath

Contents

1. Introduction 160
2. Formation of the Definitive Endoderm 162
3. The Formation and Regionalization of the Primitive Gut Tube 165
   3.1. Tube formation 165
   3.2. Formation of the enteric nervous system 167
   3.3. Patterning the early gut 168
   3.4. Role of Cdx2 168
   3.5. Role of *Hox* genes 171
   3.6. Role of Bmp signaling 172
   3.7. Role of *Hedgehog* genes 173
   3.8. Role of Wnt/planar cell polarity signaling 174
   3.9. Overview 174
4. Establishment of the Crypt–Villus Axis 175
   4.1. Role of *Hedgehog* genes 176
   4.2. Role of Wnt signaling 177
   4.3. Role of Bmps 178
5. Establishing the Stem Cell Niche and Homeostasis in the Intestinal Epithelium 179
   5.1. Role of Wnt/β-catenin signaling 179
   5.2. Role of Notch signaling 182
   5.3. Role of Eph–ephrin signaling 184
   5.4. Role of HNFα 184
6. Role of Intestinal Development Pathways in Cancer 185
Acknowledgments 186
References 186

Ludwig Institute for Cancer Research, Royal Melbourne Hospital, Parkville, Victoria, Australia and Department of Surgery, University of Melbourne, Parkville, Victoria, Australia

Current Topics in Developmental Biology, Volume 90

ISSN 0070-2153, DOI 10.1016/S0070-2153(10)90004-5

Abstract

The vertebrate intestine is a complex and highly specialized organ comprising tissues derived from all three germ layers. While a description of the morphological events underlying the consolidation and organization of the endoderm, mesoderm, and ectoderm-derived cells into a multi-layered, continuously renewing organ has been available for several decades, only recently has a strong genetic framework for this process started to emerge, and as yet it remains incomplete. This review highlights the roles played by a number of transcription factors and signaling pathways in the development of the vertebrate intestine from the moment the definitive endoderm is formed. These molecular pathways often interact with each other and play multiple roles at different stages of intestinal formation. What is currently attracting considerable attention in the field is the realization that the deregulated activities of these same pathways often play a key role in the initiation and progression of a number of complex, serious intestinal diseases, including inflammatory bowel disease and cancer.

1. Introduction

The alimentary or gastrointestinal (GI) tract is a long tubular system comprising the esophagus, stomach, small intestine (which is itself sub-divided into the duodenum, jejunum, and ileum), and the large intestine or colon, arranged in series along the rostrocaudal axis. Each successive organ exhibits distinct morphological features and expresses unique differentiated cell types that enable the tube to carry out its essential functions of ingestion, digestion, nutrient absorption, elimination, and metabolic homeostasis. In all vertebrates the inner surface of the intestine is irregular, with ridges and projections of various shapes and sizes to increase the surface area over which digestion and absorption can take place. The adult human intestinal tract is approximately 8 m in length and has a surface area typically in the region of $200\,m^2$ (the area of a tennis court) — quite an achievement for an organ that starts out early in development as a tiny ribbon of endoderm.

The mature intestine comprises four concentric layers of tissue: the mucosa, submucosa, muscularis propria, and serosa (Fig. 4.1). The inner layer is the mucosa (Fig. 4.1, inset), which comprises the intestinal epithelium, a supporting lamina propria, and the muscularis mucosae, which contains several layers of smooth muscle fibers. The endoderm-derived intestinal epithelium provides a water-tight barrier between the outside world and the body. It encloses a lumen through which all ingested material is transported and is the primary site of nutrient absorption. In mammals, the surface area of the intestinal epithelium is vastly increased by the formation

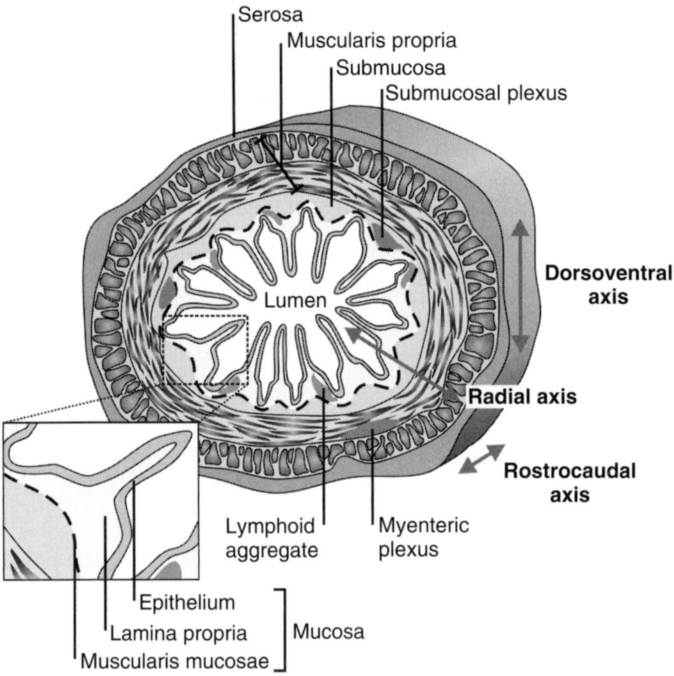

Figure 4.1 The multi-layered organization of the mature GI tract. The GI tract is a tube exhibiting rostrocaudal, dorsoventral, and radial asymmetry. It comprises four layers: the mucosa, the submucosa, the muscularis propria (comprising an inner layer of circular muscle fibers and an outer layer of longitudinal muscle fibers), and the serosa. The mucosa (inset) provides the inner lining of the tube and consists of an elaborately folded intestinal epithelium, the lamina propria, a supporting connective tissue containing lymphoid aggregates, and the muscularis mucosae, comprising a thin layer of muscle fibers. Clusters of enteric neurons are found in the submucosa (submucosal plexi) and between the two muscle layers (myenteric neurons). Figure adapted from Figure 14.1 (Burkitt *et al.*, 1993).

of elaborate structural compartments called crypts of Lieberkühn (hereafter called crypts) and villi. The crypts are goblet-shaped invaginations which penetrate the submucosa. In contrast, the finger-like villi are substantially longer and project into the lumen. These two inter-connected compartments accommodate cells exhibiting all aspects of cellular behavior including self-renewal, cell-fate determination, proliferation, differentiation, migration, and apoptosis. Multipotential stem cells are found towards the bottom of the crypts and these give rise to four distinct intestinal cell lineages and support the continuous renewal of the intestinal lining throughout life. In intimate contact with the epithelium is the mesoderm-derived lamina propria, a loose connective tissue containing α-smooth muscle actin-positive intestinal sub-epithelial myofibroblasts (ISEMFs)

(Powell *et al.*, 1999), blood and lymphatic capillaries, and large numbers of leukocytes.

The second layer is the submucosa, a mesoderm-derived, mesenchymal tissue which connects the mucosa to the muscle wall. It comprises a collagenous extracellular matrix containing larger blood and lymphatic vessels and huge numbers of immune cells (the intestine is the largest immune organ in the body). The third layer is the muscular wall or muscularis propria. It is also mesoderm-derived and contains several layers of smooth muscle arranged as inner circular and outer longitudinal muscle fibers. In concert with the neural crest (ectoderm)-derived enteric neurons, which populate the intestine in the submucosa region and between the muscle layers, the two muscle layers arranged at right angles to each other generate the peristaltic movements that propel the food along the gut. Finally, the fourth layer, or serosa, comprises a loose supporting tissue carrying the major vessels and nerves and lined by a simple squamous epithelium.

As mentioned above, the epithelial lining of the GI tract is derived from endoderm, one of the three principal germ layers. The endoderm contributes cells not only to the GI tract but also to the respiratory tract and a number of derivative organs, including the pancreas, liver, biliary tree, lung, thymus, and thyroid gland. Vertebrate endoderm organ development is an area of intense study and over the last decade studies in zebrafish and mouse, in particular, have enhanced our understanding of the genetic control of endoderm organ development.

The aim of this chapter is to document the chronological sequence of morphological events that take place during intestinal development—from the transformation of a ribbon of undifferentiated endoderm to a three-dimensional tube with rostrocaudal, dorsoventral, and radial asymmetry (Fig. 4.1), and to describe the genetic pathways that govern these events. Most information is derived from studies in mice but additional insights gleaned from studies in other model organisms, particularly zebrafish, are also included. As has been appreciated for some time, pathways governing developmental processes are frequently disrupted in a wide variety of pathological states, most notably cancer, as well as in a number of inherited syndromes. Where relevant, a discussion of the roles played by some of the intestinal development genes in the etiology of pathological states is highlighted.

2. Formation of the Definitive Endoderm

The initial specification of endoderm cells from the pluripotent cells of the epiblast in early vertebrate embryos depends on Nodal signaling (Fig. 4.2A). Nodal is a member of the transforming growth factor

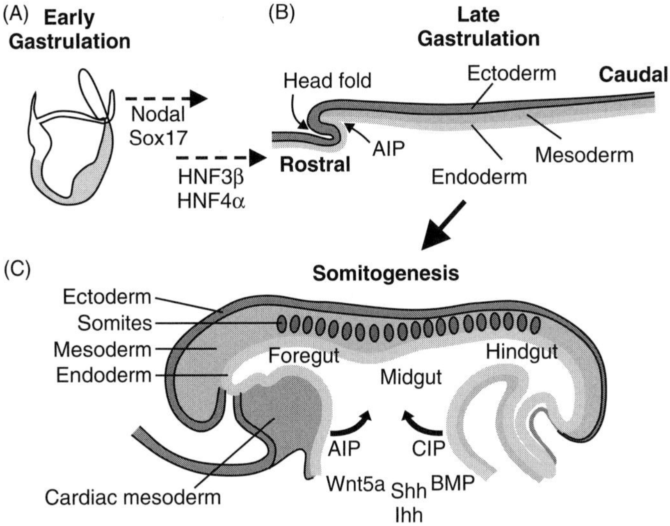

Figure 4.2 The key events during early intestinal organogenesis. (A) Nodal signaling during early gastrulation induces Sox17 expression and the specification of endoderm cells. (B) During late gastrulation, HNF3β and HNF4α contribute to the formation of a ribbon of definitive endoderm and the anterior intestinal portal (AIP) at its rostral end. Panel B adapted after Roberts 2000. (C) During somitogenesis, the endoderm layer transforms into a tube containing a lumen. This is achieved by ventrally directed movements of the AIP and the caudal intestinal portal (CIP). Following this, a proportion of the ventral endoderm cells undertake dorsolateral movements, thereby enclosing a lumen and completing the closure of the primitive gut tube. The combined activities of the Bmp, Shh/Ihh, and Wnt/PCP signaling pathways are required for this process.

(TGF)-β superfamily. A gradient of Nodal ligand activity, which in *Xenopus* is generated by the T-box transcription factor (TF), vegT (Xanthos *et al.*, 2001), stimulates the production of multiple TFs, including Gata5, Mezzo, Mixer (for reviews see Schier, 2003; Tam *et al.*, 2003; Zorn *et al.*, 2009) and, so far shown in zebrafish only, Casanova (Kikuchi *et al.*, 2001). These TFs all cooperate to stimulate the transcription of a SRY-box family TF, Sox17, which is indispensable for endoderm formation in *Xenopus* (Xanthos *et al.*, 2001), zebrafish (Aoki *et al.*, 2002a,b; Kikuchi *et al.*, 2001; Poulain *et al.*, 2002), and mice (Kanai-Azuma *et al.*, 2002). Upon specification, endodermal cells are consolidated into definitive endoderm during gastrulation, a series of complex cell movements that gives rise to the three germ layers (Fig. 4.2B).

Gastrulation movements differ markedly between fish, frog, chick, and mammalian embryos and these have been clearly described and compared in a number of reviews (Grapin-Botton *et al.*, 2000; Tam *et al.*, 2003). However, further refinement of our understanding of the precise details of these movements has been achieved recently in zebrafish (Keller *et al.*, 2008) and mice (Franklin *et al.*, 2008), where the processes are dissimilar.

In the zebrafish study (Keller *et al.*, 2008), spectacular movies generated using digital scanned laser light sheet fluorescence microscopy reveal the localization and movement of the nucleus in every cell of an entire zebrafish embryo during the first 24 h post-fertilization (hpf). After the first 4 h, when approximately 1000 pluripotent cells are generated on top of the yolk, a dynamic process called epiboly distributes the cells (blastomeres) thinly in a posterior direction over the yolk's surface. From previous studies (Alexander *et al.*, 1999), it was known that during this period (from 3.5 hpf), *sox17*-positive expression is induced in the marginal four layers of blastomeres, thereby fating these cells to become endoderm. The current study revealed that when these cells (alongside others destined to form mesoderm — a total of approximately 1550 cells) approach the "equator" (40% epiboly) they are swept inwards around the entire circumference of the embryo in a single synchronized internalization wave lasting approximately 2 h. On the dorsal side of the embryo, internalized cells become distributed along the entire future body axis. On the ventral side, the internalized ring of cells moves toward the vegetal pole. Eventually the ring closes at the vegetal pole and completes the formation of the inner cell layer or hypoblast. The observation reported here, that internalization does not continue throughout epiboly, partially revises the previous view of zebrafish gastrulation (Sepich *et al.*, 2005).

Once internalized, the endodermal cells move toward the dorsal side of the embryo in response to a gradient of Cxcl12/Cxcr4a(Sdf1) chemokine signaling (Mizoguchi *et al.*, 2008; Nair *et al.*, 2008). Cxcr4a-expressing endodermal cells move toward the dorsal side of the embryo guided by the overlying Cxcl12-expressing mesodermal cells. This results in endodermal cells converging at the midline of the embryo during late somitogenesis where they consolidate into a thin ribbon of endodermal tissue. Depletion of either Cxcl12 or Cxcr4a using antisense morpholinos disrupts integrin-dependent cell adhesion and the endoderm separates from the mesoderm. This results in anterior endodermal cells migrating too far in an anterior direction and failing to arrive at the midline at the appropriate time, resulting in the formation of bifid endodermal organs. It has been suggested that the GI bifurcations that sometimes arise during human development may be the result of defective CXCL12–CXCR4 signaling (Nair *et al.*, 2008).

Also interesting are the numerous reports of the involvement of CXCL12–CXCR4 signaling in malignancy. The CXCL12/CXCR4 axis has been shown to promote angiogenesis and the migration of tumor cells into metastatic sites by many cancers, including breast, lung, ovarian, renal, prostate cancer, and neuroblastoma (reviewed in Vandercappellen *et al.*, 2008). CXCR4-expressing tumors preferentially spread to tissues that highly express CXCL12, including the lung, liver, lymph nodes, and bone marrow and CXCL12 and CXCR4 are explicitly implicated in colorectal cancer (CRC) metastasis (Kollmar *et al.*, 2010; Ottaiano *et al.*, 2006).

A microarray study revealed that the expression of *CXCR4* by primary CRC tumors was inversely related to survival and that *CXCR4* expression was up-regulated in liver metastases compared to primary tumors (Kim *et al.*, 2005). High concentrations of CXCL12 expressed in the liver are thought to provide a specific homing target and growth-promoting environment for CXCR4-expressing CRC cells. Because interruption of the interaction between CXCL12 and its receptor may inhibit the metastatic process, the CXCR4 receptor has emerged as an important therapeutic target for cancer treatment (Vandercappellen *et al.*, 2008).

In the mouse embryo, Tam and colleagues (Franklin *et al.*, 2008; Tam *et al.*, 2007) used transgenic expression of fluorescent reporters and cell painting methods to meticulously map the origin, fate, and movement of every endoderm cell generated during gastrulation (from E7.5) through to the early organogenesis stage (E9.0; 16–19 somites). As a result of this tour-de-force achievement, the authors showed that the progenitors of all regions of the embryonic gut are present in the definitive endoderm of the early head-fold (E7.5) to early-somite stage (E8.0–E8.75) mouse embryo (Fig. 4.2A). They demonstrated that the most rostral end of the gut, the foregut, is derived from anterior endoderm that emerges early from the primitive streak (Fig. 4.2A). This process is under the control of a forkhead domain/winged helix TF, *HNF3β*, as shown by analysis of *HNF3β*-deficient mice, in which endodermal cells are specified but fail to form a foregut and midgut (Dufort *et al.*, 1998; Weinstein *et al.*, 1994). The midgut and hindgut, which together correspond to an expanse of tissue spanning the embryo from the level of somites 3–5 to the rostral end of the embryonic gut, are derived from later forming posterior endoderm cells derived from the lateral node and post-nodal cells, respectively. Interestingly, in *HNF3β*-deficent embryos the formation of the hindgut is unaffected. In both zebrafish and mouse, and indeed all vertebrates, the outcome of gastrulation and immediate post-gastrulation events is the formation of a morphologically homogeneous ribbon of endoderm tissue running along the ventral midline of the embryo (Fig. 4.2B).

3. The Formation and Regionalization of the Primitive Gut Tube

3.1. Tube formation

In mammals and the chick, the formation of a gut tube with a lumen is initiated by the ventrally directed invagination of the endoderm at its anterior and posterior ends. This gives rise to two ventral pockets, or portals, of endoderm known as the anterior intestinal portal (AIP; Fig. 4.2B, C) and the caudal intestinal portal (CIP; Fig. 4.2C). Reciprocal rostrocaudal movements of these invaginations (Fig. 4.2C) then serve to elongate the endodermal layer until they fuse in the midline of the embryo (Grapin-Botton *et al.*, 2000; Roberts, 2000). By E9.0, a proportion of the invaginated cells have also migrated dorsally and laterally to create a lumen surrounded by a tube of endoderm (for precise details of the movements of individual cells, see Franklin *et al.*, 2008) and lateral plate-derived splanchnic (visceral) mesoderm is recruited to the endoderm.

Lumen formation in zebrafish does not involve the same expansive morphological movements observed in birds and mammals. By 48 hpf the solid rod of endoderm forms a lumen by cavitation without cell death (Ng *et al.*, 2005) in a process resembling a mesenchymal to epithelial transition. It has been demonstrated that this conversion is dependent on Claudin (Cldn)15 (Bagnat *et al.*, 2007), which is a member of a family of 24 transmembrane tight junction proteins that form paracellular pores and determine the barrier properties of an epithelium. Whereas some Cldns typically increase epithelial resistance by decreasing cation permeability, others, such as Cldn2, create cation-selective pores with a concomitant decrease in resistance. Interestingly, Cldn15 can do either, depending on the complement of other Cldns in the tight junction (Zeissig *et al.*, 2007). In zebrafish, the formation of a single gut lumen is dependent on transcellular and paracellular ion transport through Cldn15-based pores between endoderm cells (Bagnat *et al.*, 2007). The asymmetric ion distribution causes the movement of fluid into spaces between the cells and the formation of multiple small lumens. As the fluid pressure builds up, these grow and start to coalesce, ultimately forming a single lumen. This process is under the control of the TF, Tcf2, which up-regulates the expression of Cldn15 and Na^+/K^+-ATPase (Bagnat *et al.*, 2007), which act together to generate an electrochemical gradient whereby Na^+/K^+-ATPase drives ion movement through Cldn15-based paracellular pores.

Since a lumen forms by a completely different mechanism in mammals, it is perhaps not surprising that intestinal tube formation in *Cldn15*$^{-/-}$ mice is unaffected. Interestingly, however, these mice develop a "megaintestine" after weaning in which the rostral part of the small intestine is approximately twice the length and diameter of its wildtype counterpart (Tamura *et al.*, 2008) and there are twice as many proliferating cells in the lower half of the crypts, the so-called transit amplifying compartment, which is characterized by uncommitted, actively proliferating cells derived from the immediately adjacent stem cells (Tamura *et al.*, 2008). Since Cldn15 appears to have no direct impact on cell proliferation, the authors speculated that the role played by Cldn15 in the formation of tight junctions and ion-conducting pores must somehow contribute to the integrity of the microenvironment and in turn contribute to the normal morphogenesis of the small intestine. Interestingly, while *Cldn15* expression has yet to be specifically implicated in inflammatory bowel disease (IBD), up-regulation of *Cldn2* expression is highly correlated with decreased intestinal epithelial barrier function in ulcerative colitis and active Crohn's disease (Zeissig *et al.*, 2007).

3.2. Formation of the enteric nervous system

The end of tube formation (E9.5 in mouse) coincides with the beginning of enteric nervous system (ENS) development. The ENS is the largest and most complex part of the peripheral nervous system and the only part capable of functioning independently of central nervous system innervation (Burzynski *et al.*, 2009). In birds and mammals, the ENS is localized into two main networks: the ganglionated, enteric plexi of the submucosa, and the myenteric plexi, located between the longitudinal and circular smooth muscle layers (Fig. 4.1).

As for all parts of the peripheral nervous system, the ENS is derived from the neural crest. In all species examined so far, ENS precursors are principally derived from neural crest cells that delaminate from the vagal region of the dorsal neural tube located between somite pairs 3–7 (Le Douarin *et al.*, 1973). These cells enter the foregut and migrate caudally in response to chemoattractant molecules expressed by the gut mesenchyme, most notably the Glial-derived neurotrophic family ligand, Gdnf (Young *et al.*, 2001). The ENS precursor cells express Gdnf co-receptors, the tyrosine kinase Ret, and a GPI-anchored receptor GFRα1, which together form a receptor complex that responds to Gdnf. The outcome of this interaction is the directed migration of ENS precursor cells along the mesenchyme in multiple chains, with individual cells ceasing migration at intervals along the way so that ultimately all regions of the gut are colonized evenly. The ENS precursors that migrate all the way to the caudal part of the hindgut spend 4–5 days on their journey; upon their arrival these cells have travelled further than any other neural crest-derived cells in the body (Anderson *et al.*, 2006; Young *et al.*, 2004). In *Ret*-deficient mice, the ENS precursors fail to arrive in the midgut by E10.5 and instead remain in the foregut where they undergo apoptosis (Durbec *et al.*, 1996). Subsequent studies have shown that Ret signaling is required not only for ENS precursor cell migration but also for their survival and proliferation (Taraviras *et al.*, 1999; Young *et al.*, 2001).

A congenital syndrome associated with *RET* mutations is Hirschsprung's disease (also known as congenital aganglionic megacolon), which affects 1 in 5000 live births (reviewed in Burzynski *et al.*, 2009). The lack of ENS innervation in the most caudal part of the large intestine in Hirschsprung's disease produces an intestinal obstruction due to tonic contraction of the aganglionic gut segment. This is manifest as an enlarged colon, or megacolon; the hallmark of the syndrome. Hirschsprung's disease is multigenic and additional mutations contributing to the etiology of the disease have been found in the *ENDOTHELIN-3* (*ET-3*) gene. In mouse, the expression of

Et-3 in the hindgut mesenchyme coincides precisely with the colonization of this region by ENS precursors (Leibl *et al.*, 1999) and subsequent studies have suggested a role for *Et-3* in maintaining their progenitor state and proliferation (Bondurand *et al.*, 2006; Landman *et al.*, 2007; Simpson *et al.*, 2007).

3.3. Patterning the early gut

The patterning of axial tissues in bilaterally symmetrical organisms (bilaterians) is orchestrated predominantly by two families of TFs known as *Hox and ParaHox* genes, which are distributed in clusters across several chromosomes in the mammalian genome. These two families of genes are descended from the same *ProtoHox* cluster identified in the common ancestors of bilaterians and cnidarians (jelly fish, sea anemones, etc.), more details of which are discussed in an earlier monograph in this series (Young *et al.*, 2009). Also in that review, the authors highlighted two phases of rostrocaudal patterning by the *Hox and ParaHox* genes during the development of axial structures: the first stage takes place during gastrulation as epiblast cells are recruited to form the three germ layers and the axis of the embryo extends. However, in the context of the endoderm, the role played by the *Hox and ParaHox* genes in this early phase of rostrocaudal patterning is not clear. In contrast, the roles played by these gene families during the second phase of rostrocaudal patterning, when the definitive endoderm is regionalized in a temporal, anterior to posterior sequence, are better understood, as discussed below. At this stage, reciprocal communication between the endoderm and the mesoderm is indispensable for normal intestinal development (Roberts, 2000).

3.4. Role of Cdx2

Cdx1, Cdx2, and *Pdx1* are *ParaHox* gene family members that have temporally and spatially distinct expression patterns in the gut during its development. For many years momentum has been gathering to support the notion that *Cdx2* functions as an intestinal "master" gene, upstream of *Hox* genes and pro-intestinal TFs that synergize to promote intestinal cell fate (Fig. 4.3). The validity of this view has recently been greatly enhanced by the publication of two new mouse models with tissue-specific and/or conditional *Cdx2* deletion that demonstrate that *Cdx2* is indispensable for intestinal development.

In mice, *Cdx2* is first expressed in the trophoectoderm of the blastocyst at E3.5 but is absent from the inner cell mass (Beck *et al.*, 1995). Its expression in the embryo proper initiates at E8.5 predominantly in the posterior (midgut and hindgut) endoderm as soon as it is formed (Beck *et al.*, 1995). Its expression in the embryo precedes that of other *ParaHox* or closely related *Hox* family members. By E12.5, its expression is extinguished

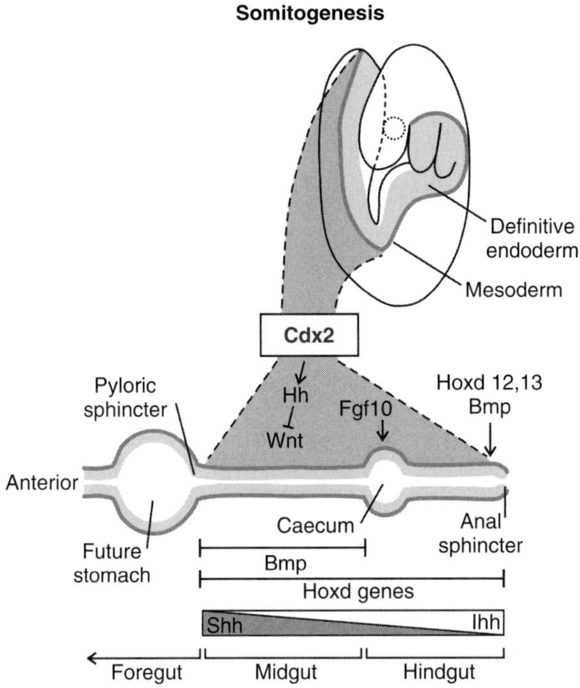

Figure 4.3 Central role of Cdx2 in the regionalization of the primitive gut tube. During somitogenesis, the gut is patterned into regions: the foregut, midgut and hindgut, the later encompassing the cecum. The intestine-specific TF, Cdx2, is expressed throughout the intestinal endoderm, from the junction of the foregut and the midgut to the anus. The activity of Cdx2 choreographs complex patterns of *Hox, Hh*, and *Wnt* gene expression at different positions along the rostrocaudal axis of the developing gut. These patterns are in turn interpreted to stimulate the expression of growth factors such as Fgf10, which induce the formation of conspicuous anatomical features such as the cecum (for Fgf10), and the pyloric and anal sphincters. (See Color Insert.)

everywhere except the intestinal endoderm (Fig. 4.3), with a clear-cut anterior boundary at the foregut–midgut junction (between the future stomach and future duodenum), and highest levels of expression at the midgut–hindgut border, where the cecum develops (Silberg *et al.*, 2000). In contrast, *Cdx1* expression is not initiated in the gut until E12.5, with highest levels of expression in the most caudal part of the hindgut (Silberg *et al.*, 2000). While *Cdx1* and *Cdx2* are functionally redundant TFs in a number of axial patterning processes, for example vertebral patterning (Savory *et al.*, 2009), they do not exhibit functional redundancy in the developing gut even though their expression patterns largely overlap in this tissue from E12.5. Thus while *Cdx1*-deficient mice are born and develop normal intestines, *Cdx2*-deficient mice die around the time of implantation.

The first evidence that *Cdx2* plays a role in intestinal development was provided by analysis of adult heterozygous *Cdx2* mice in which spontaneous inactivation of the second *Cdx2* allele led to the formation of sporadic polyp-like lesions in the colon (Chawengsaksophak *et al.*, 1997; Tamai *et al.*, 1999). These lesions originated as outpockets of embryonic gut epithelium from E11.5 onwards and were contained as benign hamartomas after birth. At the same time, areas of the cecal and colonic epithelium that retained the wildtype Cdx2 allele contained parietal cells and mucus-neck cells typical of gastric epithelium, as well as structures reminiscent of small intestinal villi (Beck *et al.*, 1999; Tamai *et al.*, 1999), suggesting that Cdx2 haploinsufficiency leads to a failure to suppress more anterior intestinal fates. A much clearer picture of the role of *Cdx2* emerged recently as a result of the creation of two independent strains of conditional *Cdx2* knock-out mice, which demonstrated the importance of Cdx2 in the specification of the colon (Gao *et al.*, 2009) and in maintaining an intestinal program of gene expression (Gao *et al.*, 2009; Grainger *et al.*, 2010).

In the first study (Gao *et al.*, 2009), *loxP* flanked *Cdx2* alleles were excised at E9.5 by the activity of Cre recombinase which was induced in a temporal and tissue-specific manner by the *cis*-regulatory elements of the *Foxa3/Hnf3γ* gene (Lee *et al.*, 2005). Defects in the growth of the mutant intestine were manifest from E14.5 and the animals died at post-natal day 1 (P1). Strikingly, *Cdx2*-deficient mice lacked all intestinal tissue distal to a malformed cecum. This failure to form a colon was accompanied by markedly abnormal small intestinal epithelium morphology: stunted villi in the duodenum and a flattened epithelium in the jejunum and ileum. There was also a substantial loss of differentiated epithelial cells (goblet cells, enterocytes, and enteroendocrine cells). In addition the smooth muscle around the rostral midgut (duodenum) was expanded. Molecular analysis of the *Cdx2*-deficient caudal midgut (future ileum) at E18.5 by immunohistochemistry, *in situ* hybridization and whole genome microarray analysis revealed loss of expression of intestine-specific genes such as *Cdx1, Hnf1α, Hnf4α, Math1, Indian Hedgehog* (*Ihh*), *Sonic Hedgehog* (*Shh*), and *Isx*. Instead, the *Cdx2*-deficient ileum expressed genes characteristic of the squamous epithelium of the oesophagus, such as *Sox2, Wnt10a, Pax9* and *Keratins 5* and *13*. The authors concluded that the expression of *Cdx2* in the definitive endoderm at the end of gastrulation is required to repress a (default) foregut differentiation program in the midgut and hindgut and is essential to establish hindgut identity. Because loss of Hh signaling in the intestinal epithelium had previously been shown to cause smooth muscle expansion (Madison *et al.*, 2005), they further speculated that the expansion of the smooth muscle layer seen in the *Cdx2*-deficient midgut is due to decreased Hh signaling in the midgut epithelium and up-regulation of Wnt ligand expression in the underlying mesenchyme (Fig. 4.3).

In the same study, interesting interactions between *Cdx2* and *Hox* genes were demonstrated as a result of qRT-PCR analysis of segments of gut corresponding to caudal foregut (future stomach), rostral midgut (future duodenum), caudal midgut (future ileum), and cecum at E12.5. Whereas some gut-enriched *Hox* genes (*Hoxa3, Hoxb3, Hoxb4, Hoxc4, Hoxd4, Hoxa5, Hoxb5, Hoxc5, Hoxa7, Hoxb7*) continued to be expressed at normal levels in the *Cdx2*-deficient gut at E12.5, *Hoxb9, Hoxc9, Hoxa13*, and *Hoxd13* were all markedly down-regulated, though this was a transient effect and normal levels were restored at E14.5 (Gao *et al.*, 2009). The fact that only selected posterior *Hox* genes and not anterior *Hox* genes are down-regulated in *Cdx2*-deficient mice reinforces the concept that *Cdx2* exerts its major influence on gene expression in the midgut and hindgut.

In the second study, the use of an inducible Villin–Cre^{ER} system produced efficient excision of a *loxP*-flanked *Cdx2* allele in the midgut (but not the hindgut) at around E14.5, resulting in death of the animals shortly after birth (Grainger *et al.*, 2010). Inspection of the small intestine revealed epithelial cells exhibiting the morphological and gene expression characteristics of gastric mucosa, rather than small intestine, providing more evidence for a profound role for Cdx2 in opposing anterior endoderm specification programs and promoting intestinal cell fate. That this is the result of cell autonomous behavior was demonstrated using chromatin immunoprecipitation assays in which Cdx2 was found to occupy the promoters of the *Cdx1, HNF1α*, and *HNF4α* genes in intestinal epithelial cells (Gao *et al.*, 2009). In zebrafish, Cdx1b is the functional equivalent of Cdx2. Knockdown of *Cdx1b* expression using antisense morpholinos also resulted in a defective intestinal epithelium including a failure to develop goblet cells (Flores *et al.*, 2008).

As alluded to above, despite its high expression in the hindgut from E12.5, identifying a role for Cdx1 in intestinal development has proved elusive. Similarly, pancreatic and duodenal homeobox factor-1 (*Pdx1*), the other *ParaHox* gene family member to be expressed in the gut endoderm (this time in the region of the presumptive duodenum from E9.5) does not appear to be essential for intestinal development. While *Pdx1*-deficient mice are apancreatic and die a few days after birth (Jonsson *et al.*, 1994; Offield *et al.*, 1996), a conditional knock-out model designed to explore Pdx1 function in the intestine from E12.5 yielded no significant phenotype (Chen *et al.*, 2009).

3.5. Role of *Hox* genes

During embryonic development, the developing gut expresses combinations of Hox genes along its rostrocaudal axis (Fig. 4.3). Though not exhaustive, the functions of *Hox* genes in patterning the mammalian gut have been explored using transgenic and knock-out mice with modified

Hox gene expression levels and gene expression profiling (microarray) analysis. Ectopic expression of *Hoxc8*, under the control of the *cis*-regulatory elements of the *Hoxa4* gene, or over-expression of *Hoxa4* itself causes intestinal patterning defects, including the formation of a megacolon from E13.5 as a result of misdirected migration of ENS progenitors (Pollock *et al.*, 1992; Tennyson *et al.*, 1998; Wolgemuth *et al.*, 1989). Deletion of all the *Hoxd* genes, or the genomic interval spanning *Hoxd genes* 4–13, severely impairs the formation of both the ileo-cecal and anal sphincters (Zakany *et al.*, 1999). However, if only *Hoxd12* or *Hoxd13* is deleted, only the morphology of the anal sphincter is affected (Kondo *et al.*, 1996).

Hoxd genes are also involved in the formation of another prominent anatomical feature in the bird and mammalian gut: the cecum, a pouch-like organ that forms at the boundary between the future small intestine and colon. At E10 all *Hoxd* genes except *Hoxd12* and *Hoxd13* (which are normally only expressed near the anus) are co-expressed in the mesenchyme in a restricted region around the posterior midgut where the cecal bud will appear (Zacchetti *et al.*, 2007). Mutant mice harboring *Hoxd*1-10 deletions fail to form a cecum. This defect resulted not from the loss of the *Hoxd*1-10 genes *per se*, but rather as a result of the ectopic (gain-of-function) expression of *Hoxd11 and 12* in the mesenchyme of the posterior midgut at E12 due to a phenomenon known as posterior prevalence (Duboule *et al.*, 1994). Ectopic expression of *Hoxd11* and *12* results in a marked reduction of both *fibroblast growth factor (Fgf)10* and *Pitx1* expression in the cecal bud region. Since the cecum is also absent in *Fgf10*-deficient mice (Burns *et al.*, 2004), these data support a model in which anterior and posterior Hox products compete to control the expression and activity of *Fgf10*, which is required for the growth of the cecum (Fig. 4.3) (Zacchetti *et al.*, 2007).

These results indicate that precise patterns of *Hoxd* gene expression in the primitive gut are required to demarcate the regions in which prominent anatomical and physiological constrictions are formed. However, our understanding of the transcriptional and post-transcriptional mechanisms involved in creating such intricate, combinatorial patterns of gene expression, remains quite rudimentary. This is clearly an area warranting further investigation.

3.6. Role of Bmp signaling

Bone morphogenetic proteins (Bmps) are secreted ligands of the Tgfβ superfamily. Together with their receptors, they are expressed widely in dynamic and partially overlapping patterns in the developing mouse from E7.5 to E10.5 (Danesh *et al.*, 2009), indicating diverse roles in morphogenesis. Among the Bmp family, Bmp7 has a unique and strikingly asymmetric expression pattern in the mesoderm on the dorsal right side of the midgut mesoderm at E10.5. Although a function has yet to be ascribed to Bmp7 at this stage of development, this expression pattern coincides with the breaking of symmetry in the embryo at the time when it is undergoing rotation and the gut tube initiates turning.

To monitor the activity of multiple Bmp ligand–receptor combinations, antibodies that recognize active, phosphorylated forms of their downstream effectors/TFs (Smad1/5/8) have proven useful. Using this approach, Smad1/5/8 phosphorylation was detected in the ventral foregut endoderm of the chick, suggesting a role for Bmps in AIP formation (Fig. 4.2C) (Smith *et al.*, 2000). Later, at E12–E13, the expression of Bmp2, BmprIb, and BmprII was detected in the midgut in the smooth muscle progenitors (Torihashi *et al.*, 2009). Bmp2-soaked latex beads applied to an *in vitro* embryoid body model that produces contractile gut-like structures equivalent to E11/12 in the mouse embryo (Torihashi *et al.*, 2006) stimulate smooth muscle differentiation, an effect that may require the cooperation of platelet-derived growth factor-A (Pdgf-A) which was also increased in response to the Bmp2 beads. Pdgf-A is known to be expressed in the developing inner circular smooth muscle and is crucial for the longitudinal smooth muscle differentiation (Torihashi *et al.*, 2009). Bmps are also involved in the formation of the pyloric sphincter (Smith *et al.*, 1999). On the whole, however, the function of Bmps in early gut development is probably not fully understood.

3.7. Role of *Hedgehog* genes

The secreted protein Shh plays a central role in a diverse array of morphogenetic processes. It is predominantly expressed in epithelia at numerous sites of epithelial–mesenchymal interactions, including the gut, and has been implicated as one of the inductive signals from the endoderm that specifies and patterns the overlying mesoderm. In mouse, two members of the Hedgehog family, *Shh* and *Ihh* are robustly co-expressed in the midgut and hindgut endoderm during somitogenesis (E8.5–E10.5). Both proteins appear first in the caudal hindgut, then in the AIP (Figs. 4.2C and 4.3), and finally in the midgut as the developing tube closes (Bitgood *et al.*, 1995). Later (E16.5), their overlapping expression patterns resolve into an opposing gradient of *Hh* expression with *Shh* expression strongest in the stomach and midgut and *Ihh* expression strongest from the hindgut to the anus (Fig. 4.3) (Kolterud *et al.*, 2009; Ramalho-Santos *et al.*, 2000). Meanwhile the Hh receptors, *Patched* (*Ptc*) *1* and *2*, and the downstream TFs, *Gli 1* and *2*, are expressed in the adjacent mesoderm (Kolterud *et al.*, 2009). $Shh^{-/-}$ and $Ihh^{-/-}$ mice die soon after birth; each displaying numerous GI tract abnormalities, including impaired smooth muscle development (demonstrating a requirement for Hh signaling in gut mesenchyme differentiation), gut malrotation, and defects in the ENS (Ramalho-Santos *et al.*, 2000; Sukegawa *et al.*, 2000).

3.8. Role of Wnt/planar cell polarity signaling

Wnt/planar cell polarity (PCP) signaling has also been implicated in early gut development. For example, the Wnt/PCP ligand, Wnt5a, which does not interfere with Wnt/β-catenin signaling, is detected in the midgut mesenchyme at E9.5, and by E12.5 its expression has spread into the hindgut (excluding the cecum). Wnt5a-deficent mice exhibit multiple defects along the GI tract; the midgut is dramatically shortened and duplicated, forming a bifurcated lumen instead of a single tube (Cervantes *et al.*, 2009). Detailed analysis revealed that Wnt5a is essential for primitive gut closure before E10.0 (Fig. 4.2C) and elongation of the midgut region from E10.5 (Cervantes *et al.*, 2009). By E18.5, the majority of Wnt5a-deficient mice display anorectal malformation in the form of a blind-ending pouch of the distal gut (Tai *et al.*, 2009).

3.9. Overview

As the sophistication of conditional knock-out mouse models increases, it is gradually becoming possible to envisage some of the epistatic relationships that exist between the various TFs and signaling pathways that impact on early intestinal development. Though available data strongly support the notion that Cdx2 is a master regulator of intestinal fate and integrity, many questions remain. In particular it is not clear yet how the signals from the various transcriptional networks and signaling pathways that Cdx2 interacts with, such as *Hox, Hnf, Hh, Bmp*, and *Wnt*, are integrated to produce the sophisticated output required to sculpt and shape the developing intestine. While the formation of some of the more striking morphological features along the rostrocaudal axis of the gut, such as the pyloric and anal sphincters and the cecum, appear to depend very heavily on an integrated output from various combinations of *Hoxd* genes, the way these signals are induced, coordinated, and interpreted remains unknown. While genetic studies in mice and other vertebrate models are already starting to make an impression, major strides forward are likely to depend on more detailed comparisons of the transcriptomes in wildtype and mutant animals than has hitherto been possible. In this regard, the advent of sensitive and quantitative RNA sequencing technologies, capable of describing the transcriptional output from increasingly discrete regions of tissue, is timely. This technology, alongside the availability of sophisticated bioinformatics approaches, including functional clustering and pathway analysis tools, will allow us to propose new transcriptional hierarchies and networks to explain how specific aspects of the intestinal developmental program are achieved. Highly tractable genetic organisms like the zebrafish will be useful alongside the mouse in putting such hypotheses to the test.

4. Establishment of the Crypt–Villus Axis

Radial asymmetry, in the form of the crypt–villus axis, is a hallmark of the mature intestinal epithelium (Fig. 4.1). As mentioned previously, in the small intestine (formed from the midgut), this axis comprises two distinct morphological but contiguous compartments, the crypts and the villi. In contrast, the large intestine or colon (formed from the hindgut) ultimately develops crypts without villi, requiring the physically separated processes of cell proliferation and differentiation to take place over a shortened axis. One of the outstanding challenges in intestinal cell biology is the identification of the molecular mechanisms that drive the formation of crypts and villi. The formation of the villi begins in mice during mid-gestation (E14.5) and proceeds in a rostrocaudal wave along the entire intestinal tract. Differential patterns of cell survival and apoptosis convert the thickened "pseudostratified" endoderm into a polarized epithelial monolayer and sculpt it into nascent folds, the precursors of the villi (Abud *et al.*, 2004, 2005). From E16.5, the spaces between the folds, the inter-villus regions, become populated with cycling cells (Abud *et al.*, 2005); these regions are destined to become the specialized proliferative compartments, the crypts. Whereas in mouse and human the formation of colonic crypts is only initiated several days after birth, in rat these structures become evident during gestation. Using transmission electron microscopy (TEM) it was shown that prior to the formation of rat colonic crypts, the pseudostratified epithelium first becomes thick enough to accommodate them. Indeed, this rapidly proliferating layer increases from 1 to 4 cells thick ($28.1 + 1.7\,\mu m$) at E16 to 6–9 cells ($68.0 + 3.4\,\mu m$) thick at E18 such that the lumen of the intestinal tube is almost occluded (Chrusch *et al.*, 1990). Shortly thereafter, groups of epithelial cells are found indented in the underlying mesenchyme, and a crypt lumen is formed that is continuous with the main lumen of the colon.

Once again these major morphological transformations are orchestrated by interactions between the endoderm and the underlying mesenchyme. The close physical relationship between these two layers was also revealed in the same TEM study of foetal rat colon (Chrusch *et al.*, 1990). From E16, mesenchymal cells become more closely abutted to epithelium and extend pseudopodia-like processes through the basement membrane to contact epithelial cells. By E19, the basement membrane at the base of the nascent crypts is thinner and the gaps have widened to $1\,\mu m$, permitting extensive epithelial–mesenchymal interactions; however, neither epithelial nor mesenchymal cells migrate through the gaps. This study supports a role for intimate epithelial–mesenchymal interactions in crypt formation and studies with mutant mice have identified the Hh, Wnt, and Bmp signaling pathways as key mediators of this reciprocal communication (Fig. 4.4).

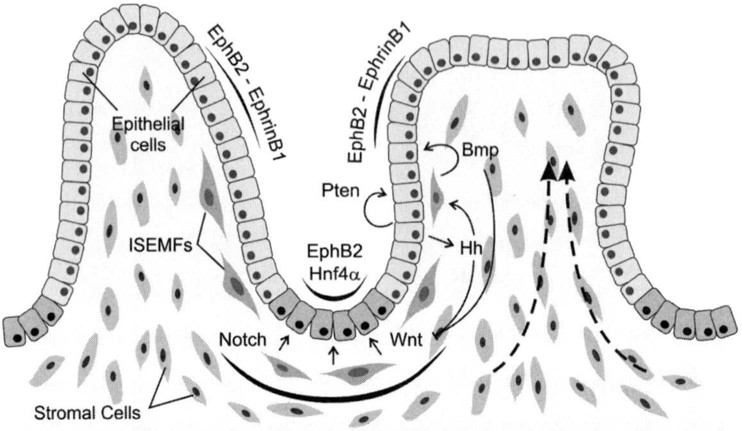

Figure 4.4 Generation of the crypt–villus axis. During mid-late gestation the endoderm, until now a pseudostratified epithelial layer, is transformed into a monolayer of epithelium that begins to elaborate folds, between which nestle the inter-villus regions (indicated by the crescent under the cells), the fore-runners of the crypts and villi. This morphological transformation is mediated by communication between the epithelial cells and cells in the underlying mesenchyme, especially the ISEMFs. Wnt signals from ISEMFs promote epithelial cell proliferation in the inter-villus region (arrows). Concurrently, epithelial cells higher up the folds secrete Hh ligands into the mesenchyme to increase Bmp expression; Bmps oppose the Wnt signal and restrict the proliferative activity of the epithelial cells to the inter-villus regions, partly through increasing the activity of Pten. The restriction of proliferative cells to the inter-villus region is reinforced by a gradient of Eph–ephrin signaling.

4.1. Role of *Hedgehog* genes

Shh$^{-/-}$ and *Ihh*$^{-/-}$ mice die soon after birth, each displaying a set of shared GI tract abnormalities (Ramalho-Santos *et al.*, 2000; Sukegawa *et al.*, 2000). In addition, *Shh*-deficient mice also exhibit intestinal metaplasia of the stomach epithelium, overproliferative villi leading to duodenal obstruction and an imperforate anus. In contrast, *Ihh*-deficient mice exhibit dramatically smaller villi in the small intestine and a dilated colon with an abnormally thin wall. Interestingly, at E16.5 *Shh* expression overlaps with cycling, bromodeoxyuridine (BrdU)-positive cells in the midgut (Kolterud *et al.*, 2009), whereas at E18.5 the expression of *Ihh* is stronger than *Shh* in cells occupying positions between the nascent villi of the midgut endoderm (Kolterud *et al.*, 2009). These data are consistent with a combined Shh–Ihh signal playing a role in proliferation and the formation of the stem cell compartment. To explore this concept further, all Hh signaling was knocked-down in the midgut endoderm of transgenic mice from E12.5 by expressing the pan-Hedgehog inhibitor, *Hhip* (Hh interacting protein) under the control of the *cis*-regulatory elements of the Villin gene (Madison *et al.*, 2005). Neonatal Hh-deficient mice developed areas of flattened villi containing highly proliferative cells, reminiscent of E14.5 pseudostratified epithelium that had failed to convert to an epithelial monolayer, while other areas developed branched villi as well as ectopic crypt-like structures on the tips of the villi. Further analysis revealed that targets of Wnt/β-catenin activity that are normally confined to the crypt compartment in wildtype mice (*c-Myc, Cdx1, Cd44*, and *Ephb2*) are all elevated and more broadly expressed in the intestinal epithelium of *Villin–Hhip* mice, suggesting that enhanced Wnt signaling from the mesenchyme to the epithelium upon loss of Hh signaling is responsible for the pro-proliferative phenotype. Of particular importance here is the behavior of the underlying ISEMFs, which are mislocalized in *Villin–Hhip* mice. While these cells are confined to the pericrytpal regions in wildtype mice, they are abnormally distributed in the villi immediately adjacent to ectopic proliferative epithelial cells in *Villin–Hhip* mice. This suggests that Hh-regulated Wnt signaling from ISEMFs is important for the proper size and location of the emerging precrypt compartment.

4.2. Role of Wnt signaling

That Wnt/β-catenin signaling plays a prominent role during the development of the crypt–villus axis was first revealed by analysis of Tcf4-deficient mice, which die shortly after birth (Korinek *et al.*, 1998). Tcf-4 is an intestine-specific member of the Tcf family of TFs that binds to β-catenin in the cell nucleus and forms a complex that can displace repressors on Wnt target genes and thereby activate their transcription. In Tcf4-deficient mice the conversion of pseudostratified endoderm into a monolayer of epithelium occurs normally in the midgut at E14.5; however, proliferating cells do not accumulate in the inter-villus regions (prospective crypt regions) and the neonatal epithelium comprises differentiated, non-cycling cells. This suggests that Wnt/β-catenin signaling plays a role in establishing the first crypts but not in villus formation. This conclusion is supported by studies exploiting transgenic mice expressing the TOP-GAL (*Tcf*)-reporter gene. In these mice, the developing intestine first exhibit Wnt/β-catenin signaling after the appearance of villi. Surprisingly, the Wnt/β-catenin activity is initiated in cells populating the growing villi, rather than in the inter-villus regions where the proliferating forerunners of intestinal crypts arise (Kim *et al.*, 2007). However, 3 days after birth the Wnt/β-catenin activity shifts from the villi to the inter-villus regions, an event which coincides with the stage at which the inter-villus regions start to invade the submucosa to form crypts. Together these results support the notion that a genetic program emanating from Tcf-4 is necessary to initiate crypt formation in the small intestine.

4.3. Role of Bmps

At E15.5 Bmp-4 expression occurs exclusively in the condensed mesenchyme underlying sites of future villus formation and inter-villus epithelial cells respond to the Bmp signal (Karlsson *et al.*, 2000). To further explore the role of Bmp signaling in the developing mouse intestine, the *cis*-regulatory elements of the fatty acid binding protein (*Fabp1*) gene were used to drive misexpression of the BMP antagonist Noggin in the epithelial cells. Noggin is a secreted cysteine knot glycoprotein, that binds with various degrees of affinity to BMP-2, -4, -5, -6, and -7 as well as GDF-5, -6 and Vg-1, and blocks their interaction with both the type I and type II BMP receptors (Groppe *et al.*, 2002). The resulting mice exhibit a profound over-proliferative phenotype in the intestinal epithelium, which encompasses the formation of numerous ectopic crypt units perpendicular to the crypt–villus axis, abnormal villus morphogenesis, stromal, and epithelial hyperplasia (Batts *et al.*, 2006). These changes suggest that Bmps play a role in restraining proliferative activity and indicate that intestinal Bmp signaling represses *de novo* crypt formation and polyp growth (Haramis *et al.*, 2004; Sancho *et al.*, 2003). Bmps are likely to achieve their function in this scenario by suppressing Wnt signaling from ISEMFs, since processes such as the duplication of stem cells, crypt fission, and increase in crypt number in the mature intestinal epithelium, which are all functions associated with Wnt signaling, are opposed by Bmp signaling (He *et al.*, 2004a). Furthermore, Bmp signaling positively regulates Pten, which increases the phosphorylation of β-catenin and enhances its rate of degradation (Persad *et al.*, 2001; Tian *et al.*, 2005). Pten is also a negative regulator of the phosphatidylinosital-3 kinase (PI3K)−Akt pathway which has been proposed to cooperate with the Wnt/ β-catenin pathway to promote intestinal epithelial cell proliferation (He *et al.*, 2004b; Tian *et al.*, 2004, 2005), although this view has recently been challenged (Ng *et al.*, 2009). Through these two actions, Bmp signaling opposes pro-proliferative Wnt/β-catenin signals. Interestingly, when gut endoderm and mesoderm co-cultures are treated with the Hh inhibitor, cyclopamine, mesodermal Bmp4 is reduced, demonstrating that Bmp4 is a target of endodermal Shh/Ihh (Sukegawa *et al.*, 2000). This is endorsed by studies in the larval frog indicating that intestinal stem cells secrete Shh and induce Bmp4 expression in the ISEMFs (Ishizuya-Oka *et al.*, 2006, 2008). While this interaction has yet to be formally demonstrated in the mammalian intestine, a hypothesis invoking a Shh → Bmp signal to inhibit Wnt/ β-catenin signaling and tightly restrict stem cell/progenitor cell proliferation to the base of the intestinal crypt is worthy of investigation.

5. Establishing the Stem Cell Niche and Homeostasis in the Intestinal Epithelium

At the time of birth, epithelial proliferation in mice remains restricted to the inter-villus regions and the more elaborate crypt structures have yet to form. This process takes place over the course of the next few weeks of post-natal life, when the epithelial cells of the inter-villus pockets penetrate into the mesenchyme to form mature crypts, which gradually increase in number by crypt fission. The mature intestinal epithelium then embarks on a process of renewal which is continuous through life. In the mouse, the epithelium turns over every 5 days.

The continuous renewal and integrity of the intestinal epithelium is achieved by a handful of stem cells positioned toward the base of each crypt in a protective niche. In the small intestine, the niche comprises stem cells, Paneth cells, and the ISEMFs that are implicated in the epithelial–mesenchymal crosstalk required to maintain the niche. Epithelial cells communicate with each other as well to determine the precise allocation of cells of the three other cell lineages and their orderly migration along the crypt–villus axis. In this context, the Notch and Eph/Ephrin pathways are the major players.

Recent reports suggest that stem cells undergo asymmetric cell division every 24 h (Barker *et al.*, 2008; Katsnelson, 2009), though this is a hotly debated topic (Barker *et al.*, 2008; Katsnelson, 2009). The two daughter cells give rise to another stem cell (self-renewal) and a progenitor cell which will leave the stem cell niche and enter the proliferative zone/transit amplifying region. In this zone, progenitor cells undergo cycles of proliferation every 12–16 h as they move upwards along the walls of the crypts. In the upper region of the crypts, migrating cells undergo differentiation and move onto the surface of a villus; each villus comprises several orderly columns of cells received from multiple crypts (Fig. 4.5). The villus accommodates differentiated cells of three distinct lineages: the abundant absorptive enterocytes, the mucin-producing goblet cells, and the relatively sparse hormone-producing enteroendocrine cells (Fig. 4.5). The fourth lineage, the Paneth cell, is unusual in that it migrates downward to the base of the crypts, where it becomes intermingled with the stem cells (Barker *et al.*, 2008; Bjerknes *et al.*, 2005). A clear and compelling picture of the dynamic nature of the mature intestinal epithelium is depicted in a movie generated by the laboratory of Hans Clevers, which is available in an online article (Katsnelson, 2009) at http://www.the-scientist.com/2009/07/1/51/1/#video.

5.1. Role of Wnt/β-catenin signaling

The continuous production of cells in the intestinal epithelium is maintained by a morphogen-like gradient of Wnt/β-catenin signaling

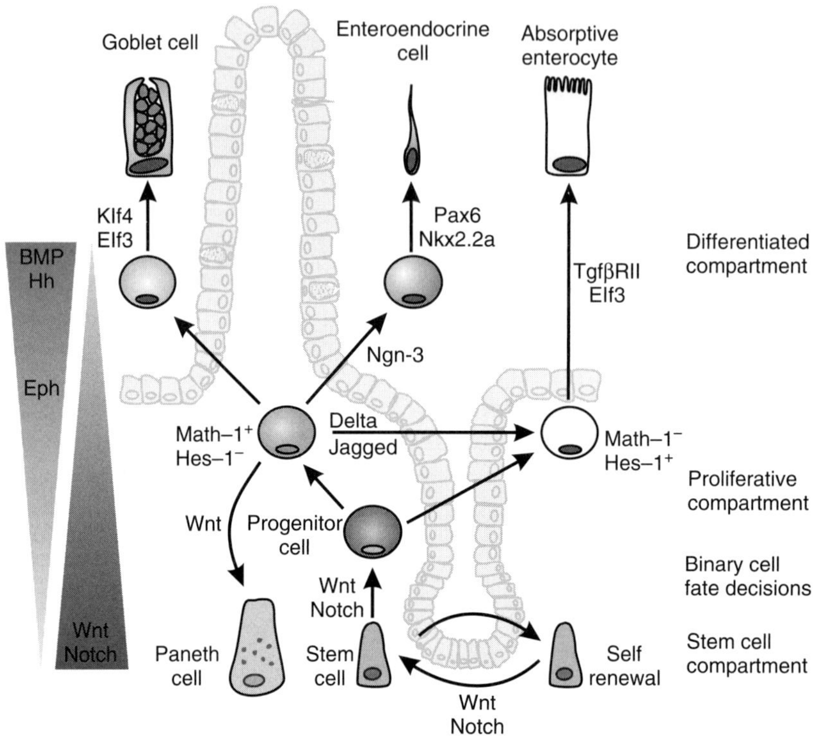

Figure 4.5 Organization of the intestinal epithelium and the crypts of Lieberkühn and cell lineage determination in the intestinal epithelium. Stem cells reside at or close to the base of the crypts, either between or just above the Paneth cells. Above the stem cells are the rapidly dividing progenitor/transit amplifying cells. Four different cell lineages (three secretory and one absorptive) arise from a common multipotential progenitor cell through a Notch-mediated binary cell fate decision. The three classes of secretory cells are Paneth cells, which secrete antimicrobial proteins, goblet cells containing thece full of mucus, and the much rarer enteroendocrine cells, which secrete various gut hormones. The absorptive cells have an elaborate brush border on their apical surface. Progenitor cells that receive a Notch signal (Delta/Jagged) from their neighbor(s) turn on the expression of the Hairy and enhancer of split-1 (Hes-1) bHLH TF. In turn, Hes-1 antagonizes Math-1, also a bHLH TF, which is required for commitment toward secretory lineages. Thus the Notch signal-receiving cell (Hes-1$^+$/Math-1$^-$) differentiates into an absorptive cell, rather than a secretory cell. In contrast, cells expressing Notch ligands (Hes-1$^-$/Math-1$^+$) escape Notch activation and differentiate into secretory cells. Once this secretory versus absorptive cell fate decision has been made, further TFs are required to promote terminal differentiation of the four cell lineages, some examples of which are shown. Neurogenin-3 (also a bHLH TF), Nkx2.2a, and Pax6 all promote enteroendocrine differentiation while the Kruppel-like TF, Klf-4, is required for goblet cell differentiation (Zheng *et al.*, 2009). An Ets-family TF, Elf-3, activates TgfβRII expression and thereby promotes terminal differentiation of goblet cells and absorptive cells (Flentjar *et al.*, 2007; Ng *et al.*, 2002). Epithelial turnover is maintained by a gradient of Wnt/β-catenin signaling, which is highest at the base of the crypts and tapers off toward the top of the crypts and villi. Other prominent signaling pathways (Bmp and Hh) are also established across the crypt/villus axis to oppose the pro-proliferative force of Wnt signaling and permit cell differentiation. Eph–ephrin signaling prevents intermingling of the proliferating and differentiating cells.

(Ireland *et al.*, 2004; Pinto *et al.*, 2003; Sansom *et al.*, 2004; van de Wetering *et al.*, 2002), which is highest at the base of the crypts where the stem cells reside, and tapers off toward the crypt–villus junction where fully differentiated cells are found (Fig. 4.5). Wnt/β-catenin signaling is not only the driving force in the proliferative compartment, but its activity at the crypt base also controls stem cell renewal and promotes the terminal differentiation of Paneth cells, which are thought to protect the stem cell niche by secreting antimicrobial molecules such as lysozyme, cryptidins/defensins (Fig. 4.5). Wnt/β-catenin signaling generates an opposing gradient of Eph–ephrin signaling along the established crypt–villus axis which functions to restrict the intermingling of proliferative and differentiated cells. Cells occupying positions at different levels along the adult crypt axis are characterized by different expression levels of EphB and Ephrin-B proteins such that the migratory behavior of cells is tightly linked to their proliferation and differentiation programs (Batlle *et al.*, 2002). Other prominent signaling pathways (Bmp and Hh) also establish reciprocal gradients along the crypt/villus axis to oppose the pro-proliferative force of Wnt signaling and permit cell differentiation. The role of Wnt/β-catenin signaling in controlling cell proliferation and homeostasis in the mature intestinal epithelium in mice is documented extensively in multiple reviews (for example see, van der Flier and Clevers, 2009; Verzi *et al.*, 2008). Accordingly, the discussion here will be limited to recent developments which have transformed our understanding of the molecular nature of the intestinal stem cell, a previously enigmatic subject.

Rapid progress in this field resulted from a microarray analysis of human CRC cell lines harboring endogenous adenomatous polyposis coli (APC) mutations and stably expressing inducible dominant-negative versions of TCF1 or TCF4, which led to the identification of a Wnt/TCF signaling "gene signature" comprising 80 Wnt-responsive genes (van der Flier *et al.*, 2007). Of these 80 genes, 17 exhibited an expression pattern restricted to the crypts (and adenomas) in a mouse intestinal adenoma model, and the expression of one of these, Lgr5/Gpr49, was found only in columnar cells occupying positions at the base of the crypts wedged in between Paneth cells (Cheng *et al.*, 1974). Lineage tracing experiments in mice harboring an inducible Lgr5 knock-in allele encoding GFP and Cre recombinase demonstrated that $Lgr5^{+}/GFP^{+}$ cells are multipotent, capable of generating all the cell lineages in the intestinal epithelium (Barker *et al.*, 2007, 2008). Moreover, single $Lgr5^{+}/GFP^{+}$ cells were capable of generating an entire intestinal crypt *in vitro* under precisely defined culture conditions designed to provide components expected to be secreted by mesenchymal cells (including ISEMFs) in the stem cell niche, including a Notch agonist peptide, R-spondin1 (Wnt agonist), and Noggin (Bmp antagonist) (Sato *et al.*, 2009). This is a remarkable observation in view of the long-held conviction that intestinal stem cells require positional cues from their highly specialized niche environment to generate radially asymmetrical crypts. Moreover, the ability to manipulate the behavior and differentiation of intestinal stem cells in this fashion undoubtedly has considerable implications for studies directed toward inducing endoderm-like tissues *in vitro*; it is likely that the strategies employed above may be of value in the future for the repair/regeneration of endodermal organs ravaged by diseases such as diabetes and IBD.

The next development was the identification of a gene expression signature for essentially pure (fluorescence activated cell sorted) $Lgr5^+/GFP^{hi}$ stem cells. Several of the genes identified, including *Achete scute-like 2* (*Ascl2*), which encodes a basic helix–loop–helix (bHLH) TF, were found to be Wnt target genes. *Ascl2* is of particular interest because its expression in the intestinal epithelium was found to be exquisitely restricted to $Lgr5^+$ stem cells (van der Flier *et al.*, 2009). Compound mutant mice harboring conditional knock-out alleles of *Ascl2* and a β-napthoflavone-inducible Ah-Cre transgene were generated and within a few days of inducing Cre activity (in a proportion of crypts), $Lgr5^+$stem cells disappeared. While the proliferative cells in the transit amplifying compartments of the recombined crypts continued to proliferate and provide the villi with differentiated cells for a few days, the failure of the stem cell-depleted crypts to replace them, coupled with their lack of inherent self-renewal capacity, caused increased apoptosis in this compartment. Conversely, mice harboring a *Villin–Ascl2* transgene to drive intestine-specific *Ascl2* expression from E14.5, developed crypt hyperplasia and ectopic crypts on villi (van der Flier *et al.*, 2009). These studies not only highlight the key role of *Ascl2* in determining intestinal stem cell fate but also herald an ability to provide a detailed characterization of intestinal stem cells that was unthinkable just a few years ago.

5.2. Role of Notch signaling

Notch and its ligands (Delta and Serrate/Jagged family members) are transmembrane proteins that control cell fate decisions by the exchange of inhibitory signals between cells that are in contact with each other. In the intestinal epithelium, Notch pathway components are expressed in the stem cells and crypt epithelial cells. In mouse and zebrafish mutants harboring mutations in Notch pathway components, or upon application of Notch inhibitors, such as γ-secretase inhibitors, the intestinal epithelium becomes engorged with mucin-producing goblet cells, indicating that cells that evade the Notch signal adopt a secretory cell fate (Crosnier *et al.*, 2005). Conversely, Notch signaling is indispensable for the differentiation of the enterocytes, the cells responsible for nutrient absorption and which constitute the bulk of the intestinal epithelium. The Notch-mediated choice between adopting an absorptive fate versus a secretory fate is illustrated schematically in Fig. 4.5. This figure depicts a simple model for determination of cell fate in the intestinal epithelium based largely on observations reported for Hes1$^{-/-}$ and Math$^{-/-}$ mice. Whereas Math$^{-/-}$ mice lack goblet and enteroendocrine cells (Yang *et al.*, 2001), Hes$^{-/-}$ mice display an opposite phenotype and display an excess of secretory cells over absorptive cells (Jensen *et al.*, 2000). Cells receiving a Notch signal are induced to express Hes-1, a bHLH transcriptional repressor. One of the targets of Hes-1 is Math-1. Cells without Math-1 expression remain as stem cells or can differentiate into absorptive cells, whereas cells expressing Math-1 become committed to a secretory cell fate (Fig. 4.5) (Crosnier *et al.*, 2005; Yang *et al.*, 2001). Notch signaling also suppresses Kruppel-like factor (Klf)-4 which is a C_2H_2 zinc-finger containing TF that is highly expressed in the GI tract and required for cells to adopt a goblet cell fate (Zheng *et al.*, 2009).

In addition to determining cell fate in the intestinal epithelium, Notch signaling also stimulates intestinal epithelial cell proliferation (Fre *et al.*, 2005, 2009; Rodilla *et al.*, 2009; van Es *et al.*, 2005). When the Villin–Cre system was used to induce the expression of a constitutively active version of the Notch receptor (N1ic) in intestinal epithelial cells from E14.5, the differentiation of the secretory cell lineages was inhibited as expected. In addition, N1ic over-expression in the intestinal epithelium resulted in a striking expansion of the population of proliferating cells in the inter-villus regions. Staining for proliferative markers (Ki67 and BrdU accumulation) revealed cycling cells distributed all along the villus axis (Fre *et al.*, 2005). Conversely, the conditional removal of the common Notch pathway TF, Csl/Rbj-2 results in mice with an intestinal phenotype characterized by an expansion of the post-mitotic goblet cell population at the expense of proliferative cells (van Es *et al.*, 2005). However, the expanded proliferative activity of intestinal epithelial cells observed in N1ic over-expressing mice is extinguished if these mice are crossed onto a Tcf4 null background; indeed the phenotype of the compound mice was exactly the same as that of Tcf4-deficient mice (Fre *et al.*, 2009). This indicates that for Notch signals to exert a pro-proliferative effect, they must be integrated with Wnt/Tcf4 signaling (Fre *et al.*, 2009). In a satisfying extension of the same study, the authors uncovered a striking synergy between activation of Notch and Wnt/β-catenin signaling in promoting adenoma formation of compound Apc$^{min/+}$; Villin–CreN1ic mice (Fre *et al.*, 2009). The pro-proliferative activity of Notch signaling appears to depend on the up-regulation of Jagged ligand expression in response to Wnt/β-catenin signaling, though its impact on cell lineage determination does not (Rodilla *et al.*, 2009). Interestingly, tumors taken from familial adenomatous polyposis patients, who harbor APC mutations and are highly predisposed to colon cancer development, show over-expression of Jagged ligand (Rodilla *et al.*, 2009). A more comprehensive discussion of the role of Notch in the intestine is available in a number of excellent recent reviews (Barker *et al.*, 2008; Crosnier *et al.*, 2006; van der Flier and Clevers 2009; Zorn *et al.*, 2009).

5.3. Role of Eph–ephrin signaling

Like Notch signaling, interactions between membrane-bound Eph and Ephrin molecules are bidirectional cell–cell communication events that occur between neighboring cells in contact with each other. Unlike Delta-Notch signaling, which controls binary cell fate decisions, the ligation and clustering of Eph–ephrin complexes dictate a switch between attractive and repulsive cellular behaviors that impact on the precise positioning of cells and their pathways of migration (Pasquale, 2005; Vearing *et al.*, 2005; Wimmer-Kleikamp *et al.*, 2004). This is thought to involve interactions with the Rho and Ras families of small GTPases, which in turn impact on the cytoskeleton (Coulthard *et al.*, 2002).

The possible role of Eph–ephrin signaling in intestinal development was first suggested by expression profiling experiments showing that components of this pathway are targets of Wnt/β-catenin signaling (Batlle *et al.*, 2002). Subsequently, EphB2, EphB3, and Ephrin-B1 were found to be expressed in the inter-villus regions of neonatal mice, with restricted and complementary expression domains (Batlle *et al.*, 2002). In particular, EphB receptor expression alone is found in proliferative cells at the base of the inter-villus pockets while EphB2 and Ephrin-B1 are co-expressed in cells adjacent to the proliferative zone (Fig. 4.4). In *EphB3* null mice, the distribution of proliferative cells in the inter-villus regions is unaffected; however, Paneth cells no longer follow a downward migratory path and are found in positions further up the growing villus. This is not the case, however, in compound *EphB2/EphB3* mutant mice where the number of proliferating cells is substantially decreased and differentiated cells are found occupying positions in the proliferative zone, suggesting that EphB receptor expression in wildtype proliferative cells repels the downward migration of differentiated cells (Batlle *et al.*, 2002).

The ability of EphB2 to influence both cell positioning and cell proliferation in the intestinal epithelium was recently shown to be mediated by independent signaling pathways (Genander *et al.*, 2009). Whereas EphB2-regulated cell positioning was found to be unconnected to its kinase activity and involved increased transcription of both the catalytic and regulatory sub-units of PI3K, the tyrosine kinase activity of EphB2 was integral to its ability to promote mitosis, and was mediated by post-transcriptional events that culminated in a c-Abl-dependent stabilization of Cyclin D1 protein levels (Genander *et al.*, 2009).

5.4. Role of HNFα

Although a role for hepatocyte nuclear factor (HNF)-4α in shaping the crypt–villus axis is only just emerging, the likelihood that this protein may be key player in the development of human ulcerative colitis (Barrett *et al.*,

2009) has attracted renewed attention in this nuclear hormone TF. It has been known for some time that HNF-4α plays an indispensable role during early development; Hnf4α-deficient mice die before E10.5 as a result of disrupted gastrulation (Chen *et al.*, 1994). Accordingly, to investigate its function in the development of the digestive system, a number of conditional mouse models were generated. Intestine-specific knock-out mice were created by the excision of *loxP* flanked *Hnf4α* alleles by Cre recombinase induced at E9.5 by the *cis*-regulatory elements of the *Foxa3* gene (Lee *et al.*, 2005). These mice die perinatally, but colonic tissue recovered from late development embryos (E18.5) shows reduced epithelial cell proliferation, an absence of inter-villus regions and defective goblet cell maturation (Garrison *et al.*, 2006). Further work demonstrated that during mid-gestation Hnf4α is essential for the expression of a multitude of genes encoding cell junction and adhesion proteins (Battle *et al.*, 2006), indicating a role in the development of epithelial apicobasal polarity (Fig. 4.4).

To explore the impact of Hnf4α-deficiency on intestinal barrier function, *loxP* flanked *Hnf4α* alleles were excised at E14.5 by the activity of Cre recombinase induced by the *cis*-regulatory elements of the *Villin* gene. Inducing recombination later in development circumvented the perinatal lethality. Following dextran sodium sulfate exposure, adult mice showed increased epithelial permeability and markedly more severe colitis than their wild-type littermates (Ahn *et al.*, 2008). This observation strongly resonates with the results of a recent genome-wide association study which identified a highly significant association between a single nucleotide polymorphism in the HNF4α gene and ulcerative colitis (Barrett *et al.*, 2009).

6. Role of Intestinal Development Pathways in Cancer

This review has highlighted a few signaling pathways that play critical roles during vertebrate intestinal development, most notably Wnt/β-catenin, Hh, Eph, Notch, and Bmp signaling. In the normal adult intestinal epithelium, these pathways continue to interact with each other to maintain tight control over the opposing processes of proliferation and differentiation along the crypt–villus axis. However, it is clear that with the passage of time, specific components of these pathways are particularly susceptible to mutations that predispose to aberrant activity, the consequences of which can be catastrophic for the orderly behavior of the cell. In particular, mutations in one or more components of the Wnt/β-catenin signaling pathway that result in elevated TCF4/β-catenin co-transcriptional activity are found in almost all instances of CRC. It is widely accepted that nonsense mutations encoding truncated versions of the tumor suppressor protein, APC are integral to at least 80% of all colon cancer cases (Cottrell *et al.*, 1992; Powell *et al.*, 1992). APC, alongside AXIN, casein kinase 1 (CK1), and glycogen synthase kinase (GSK)-3β are key participants in the so-called "β-catenin destruction complex" which restrains the transcriptional activity of β-catenin by phosphorylating the molecule on Ser^{33} and Ser^{37}, thereby targeting it for degradation by the proteasomal pathway (Orford *et al.*, 1997). The importance of this regulatory mechanism is reiterated by the fact that in the remaining 20% of CRC cases that do not harbor *APC* mutations, other mutations impacting on Wnt/β-catenin signaling are invoked. These include missense mutations in *CTNNB1* that encode versions of β-catenin that cannot be phosphorylated (Polakis, 2000). Mutations in the Wnt components AXIN1, AXIN2, and TCF4 have been found in microsatellite-unstable colon cancers, though it is not proven in every case that these changes are functional (Segditsas *et al.*, 2006). In addition, aberrant HH, EPH, NOTCH, TGFβ, and BMP signaling are all strongly associated with the development of multiple cancers, including cancers of the GI tract. Although tremendous effort is being directed to the development of drugs to target these pathways in one way or another, currently there are no strategies available to successfully combat their impact on cell behavior once they have become perturbed. However, as we strive for further refinement of our mechanistic understanding of the roles these pathways play during intestinal organogenesis, we are likely to produce new insights into the molecular players involved and the critical interactions between them, and this in turn may suggest new targets for cancer therapy.

ACKNOWLEDGMENTS

I apologize to the many authors who have contributed knowledge to our understanding of the genetic control of intestinal development whose work is not cited here. I thank Janna Taylor for figure preparation and Tony Burgess, Matthias Ernst, Tanya de Jong-Curtain, and Yeliz Rifat for helpful comments on the manuscript. Work in my laboratory on the genetic basis of intestinal development and colon cancer is funded by the National Health and Medical Research Council (NHMRC), Australia through Program Grant 487922, and an NHMRC Senior Research Fellowship.

REFERENCES

Abud, H. E. *et al.* (2005). Growth of intestinal epithelium in organ culture is dependent on EGF signalling. *Exp. Cell Res.* **303**, 252–262.

Abud, H. E. and Heath, J. K. (2004). Detecting apoptosis during the formation of polarized intestinal epithelium in organ culture. *Cell Death Differ.* **11**, 788–789.

Ahn, S. H. *et al.* (2008). Hepatocyte nuclear factor 4alpha in the intestinal epithelial cells protects against inflammatory bowel disease. *Inflamm. Bowel Dis.* **14**, 908–920.

Alexander, J. and Stainier, D.Y.R. (1999). A molecular pathway leading to endoderm formation in zebrafish. *Curr. Biol.* **9**, 1147–1157.

Anderson, R. B. *et al.* (2006). Neural crest and the development of the enteric nervous system. *Adv. Exp. Med. Biol.* **589**, 181–196.

Aoki, T. O. *et al.* (2002a). Molecular integration of casanova in the nodal signalling pathway controlling endoderm formation. *Development* **129**, 275–286.

Aoki, T. O. *et al.* (2002b). Regulation of nodal signalling and mesendoderm formation by TARAM-A, a TGF[beta]-related type I receptor. *Dev. Biol.* **241**, 273–288.

Bagnat, M. *et al.* (2007). Genetic control of single lumen formation in the zebrafish gut. *Nat. Cell Biol.* **9**, 954–960.

Barker, N. *et al.* (2007). Identification of stem cells in small intestine and colon by marker gene lgr5. *Nature* **449**, 1003–1007.

Barker, N. *et al.* (2008). The intestinal stem cell. *Genes Dev.* **22**, 1856–1864.

Barrett, J. C. *et al.* (2009). Genome-wide association study of ulcerative colitis identifies three new susceptibility loci, including the HNF4A region. *Nat. Genet.* **41**, 1330–1334.

Batlle, E. *et al.* (2002). Beta-catenin and TCF mediate cell positioning in the intestinal epithelium by controlling the expression of EphB/ephrinB. *Cell* **111**, 251–263.

Battle, M. A. *et al.* (2006). Hepatocyte nuclear factor 4alpha orchestrates expression of cell adhesion proteins during the epithelial transformation of the developing liver. *Proc. Natl. Acad. Sci. USA* **103**, 8419–8424.

Batts, E. B. *et al.* (2006). Bmp signaling is required for intestinal growth and morphogenesis. *Dev. Dyn.* **235**, 1563–1570.

Beck, F. *et al.* (1995). Expression of cdx-2 in the mouse embryo and placenta: possible role in patterning of the extra-embryonic membranes. *Dev. Dyn.* **204**, 219–227.

Beck, F. *et al.* (1999). Reprogramming of intestinal differentiation and intercalary regeneration in cdx2 mutant mice. *Proc. Natl. Acad. Sci. USA* **96**, 7318–7323.

Bitgood, M. J. and McMahon, A. P. (1995). Hedgehog and Bmp genes are coexpressed at many diverse sites of cell–cell interaction in the mouse embryo. *Dev. Biol.* **172**, 126–138.

Bjerknes, M. and Cheng, H. (2005). Gastrointestinal stem cells. II. Intestinal stem cells. *Am. J. Physiol. Gastrointest. Liver Physiol.* **289**, G381–G387.

Bondurand, N. *et al.* (2006). Maintenance of mammalian enteric nervous system progenitors by SOX10 and endothelin 3 signalling. *Development* **133**, 2075–2086.

Burkitt, H. G. *et al.* (1993). *Wheater's Functional Histology*. Churchill Livingstone (Edinburgh and Melbourne).

Burns, R. C. *et al.* (2004). Requirement for fibroblast growth factor 10 or fibroblast growth factor receptor 2-IIIb signaling for cecal development in mouse. *Dev. Biol.* **265**, 61–74.

Burzynski, G. *et al.* (2009). Genetic model system studies of the development of the enteric nervous system, gut motility and Hirschsprung's disease. *Neurogastroenterol. Motil.* **21**, 113–127.

Cervantes, S. *et al.* (2009). Wnt5a is essential for intestinal elongation in mice. *Dev. Biol.* **326**, 285–294.

Chawengsaksophak, K. *et al.* (1997). Homeosis and intestinal tumours in cdx2 mutant mice. *Nature* **386**, 84–87.

Chen, W. S. *et al.* (1994). Disruption of the HNF-4 gene, expressed in visceral endoderm, leads to cell death in embryonic ectoderm and impaired gastrulation of mouse embryos. *Genes Dev.* **8**, 2466–2477.

Chen, C. *et al.* (2009). Pdx1 inactivation restricted to the intestinal epithelium in mice alters duodenal gene expression in enterocytes and enteroendocrine cells. *Am. J. Physiol. Gastrointest. Liver Physiol.* **297**, G1126–G1137.

Cheng, H. and Leblond, C. P. (1974). Origin, differentiation and renewal of the four main epithelial cell types in the mouse small intestine. I. Columnar cell. *Am. J. Anat.* **141**, 461–479.

Chrusch, C. *et al.* (1990). Morphogenesis as a parallel of invasion: the epithelial–mesenchymal boundary and basal lamina in foetal rat colon. *Clin. Exp. Metastasis* **8**, 121–127.

Cottrell, S. *et al.* (1992). Molecular analysis of APC mutations in familial adenomatous polyposis and sporadic colon carcinomas. *Lancet* **340**, 626–630.

Coulthard, M. G. *et al.* (2002). The role of the Eph–ephrin signalling system in the regulation of developmental patterning. *Int. J. Dev. Biol.* **46**, 375–384.
Crosnier, Cc. *et al.* (2005). Delta-Notch signalling controls commitment to a secretory fate in the zebrafish intestine. *Development* **132**, 1093–1104.
Crosnier, Cc. *et al.* (2006). Organizing cell renewal in the intestine: stem cells, signals and combinatorial control. *Nat. Rev. Genet.* **7**, 349–359.
Danesh, S. M. *et al.* (2009). BMP and BMP receptor expression during murine organogenesis. *Gene Expr. Patterns* **9**, 255–265.
Duboule, D. and Morata, G. (1994). Colinearity and functional hierarchy among genes of the homeotic complexes. *Trends Genet.* **10**, 358–364.
Dufort, D. *et al.* (1998). The transcription factor HNF3beta is required in visceral endoderm for normal primitive streak morphogenesis. *Development* **125**, 3015–3025.
Durbec, P. L. *et al.* (1996). Common origin and developmental dependence on c-ret of subsets of enteric and sympathetic neuroblasts. *Development* **122**, 349–358.
Flentjar, N. *et al.* (2007). TGF-betaRII rescues development of small intestinal epithelial cells in Elf3-deficient mice. *Gastroenterology* **132**, 1410–1419.
Flores, C. F. *et al.* (2008). Intestinal differentiation in zebrafish requires Cdx1b, a functional equivalent of mammalian Cdx2. *Gastroenterology* **135**, 1665–1675.
Franklin, V. *et al.* (2008). Regionalisation of the endoderm progenitors and morphogenesis of the gut portals of the mouse embryo. *Mech. Dev.* **125**, 587–600.
Fre, S. *et al.* (2005). Notch signals control the fate of immature progenitor cells in the intestine. *Nature* **435**, 964–968.
Fre, S. *et al.* (2009). Notch and Wnt signals cooperatively control cell proliferation and tumorigenesis in the intestine. *Proc. Natl. Acad. Sci. USA* **106**, 6309–6314.
Gao, N. *et al.* (2009). Establishment of intestinal identity and epithelial–mesenchymal signaling by Cdx2. *Dev. Cell* **16**, 588–599.
Garrison, W. D. *et al.* (2006). Hepatocyte nuclear factor 4alpha is essential for embryonic development of the mouse colon. *Gastroenterology* **130**, 1207–1220.
Genander, M. *et al.* (2009). Dissociation of EphB2 signaling pathways mediating progenitor cell proliferation and tumor suppression. *Cell* **139**, 679–692.
Grainger, S. *et al.* (2010). Cdx2 regulates patterning of the intestinal epithelium. *Dev. Biol.* **339**, 155-165.
Grapin-Botton, A. and Melton, D. A. (2000). Endoderm development: from patterning to organogenesis. *Trends Genet.* **16**, 124–130.
Groppe, J. *et al.* (2002). Structural basis of BMP signalling inhibition by the cystine knot protein Noggin. *Nature* **420**, 636–642.
Haramis, A.-P.G. *et al.* (2004). De novo crypt formation and juvenile polyposis on BMP inhibition in mouse intestine. *Science* **303**, 1684–1686.
He, X. C. *et al.* (2004a). BMP signaling inhibits intestinal stem cell self-renewal through suppression of Wnt-[beta]-catenin signaling. *Nat. Genet.* **36**, 1117–1121.
He, X. C. *et al.* (2004b). BMP signaling inhibits intestinal stem cell self-renewal through suppression of Wnt-beta-catenin signaling. *Nat. Genet.* **36**, 1117–1121.
Ireland, H. *et al.* (2004). Inducible Cre-mediated control of gene expression in the murine gastrointestinal tract: effect of loss of beta-catenin. *Gastroenterology* **126**, 1236–1246.
Ishizuya-Oka, A. *et al.* (2006). Shh/BMP-4 signaling pathway is essential for intestinal epithelial development during Xenopus larval-to-adult remodeling. *Dev. Dyn.* **235**, 3240–3249.
Ishizuya-Oka, A. and Hasebe, T. (2008). Sonic hedgehog and bone morphogenetic protein-4 signaling pathway involved in epithelial cell renewal along the radial axis of the intestine. *Digestion* **77**(Suppl 1), 42–47.
Jensen, J. *et al.* (2000). Control of endodermal endocrine development by Hes-1. *Nat. Genet.* **24**, 36–44.

Jonsson, J. *et al.* (1994). Insulin-promoter-factor 1 is required for pancreas development in mice. *Nature* **371**, 606–609.

Kanai-Azuma, M. *et al.* (2002). Depletion of definitive gut endoderm in Sox17-null mutant mice. *Development* **129**, 2367–2379.

Karlsson, L. *et al.* (2000). Abnormal gastrointestinal development in PDGF-A and PDGFR-(alpha) deficient mice implicates a novel mesenchymal structure with putative instructive properties in villus morphogenesis. *Development* **127**, 3457–3466.

Katsnelson, A. (2009). Gut Churning. *Scientist* **23**, 51.

Keller, P. J. *et al.* (2008). Reconstruction of zebrafish early embryonic development by scanned light sheet microscopy. *Science* **322**, 1065–1069.

Kikuchi, Y. *et al.* (2001). Casanova encodes a novel Sox-related protein necessary and sufficient for early endoderm formation in zebrafish. *Genes Dev.* **15**, 1493–1505.

Kim, J. *et al.* (2005). Chemokine receptor CXCR4 expression in colorectal cancer patients increases the risk for recurrence and for poor survival. *J. Clin. Oncol.* **23**, 2744–2753.

Kim, B.-M. *et al.* (2007). Phases of canonical Wnt signaling during the development of mouse intestinal epithelium. *Gastroenterology* **133**, 529–538.

Kollmar, O. *et al.* (2010). CXCR4 and CXCR7 regulate angiogenesis and CT26.WT tumor growth independent from SDF-1. *Int. J. Cancer* **126**, 1302–1315.

Kolterud, Å. *et al.* (2009). Paracrine hedgehog signaling in stomach and intestine: new roles for hedgehog in gastrointestinal patterning. *Gastroenterology* **137**, 618–628.

Kondo, T. *et al.* (1996). Function of posterior HoxD genes in the morphogenesis of the anal sphincter. *Development* **122**, 2651–2659.

Korinek, V. *et al.* (1998). Depletion of epithelial stem-cell compartments in the small intestine of mice lacking Tcf-4. *Nat. Genet.* **19**, 379–383.

Landman, K. A. *et al.* (2007). Mathematical and experimental insights into the development of the enteric nervous system and Hirschsprung's disease. *Dev. Growth Differ.* **49**, 277–286.

Le Douarin, N. M. and Tellet, M. A. (1973). The migration of neural crest cells to the wall of the digestive tract in avian embryo. *J. Embryol. Exp. Morphol.* **30**(1), 31–48.

Lee, C. S. *et al.* (2005). The initiation of liver development is dependent on Foxa transcription factors. *Nature* **435**, 944–947.

Leibl, M. A. *et al.* (1999). Expression of endothelin 3 by mesenchymal cells of embryonic mouse caecum. *Gut* **44**, 246–252.

Madison, B. B. *et al.* (2005). Epithelial hedgehog signals pattern the intestinal crypt–villus axis. *Development* **132**, 279–289.

Mizoguchi, T. *et al.* (2008). Sdf1/Cxcr4 signaling controls the dorsal migration of endodermal cells during zebrafish gastrulation. *Development* **135**, 2521–2529.

Nair, S. and Schilling, T. F. (2008). Chemokine signaling controls endodermal migration during zebrafish gastrulation. *Science* **322**, 89–92.

Ng, A. Y. *et al.* (2002). Inactivation of the transcription factor Elf3 in mice results in dysmorphogenesis and altered differentiation of intestinal epithelium. *Gastroenterology* **122**, 1455–1466.

Ng, A.N.Y. *et al.* (2005). Formation of the digestive system in zebrafish: III. Intestinal epithelium morphogenesis. *Dev. Biol.* **286**, 114–135.

Ng, S. S. *et al.* (2009). Phosphatidylinositol 3-kinase signaling does not activate the Wnt cascade. *J. Biol. Chem.* **284**, 35308–35313.

Offield, M. F. *et al.* (1996). PDX-1 is required for pancreatic outgrowth and differentiation of the rostral duodenum. *Development* **122**, 983–995.

Orford, K. *et al.* (1997). Serine phosphorylation-regulated ubiquitination and degradation of beta-catenin. *J. Biol. Chem.* **272**, 24735–24738.

Ottaiano, A. *et al.* (2006). Overexpression of both CXC chemokine receptor 4 and vascular endothelial growth factor proteins predicts early distant relapse in stage II–III colorectal cancer patients. *Clin. Cancer Res.* **12**, 2795–2803.

Pasquale, E. B. (2005). Eph receptor signalling casts a wide net on cell behaviour. *Nat. Rev. Mol. Cell Biol.* **6**, 462–475.

Persad, S. *et al.* (2001). Tumor suppressor Pten inhibits nuclear accumulation of beta-catenin and T cell/lymphoid enhancer factor 1-mediated transcriptional activation. *J. Cell Biol.* **153**, 1161–1174.

Pinto, D. *et al.* (2003). Canonical Wnt signals are essential for homeostasis of the intestinal epithelium. *Genes Dev.* **17**, 1709–1713.

Polakis, P. (2000). Wnt signaling and cancer. *Genes Dev.* **14**, 1837–1851.

Pollock, R. A. *et al.* (1992). Altering the boundaries of Hox3.1 expression: evidence for antipodal gene regulation. *Cell* **71**, 911–923.

Poulain, M. and Lepage, T. (2002). Mezzo, a paired-like homeobox protein is an immediate target of Nodal signalling and regulates endoderm specification in zebrafish. *Development* **129**, 4901–4914.

Powell, S. M. *et al.* (1992). APC mutations occur early during colorectal tumorigenesis. *Nature* **359**, 235–237.

Powell, D. W. *et al.* (1999). Myofibroblasts. II. Intestinal subepithelial myofibroblasts. *Am. J. Physiol. Cell Physiol.* **277**, C183–C201.

Ramalho-Santos, M. *et al.* (2000). Hedgehog signals regulate multiple aspects of gastro-intestinal development. *Development* **127**, 2763–2772.

Roberts, J. R. (2000). Molecular mechanisms of development of the gastrointestinal tract. *Dev. Dyn.* **219**, 109–120.

Rodilla, V. *et al.* (2009). Jagged1 is the pathological link between Wnt and Notch pathways in colorectal cancer. *Proc. Natl. Acad. Sci. USA* **106**, 6315–6320.

Sancho, E. *et al.* (2003). Live and let die in the intestinal epithelium. *Curr. Opin. Cell Biol.* **15**, 763–770.

Sansom, O. J. *et al.* (2004). Loss of Apc in vivo immediately perturbs Wnt signaling, differentiation, and migration. *Genes Dev.* **18**, 1385–1390.

Sato, T. *et al.* (2009). Single Lgr5 stem cells build crypt–villus structures in vitro without a mesenchymal niche. *Nature* **459**, 262–265.

Savory, J.G.A. *et al.* (2009). Cdx1 and Cdx2 are functionally equivalent in vertebral patterning. *Dev. Biol.* **330**, 114–122.

Schier, A. F. (2003). Nodal signaling in vertebrate development. *Annu. Rev. Cell Dev. Biol.* **19**, 589–621.

Segditsas, S. and Tomlinson, I. (2006). Colorectal cancer and genetic alterations in the Wnt pathway. *Oncogene* **25**, 7531–7537.

Sepich, S. S. *et al.* (2005). Initiation of convergence and extension movements of lateral mesoderm during zebrafish gastrulation. *Dev. Dyn.* **234**, 279–292.

Silberg, D. G. *et al.* (2000). Cdx1 and Cdx2 expression during intestinal development. *Gastroenterology* **119**, 961–971.

Simpson, M. J. *et al.* (2007). Cell proliferation drives neural crest cell invasion of the intestine. *Dev. Biol.* **302**, 553–568.

Smith, D. M. *et al.* (2000). Roles of BMP signaling and Nkx2.5 in patterning at the chick midgut–foregut boundary. *Development* **127**, 3671–3681.

Smith, D. M. and Tabin, C. J. (1999). Developmental biology: BMP signalling specifies the pyloric sphincter. *Nature* **402**, 748–749.

Sukegawa, A. *et al.* (2000). The concentric structure of the developing gut is regulated by sonic hedgehog derived from endodermal epithelium. *Development* **127**, 1971–1980.

Tai, C. C. *et al.* (2009). Wnt5a knock-out mouse as a new model of anorectal malformation. *J. Surg. Res.* **156**, 278–282.

Tam, P.P.L. *et al.* (2003). Early endoderm development in vertebrates: lineage differentiation and morphogenetic function. *Curr. Opin. Genet. Dev.* **13**, 393–400.

Tam, P.P.L. *et al.* (2007). Sequential allocation and global pattern of movement of the definitive endoderm in the mouse embryo during gastrulation. *Development* **134**, 251–260.

Tamai, Y. *et al.* (1999). Colonic hamartoma development by anomalous duplication in Cdx2 knockout mice. *Cancer Res.* **59**, 2965–2970.

Tamura, A. *et al.* (2008). Megaintestine in claudin-15-deficient mice. *Gastroenterology* **134**, 523–534.

Taraviras, S. *et al.* (1999). Signalling by the RET receptor tyrosine kinase and its role in the development of the mammalian enteric nervous system. *Development* **126**, 2785–2797.

Tennyson, M. T. *et al.* (1998). Fetal development of the enteric nervous system of transgenic mice that overexpress the *Hoxa-4* gene. *Dev. Dyn.* **211**, 269–291.

Tian, Q. *et al.* (2004). Proteomic analysis identifies that 14-3-3zeta interacts with beta-catenin and facilitates its activation by Akt. *Proc. Natl. Acad. Sci. USA* **101**, 15370–15375.

Tian, Q. *et al.* (2005). Bridging the BMP and Wnt pathways by PI3 kinase/Akt and 14-3-3zeta. *Cell Cycle* **4**, 215–6.

Torihashi, S. *et al.* (2006). Gut-like structures from mouse embryonic stem cells as an in vitro model for gut organogenesis preserving developmental potential after transplantation. *Stem Cells* **24**, 2618–2626.

Torihashi, S. *et al.* (2009). The expression and crucial roles of BMP signaling in development of smooth muscle progenitor cells in the mouse embryonic gut. *Differentiation* **77**, 277–289.

van de Wetering, M. *et al.* (2002). The beta-catenin/TCF-4 complex imposes a crypt progenitor phenotype on colorectal cancer cells. *Cell* **111**, 241–250.

van der Flier, L. G. and Clevers, H. (2009). Stem cells, self-renewal, and differentiation in the intestinal epithelium. *Annu. Rev. Physiol.* **71**, 241–260.

van der Flier, L. G. *et al.* (2007). The intestinal Wnt/TCF signature. *Gastroenterology* **132**, 628–632.

van der Flier, L. G. *et al.* (2009). Transcription factor achaete scute-like 2 controls intestinal stem cell fate. *Cell* **136**, 903–912.

van Es, J. H. *et al.* (2005). Notch/gamma-secretase inhibition turns proliferative cells in intestinal crypts and adenomas into goblet cells. *Nature* **435**, 959–963.

Vandercappellen, J. *et al.* (2008). The role of CXC chemokines and their receptors in cancer. *Cancer Lett.* **267**, 226–244.

Vearing, C. J. and Lackmann, M. (2005). Eph receptor signalling; dimerisation just isn't enough. *Growth Factors* **23**, 67–76.

Verzi, M. P. and Shivdasani, R. A. (2008). Wnt signaling in gut organogenesis. *Organogenesis* **4**, 87–91.

Weinstein, D. C. *et al.* (1994). The winged-helix transcription factor HNF-3[beta] is required for notochord development in the mouse embryo. *Cell* **78**, 575–588.

Wimmer-Kleikamp, S. H. *et al.* (2004). Recruitment of Eph receptors into signaling clusters does not require ephrin contact. *J. Cell Biol.* **164**, 661–666.

Wolgemuth, B. R., Mostoller, M. P., Brinster, R. L. and Palmiter, R. D. (1989). Transgenic mice overexpressing the mouse homeobox-containing gene Hox-1.4 exhibit abnormal gut development. *Nature* **337**, 464–467.

Xanthos, J. B. *et al.* (2001). Maternal VegT is the initiator of a molecular network specifying endoderm in Xenopus laevis. *Development* **128**, 167–180.

Yang, Q. *et al.* (2001). Requirement of math1 for secretory cell lineage commitment in the mouse intestine. *Science* **294**, 2155–2158.

Young, H. M. *et al.* (2001). GDNF is a chemoattractant for enteric neural cells. *Dev. Biol.* **229**, 503–516.

Young, H. M. *et al.* (2004). Dynamics of neural crest-derived cell migration in the embryonic mouse gut. *Dev. Biol.* **270**, 455–473.

Young, T. and Deschamps, J. (2009). Chapter 8 Hox, Cdx, and anteroposterior patterning in the mouse embryo. *Curr. Top. Dev. Biol.* **88**, 235–255.

Zacchetti, G. *et al.* (2007). Hox gene function in vertebrate gut morphogenesis: the case of the caecum. *Development* **134**, 3967–3973.

Zakany, J. and Duboule, D. (1999). Hox genes and the making of sphincters. *Nature* **401**, 761–762.

Zeissig, S. *et al.* (2007). Changes in expression and distribution of claudin 2, 5 and 8 lead to discontinuous tight junctions and barrier dysfunction in active Crohn's disease. *Gut* **56**, 61–72.

Zheng, H. *et al.* (2009). KLF4 gene expression is inhibited by the notch signaling pathway that controls goblet cell differentiation in mouse gastrointestinal tract. *Am. J. Physiol. Gastrointest. Liver Physiol.* **296**, G490–G498.

Zorn, A. M. and Wells, J. M. (2009). Vertebrate endoderm development and organ formation. *Annu. Rev. Cell Dev. Biol.* **25**, 221–251.

CHAPTER FIVE

Kidney Development: Two Tales of Tubulogenesis

Melissa Little, Kylie Georgas, David Pennisi, *and* Lorine Wilkinson

Contents

1. Introduction: How You Get a Kidney 194
2. The First Tale of Tubulogenesis: a Branching Tree 197
   2.1. Gdnf/Ret signaling initiates and guides UB outgrowth 197
   2.2. Regulation of UB initiation, position, width, and number 199
   2.3. A di/trichotomous branching tree—suppression and induction of branching 200
   2.4. What is there yet to know? 202
3. Tubulogenesis *Via* MET—Tube One Induces Tube Two 203
   3.1. CM as a population of nephron progenitors 205
   3.2. Nephron progenitors and cessation of nephrogenesis 207
4. Patterning the Resulting Tubules 207
   4.1. Notch signaling in the patterning of the proximal nephron 210
   4.2. Patterning of the LH and DTs 211
   4.3. Patterning of UT versus stalk and differentiation of the ureter–pelvic junction 212
5. Moving From Structure to a Functional Filter 213
   5.1. Patterning and vascularization of the developing RC 213
   5.2. Maturation of the RC 216
6. Disruptions to Kidney Tubulogenesis in Human Disease 217
   6.1. Consequences of defects in branching morphogenesis 218
   6.2. Dysplasia and defects in MET in humans 218
   6.3. Nephron number and renal disease 219
7. Conclusion 220
Acknowledgments 221
References 221

Institute for Molecular Bioscience, The University of Queensland, St. Lucia, Australia

Current Topics in Developmental Biology, Volume 90
ISSN 0070-2153, DOI 10.1016/S0070-2153(10)90005-7

Abstract

The mammalian kidney may well be one of the most complex organs of postnatal life. Each adult human kidney contains on average more than one million functional filtration units, the nephrons, residing within a specialized cellular interstitium. Each kidney also contains over 25 distinct cell types, each of which must be specifically aligned with respect to each other to ensure both normal development and ultimately, normal renal function. Despite this complexity, the development of the kidney can be simplistically described as the coordinate formation of two distinct sets of tubules. These tubules develop cooperatively with each other in time and space, yet represent two distinct but classical types of tubulogenesis. The first of these tubules, the ureteric bud, forms as an outgrowth of another epithelial tube, the nephric duct, and undergoes extensive branching morphogenesis to create the collecting system of the kidney. The second tubules are the nephrons themselves which arise *via* a mesenchyme-to-epithelial transition induced by the first set of tubules. These tubules never branch, but must elongate to become intricately patterned and functionally segmented tubules. The molecular drivers for these two tales of tubulogenesis include many gene families regulating tubulogenesis and branching morphogenesis in other organs; however, the individual players and codependent interrelationships between a branched and non-branched tubular network make organogenesis in the kidney unique. Here we review both what is known and remains to be understood in kidney tubulogenesis.

1. Introduction: How You Get a Kidney

The adult mammalian kidney is derived from the intermediate mesoderm (IM) located lateral to the somites. Referred to as the metanephros, this is the final of three embryonic excretory organs that develop in a temporally and spatially distinct order from the rostral to caudal end of the embryo; the pronephros, mesonephros, and metanephros. All three of these excretory structures empty into the primary nephric duct (ND) or Wolffian duct, which also arises from the IM. The pronephros is the earliest to arise and is functional during development only in lower vertebrates. It consists of simple tubules that empty into the ND. The mesonephros develops more caudally as the pronephros degenerates and comprises more complex ducts, consisting of a glomerulus and tubule, a proportion of which independently enter the ND. The mesonephros is functional in adult higher fishes and amphibians, but only during embryogenesis in mammals. The metanephros, the final of the three embryonic organs, forms the permanent adult kidney in mammals.

The metanephros is derived from two key tissues, the ureteric bud (UB) and a region of spatially defined IM termed the metanephric mesenchyme

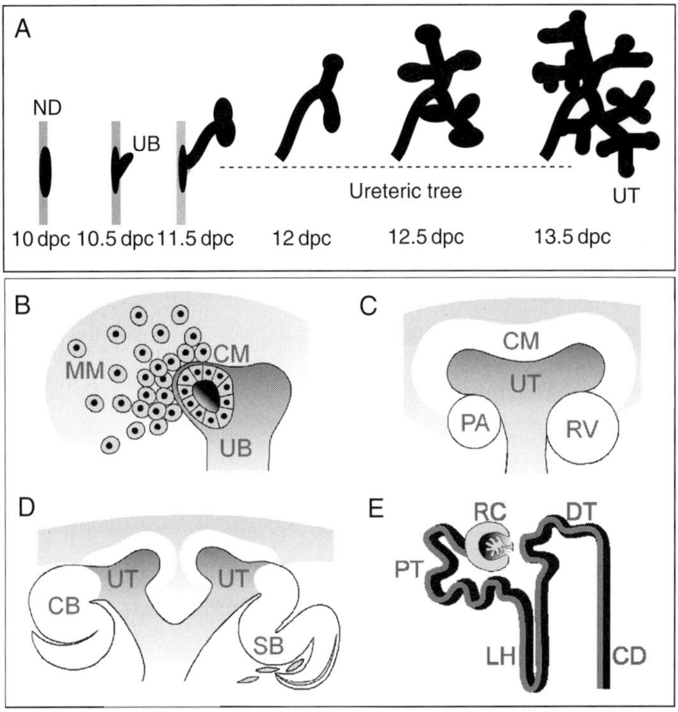

Figure 5.1 Overview of tubulogenesis during kidney development. (A) The first tale of tubulogenesis: branching morphogenesis. From left to right, a swelling of the ND occurs in response to GDNF secretion from the presumptive kidney mesenchyme. This forms a single branch, the UB. The bud elongates before forming an ampulla (11.5 dpc) and undergoing the first dichotomous branching event. This forms two UTs. Each tip then branches again, with the ureteric tree as a whole progressively invading and inducing the surrounding mesenchyme. (B)–(E). The second tale of tubulogenesis: formation of individual nephros *via* MET and elongation. (B) Around the tip of the branching UB, the MM condenses to form a CM. These CM cells self renew throughout kidney development to supply the cells required for nephron formation. (C) CM proximal to the UT and immediately adjacent to its basement membrane condenses to form a PA. Cells of the PA then undergo changes in cell adhesion, morphology, and extracellular matrix production to produce a RV with a polarized epithelium (MET). The formation of one RV is equivalent to the birth of one nephron as all components of the nephron from glomerulus to DTs arise from this ball of cells. Fusion also occurs between the late RV and the UT to form a contiguous tubule. (D) Cells of the RV proliferate and segment, forming intermediate stages referred to as CB and SB. Endothelial cells are attracted into the cleft at the proximal end of the SSB to begin to form the glomerular vasculature. (E) Elongation, including patterning and segmentation, proceeds in the absence of branching to form the mature nephron compartmentalized into PT, LH, DT, and CD.

(MM). The formation of the UB is the initiating tubulogenic event that starts off kidney development (Fig. 5.1A). Around 10 days *post coitum* (dpc) in the mouse, a region of the ND adjacent to the mesenchyme of the presumptive kidney swells due to cellular proliferation and migration of cells into that region from elsewhere in the ND. This marks the site of budding from which the UB will appear. Paired epithelial buds arise from the ND at 10.5 dpc (approximately five weeks in the human) in response to signals from the adjacent MM. The same signals attract each UB to grow towards the MM while mesenchymal inhibitors discourage branching until each UB reaches the target tissue. Once this has happened, the tip of each UB expands to form an ampulla from which two branches are formed (Fig. 5.1A; 11.5 dpc). This is the commencement of extensive UB branching which will continue for the duration of kidney development and ultimately form the collecting system of the kidney. The branched tree remains connected back to the ND with the region falling outside of the kidney forming the ureter. Just as the initial outgrowth is regulated such that there is only one UB, the pattern of subsequent branching is controlled such that all branching events arise from a terminal ampulla and not as side branching. This is quite different to the initial domain form of branching that occurs in the lung (Metzger *et al.*, 2008). This process is controlled by expression of genes both in the MM, acting to guide and pattern the UB, as well as cell-autonomous gene expression within the ureteric tree itself.

The second tubulogenic event key to kidney development is distinctly different. On this occasion, the tubules in question do not arise from an existing tubule but develop *via* a mesenchyme-to-epithelial transition (MET) of the MM surrounding the branching ureteric compartment (Fig. 5.1B–E). This event occurs at the tips of that branching ureteric epithelium and results in the formation of the nephrons. As the UB grows and branches within the uninduced MM, signals from the UB induce that mesenchyme to condense around the nascent tips of the UB (Fig. 5.1B). This condensing MM is referred to as cap mesenchyme (CM), and this CM is regarded as a self-renewing progenitor population. At a specific location in relation to the ureteric tip (UT), the CM ceases to self-renew and undergoes a progressive MET involving an initial pretubular aggregate (PA) stage followed by the formation of an epithelial renal vesicle (RV). This always occurs in the "armpit" of the terminal UT alongside the adjacent basement membrane of the UT (Fig. 5.1C). Each of these epithelialization events (RV formation) represents the "endowment" of a nephron (Fig. 5.1C). It is at this time that these neo-tubules fuse with the adjacent UTs to ultimately form contiguous uriniferous tubules with the collecting ducts (CDs), thereby allowing the urinary filtrate to drain from the kidney to the bladder *via* the ureter.

The tubules that make up the nephrons are many and distinct from the ureteric compartment in that they never branch. However, they do bend and twist and elongate into an intricate but tightly patterned structure that ultimately segments into regions with very distinct functions. Just as the first tubules rely on the MM that will form the second tubules, the second tubules will not form without active survival and differentiation signals from the first. Genes expressed by the invading UB regulate the survival of the MM. Failure of UB outgrowth invariably leads to apoptosis of the MM and consequent renal agenesis, while ectopic UB outgrowth leads to dysplasia, such as hydroureter and duplex kidney. The MET events that form these second tubules also depend upon signals from the ureteric compartment, notably secreted Wnt ligands. It is also likely that the complex patterning and segmentation that follows is also regulated by signals from the branched tree network that has and will form. Conversely, branching morphogenesis must be controlled spatially and temporally to ensure a functional organ and perturbations to this process affect not only branching but subsequent nephron formation. Herein lies an exquisite example of reciprocal signaling and morphogenetic interdependence. If you disrupt one of these processes, you disrupt the other as well. This has on occasion made it difficult to separate cause and effect when examining what genes regulate kidney development. In this review we will begin by separately discussing these two tales of tubulogenesis in turn, highlighting the molecular pathways involved in tubule formation and patterning. We will then discuss how nephron segmentation and functionalization occurs, how this understanding relates to human disease, and what there is left to understand.

2. The First Tale of Tubulogenesis: A Branching Tree

2.1. Gdnf/Ret signaling initiates and guides UB outgrowth

Initial budding of the UB from the ND is a classical example of branching morphogenesis. In many organs, the initiation of tubule outgrowth and subsequent branching morphogenesis involves a response by the existing epithelium to a local mesenchymal growth factor signal. This is often a receptor-tyrosine kinase ligand, such as epidermal growth factor (EGF), fibroblast growth factor (FGF), or transforming growth factor alpha (TGFα), which promotes cellular proliferation and migration. In the *Drosophila* airways, for example, the expression of FGF by the surrounding mesenchyme signaling to the FGF receptor in the epithelium results in formation of a tip. There is competition for this signal such that the cells with the highest level of FGF signaling become the first tip, in turn inhibiting the generation of additional tips nearby. Similarly, the regulation of branch position, number, width, and branching pattern in lung, prostate, and mammary gland also involve receptor tyrosine kinases and their ligands. This is also the case for the kidney where UB outgrowth is initiated when the soluble chemotactic growth factor, glial cell line-derived neurotrophic factor (Gdnf) binds to the tyrosine kinase receptor Ret (Fig. 5.2). The requirement for Gdnf/Ret signaling in UB initiation and subsequent branching morphogenesis is central to kidney development and was established and elaborated upon *via* manipulation of

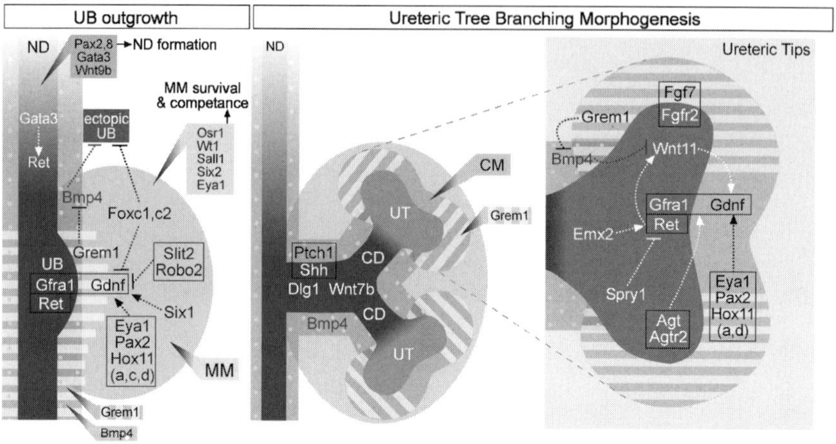

Figure 5.2 Genes regulating ureteric bud outgrowth and branching morphogenesis. This schematic illustrates key genes involved in the ureteric bud initiation, outgrowth (left), and branching events (right). This includes genes expresson within regions of the MM, ND and forming ureteric tree, where the expression of multiple genes can be seen in overlapping domains. The *Bmp4* expression domain (spotted region) is restricted to the mesenchyme surrounding the entire length of the ND epithelium and the emerging UB (left), and later is maintained in the mesenchyme (stroma/interstitium) surrounding the epithelium of the branching ureteric tree (right). The Bmp4 antagonist *Grem1* (striped region) is expressed in an overlapping subdomain of the mesenchyme surrounding the UB (left) and continues to be maintained in the CM surrounding the UTs (right). In these regions Grem1 acts to suppress Bmp4 activity that inhibits branching and maintains UB tip outgrowth. Genes expressed by the UB are shown as white text and mesenchymal genes are in black text. Genes encoding proteins that directly interact in a complex are boxed. See text (Section 2) for details.

receptor and ligand, as well as subsequent signaling in the mouse (Moore *et al.*, 1996; Pichel *et al.*, 1996; Sánchez *et al.*, 1996; Schuchardt *et al.*, 1996). *Gdnf* is expressed in the IM/nephrogenic cord from 8.5 dpc in the mouse. This region of expression becomes progressively restricted, such that by 10–10.5 dpc it defines the location of the presumptive metanephros. The *Ret* receptor tyrosine kinase is expressed by all cells in the ND. However, ligand binding leads to an upregulation of *Ret* on a subset of cells closest to the source of ligand, thereby rendering these cells most competent to respond to further Gdnf signaling. This dictates the position of the UB. The co-receptor for Gdnf, *Gfrα1*, is expressed by both MM and ND and appears to assist in Ret signaling (Costantini and Shakya, 2006). The proliferation of the responding cells, which express *Ret*, facilitates the migration of the UB towards the MM. Along with *Ret*, the UB tip expresses *Wnt11* and this spatial expression persists in all subsequent branch tips. A positive feedback loop between *Gdnf, Ret*, and *Wnt11* exists such that the more the Gdnf signaling, the higher the expression of *Wnt11* and *Ret* (Fig. 5.2).

Advanced real-time imaging and genetically tagged UB cells have indicated that Ret has a direct role in defining the UTs of this branching network. The UB cells with mutated Ret preferentially reside in the UB trunk and are excluded from the UB tips (Shakya *et al.*, 2005). In an elegant series of experiments, Chi *et al.* (2009a) found that cells in the ND fated to form the primary UB outgrowth form a pseudostratified epithelium in a Ret-independent manner, but ND cells able to express Ret compete with mutant ND cells and selectively converge to form UB tips. Conversely, ND cells lacking *Spry1*, an inhibitor of Ret-signaling, out-compete normal ND cells in contributing to the UB tips. Attraction between Ret-expressing UB tip cells is proposed to keep them at the UB tips, partly accounting for the UB tip-specific expression (Chi *et al.*, 2009b).

2.2. Regulation of UB initiation, position, width, and number

It is critical that only a single UB arises from the ND rather than many, and that it is located at the right place. This is dictated by the rostro-caudal location of *Gdnf* expression in the IM, which is in turn regulated by the forkhead/winged helix transcription factors, *Foxc1* and *Foxc2*. *Foxc1/c2* are expressed by the MM at 10.5 dpc. *Foxc1* homozygous null and *Foxc1/c2* compound heterozygotes display a rostral expansion of *Gdnf* and *Eya1* expression in the IM, leading to ectopic UB formation, variably penetrant hydroureter, and duplex kidney (Kume *et al.*, 2000). Likewise, loss of normal mesenchymal expression of *Robo2* or its ligand, Slit2 results in an anterior expansion of Gdnf and supernumerary UB formation (Grieshammer *et al.*, 2004). As well as being regulated by mesenchymal gene expression, UB outgrowth is also modified by the expression of transcription factors in the ND. *Gata3* inactivation in the ND results in ectopic and supernumerary ureter budding, implicating *Gata3* in the transcriptional regulation of *Ret* expression. This appears to be independent of its role in Ret activation, stemming instead from mediation of β-catenin to prevent premature differentiation of ND cells (Grote *et al.*, 2008).

The UB initiation is clearly dependent upon the formation of the MM itself. The loss of mesenchymal genes upstream of *Gdnf* results in a failure of UB growth and invasion. Gene ablation of *Eya1* or *Six1*, normally expressed by the MM at 10.5 dpc, leads to loss of *Gdnf* in the MM, failure of UB outgrowth, and apoptosis of the MM by 12.5 dpc (Xu *et al.*, 2003). Loss of Wt1 results in apoptosis of the MM (Kreidberg *et al.*, 1993). Pax2 also plays a role in initiation and maintenance of Gdnf/Ret signaling (Clarke *et al.*, 2006). In *Eya1* mutants, *Six2* is also absent suggesting it is downstream of *Eya1*, however, *Pax2* and *Wt1* are expressed normally in the MM. Similarly, ablation in mice of the paralogous *Hox11* genes, *Hoxa11*, *Hoxc11* and *Hoxd11*, which are also expressed in the MM, results in a loss of UB induction and apoptosis of MM (Patterson *et al.*, 2001; Wellik *et al.*, 2002). The Hox11 homeodomain proteins have been proposed to complex with Eya1 and Pax2 to directly upregulate *Six2* and *Gdnf* expression.

Indeed, loss of the *Hox11* genes results in a loss of expression of both *Gdnf* and *Six2* (Gong *et al.*, 2007). Six1 induces expression of *Sall1* in the MM and is also required for invasion of the UB into the MM. Hence, deletion of either gene results in a failure of the UB to invade the mesenchyme with subsequent apoptosis of the MM (Chai *et al.*, 2006)

In many branching organs, there is inhibition of further branching around an advancing tip to regulate both total branch/tip number and pattern. This can involve the production of inhibitors of branching by the tip itself or the surrounding mesenchyme. In *Drosophila*, the tip cells of the extending tracheal network secrete notch to prevent creation of additional tips nearby. In the mammary gland, expression of TGFβ by the cells at the tip of the epithelial ducts prevents branching of the cells in the adjacent trunk. In the kidney, once the primary UB outgrowth is initiated, ectopic budding from the ND is suppressed by inhibitors expressed along the ND. This includes *Bmp4* that is expressed in the mesenchyme along the length of the ND and in the mesenchyme around the outgrowing UB, as well as around subsequent branches (Fig. 5.2). *Bmp4* inhibits UB branching by down-regulating the level of *Wnt11* expression in the trunk of the outgrowing UB. As the UB continues to grow towards the MM, *Bmp4* expression extends along the mesenchyme adjacent to the UB trunk, again preventing subsequent branching to ensure UB elongation (Miyazaki *et al.*, 2000). This inhibition of branching is relieved by the expression of the *Bmp4* antagonist, Gremlin (*Grem1*), in the mesenchyme surrounding both the emerging UB and subsequent UB tips (Michos *et al.*, 2007) (Fig. 5.2). The combination of a positive guidance and positional cue from the MM (Gdnf) and the relief of Bmp4 suppression of outgrowth (Grem1) ensure the correct positioning and number of UBs. While *Bmp4-/-* mice are early embryonic lethal, *Bmp4* +/- mice display anteriorly misplaced or ectopic UBs resulting in megaureter or duplex collecting systems, in addition to glomerular cysts (Miyazaki *et al.*, 2000; Ueda *et al.*, 2008). Bmp4 thereby regulates the anterior–posterior axis of the developing kidney.

2.3. A di/trichotomous branching tree—suppression and induction of branching

Although not as fully documented as the developing lung (Metzger *et al.*, 2008), the pattern of UB branching is highly regulated to ensure the formation of an organ of the appropriate size and shape. Once the UB has extended from the ND and reaches the MM, a series of dichotomous or trichotomous branching events generate the CD network of the kidney. Many of the same pathways involved in the induction of the initial UB outgrowth continue to be involved in this process. Gdnf/Ret signaling remains critical and is regulated and modified by members of the bone morphogenetic protein (BMP), FGF, and TGFβ superfamily members as well as retinoic acid signaling from the stroma (Batourina *et al.*, 2001; Mendelsohn *et al.*, 1999). However, surprisingly little is known about the consequence of Ret signaling. Lu *et al.* (2009) have recently shown that the transcription factors Etv4 and Etv5 are upregulated in response to Ret signaling, in turn leading to upregulation of Cxcr4, Myb, Met, and Mmp14 from the tips of the ureteric tree. The latter may play a role in facilitating the extension of the growing tip into the surrounding extracellular matrix. As described later, Met is involved in regulation of nephron number. While the centrality of the Gdnf/Ret relationship has long been a given in the field, a recent study investigated whether branching morphogenesis could proceed in the absence of this signaling pathway. Here, Michos *et al.* (2010) examined the ability of mice lacking either Gdnf or Ret together with the branching inhibitor sprouty1 to form a kidney. Both Gdnf-/-, Spry1-/- and Ret-/-;Spry1-/- mice showed substantial kidney development, including the formation of a branching CD system, however disturbances to the normal pattern of branch spacing and angle were still apparent. This partial rescue phenotype was due to the expression of FGF10 from the MM, which also feeds into Etv4/5 to elicit branching. Hence, removal of even one copy of FGF10 caused renal agenesis and on this occasion, together with a loss of downstream signaling *via* Etv4/5. Hence, it is a redundant receptor tyrosine kinase signal that is central for branching rather than GDNF/Ret per se.

Involvement of FGFs in this process extends beyond Fgf10. MM expression of *Fgf7* and UT expression of Fgfr2 isoform IIIb are also required for optimal branching, the perturbation of which leads to a reduction in total nephron number (Qiao *et al.*, 1999; Revest *et al.*, 2000). This high affinity receptor can bind a range of FGF ligands and regulates UB stalk diameter (Zhao *et al.*, 2004). Fgf10, another ligand for this receptor, is also required for normal kidney UB branching, with mutants displaying a smaller collecting system (Ohuchi *et al.*, 2000). Two other high-affinity FGF receptors, Fgfr1 and Fgfr2, have also been implicated in kidney development, acting in a redundant manner in both the MM (Poladia *et al.*, 2006) and the developing ureteric tree (Zhao *et al.*, 2004). Angiotensin II, one of several physiologically active peptide products of the angiotensinogen gene, *Agt*, increases ureteric branching and cell proliferation in the tip *via* the angiotensin II receptor, type 2 *(Agt2r)*, partially through indirect upregulation of the Gdnf/Ret pathway (Yosypiv *et al.*, 2008). As in the initial phases of UB outgrowth, Hoxa11 and Hoxd11 are involved in branching. In *Hoxa11/Hoxd11* double mutant kidneys, expression of *Wnt11, Emx2,* and *Ret* was lost in the UB tips (Patterson *et al.*, 2001). Of note, UB branching was more severely affected in the mid-ventral part of *Hoxa11d11/-* kidneys, suggesting that these homeotic transcription factors regulate dorso-ventral and antero-posterior patterning of the kidney (Patterson *et al.*, 2001).

While at first blush it seems that continued branching morphogenesis during kidney development involves the same players as the initiation of UB outgrowth, this is not always the case. *Emx2* is essential for early stages of subsequent branching but not initial UB outgrowth (Miyamoto *et al.*, 1997). In *Emx2* mutants, *Pax2* expression was initially induced in the MM but greatly diminished in the UB, and *Ret* expression in the UB was greatly reduced. The MM expression of *Wt1* appeared normal in these mutants; however, *Gdnf* expression in the MM was initiated but not maintained resulting in complete agenesis of the kidneys and ureters as well as the gonads and reproductive tracts (Miyamoto *et al.*, 1997). Other pathways that seemed to have only subtle roles, such as Hgf/Met signaling, become critical in the presence of defects in other pathways. *Hgf* is expressed by the MM and its receptor, *Met,* by both the UB and mesenchyme (Woolf *et al.*, 1995). While regarded until recently as not essential in the kidney, loss of Hgf/Met signaling results in a postnatal deficit in glomerular number (Ishibe *et al.*, 2009) which is only partially compensated for by upregulation of Egfr (Ishibe *et al.*, 2009). The recent finding that Met is downstream of Fgf10, which can replace Gdnf in initiating and sustaining branching (Michos *et al.*, 2010), suggests a more important role for Hgf/Met under some circumstances.

It would appear from what has been discussed to date that the whole formation of the ureteric tree cannot come about without signals from the surrounding mesenchyme. Surprisingly, isolated rat UB epithelium can grow and branch in the absence of MM in *ex vivo* culture systems if provided with Gdnf and other MM-derived secreted factors. This suggests that there is no need for direct contact between the UB and the MM (Qiao *et al.*, 1999). However, direct cellular contact between these two tissues is required for the appropriate elongation of the UB stalk (Shah *et al.*, 2009). Recombinant FGFs (in addition to factors described above) promote the growth of isolated UB epithelia, although to varying extents. Fgf1 and Fgf10 induce UB to form branched structures with elongated stalks and distinct tip cells, whilst Fgf2 and Fgf7 induce branching with less elongated stalks and less distinct tip cells (Qiao *et al.*, 2001). The complexity of the dynamic interaction between the heterogeneous UB and the (equally) heterogeneous MM is further highlighted by explant recombination studies showing that fresh MM can induce older UBs to continue branching (Shah *et al.*, 2009). This study also revealed plasticity in UB patterning whereby MM can correct UB branching defects induced by recombinant Fgf7 or heregulin (neuregulin 1, *Nrg1*) (Shah *et al.*, 2009).

2.4. What is there yet to know?

While our understanding of branching morphogenesis is increasing, what is left to learn is not always easy to predict. This is in large part due to the fact that visualization of the ureteric tree in 3D is limited to the duration of explants cultures, which is rarely more than 4–5 days. More critically, such explants grow as flat structures unlike the *in vivo* organ itself. Later stages of branching may well involve considerable remodeling and reshaping of the tree in order to accommodate the elongating tubules of the forming nephrons. Indeed, it is likely that these nephrons drive such reshaping. Improvements in imaging capacity will begin to make inroads into the advanced patterning of the ureteric tree. The application of optical projection tomography (Short *et al.*, 2010) will allow us to learn whether the branching pattern is as tightly regulated as that for lung development as well as allowing the quantitative assessment of changes in branch number, tip number, ureteric volume and shape, branchpoints and angles in whole organs as opposed to explants. This will be assisted by the new generation of fluorescently tagged mice, such as the Hoxb7-myrVenus mice (Chi *et al.*, 2009b), which will assist in examining the individual shape of cells within the ureteric tree during development.

3. Tubulogenesis *Via* MET–Tube One Induces Tube Two

In many organs, including the lung and the mammary gland, epithelial branching morphogenesis is accompanied by the terminal specialization of the tubular tips to form the functional units. In case of the lung, these are the alveoli, whereas in the breast it is the milk-producing epithelium. In the kidney, unlike these other organs, it is not specialization of the tip but the induction of a second tubulogenetic event at each of those tips that is required to form the functional units of excretion. Hence, tube one induces tube two and this occurs at the tip of every branch. Unlike the first tubulogenic event, this does not just mean outgrowth of a new tube from those tips but the induction of the mesenchyme around the tip to become a new tube. In contrast to mesenchymal cells, epithelial cells are polarized, with an apical side that borders a lumen, and a basal side that generates a basement membrane. Specialized lateral cell contacts are formed including adherens and tight junctions. Hence, an MET is required. The cells required to undergo this MET are the MM-derived CM cells that lie in close apposition with the tips of the ureteric tree. That a primary inductive event from the UB is essential for initiating MET in the kidney has been shown in classical experiments in which the UB was dissected away from the mesenchyme. Without the UB, the MM fails to form epithelial tubules and rapidly degenerates (Grobstein, 1956). The search for the epithelialization signal has been a major focus of the field. A number of other tissue types, including embryonic brain and spinal cord, can induce MET and tubulogenesis when placed in contact with the MM. This indicated that the factors responsible were not specific to UB, but that cell–cell contact was critical for induction of MET. In experiments using microporous filters to separate the tissues, induction only occurred when cells from the inducing tissue were able to extend processes through the pores (Sariola *et al.*, 1989). The induction of MET has now been shown to involve the Wnt family of secreted glycoproteins (Carroll *et al.*, 2005; Stark *et al.*, 1994) and the requirement for cell–cell contact is thought to be due to strong binding of the Wnt proteins to cell surface by heparan sulfate proteoglycans (Bullock *et al.*, 1998; Kispert *et al.*, 1996).

Just as Gdnf/Ret signaling has been central to branching, Wnt signaling is key to nephron endowment, with both Wnt4 and Wnt9b being involved in MET. It is the expression of *Wnt9b* from the UT that is thought to initiate the MET event (Carroll *et al.*, 2005) and this in turn results in the expression of *Wnt4* in the PA (Stark *et al.*, 1994) (Fig. 5.3). The transcription factor *Emx2*, expressed in UB, is also critical for inducing *Wnt4* expression (Miyamoto *et al.*, 1997). Interestingly, contact with cultured *Wnt4*-expressing cells can initiate MET in isolated MM, bypassing the requirement for Wnt9b signaling and suggesting that after initial induction, Wnt4 acts in a cell-autonomous fashion (Kispert *et al.*, 1998). Wnt signaling in MET is thought to be dependent upon canonical Wnt signaling involving β-catenin, while later epithelial differentiation and polarity may be reliant on the non-canonical Wnt/PCP (planar cell polarity) pathway,

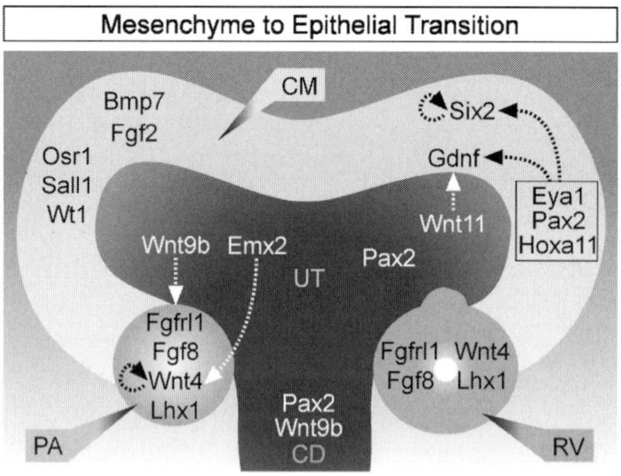

Figure 5.3 Genes regulating nephron endowment *via* MET. A specific region of the CM on the medullary or underside of the UT (armpit region) receives signals from the UT to form a PA, which undergoes a MET to form a RV. This schematic illustrates the expression domains of genes involved in this event, including signals from the UT and gene expression within the CM. See text (Section 3) for details.

abnormalities in which can lead to tubular cyst formation (Schmidt-Ott and Barasch, 2008). Conditional β-catenin (*Ctnnb1*) gene deletion in the mouse MM leads to a phenotype similar to *Wnt9b* and *Wnt4* mutants, including a loss of MET and formation of only a few nephrons (Park *et al.*, 2007). The transcription factor gene *Lhx1* is expressed in aggregates slightly later than *Wnt4* and is required for further development of the aggregate (Kobayashi *et al.*, 2005). Although regarded as critical, Wnt signaling is not the only contributor to MET; FGFs are another growth factor family required. *Fgf8* is expressed earlier than *Wnt4* and is necessary for the expression of both *Wnt4* and *Lhx1*. After conditional *Fgf8* gene deletion, PAs form but neither *Wnt4* nor *Lhx1* are expressed and nephron formation does not progress, leading to severe renal hypopasia. Ffg8 may act in conjunction with Wnt4 to induce the expression of *Lhx1* (Grieshammer *et al.*, 2005; Perantoni *et al.*, 2005). Similarly, deletion of the novel FGF receptor-like 1 (*FGFrl1*) gene, which is normally expressed in the RV, leads to severe renal dysgenesis due to a failure to initiate MET (Gerber *et al.*, 2009). *Fgfrl1-/-* mice do not express any of the mesenchymal genes essential for MET, including *Fgf8, Pax8, Wnt4* and *Lhx1,* placing *Fgfrl1* upstream of these genes in the transcriptional hierarchy (Fig. 5.3). Although the mice express UB genes including *Wnt9b* and *Wnt11*, they also display a ureteric branching defect, possibly the result of a failure in the reciprocal signaling between the differentiating nephrons and the UB. As Fgf8 is required to initiate Wnt4 signaling, it has been speculated that Fgfrl1 may modulate a cooperative Fgf–Wnt9b signal to initiate MET (Gerber *et al.*, 2009).

This entire cascade continues to place Wnt9b signaling at its apex and the key initiating signal. The dilemma with this model is the ubiquitous expression of Wnt9b throughout the ureteric epithelium. Given that the whole tree expresses this gene, why does MET only occur in the armpit of the terminal branches of the ureteric tree? One could predict that this will be the site of the highest level of Wnt9b signaling, but this seems simplistic. There is another Wnt signal emanating from the UT—Wnt11. Most studies have focused on the role of Wnt11 in reinforcing the expression of Gdnf and Ret. Coincident expression of these two Wnt proteins may be important, although the phenotype of Wnt11 knockout mice suggests a primary role in branching rather than MET (Majumdar *et al.*, 2003). Given the specific location for the MET event, one can also speculate as to whether self-renewal of the CM occurs in a different parts of the CM or whether CM cells must actively migrate into the armpit to undergo MET. If so, the signal for such migration is not known.

3.1. CM as a population of nephron progenitors

Cap mesenchyme is defined as the region of the MM closest to each UT and to distinguish these MM-derived cells from others within the stroma that will not form nephrons. However, whether this is a homogeneous or heterogeneous compartment remains debatable. Transient lineage tracing does suggest that once the UB first enters the MM field, the cells that are going to contribute to the CM become committed to that fate and no further cells are contributed by the MM to the CM from that time onwards (Kobayashi *et al.*, 2008). Based on fluorescence-activated cell sorting (FACS) analysis of 11.5 dpc kidneys from Six2GFP mice, which facilitates the isolation of CM, the starting population is around 10,000 cells. This number reaches approximately 180,000 cells by 19.5 dpc (Kobayashi *et al.*, 2008). Given that both branching and nephron induction proceed for the duration of kidney development, the CM must undergo self-renewal to both generate nephrons and increase in number. Hence, the CM is regarded as the nephron progenitor population. *Six2* is not only a good marker of CM but has been shown to be an essential gene for CM self-renewal, simultaneously suppressing differentiation. *Six2* null mice display ectopic nephron formation and a failure to renew the CM cell population. This is thought to result in CM exhaustion and consequential renal hyperplasia. *Six2* gain-of-function in explants inhibits MET, suggesting Six2 is essential for the maintenance of an undifferentiated nephron progenitor population (Kobayashi *et al.*, 2008; Self *et al.*, 2006).

Many other genes expressed in the CM are key MM transcription factors, including *Osr1, Sall1, Wt1, Hoxa11,* and *Eya1.* If such genes are not purely directing branching, it is reasonable to assume that their expression in the CM is related to nephron endowment and drives self-renewal (as for Six2), survival, or commitment and competence to undergo MET. On some occasions, this has been investigated. While also expressed in the ureteric epithelium, *Pax2* expression in the CM is critical for the commitment of the MM to a CM state. *Pax2* expression persists in the tubules of the developing nephrons, although is restricted from the developing glomeruli possibly due to transcriptional repression by Wt1 (Ryan *et al.*, 1995). A tripartite complex between Hoxa11, Pax2, and Eya1 positively regulates the expression of *Six2* and *Gdnf* in the CM (Gong *et al.*, 2007). Hence this complex not only ensures CM survival but also continued UB outgrowth.

The survival of the MM itself is required for the formation of the CM. The MM appears programmed to undergo apoptosis if it does not receive the appropriate survival signals from the ingrowing UB. *Bmp7* also appears necessary for proliferation and survival of this uninduced mesenchyme population. After 14.5 dpc in *Bmp7* null mice, kidney development halts and massive apoptosis occurs resulting in renal hypoplasia (Simic and Vukicevic, 2005). Dudley *et al.* (1999) have shown that a combination of Bmp7 and Fgf2 is critical for MM survival. Loss of the early MM transcription factor *Wt1*, results in a lack of UB induction and a subsequent apoptosis of the MM (Kreidberg *et al.*, 1993). In fact, an MM lacking *Wt1* is unable to respond to wild-type UB or heterologous inducing tissues such as spinal cord, suggesting that the role of Wt1 in MET is cell-autonomous and not dependent on UB invasion. Hence Wt1 has been proposed to be involved in the survival and competence of the MM (Donovan *et al.*, 1999).

3.2. Nephron progenitors and cessation of nephrogenesis

It is the elegant lineage tracing studies of Kobayashi *et al.* (2008) and Boyle *et al.* (2008) that have clearly demonstrated that all elements of the nephron are derived from the $Six2^{+}Cited1^{+}$ CM. This mechanism of tubulogenesis does not occur in the postnatal kidney, even after injury (Humphreys *et al.*, 2008). Indeed, it has been known for decades that there are no new nephrons endowed in the kidney after the end of development (Hartman *et al.*, 2007). Hence, the CM is a transient progenitor population. The last MET events occur in the human at 36 weeks of gestation and in the mouse in the first few days of postnatal life (Hartman *et al.*, 2007). We do not know what triggers the cessation of nephrogenesis and the loss of the CM compartment. However, as this does not involve active CM apoptosis or the loss of the inducing signal from the ureteric compartment, it is assumed that any remaining CM commits to nephron formation and hence is lost. Whether this requires an active trigger or the loss of a self-renewal signal and whether this could be prevented or delayed is not known. Another large hole in our understanding of this process is exemplified by the fact that it has never been possible to successfully culture isolated CM that can retain nephron-forming potential. This supports a view that the survival of the CM phenotype is reliant upon a highly regulated niche. Our inability to culture CM may, therefore, stem from a lack of understanding of the precise stromal, UT, autocrine, and potentially renal capsule signals that combine to maintain this unique niche *in vivo*. Until this is better understood, the concept of *de novo* kidney bioengineering based on developmental paradigms will not succeed.

4. Patterning the Resulting Tubules

The formation of an RV is only a beginning in terms of this tubulogenic event. The resulting epithelial ball needs to elongate, bend, segment/pattern, and differentiate to generate a functional, mature nephron within the adult kidney. Ultimately, each nephron is segmented and patterned along its proximal–distal axis into a renal corpuscle (RC) followed by proximal tubules (PTs) (convoluted and straight), loop of Henle (LH), distal tubules (DTs) and connecting tubule (CT) segments, respectively (Fig. 5.1E). These are connected to the CDs to form a continuous uriniferous tubule (Georgas *et al.*, 2008; Kriz and Bankir, 1988; Little *et al.*, 2007). Individual tubular segments can be distinguished at the molecular, cellular and anatomical levels, with each segment ultimately conferred a specific functional role in the regulation of electrolyte and water balance. While this segmentation of the nephron is critical for normal function, and the absence of particular nephron segments can result in a variety of diseases in humans (Allanson *et al.*, 1992), the molecular programs governing nephron segmentation and proximal–distal polarity are still poorly elucidated.

Proximal–distal patterning along the length of the nephron begins very early in the life of a nephron. This is readily evident at the histological level as early as the S-shaped body (SB) stage which can be divided into distinct proximal, medial, and distal (with respect to the position of the adjacent UT) segments. The proximal segment is further organized into two distinct epithelial layers, the parietal (Bowman's capsule) and visceral (podocyte) layers which, together with the invading endothelial and mesangial cells, will become the RC. This proximal–distal patterning of the SSB is reflected in restricted expression domains of known marker genes (Dressler, 2006; Georgas *et al.*, 2009; Kopan *et al.*, 2007). Although histologically indistinct, differential gene expression suggests that proximal and distal domains exist as early as the RV stage (Georgas *et al.*, 2009; Kopan *et al.*, 2007) (Fig. 5.4). This includes proximal–distal differences in the expression of genes novel to kidney development (*Tmem100, Papss2, Greb1, Pcsk9*, and *Ccdc86*) and components of pathways with established roles in kidney development. For example, members of the Notch (*Jag1, Dll1*), BMP (*Bmp2*), and Wnt (*Wnt4, Dkk1*, and *Ccnd1*) signaling pathways are all polarized to the distal RV.

A comparison of the location of gene expression in the RV with the expression pattern of the same genes in later stages of nephron development suggests that the precursors of every segment of the mature nephron are already distinguishable within the RV. *Lhx1, Pou3f3, Dll1* (distal RV), and *Wt1* (proximal RV) were among the first molecules identified with obvious polarity of expression in the RV (Georgas *et al.*, 2008; Kobayashi *et al.*, 2005; Nakai *et al.*, 2003) (Fig. 5.4). All of these genes are critical for specification of various nephron segments. *Wt1* expression is polarized in the proximal RV, restricted to the proximal SSB and maintained in the podocytes throughout all stages of nephrogenesis, including the adult (Georgas *et al.*, 2008, 2009; Pritchard-Jones 1999). Here *Wt1* is critical for the specification of the podocyte layer of the RC. *Wnt4* and *Lhx1* are expressed as early as the PA. Consequently when these genes are disrupted, nephrogenesis is blocked very early. *Wnt4* mutants develop only PAs (Kispert *et al.*, 1998; Stark *et al.*, 1994). *Lhx1* conditional mutants (Rarb2Cre; Lim1) develop RVs

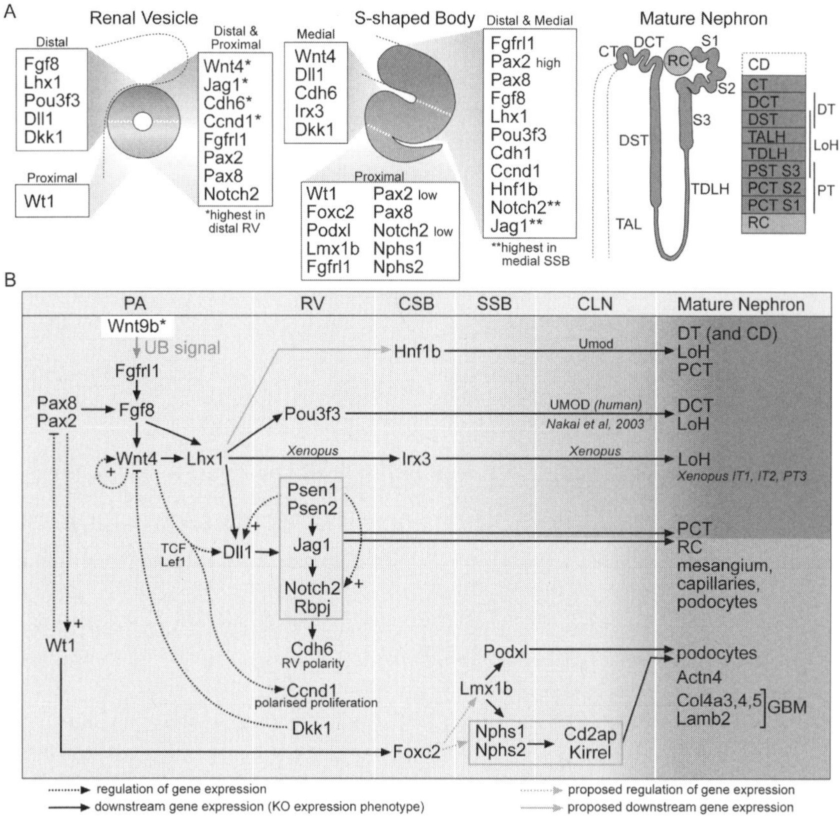

Figure 5.4 Summary of genes involved in nephron segmentation. (A) The expression patterns of polarized genes along the proximal–distal axis of the renal vesicle and SB. The ureteric epithelium is represented by a dotted black line. The mature nephron segments are shown (right); with CD derived from the ureteric epithelium and nephron segments derived from the CM; CT, DCT, distal straight tubule, thin ascending loop of Henle; thin descending loop of Henle; proximal straight tubule, PCT and RC. (B) The transcriptional hierarchy of genes governing nephron patterning. Genes are shown in order of their gene expression from left to right; in PA, RV, CSB, SSB, and capillary loop nephron. The heirarchy of specification within the mature nephron are indicated (right). This has been determined from the genetic analysis of mutant mice, see text (Section 4-5) for details. (See Color Insert.)

that express *Wnt4, Pax8*, and *Fgf8*; however, these RVs are not correctly patterned and they do not show regional expression of the polarized downstream genes *Pou3f3* and *Dll1* (Kobayashi *et al.*, 2005). The expression pattern of *Lhx1* suggests that it is required only for the tubular segments of the nephron that are likely derived from the distal RV.

Evidence that this is the case comes from chimeric experiments using *Lhx1* knockout cells where these cells only contribute to the proximal segment of the SSB (Kobayashi *et al.*, 2005). *Dkk1*, an inhibitor of canonical Wnt signaling, shows expression restricted to the distal RV in the early nephron, and medial SSB; however, the details of its role in nephrogenesis are currently unknown (Georgas *et al.*, 2009). Of note, *Dkk1* expression in the distal RV may act to lessen the canonical signal from the Wnt9b in the adjacent UT. CyclinD1 (*Ccnd1*), a cell cycle regulator that is also likely to be involved in the canonical Wnt pathway as a target of TCF/Lef signaling, is also differentially expressed in the distal RV. Indeed, *Ccnd1* expression indicates preferential proliferation of cells within the distal RV that may aid the fusion of the developing nephron to the ureteric epithelium (Georgas *et al.*, 2009).

4.1. Notch signaling in the patterning of the proximal nephron

A proximal segment of the elongating nephron becomes highly convoluted but stays within the cortex of the kidney to form the PTs. To get a tube to bend requires differential changes in cell shape, such as apical constriction to cause a curve, and/or the provision of external pressures to force a tube to turn rather than grow in one direction. How this occurs is not known, but we do know that patterning of the PTs requires the notch signaling pathway (Kopan *et al.*, 2007) (Fig. 5.4). *Notch1, Notch2*, and the ligands *Dll1* and *Jag1* are all expressed within the RV (Chen and Al-Awqati, 2005). The loss of *Notch2* and *Jag1* in mutant mice results in the loss of proximal nephron segments, including the proximal convoluted tubules (PCT) and also the capillaries and mesangium of RCs, the latter resulting from a role of notch signaling in vascular development (McCright *et al.*, 2001; 2002). Podocytes develop in these hypomorphic *Notch2* kidneys, although their organization is abnormal, indicating that a small amount of notch signaling can initiate their development. However, in the *Notch2* null allele (*Pax3*-Cre conditional loss of *Notch2* in the MM), podocytes do not develop and the nephrons are severely truncated with only DTs identified (Cheng and Kopan 2005, Cheng *et al.*, 2007). In these *Notch2*-deficient kidneys, RVs develop normally and are polarized (normal *Jag1, Dll1*, and *Lhx1* regional expression), however, by the SSB stage, segmentation is defective with the *Wt1*-expressing podocyte precursors and *Cdh6*-expressing proximal convoluted tubule precursors are absent (Cheng *et al.*, 2007). *Cdh6* is, therefore, downstream of notch signaling. A marker of the medial SSB, *Cdh6* is also differentially polarized to the distal RV. Although *Cdh6* mutant mice show no renal dysfunction they have reduced nephron number due to postnatal necrosis, which is likely to be the result of abnormal fusion of nephrons to the ureteric epithelium, possibly as a result of a delay in MET at the distal RV (Georgas *et al.*, 2009; Mah *et al.*, 2000). Although ectopic activation of *Notch1* in nephron precursors can lead to a proximalization of the nephron, unlike *Notch2*, it is not sufficient for the development of proximal cell fates (Cheng *et al.*, 2007). In *Lhx1* mutant nephrons that are blocked at the RV stage, the notch ligand *Dll1* is altered and as a result notch signaling is impaired. This suggests that *Lhx1* is an upstream regulator of notch signaling in the initiation of PT fate. While *Dll1* is restricted to the distal RV and medial segment of the SSB, its role in nephron segmentation is unclear as *Dll1* mutants do not develop a phenotype (Cheng *et al.*, 2007; McCright *et al.*, 2001). Notch signaling occurs when notch is cleaved and the notch intracellular domain (NICD) is translocated to the nucleus. The NICD then complexes with the transcription factor Rbpj, and together with cofactors, activates the transcription of downstream notch target genes. Conditional removal of *Rbpj* in the MM leads to a similar phenotype to the *Notch2* null, resulting in nephrons lacking both PTs and podocytes (Cheng *et al.*, 2007). Notch signaling is also modulated by two presenilins, *Psen1* and *Psen2*, which are essential for the proteolytic activation of notch protein. These genes are also known to function in proximal segmentation of the nephron. In a mutant mouse affecting both *Psen1* and *Psen2*, RVs develop, however, *Jag1* and *Dll1* are absent and they do not progress into a correctly patterned SSB stage. Nephrons that do develop into tubules have a DT phenotype and completely lack PTs and RCs, including podocytes. This is again consistent with specification of the proximal nephron by the notch pathway (Wang *et al.*, 2003).

4.2. Patterning of the LH and DTs

Unlike other portions of the nephron, the LH is a long, straight, tubular segment that lies parallel to the orientation of the CDs and descends from the cortex or medulla into the papilla of the kidney. In other organs, tubule elongation can involve convergent extension and/or cellular proliferation. The plane of cell division or the presence of convergent extension within the tubule will determine the diameter of each segment. Karner *et al.* (2009) propose that the patterning of the LH arises in response to the secretion of Wnt9b from the CDs promoting convergent extension. This would ensure thinning and elongation of the tubule at this point. They did demonstrate that this was not mediated by canonical signaling, which is interesting given the accepted role for a Wnt9b-mediated canonical signal in nephron induction. Three members of the Iroquois (Irx) gene family, *Irx1, Irx2* and *Irx3,* are expressed in the mouse nephron from the comma-shaped body (CB) stage. Their restricted pattern of expression in both the embryonic and adult mouse nephron suggests that they may be involved in patterning the LH

(Reggiani *et al.*, 2007). While there is no formal evidence for this in mouse, gain- and loss-of-function studies in *Xenopus* have shown that the Irx genes, while not required for the initial activation of early kidney genes, are required early for maintenance of the pronephric anlage and later segmentation of the nephron (Alarcon *et al.*, 2008; Reggiani *et al.*, 2007). *Irx1* and *Irx3* are necessary and sufficient for the formation of the pronephric tubule intermediate segments, these being regarded as equivalent to the LH in the mammalian nephron (Alarcon *et al.*, 2008; Reggiani *et al.*, 2007). Given the high degree of similarity in nephron segmentation and gene patterning between the *Xenopus* pronephros and mouse nephron (Raciti *et al.*, 2008), it can be inferred that this role for *Irx* genes in LH patterning may extend to mammals.

A POU-domain transcription factor that is polarized in expression from RV, Pou3f3 (Brn1), is also required for the differentiation of the distal nephron [LH, distal convoluted tubules (DCTs) and macula densa] (Nakai *et al.*, 2003). In *Pou3f3* mutant mice, the DT and LH segments are absent and the LH marker, Umod, cannot be detected. *Umod* encodes the glycoprotein uromodulin, and its expression is widely used as marker of the adult DT and LH segments, where it is expressed from the capillary loop stage (Bachmann *et al.*, 1990). *Umod* knockout mice do not display a phenotype (Raffi *et al.*, 2006), however, in humans, *UMOD* mutations result in hyperuricemic nephropathy, familial juvenile and medullary cystic kidney disease type 2 (MCKD2) (Hart *et al.*, 2002; Turner *et al.*, 2003). This suggests that at least in humans UMOD is likely to function in either the terminal differentiation and/or maintained function of the distal and LH nephron segments.

4.3. Patterning of UT versus stalk and differentiation of the ureter–pelvic junction

Tubular patterning and maturation also occurs in the branching ureteric compartment, the branches of which differentiate into CDs that empty into the pelvis and the initial stalk of which becomes the ureter. Again, spatially regulated differentiation of these segments ensures their capacity to play a distinct functional role. The CDs are highly water-permeable as they represent the last site of NaCl and water reabsorption for the kidney. This requires the expression of the water channel aquaporin 2 (*Aqp2*). In contrast, the pelvic lining and the ureter are water-tight conduits for urine which express a number of uroplakin proteins within specialized epithelial cells called umbrella cells. Ureter function also requires the formation of a musculature to facilitate peristalsis. How differentiation in these distinct phenotypes occurs is poorly understood; however, hedgehog signaling is important. The hedgehog ligand Shh, secreted from the ureteric stalk and medullary ureteric epithelium, signals *via* the Ptch1 receptor in the surrounding mesenchyme to induce the production of smooth muscle (Fig. 5.2). A similar process is important in the development of the bladder and urethral musculature. Bmp4 continues to be secreted by a mesenchymal population along the inner CDs and plays a role in establishing the ureteropelvic junction that demarcates the CD system from the epithelia of the pelvis and ureter. This is exemplified by the Bmp4 induction of genes selectively expressed by the ureter, such as the uroplakins (Brenner-Anantharam *et al.*, 2007). Bmp4 is also involved in the recruitment of smooth muscle to the ureter (Wang *et al.*, 2009). The transcription factor Thsz3 is downstream of Shh and BMP4 signaling and is also required for the normal differentiation of ureteric smooth muscle (Caubit *et al.*, 2008). Defects in Tbx18 also disrupt ureteral mesenchyme development (Airik *et al.*, 2006). Targeted deletion of the Disc-large homolog 1 (*Dlgh1*), which is expressed in the developing ureter, causes a loss of ureteral stromal cells and disorganized ureteral smooth muscle cells. This results in impaired ureteral peristalsis and hydronephrosis (Mahoney *et al.*, 2006). The patterning and segmentation of the nephrons is also influenced by the demarcation between the UB tip and stalk. For example, while expression of *Ret, Gfrα1*, and *Wnt11* are confined to the advancing tips/ampullae, *Wnt7b* expression becomes restricted to UB stalk and plays a key role in growth (and extension) of the adjacent developing loops of Henle by regulating the plane of epithelial cell division (Yu *et al.*, 2009).

5. Moving From Structure to a Functional Filter

5.1. Patterning and vascularization of the developing RC

Much of what has been discussed to date relates to early kidney development. As mutations affecting these early events mask roles for genes in later stages of differentiation, in some ways we know less about this later critical stage of development. Ultimately the kidney is a filter of blood; hence vascularization of the RC at the proximal end of the nephron is one of the most critical events in kidney development. This occurs as the presumptive glomeruli are being patterned and is regulated by the cells of that RC. By the CSB stage, RC patterning is evident as a distinct cleft (vascular cleft) between the basal aspects of the proximal epithelia (Fig. 5.1). *Foxc2* is the earliest known specific podocyte marker, with expression beginning in presumptive podocytes at the comma-stage, before a morphological distinction between visceral and parietal cells can be seen (Takemoto *et al.*, 2006). The outermost layer of epithelia at the proximal end of the comma-shape body begins to flatten and elongate to form the parietal epithelia, while the corresponding internal layer, the visceral layer, remains polygonal. Distinction of these epithelial layers is observed at the SSB stage and coincides with expression of the podocyte markers *Lmx1b, Podxl, Nphs1*, and *Nphs2* in the visceral layer (Takemoto *et al.*, 2006). The subsequent separation of the parietal (Bowman's capsule) and visceral (podocyte) epithelia forms the Bowman's space. The functional and structural phenotype of the visceral and parietal epithelium of the RC diverges dramatically from this point in time, yet there are very few early differences in their pattern of gene expression. Wt1 is upregulated in both presumptive podocytes and parietal epithelium, while Pax2 expression is lost (Ryan *et al.*, 1995). In mature glomeruli, Wt1 expression is maintained only in the podocyte, where it plays a key role in maintenance of the podocyte phenotype. Podocyte-specific expression of a dominant-negative form of *Wt1*, found in Denys–Drash syndrome, results in mesangial sclerosis and podocyte foot process abnormalities (Patek *et al.*, 1999). Given the requirement for podocytes to attach closely to the glomerular basement membrane (GBM) for normal renal function and the lack of evidence for cell division in these cells, it has been assumed that mature podocytes have little capacity for self-renewal. However, recent work has shown in both mice and humans that new podocytes are recruited from the parietal epithelial cells at the urinary pole of the glomerulus (Appel *et al.*, 2009; Ronconi *et al.*, 2009), reflecting the common origin of these two cell types.

The forming podocytes express Vegfa. Endothelial progenitors (angioblasts) within the stroma express the Vegfa receptor, Flk1, and migrate into the cleft in response to the podocyte secretion of Vegfa. The origin of these angioblasts has not been proven, but they may derive from the MM and/or extrarenal cells (Sariola, 1985). Initially angioblasts aggregate to form solid precapillary cords, but develop a lumen over several days by Tgfβ1-induced apoptosis of superfluous endothelial cells (Ballerman, 2005). As endothelial cells migrate from the stroma into the cleft they begin to express Pdgfb which signals *via* binding to the receptor Pdgfrβ expressed on mesangial cell precursors. These mesangial cells are pericyte-like vascular support cells that share the same origin as the vascular smooth muscle cells and the pericytes lining the extraglomerular vasculature (Lindahl *et al.*, 1998; Takemoto *et al.*, 2006). Mesangial precursors appear to be derived from the MM and reside in the stroma. The capillary loop stage of glomerulogenesis marks the point at which the developing glomerulus contains a single capillary, continuous with an afferent and efferent arteriole. Further development of the glomerular capillary to form the complex capillary network requires branching of the single capillary loop and expansion of the mesangium. The capillary network is thought to form through either sprouting from the capillary loop or, more likely, intussusceptive angiogenesis, a process whereby endothelium forms contacts or pillars within the lumen that expand longitudinally to cause vessel splitting (Burri *et al.*, 2004). Mesangial cells are involved in providing structural support to the newly formed vessels by synthesizing the surrounding extracellular matrix (ECM) (Betsholtz *et al.*, 2001). Loss of mesangial cells, as in mesangiolysis, result in dilation of glomerular capillaries.

Podocyte-specific gain- and loss-of-function mutants have shown that the dose of Vegfa is critically important in glomerular vascular development. Too little or no podocyte expression of Vegfa causes severe glomerular vascular defects specifically involving recruitment and morphology of endothelial cells. Overexpression of Vegfa in podocytes leads to collapsing glomerulopathy possibly resulting from hyper-perfusion of endothelial cells (Eremina *et al.*, 2003). These studies have shown that gene dosage of *Vegfa* is critical in the maintenance of the endothelial fenestrae, through which filtration occurs. Furthermore, Vegfa or Vegfa receptor blocking molecules now used as anti-angiogenic drugs in the treatment of cancers result in proteinuria (Launay-Vacher *et al.*, 2009). Bioavailability of Vegfa is regulated by alternate splicing of the heparan-binding domain. Isoforms (Vegfa 164, 188) containing the binding domain are sequestered to heparan sulfate proteoglycans in the ECM, while the isoform lacking the heparan-binding domain, Vegfa 120, is freely soluble. Another molecule that has been shown to regulate Vegfa secretion from the podocyte is Crim1 (Wilkinson *et al.*, 2007). *Crim1* mutant adult mice develop a progressive renal disease characterized by a leaky GBM, interstitial fibrosis, and glomerular cysts (Wilkinson *et al.*, 2009). Crim1 is a cell surface protein expressed in the podocytes that sequesters Vegfa at the cell surface, regulating its release or spatial localization. In *Crim1* hypomorphic mouse embryos, increased Vegfa signaling within the glomerular endothelium results from a disruption to Vegfa tethering to the podocytes and this leads to aberrant glomerular capillary development (Wilkinson *et al.*, 2007). Interestingly, in these mouse models, the primary defect is in endothelial cells; however, podocyte architecture is also severely affected, underscoring the importance of endothelial cell integrity to maintenance of other components of the glomerular filtration barrier (GFB). Development of the mesangium and the coincident formation of the glomerular capillary network are reliant on expression of Pdgfb and Pdgfrβ. Genetic ablation of *Pdgfb* or *Pdgfrβ* results in defective recruitment of mesangial cells into the developing glomerulus and the glomerular vasculature develops as one or a few dilated capillary loops (Lindahl *et al.*, 1998). Like Vegfa, Pdgfb also contains a heparan-binding motif that appears to be necessary for regulation of extra-cellular localization (Lindblom *et al.*, 2003). Correct dosage of podocyte expression of *Bmp4* during podocyte differentiation is also essential for proper glomerular tuft development. *Bmp4* heterozygous null mutants develop glomerular cysts and collapsed capillary tufts. Likewise podocyte-specific expression of Noggin, a BMP antagonist, resulted in a similar glomerular phenotype as well as poorly developed tubules and reduced numbers of tubules, while podocyte overexpression of *Bmp4* resulted in a the absence of endothelium in the tuft (Ueda *et al.*, 2008).

5.2. Maturation of the RC

As development progresses beyond the capillary loop stage, podocytes and endothelial cells contribute to the GBM. As they mature during the capillary loop stage, the podocytes develop highly specialized extensions called foot processes that adhere to the GBM *via* cytoskeletal elements. Adherens-like cell–cell junctions called the slit diaphragm develop between interdigitating podocyte foot processes. Endothelial cells become flattened against the GBM and form fenestrae. Endothelial cells, podocytes, and GBM together form the selectively permeable GFB. The proteins Kreisler and Glepp1 are involved in podocyte differentiation and maturation. Mutations of these genes in mice lead to poorly formed podocyte foot processes and in the case of Kreisler, proteinuria (Sadl *et al.*, 2002; Wang *et al.*, 2000).

Both the slit diaphragm and foot process attachment of podocytes to the GBM are critical in maintaining the integrity of the GFB. Disruption of these structures through gene deletion, mutation, disease, or environmental factors can lead to foot process effacement, leakiness of the GFB, and proteinuria. Mutations in the genes coding for the slit diaphragm proteins Nephrin (NPHS1) and Podocin (NPHS2) in humans lead to congenital nephrotic syndrome, characterized by proteinuria and renal failure (Koziell *et al.*, 2002). Knockout of Podocin in the mouse leads to defects in glomerular capillary formation and mesangial cell recruitment, underlying the codependence of all the components of the glomerular tuft for normal development. CD2AP associates with Nephrin in the slit diaphragm and is linked with focal and segmental glomerular sclerosis (FSGS), also characterized by proteinuria and progressive kidney disease (Kim *et al.*, 2003). Alpha-actinin-4 (ACTN4) is a podocyte cytoskeletal protein that has been linked to an autosomal dominant form of FSGS (Kaplan *et al.*, 2000). Lmx1b, a Lim-homeodomain transcription factor, is necessary for podocyte differentiation from the SSB stage and later for maintenance of the slit diaphragm (Chen *et al.*, 1998). Mutations in this gene in humans results in Nail Patella syndrome, a syndrome which frequently includes glomerulopathy (Dreyer *et al.*, 1998). Other proteins involved in the slit diaphragm but not yet known to be associated with human disease include FAT1, Tight junction protein (ZO-1), and P-cadherin (Cadherin 3) (Pavenstadt *et al.*, 2003). Attachment of podocytes to the underlying GBM involves the transmembrane receptors integrin $\alpha3\beta1$ and dystroglycan. These proteins bind GBM matrix molecules such as collagen, laminin, agrin, and perlecan *via* their extracellular domain and are coupled to the actin cytoskeleton intracellularly. Genetic ablation of integrin $\alpha3$ (*Itga3*) (Kreidberg *et al.*, 1996) or podocyte specific loss of Integrin β1 (*Itgb1*) (Pozzi *et al.*, 2008) results in neonatal proteinuria and early renal failure.

The GBM is synthesized as two separate layers produced by the endothelial cells and the maturing podocytes. These layers fuse during maturation of the glomerulus. Collagen IV and laminins are the major protein contributors to the GBM (Miner, 2005). Composition of the GBM changes with maturity of the glomerulus as collagen IV, and laminins undergo isoform shifts during development (Miner, 1999; St John and Abrahamson, 2001). Col4a3, Col4a4, and Col4a5 are the mature collagens of the mature GBM. Alport syndrome, which is characterized by adult onset nephritis progressing to renal failure, results from mutations in any one of the collagen IV isoforms, *COL4A3, COL4A4*, or *COL4A5*. This prevents the usual isoform transition, resulting in retention of early COL4A1 and COL4A2 isoforms and leads to a disruption of the COL4A3–4–5 collagen network in the GBM (Heidet and Gubler, 2009). Mutations in either *COL4A3* or *COL4A4* also result in benign familial hematuria that is a non-progressive nephropathy due to a thinning of the GBM (Longo *et al.*, 2002). The negatively charged heparan sulfate proteoglycans are an important component of the GBM matrix particularly in conferring charge selectivity to the GFB. The podocyte-derived sialoprotein, Podocalyxin, also contributes to the overall negative charge on the GBM (Pavenstadt *et al.*, 2003).

Coincident with the specialization of the podocyte at the capillary loop stage is a loss in mitotic activity including a loss of cell cycle promoters and upregulation of cell cycle inhibiters such as $p27^{Kip1}$ and $p57^{Kip2}$. Under normal conditions podocytes do not proliferate beyond this stage, however, expression of cell cycle proteins may be reversed in some diseases such as collapsing glomerulopathy and HIV-associated nephropathy leading to podocyte de-differentiation, proliferation, foot effacement, and loss of barrier function (Marshall and Shankland, 2007).

6. Disruptions to Kidney Tubulogenesis in Human Disease

Much of the analysis of mammalian kidney development has been performed in mice. However, the association of gene mutations with defects in kidney development in humans has both validated the results observed in mouse and extended our understanding of the roles these genes play. Defects of genes critical at almost any stage of kidney morphogenesis can affect ultimate renal function. Here we have reviewed the literature again by linking diseases with disruption to one or other tubulogenic event.

6.1. Consequences of defects in branching morphogenesis

Congenital anomalies of the kidney and urinary tract (CAKUT) occur in 1 in 500 humans (Schedl, 2007), ultimately constituting approximately 20–30% of all prenatal anomalies and representing a major cause of renal failure in infants and children. CAKUT can include a wide variety of defects (aberrant ureteric budding, duplicated collecting systems, duplex kidneys, megaureter, abnormal glomerulogenesis, papillary hypoplasia, and hydronephrosis), but most frequently defects include vesicoureteric reflux (VUR; improper insertion of the ureters into the bladder causing urine to flush back into the kidney) and renal hypoplastic dysplasia (1 in 1000 live births) (Winyard and Chitty, 2008). While the genetic basis of CAKUT is complex with variable penetrance, a common determinant appears to be disturbances in budding of the UB from the ND (Ichikawa *et al.*, 2002). Hence, one would imagine mutations in genes involved in this process. Indeed, mutations in *RET* in humans have been linked to primary VUR (Yang *et al.*, 2008). Components of the renin–angiontensin system, later critical in the role of the kidney in regulation of blood pressure, are also disrupted in some cases of CAKUT (Pope *et al.*, 1999; Rigoli *et al.*, 2004; Yosypiv *et al.*, 2008), implicating these genes in UB branching. Loss of *Robo2* in mice leads to additional UBs and ultimately VUR (Lu *et al.*, 2007). Similarly, *ROBO2* mutations underlie 5.1% of CAKUT in humans (Bertoli-Avella *et al.*, 2008). A number of other gene mutations have been associated with dominant renal hypodysplastic disease (RHD) without overt VUR or megaureter, usually presenting as part of a syndrome. These include mutations in HNF1β (*TCF2*) (renal cysts and mature onset diabetes of the young—MODY5), *PAX2* (renal–coloboma syndrome), *EYA1* and *SIX1* (branchio-oto-renal syndrome; craniofacial, hearing, and kidney defects), and *SALL1* (Townes–Brocks syndrome; multiorgan syndrome with renal hypoplasia) (Abdelhak *et al.*, 1997; Ruf *et al.*, 2004; Weber *et al.*, 2006). All of these genes are expressed in the MM but affect the ability of that tissue to stimulate UB outgrowth. Of patients with RHD, 15% show mutations in *TCF2* or *PAX2*, while *EYA1, SALL1*, and *SIX1* mutations are less frequent. One of the largest studies investigating the genetic basis of CAKUT and involving a cohort of >250 patients (Weber *et al.*, 2006), has also identified mutations in the mesenchyme-expressed *BMP4* and *SIX2* associated with renal hypodysplasia (Weber *et al.*, 2008). In conclusion, the establishment of this first tubulogenic event is critical both to the initiation of kidney development as well as the determination of overall organ size, shape, and function.

6.2. Dysplasia and defects in MET in humans

Renal dysplasias not in association with clear ureteric defects can also arise from disruption to the second tubulogenic event, nephron endownment.

Conversely, nephron endowment can be reduced both by a reduction in MET as well as a reduction in UB branching as the two processes are codependent and intimately linked. Certainly in humans, mutations in *PAX2* are associated with renal–coloboma syndrome, a congenital defect characterized by renal hypoplasia, variably penetrant VUR, and optic nerve colobomas (Sanyanusin *et al.*, 1995). Mutations in *WNT4* also result in SERKAL syndrome, which includes renal hypoplasia. Rare cases of unilateral renal agenesis have been reported in isolated patients with *WNT4* and *PAX8* mutations (Biason-Lauber *et al.*, 2004; Meeus *et al.*, 2004). Point mutations, aberrant splicing, or large deletions are prevalent in the *WT1* gene in many human syndromes such as Denys–Drash syndrome, WAGR (Wilms–Aniridia–Genital Anomoloies–Retardation), Frasier syndrome, and in Wilms' tumor, one of the most common childhood solid tumors. These syndromes frequently involve genital defects as well as renal dysplasia and an increased risk for developing Wilms' tumor. Wilms' tumor is a malignant embryonal tumor that is thought to develop from foci of persistant embryonic structures called nephrogenic rests (Khoury, 2005). Tumors most often contain tubular- and glomerular- like structures that recapitulate embryonic renal development; however, non-renal tissues such as skeletal muscle, bone, and neural tissue may also be found (Beckwith, 1998; Khoury, 2005). One interpretation of this triphasic phenotype is that the *WT1* defects result in aberrant fates for the MM with some attempting unsuccessful MET and other cells reverting to heterologous mesenchymal fates befitting an earlier state of IM development. With respect to genes involved in nephron patterning and segmentation, two forms of Alagille syndrome, a multisystem developmental disorder with severe renal disease, result when either *NOTCH2* (ALGS2) or *JAG1* (ALGS1) is mutated (Li *et al.*, 1997; McDaniell *et al.*, 2006; Oda *et al.*, 1997). The defects observed here include cystic or dysplastic kidneys with renal tubular acidosis leading to renal insufficiency and hypertension.

6.3. Nephron number and renal disease

Each human kidney contains between 300,000 and 1.8 million nephrons, all of which are endowed before birth at around 36 weeks of gestation (Hoy *et al.*, 2008; Nyengaard *et al.*, 1992). This dramatic variability in nephron number is a direct result of both genetic insults, many of which have been discussed in the context of the mouse, and environmental factors affecting the process of nephron endowment *in utero*. Even a slowing of the process due to intrauterine growth retardation appears to reduce the overall number of functional nephrons. There is even some evidence that some indigenous populations, including the Australian Aboriginal, have a lower average nephron number. This may stem from common environmental disruptors or an inherited genetic trait more common in this population. What is becoming clear is that while reduced nephron number may not be apparent at birth, there is evidence linking reduced overall nephron number with renal failure later in life due to an increased risk of renal disease and hypertension (reviewed in Zandi-Nejad *et al.*, 2006). Low birth weight is the strongest indicator for a reduced nephron number and this can arise due to a variety of intrauterine insults, including acute or prolonged hypoxia, preeclampsia, placental insufficiency, recreational drug use, smoking, stress, and malnutrition. How these insults translate to a reduction in MET is not yet clear, although animal studies suggest that reduced nephron number in response to intrauterine exposure to dexamethasone may be due to reduced branching (Singh *et al.*, 2007). It will be important to functionally link these environmental effectors with molecular pathways to begin to intervene.

7. Conclusion

In summary, this review describes the state of play in our understanding of tubulogenesis in the kidney. However, what has been soundly overlooked in these discussions has been the regulation of these processes by the surrounding renal intersistium, some of which is MM derived and some of which presumably migrates into the kidney during development. This includes the interstitial fibroblasts, major blood vessels, lymphatics and nerves, as well as resident tissue macrophages which are known to be present within the MM as early as 12 dpc in mouse (Rae *et al.*, 2007). The nature of the relationship between the interstitial microenvironment and the tubular nephrons is poorly understood. As a result, our understanding of why ischemic, toxic, or genetic disturbances lead to chronic fibrosis and loss of glomerular and tubular function also remains unclear. Indeed, a great deal remains to be understood. At birth, a human kidney is approximately 5 cm in length but will grow to a final length of 13 cm without any increase in nephron number. This continues to require tubular elongation and maturation as well as considerable remodeling of the interstitial compartment, none of which is understood. Disruption to tubule elongation and tubule cell orientation results in the formation of cystic conditions including polycystic kidney disease and nephronopthisis both during and after birth. This, together with even subtle changes in overall nephron number will strongly influence renal function throughout life. And so the final phases of this two-fold tale of tubulogenesis remains incomplete. The challenge lies in integrating what we understand about the early stages of kidney development with the finishing off and ongoing function of the organ. This will require strong integration between developmental biology and experimental nephrology. Such an understanding will undoubtedly assist in the diagnosis and treatment of renal disease as well as perhaps one day allow the regulation or reinitiation of nephrogenesis.

ACKNOWLEDGMENTS

ML is a Principal Research Fellow of the National Health and Medical Research Council (NHMRC) of Australia. Her research is supported by the National Institutes of Health (DK070136), the NHMRC (ID455972), and the Australian Stem Cell Centre.

REFERENCES

Abdelhak, S., Kalatzis, V., Heilig, R., Compain, S., Samson, D., Vincent, C., Weil, D., Cruaud, C., Sahly, I., Leibovici, M., Bitner-Glindzicz, M., Francis, M., *et al.* (1997). A human homologue of the drosophila eyes absent gene underlies branchio-oto-renal (BOR) syndrome and identifies a novel gene family. *Nat. Genet.* **15**, 157–164.

Airik, R., Bussen, M., Singh, M. K., Petry, M., and Kispert, A. (2006). Tbx18 regulates the development of the ureteral mesenchyme. *J. Clin. Invest.* **116**(3), 663–674.

Alarcon, P., Rodriguez-Seguel, E., Fernandez-Gonzalez, A., Rubio, R., and Gomez-Skarmeta, J. L. (2008). A dual requirement for iroquois genes during xenopus kidney development. *Development* **135**(19), 3197–3207.

Allanson, J. E., Hunter, A. G., Mettler, G. S., and Jimenez, C. (1992). Renal tubular dysgenesis: A not uncommon autosomal recessive syndrome: A review. *Am. J. Med. Genet.* **43**(5), 811–814.

Appel, D., Kershaw, D. B., Smeets, B., Yuan, G., Fuss, A., Frye, B., Elger, M., Kriz, W., Floege, J., and Moeller, M. J. (2009). Recruitment of podocytes from glomerular parietal epithelial cells. *J. Am. Soc. Nephrol.* **20**(2), 333–343.

Bachmann, S., Metzger, R., and Bunnemann, B. (1990). Tamm-horsfall protein-mRNA synthesis is localized to the thick ascending limb of henle's loop in rat kidney. *Histochemistry* **94**(5), 517–523.

Ballermann, B. J. (2005). Glomerular endothelial cell differentiation. *Kidney Int.* **67**, 1668–1671.

Batourina, E., Gim, S., Bello, N., Shy, M., Clagett-Dame, M., Srinivas, S., Costantini, F., and Mendelsohn, C. (2001). Vitamin A controls epithelial/mesenchymal interactions through ret expression. *Nat. Genet.* **27**, 74–78.

Beckwith, J. B. (1998). Nephrogenic rests and the pathogenesis of wilms tumor: Developmental and clinical considerations. *Am. J. Med. Genet.* **79**, 268–273.

Bertoli-Avella, A. M., Conte, M. L., Punzo, F., de Graaf, B. M., Lama, G., La Manna, A., Polito, C., Grassia, C., Nobili, B., Rambaldi, P. F., Oostra, B. A., and Perrotta, S. (2008). ROBO2 gene variants are associated with familial vesicoureteral reflux. *J. Am. Soc. Nephrol.* **19**(4), 825–831.

Betsholtz, C., Karlsson, L., Lindahl, P. (2001) Developmental roles of platelet-derived growth factors. *Bioessays* **23**(6), 494–507.

Biason-Lauber, A., Konrad, D., Navratil, F., and Schoenle, E. J. (2004). A WNT4 mutation associated with mullerian-duct regression and virilization in a 46,XX woman. *N. Engl. J. Med.* **351**(8), 792–798.

Boyle, S., Misfeldt, A., Chandler, K. J., Deal, K. K., Southard-Smith, E. M., Mortlock, D. P., Baldwin, H. S., and de Caestecker, M. (2008). Fate mapping using cited1-CreERT2 mice demonstrates that the cap mesenchyme contains self-renewing progenitor cells and gives rise exclusively to nephronic epithelia. *Dev. Biol.* 313(1), 234–245.

Brenner-Anantharam, A., Cebrian, C., Guillaume, R., Hurtado, R., Sun, T. T., and Herzlinger, D. (2007). Tailbud-derived mesenchyme promotes urinary tract segmentation via BMP4 signaling. *Development* **134**, 1967–1975.

Bullock, S. L., Fletcher, J. M., Beddington, R. S., and Wilson, V. A. (1998). Renal agenesis in mice homozygous for a gene trap mutation in the gene encoding heparan sulfate 2-sulfotransferase. *Genes Dev.* **12**(12), 1894–1906.

Burri, P. H., Hlushchuk, R., and Djonov, V. (2004). Intussusceptive angiogenesis: Its emergence, its characteristics, and its significance. *Dev. Dyn.* **231**, 474–488.

Carroll, T. J., Park, J. S., Hayashi, S., Majumdar, A., and McMahon, A. P. (2005). Wnt9b plays a central role in the regulation of mesenchymal to epithelial transitions underlying organogenesis of the mammalian urogenital system. *Dev. Cell* **9**(2), 283–292.

Caubit, X., Lye, C. M., Martin, E., Coré, N., Long, D. A., Vola, C., Jenkins, D., Garratt, A. N., Skaer, H., Woolf, A. S., and Fasano, L. (2008). Teashirt 3 is necessary for ureteral smooth muscle differentiation downstream of SHH and BMP4. *Development* **135**(19), 3301–3310.

Chai, L., Yang, J., Di, C., Cui, W., Kawakami, K., Lai, R., and Ma, Y. (2006). Transcriptional activation of the SALL1 by the human SIX1 homeodomain during kidney development. *J. Biol. Chem.* **281**(28), 18918–18926.

Chen, L., and Al-Awqati, Q. (2005). Segmental expression of Notch and Hairy genes in nephrogenesis. *Am. J. Physiol. Renal Physiol.* **288**(5), F939–F952.

Chen, H., Lun, Y., Ovchinnikov, D., Kokubo, H., Oberg, K. C., Pepicelli, C. V., Gan, L., Lee, B., and Johnson., R. L. (1998). Limb and kidney defects in Lmx1b mutant mice suggest an involvement of LMX1B in human nail patella syndrome. *Nat. Genet.* **19**, 51–55.

Cheng, H. T., Kim, M., Valerius, M. T., Surendran, K., Schuster-Gossler, K., Gossler, A., McMahon, A. P., and Kopan, R. (2007). Notch2, but not Notch1, is required for proximal fate acquisition in the mammalian nephron. *Development* **134**(4), 801–811.

Cheng, H. T., and Kopan, R. (2005). The role of Notch signaling in specification of podocyte and proximal tubules within the developing mouse kidney. *Kidney Int.* **68**(5), 1951–1952.

Chi, X., Hadjantonakis, A. K., Wu, Z., Hyink, D., and Costantini, F. (2009). A transgenic mouse that reveals cell shape and arrangement during ureteric bud branching. *Genesis* **47**(2), 61–66.

Chi, X., Michos, O., Shakya, R., Riccio, P., Enomoto, H., Licht, J. D., Asai, N., Takahashi, M., Ohgami, N., Kato, M., Mendelsohn, C., and Costantini, F. (2009). Ret-dependent cell rearrangements in the Wolffian duct epithelium initiate ureteric bud morphogenesis. *Dev. Cell* **17**, 199–209.

Clarke, J. C., Patel, S. R., Raymond, R. M.Jr., Andrew, S., Robinson, B. G., Dressler, G. R., and Brophy, P. D. (2006). Regulation of c-Ret in the developing kidney is responsive to Pax2 gene dosage. *Hum. Mol. Genet.* **15**, 3420–3428.

Costantini, F., and Shakya, R. (2006). GDNF/Ret signaling and the development of the kidney. *BioEssays* **28**, 117–127.

Donovan, M. J., Natoli, T. A., Sainio, K., Amstutz, A., Jaenisch, R., Sariola, H., and Kreidberg., J. A. (1999). Initial differentiation of the metanephric mesenchyme is independent of WT1 and the ureteric bud. *Dev. Genet.* **24**, 252–262.

Dressler, G. R. (2006). The cellular basis of kidney development. *Ann. Rev. Cell Dev. Biol.* **22**, 509–529.

Dreyer, S. D., Zhou, G., Baldini, A., Winterpacht, A., Zabel, B., Cole, W., Johnson, R. L., and Lee, B. (1998). Mutations in LMX1B cause abnormal skeletal patterning and renal dysplasia in nail patella syndrome. *Nat. Genet.* **19**, 47–50.

Dudley, A. T., Godin, R. E., and Robertson, E. J. (1999). Interaction between FGF and BMP signaling pathways regulates development of metanephric mesenchyme. *Genes Dev.* **13**(12), 1601–1613.

Eremina, V., Sood, M., Haigh, J., Nagy, A., Lajoie, G., Ferrara, N., Gerber, H. P., Kikkawa, Y., Miner, J. H., and Quaggin., S. E. (2003). Glomerular-specific alterations of VEGF-A expression lead to distinct congenital and acquired renal diseases. *J. Clin. Invest.* **111**, 707–716.

Georgas, K., Rumballe, B., Valerius, M. T., Chiu, H. S., Thiagarajan, R. D., Lesieur, E., Aronow, B. J., Brunskill, E. W., Combes, A. N., Tang, D., Taylor, D., Grimmond, S. M., *et al.* (2009). Analysis of early nephron patterning reveals a role for distal RV proliferation in fusion to the ureteric tip via a cap mesenchyme-derived connecting segment. *Dev. Biol.* **332**(2), 273–286.

Georgas, K., Rumballe, B., Wilkinson, L., Chiu, H. S., Lesieur, E., Gilbert, T., and Little, M. H. (2008). Use of dual section mRNA in situ hybridisation/immunohistochemistry to clarify gene expression patterns during the early stages of nephron development in the embryo and in the mature nephron of the adult mouse kidney. *Histochem. Cell Biol.* **130**(5), 927–942.

Gerber, S. D., Steinberg, F., Beyeler, M., Villiger, P. M., and Trueb, B. (2009). The murine Fgfrl1 receptor is essential for the development of the metanephric kidney. *Dev. Biol.* **335**(1), 106–119.

Gong, K. Q., Yallowitz, A. R., Sun, H., Dressler, G. R., and Wellik, D. M. (2007). A Hox-Eya-Pax complex regulates early kidney developmental gene expression. *Mol. Cell. Biol.* **27**(21), 7661–7668.

Grieshammer, U., Cebrian, C., Ilagan, R., Meyers, E., Herzlinger, D., and Martin, G. R. (2005). FGF8 is required for cell survival at distinct stages of nephrogenesis and for regulation of gene expression in nascent nephrons. *Development* **132**(17), 3847–3857.

Grieshammer, U., Le, M., Plump, A. S., Wang, F., Tessier-Lavigne, M., and Martin, G. R. (2004). SLIT2-mediated ROBO2 signaling restricts kidney induction to a single site. *Dev. Cell* **6**, 709–717.

Grobstein, C. (1956). Trans-filter induction of tubules in mouse metanephrogenic mesenchyme. *Exp. Cell Res.* **10**(2), 424–440.

Grote, D., Boualia, S. K., Souabni, A., Merkel, C., Chi, X., Costantini, F., Carroll, T., and Bouchard, M. (2008). Gata3 acts downstream of beta-catenin signaling to prevent ectopic metanephric kidney induction. *PLoS Genet.* **4**, e1000316.

Hart, T. C., Gorry, M. C., Hart, P. S., Woodard, A. S., Shihabi, Z., Sandhu, J., Shirts, B., Xu, L., Zhu, H., Barmada, M. M., and Bleyer, A. J. (2002). Mutations of the UMOD gene are responsible for medullary cystic kidney disease 2 and familial juvenile hyperuricaemic nephropathy. *J. Med. Genet.* **39**, 882–892.

Hartman, H. A., Lai, H. L., and Patterson, L. T. (2007). Cessation of renal morphogenesis in mice. *Dev. Biol.* **310**, 379–387.

Heidet, L., and Gubler, M. C. (2009). The renal lesions of Alport syndrome. *J. Am. Soc. Nephrol.* **20**(6), 1210–1215.

Hoy, W. E., Bertram, J. F., Denton, R. D., Zimanyi, M., Samuel, T., and Hughson, M. D. (2008). Nephron number, glomerular volume, renal disease and hypertension. *Curr. Opin. Nephrol. Hypertens.* **17**(3), 258–265.

Humphreys, B. D., Valerius, M. T., Kobayashi, A., Mugford, J. W., Soeung, S., Duffield, J. S., McMahon, A. P., and Bonventre, J. V. (2008). Intrinsic epithelial cells repair the kidney after injury. *Cell Stem Cell* **2**(3), 284–291.

Ichikawa, I., Kuwayama, F., Pope, J. C.4th, Stephens, F. D., and Miyazaki, Y. (2002). Paradigm shift from classic anatomic theories to contemporary cell biological views of CAKUT. *Kidney Int.* **61**, 889–898.

Ishibe, S., Karihaloo, A., Ma, H., Zhang, J., Marlier, A., Mitobe, M., Togawa, A., Schmitt, R., Czyczk, J., Kashgarian, M., Geller, D. S., Thorgeirsson, S. S., *et al.* (2009). Met and the epidermal growth factor receptor act cooperatively to regulate final nephron number and maintain collecting duct morphology. *Development* **136**, 337–345.

Kaplan, J. M., Kim, S. H., North, K. N., Rennke, H., Correia, L. A., Tong, H. -Q., Mathis, B. J., Rodriguez-Perez, J. -C., Allen, P. G., Beggs, A. H., and Pollak, M. R. (2000). Mutations in ACTN4, encoding alpha-actinin-4, cause familial focal segmental glomerulosclerosis. *Nat. Genet.* **24**, 251–256.

Karner, C. M., Chirumamilla, R., Aoki, S., Igarashi, P., Wallingford, J. B., and Carroll, T. J. (2009). Wnt9b signaling regulates planar cell polarity and kidney tubule morphogenesis. *Nat. Genet.* **41**, 793–799.

Khoury, J. D. (2005). Nephroblastic neoplasms. *Clin. Lab. Med.* **25**, 341–361.

Kim, J. M., Wu, H., Green, G., Winkler, C. A., Kopp, J. B., Miner, J. H., Unanue, E. R., and Shaw, A. S. (2003). CD2-associated protein haploinsufficiency is linked to glomerular disease susceptibility. *Science* **300**, 1298–1300.

Kispert, A., Vainio, S., Shen, L., Rowitch, D. H., and McMahon., A. P. (1996). Proteoglycans are required for maintenance of Wnt-11 expression in the ureter tips. *Development* **122**, 3627–3637.

Kispert, A., Vainio, S., and McMahon A. P. (1998). Wnt-4 is a mesenchymal signal for epithelial transformation of metanephric mesenchyme in the developing kidney. *Development* **125**(21), 4225–4234.

Kobayashi, A., Kwan, K. M., Carroll, T. J., McMahon, A. P., Mendelsohn, C. L., and Behringer, R. R. (2005). Distinct and sequential tissue-specific activities of the LIM-class homeobox gene Lim1 for tubular morphogenesis during kidney development. *Development* **132**, 2809–2823.

Kobayashi, A., Valerius, M. T., Mugford, J. W., Carroll, T. J., Self, M., Oliver, G., and McMahon, A. P. (2008). Six2 defines and regulates a multipotent self-renewing nephron progenitor population throughout mammalian kidney development. *Cell Stem Cell* **3**(2), 169–181.

Kopan, R., Cheng, H. T., and Surendran, K. (2007). Molecular insights into segmentation along the proximal–distal axis of the nephron. *J. Am. Soc. Nephrol.* **18**(7), 2014–2020.

Koziell, A., Grech, V., Hussain, S., Lee, G., Lenkkeri, U., Tryggvason, K., and Scambler, P. (2002). Genotype/phenotype correlations of NPHS1 and NPHS2 mutations in nephrotic syndrome advocate a functional inter-relationship in glomerular filtration. *Hum. Mol. Genet.* **11**, 379–388.

Kreidberg, J. A., Donovan, M. J., Goldstein, S. L., Rennke, H., Shepherd, K., Jones, R. C., and Jaenisch, R. (1996). Alpha 3 beta 1 integrin has a crucial role in kidney and lung organogenesis. *Development* **122**, 3537–3547.

Kreidberg, J. A., Sariola, H., Loring, J. M., Maeda, M., Pelletier, J., Housman, D., and Jaenisch, R. (1993). WT-1 is required for early kidney development. *Cell* **74**(4), 679–691.

Kriz, W., and Bankir, L. (1998). A standard nomenclature for structure of the kidney. The Renal Commission of the International Union of Physiological Sciences (IUPS). *Anat. Embryol. (Berl)* **178**(2), N1–N8.

Kume, T., Deng, K., and Hogan, B. L. (2000). Murine forkhead/winged helix genes Foxc1 (Mf1) and Foxc2 (Mfh1) are required for the early organogenesis of the kidney and urinary tract. *Development* **127**, 1387–1395.

Launay-Vacher, V., Ayllon, J., Janus, N., Medioni, J., Deray, G., Isnard-Bagnis, C., and Oudard., S. (2009). Evolution of renal function in patients treated with antiangiogenics after nephrectomy for renal cell carcinoma. *Urol. Oncol.* Epub.

Li, L., Krantz, I. D., Deng, Y., Genin, A., Banta, A. B., Collins, C. C., Qi, M., Trask, B. J., Kuo, W. L., Cochran, J., Costa, T., Pierpont, M. E., *et al.* (1997). Alagille syndrome is caused by mutations in human Jagged1, which encodes a ligand for Notch1. *Nat. Genet.* **16**(3), 243–251.

Lindahl, P., Hellstrom, M., Kalen, M., Karlsson, L., Pekny, M., Pekna, M., Soriano, P., and Betsholtz., C. (1998). Paracrine PDGF-B/PDGF-Rbeta signaling controls mesangial cell development in kidney glomeruli. *Development* **125**, 3313–3322.

Lindblom, P., Gerhardt, H., Liebner, S., Abramsson, A., Enge, M., Hellstrom, M., Backstrom, G., Fredriksson, S., Landegren, U., Nystrom, H. C., Bergstrom, G., Dejana, E., *et al.* (2003). Endothelial PDGF-B retention is required for proper investment of pericytes in the microvessel wall. *Genes Dev.* **17**, 1835–1840.

Little, M. H., Brennan, J., Georgas, K., Davies, J. A., Davidson, D. R., Baldock, R. A., Beverdam, A., Bertram, J. F., Capel, B., Chiu, H. S., *et al.* (2007). A high-resolution anatomical ontology of the developing murine genitourinary tract. *Gene Exp. Patterns* **7**(6), 680–699.

Longo, I., Porcedda, P., Mari, F., Giachino, D., Meloni, I., Deplano, C., Brusco, A., Bosio, M., Massella, L., Lavoratti, G., Roccatello, D., Frascá, G., *et al.* (2002). COL4A3/COL4A4 mutations: From familial hematuria to autosomal-dominant or recessive Alport syndrome. *Kidney Int.* **61**(6), 1947–1956.

Lu, B. C., Cebrian, C., Chi, X., Kuure, S., Kuo, R., Bates, C. M., Arber, S., Hassell, J., MacNeil, L., Hoshi, M., Jain, S., Asai, N., *et al.* (2009). Etv4 and Etv5 are required downstream of GDNF and Ret for kidney branching morphogenesis. *Nat. Genet.* **41**(12), 1295–1302.

Lu, W., van Eerde, A. M., Fan, X., Quintero-Rivera, F., Kulkarni, S., Ferguson, H., Kim, H. G., Fan, Y., Xi, Q., Li, Q. G., Sanlaville, D., Andrews, W., *et al.* (2007). Disruption of ROBO2 is associated with urinary tract anomalies and confers risk of vesicoureteral reflux. *Am. J. Hum. Genet.* **80**, 616–632.

Mah, S. P., Saueressig, H., Goulding, M., Kintner, C., and Dressler, G. R. (2000). Kidney development in cadherin-6 mutants: Delayed mesenchyme-to-epithelial conversion and loss of nephrons. *Dev. Biol.* **223**(1), 38–53.

Mahoney, Z. X., Sammut, B., Xavier, R. J., Cunningham, J., Go, G., Brim, K. L., Stappenbeck, T. S., Miner, J. H., and Swat, W. (2006). Discs-large homolog 1 regulates smooth muscle orientation in the mouse ureter. *Proc. Natl. Acad. Sci. USA* **103**, 19872–19877.

Majumdar, A., Vainio, S., Kispert, A., McMahon, J., and McMahon, A. P. (2003). Wnt11 and Ret/Gdnf pathways cooperate in regulating ureteric branching during metanephric kidney development. *Development* **130**(14), 3175–3185.

Marshall, C. B., and Shankland, S. J. (2007). Cell cycle regulatory proteins in podocyte health and disease. *Nephron Exp. Nephrol.* **106**, e51–e59.

McCright, B., Gao, X., Shen, L., Lozier, J., Lan, Y., Maguire, M., Herzlinger, D., Weinmaster, G., Jiang, R., and Gridley, T. (2001). Defects in development of the kidney, heart and eye vasculature in mice homozygous for a hypomorphic Notch2 mutation. *Development* **128**(4), 491–502.

McCright, B., Lozier, J., and Gridley, T. (2002). A mouse model of Alagille syndrome: Notch2 as a genetic modifier of Jag1 haploinsufficiency. *Development* **129**(4), 1075–1082.

McDaniell, R., Warthen, D. M., Sanchez-Lara, P. A., Pai, A., Krantz, I. D., Piccoli, D. A., and Spinner, N. B. (2006). NOTCH2 mutations cause Alagille syndrome, a heterogeneous disorder of the Notch signaling pathway. *Am. J. Hum. Genet.* **79**, 169–173.

Meeus, L., Gilbert, B., Rydlewski, C., Parma, J., Roussie, A. L., Abramowicz, M., Vilain, C., Christophe, D., Costagliola, S., and Vassart, G. (2004). Characterization of a novel loss of function mutation of PAX8 in a familial case of congenital hypothyroidism with in-place, normal-sized thyroid. *J. Clin. Endocrinol. Metab.* **89**(9), 4285–4291.

Mendelsohn, C., Batourina, E., Fung, S., Gilbert, T., and Dodd, J. (1999). Stromal cells mediate retinoid-dependent functions essential for renal development. *Development* **126**, 1139–1148.

Metzger, R. J., Klein, O. D., Martin, G. R., and Krasnow, M. A. (2008). The branching programme of mouse lung development. *Nature* **453**(7196), 745–750.

Michos, O., Cebrian, C., Hyink, D., Grieshammer, U., Williams, L., D'Agati, V., Licht, J. D., Martin, G. R., and Costantini, F. (2010). Kidney development in the absence of Gdnf and Spry1 requires Fgf10. *PLoS Genet.* **6**(1), e1000809.

Michos, O., Goncalves, A., Lopez-Rios, J., Tiecke, E., Naillat, F., Beier, K., Galli, A., Vainio, S., and Zeller, R. (2007). Reduction of BMP4 activity by gremlin 1 enables ureteric bud outgrowth and GDNF/WNT11 feedback signalling during kidney branching morphogenesis. *Development* **134**, 2397–2405.

Miner, J. H. (1999). Renal basement membrane components. *Kidney Int.* **56**, 2016–2024.

Miner, J. H. (2005). Building the glomerulus: A matricentric view. *J. Am. Soc. Nephrol.* **16**, 857–861.

Miyamoto, N., Yoshida, M., Kuratani, S., Matsuo, I., and Aizawa, S. (1997). Defects of urogenital development in mice lacking Emx2. *Development* **124**, 1653–1664.

Miyazaki, Y., Oshima, K., Fogo, A., Hogan, B. L., and Ichikawa, I. (2000). Bone morphogenetic protein 4 regulates the budding site and elongation of the mouse ureter. *J. Clin. Invest.* **105**, 863–873.

Moore, M. W., Klein, R. D., Farinas, I., Sauer, H., Armanini, M., Phillips, H., Reichardt, L. F., Ryan, A. M., Carver-Moore, K., and Rosenthal, A. (1996). Renal and neuronal abnormalities in mice lacking GDNF. *Nature* **382**, 76–79.

Nakai, S., Sugitani, Y., Sato, H., Ito, S., Miura, Y., Ogawa, M., Nishi, M., Jishage, K., Minowa, O., and Noda, T. (2003). Crucial roles of Brn1 in distal tubule formation and function in mouse kidney. *Development* **130**(19), 4751–4759.

Nyengaard, J. R., Bendtsen, T. F., (1992). Glomerular number and size in relation to age, kidney weight, and body surface in normal man. *Anat. Rec.* **232**(2), 194–201.

Oda, T., Elkahloun, A. G., Pike, B. L., Okajima, K., Krantz, I. D., Genin, A., Piccoli, D. A., Meltzer, P. S., Spinner, N. B., Collins, F. S., and Chandrasekharappa, S. C. (1997). Mutations in the human Jagged1 gene are responsible for Alagille syndrome. *Nat. Genet.* **16**, 235–242.

Ohuchi, H., Hori, Y., Yamasaki, M., Harada, H., Sekine, K., Kato, S., and Itoh, N. (2000). FGF10 acts as a major ligand for FGF receptor 2 IIIb in mouse multi-organ development. *Biochem. Biophys. Res. Commun.* **277**, 643–649.

Park, J. S., Valerius, M. T., and McMahon, A. P. (2007). Wnt/beta-catenin signaling regulates nephron induction during mouse kidney development. *Development* **134**, 2533–2539.

Patek, C. E., Little, M. H., Fleming, S., Miles, C., Charlieu, J. -P., Clarke, A. R., Miyagawa, K., Christie, S., Doig, J., Harrison, D. J., Porteous, A. J., Brookes, D., *et al.* (1999). A zinc finger truncation of murine *WT1* results in the characteristic urogenital abnormalities of Denys–Drash syndrome. *Proc. Natl. Acad. Sci. USA* **96**(6), 2931–2936.

Patterson, L. T., Pembaur, M., and Potter, S. S. (2001). Hoxa11 and Hoxd11 regulate branching morphogenesis of the ureteric bud in the developing kidney. *Development* **128**, 2153–2161.

Pavenstadt, H., Kriz, W., and Kretzler, M. (2003). Cell biology of the glomerular podocyte. *Physiol. Rev.* **83**, 253–307.

Perantoni, A. O., Timofeeva, O., Naillat, F., Richman, C., Pajni-Underwood, S., Wilson, C., Vainio, S., Dove, L. F., and Lewandoski, M. (2005). Inactivation of FGF8 in early mesoderm reveals an essential role in kidney development. *Development* **132**(17), 3859–3871.

Pichel, J. G., Shen, L., Sheng, H. Z., Granholm, A. C., Drago, J., Grinberg, A., Lee, E. J., Huang, S. P., Saarma, M., Hoffer, B. J., Sariola, H., and Westphal, H. (1996). Defects in enteric innervation and kidney development in mice lacking GDNF. *Nature* **382**, 73–76.

Poladia, D. P., Kish, K., Kutay, B., Hains, D., Kegg, H., Zhao, H., and Bates, C. M. (2006). Role of fibroblast growth factor receptors 1 and 2 in the metanephric mesenchyme. *Dev. Biol.* **291**, 325–339.

Pope, J. C.4th, Brock, J. W.3rd, Adams, M. C., Stephens, F. D., and Ichikawa, I. (1999). How they begin and how they end: Classic and new theories for the development and deterioration of congenital anomalies of the kidney and urinary tract, CAKUT. *J. Am. Soc. Nephrol.* **10**(9), 2018–2028.

Pozzi, A., Jarad, G., Moeckel, G. W., Coffa, S., Zhang, X., Gewin, L., Eremina, V., Hudson, B. G., Borza, D. B., Harris, R. C., Holzman, L. B., Phillips, C. L., *et al.* (2008). Beta1 integrin expression by podocytes is required to maintain glomerular structural integrity. *Dev. Biol.* **316**, 288–301.

Pritchard-Jones, K. (1999). The Wilms tumour gene, WT1, in normal and abnormal nephrogenesis. *Pediatr. Nephrol.* **13**, 620–625.

Qiao, J., Bush, K. T., Steer, D. L., Stuart, R. O., Sakurai, H., Wachsman, W., and Nigam, S. K. (2001). Multiple fibroblast growth factors support growth of the ureteric bud but have different effects on branching morphogenesis. *Mech. Dev.* **109**, 123–135.

Qiao, J., Uzzo, R., Obara-Ishihara, T., Degenstein, L., Fuchs, E., and Herzlinger., D. (1999). FGF-7 modulates ureteric bud growth and nephron number in the developing kidney. *Development* **126**, 547–554.

Raciti, D., Reggiani, L., Geffers, L., Jiang, Q., Bacchion, F., Subrizi, A. E., Clements, D., Tindal, C., Davidson, D. R., Kaissling, B., and Brandli, A. W. (2008). Organization of the pronephric kidney revealed by large-scale gene expression mapping. *Genome Biol.* **9**, R84.

Rae, F., Woods, K., Sasmono, T., Campanale, N., Taylor, D., Ovchinnikov, D. A., Grimmond, S. M., Hume, D. A., Ricardo, S. D., and Little, M. H. (2007). Characterisation and trophic functions of murine embryonic macrophages based upon the use of a Csf1r-EGFP transgene reporter. *Dev. Biol.* **308**(1), 232–246.

Raffi, H., Bates, J. M., Laszik, Z., and Kumar., S. (2006). Tamm-Horsfall protein knockout mice do not develop medullary cystic kidney disease. *Kidney Int.* **69**, 1914–1915.

Reggiani, L., Raciti, D., Airik, R., Kispert, A., and Brandli, A. W. (2007). The prepattern transcription factor Irx3 directs nephron segment identity. *Genes Dev.* **21**, 2358–2370.

Revest, J. M., Spencer-Dene, B., Kerr, K., De Moerlooze, L., Rosewell, I., and Dickson, C. (2001). Fibroblast growth factor receptor 2-IIIb acts upstream of Shh and Fgf4 and is required for limb bud maintenance but not for the induction of Fgf8, Fgf10, Msx1, or Bmp4. *Dev. Biol.* **231**, 47–62.

Rigoli, L., Chimenz, R., di Bella, C., Cavallaro, E., Caruso, R., Briuglia, S., Fede, C., and Salpietro, C. D. (2004). Angiotensin-converting enzyme and angiotensin type 2 receptor gene genotype distributions in Italian children with congenital uropathies. *Pediatr. Res.* **56**, 988–993.

Ronconi, E., Sagrinati, C., Angelotti, M. L., Lazzeri, E., Mazzinghi, B., Ballerini, L., Parente, E., Becherucci, F., Gacci, M., Carini, M., Maggi, E., Serio, M., *et al.* (2009). Regeneration of glomerular podocytes by human renal progenitors. *J. Am. Soc. Nephrol.* **20**(2), 322–332.

Ruf, R. G., Xu, P. X., Silvius, D., Otto, E. A., Beekmann, F., Muerb, U. T., Kumar, S., Neuhaus, T. J., Kemper, M. J., Raymond, R. M. Jr., Brophy, P. D., Berkman, J., *et al.* (2004). SIX1 mutations cause branchio-oto-renal syndrome by disruption of EYA1–SIX1–DNA complexes. *Proc. Natl. Acad. Sci. USA* **101**, 8090–8095.

Ryan, G., Steele-Perkins, V., Morris, J. F., Rauscher, F. J.3rd, and Dressler, G. R. (1995). Repression of Pax-2 by WT1 during normal kidney development. *Development* **121**(3), 867–875.

Sadl, V., Jin, F., Yu, J., Cui, S., Holmyard, D., Quaggin, S., Barsh, G., and Cordes, S. (2002). The mouse Kreisler (Krml1/MafB) segmentation gene is required for differentiation of glomerular visceral epithelial cells. *Dev. Biol.* **249**, 16–29.

Sanyanusin, P., Schimmenti, L. A., McNoe, L. A., Ward, T. A., Pierpont, M. E., Sullivan, M. J., Dobyns, W. B., and Eccles, M. R. (1995). Mutation of the PAX2 gene in a family with optic nerve colobomas, renal anomalies and vesicoureteral reflux. *Nat. Genet.* **9**, 358–364.

Sariola, H. (1985). Interspecies chimeras: An experimental approach for studies on embryonic angiogenesis. *Med. Biol.* **63**, 43–65.

Sariola, H., Ekblom, P., and Henke-Fahle, S. (1989). Embryonic neurons as in vitro inducers of differentiation of nephrogenic mesenchyme. *Dev. Biol.* **132**, 271–281.

Schedl, A. (2007). Renal abnormalities and their developmental origin. *Nat. Rev. Genet.* **8**, 791–802.

Schmidt-Ott, K. M., and Barasch., J. (2008). WNT/beta-catenin signaling in nephron progenitors and their epithelial progeny. *Kidney Int.* **74**, 1004–1008.

Schuchardt, A., D'Agati, V., Pachnis, V., and Costantini, F. (1996). Renal agenesis and hypodysplasia in ret-k- mutant mice result from defects in ureteric bud development. *Development* **122**, 1919–1929.

Self, M., Lagutin, O. V., Bowling, B., Hendrix, J., Cai, Y., Dressler, G. R., and Oliver, G. (2006). Six2 is required for suppression of nephrogenesis and progenitor renewal in the developing kidney. *EMBO J.* **25**, 5214–5228.

Shah, M. M., Tee, J. B., Meyer, T., Meyer-Schwesinger, C., Choi, Y., Sweeney, D. E., Gallegos, T. F., Johkura, K., Rosines, E., Kouznetsova, V., Rose, D. W., Bush, K. T., *et al.* (2009). The instructive role of metanephric mesenchyme in ureteric bud patterning, sculpting, and maturation and its potential ability to buffer ureteric bud branching defects. *Am. J. Physiol. Renal Physiol.* **297**, F1330–F1341.

Shakya, R., Watanabe, T., and Costantini, F. (2005). The role of GDNF/Ret signaling in ureteric bud cell fate and branching morphogenesis. *Dev. Cell* **8**, 65–74.

Short, K. M., Hodson, M. J., and Smyth, I. M. (2010). Tomographic quantification of branching morphogenesis and renal development. *Kidney Int.* **77**(12), 1132–1139.

Simic, P., and Vukicevic., S. (2005). Bone morphogenetic proteins in development and homeostasis of kidney. *Cytokine Growth Factor Rev.* **16**, 299–308.

Singh, R. R., Moritz, K. M., Bertram, J. F., and Cullen-McEwen, L. A. (2007). Effects of dexamethasone exposure on rat metanephric development: In vitro and in vivo studies. *Am. J. Physiol. Renal Physiol.* **293**(2), F548–F554.

St John, P. L., and Abrahamson., D. R. (2001). Glomerular endothelial cells and podocytes jointly synthesize laminin-1 and -11 chains. *Kidney Int.* **60**, 1037–1046.

Stark, K., Vainio, S., Vassileva, G., and McMahon, A. P. (1994). Epithelial transformation of metanephric mesenchyme in the developing kidney regulated by Wnt-4. *Nature* **372**(6507), 679–683.

Sánchez, M. P., Silos-Santiago, I., Frisén, J., He, B., Lira, S. A., and Barbacid, M. (1996). Renal agenesis and the absence of enteric neurons in mice lacking GDNF. *Nature* **382**, 70–73.

Takemoto, M., He, L., Norlin, J., Patrakka, J., Xiao, Z., Petrova, T., Bondjers, C., Asp, J., Wallgard, E., Sun, Y., Samuelsson, T., Mostad, P., *et al.* (2006). Large-scale identification of genes implicated in kidney glomerulus development and function. *EMBO J.* **25**, 1160–1174.

Turner, J. J., Stacey, J. M., Harding, B., Kotanko, P., Lhotta, K., Puig, J. G., Roberts, I., Torres, R. J., and Thakker, R. V. (2003). UROMODULIN mutations cause familial juvenile hyperuricemic nephropathy. *J. Clin. Endocrinol. Metabol.* **88**, 1398–1401.

Ueda, H., Miyazaki, Y., Matsusaka, T., Utsunomiya, Y., Kawamura, T., Hosoya, T., and Ichikawa, I. (2008). Bmp in podocytes is essential for normal glomerular capillary formation. *J. Am. Soc. Nephrol.* **19**, 685–694.

Wang, G. J., Brenner-Anantharam, A., Vaughan, E. D., and Herzlinger, D. (2009). Antagonism of BMP4 signaling disrupts smooth muscle investment of the ureter and ureteropelvic junction. *J. Urol.* **181**, 401–407.

Wang, P., Pereira, F. A., Beasley, D., and Zheng, H. (2003). Presenilins are required for the formation of comma- and S-shaped bodies during nephrogenesis. *Development* **130**(20), 5019–5029.

Wang, R., St John, P. L., Kretzler, M., Wiggins, R. C., and Abrahamson., D. R. (2000). Molecular cloning, expression, and distribution of glomerular epithelial protein 1 in developing mouse kidney. *Kidney Int.* **57**, 1847–1859.

Weber, S., Moriniere, V., Knuppel, T., Charbit, M., Dusek, J., Ghiggeri, G. M., Jankauskiene, A., Mir, S., Montini, G., Peco-Antic, A., Wuhl, E., Zurowska, A. M., *et al.* (2006). Prevalence of mutations in renal developmental genes in children with renal hypodysplasia: Results of the ESCAPE study. *J. Am. Soc. Nephrol.* **17**, 2864–2870.

Weber, S., Taylor, J. C., Winyard, P., Baker, K. F., Sullivan-Brown, J., Schild, R., Knuppel, T., Zurowska, A. M., Caldas-Alfonso, A., Litwin, M., Emre, S., Ghiggeri, G. M., *et al.*

(2008). SIX2 and BMP4 mutations associate with anomalous kidney development. *J. Am. Soc. Nephrol.* **19**, 891–903.

Wellik, D. M., Hawkes, P. J., and Capecchi, M. R. (2002). Hox11 paralogous genes are essential for metanephric kidney induction. *Genes Dev.* **16**, 1423–1432.

Wilkinson, L., Gilbert, T., Kinna, G., Ruta, L. A., Pennisi, D., Kett, M., and Little, M. H. (2007). Crim1KST264/KST264 mice implicate Crim1 in the regulation of vascular endothelial growth factor-A activity during glomerular vascular development. *J. Am. Soc. Nephrol.* **18**, 1697–1708.

Wilkinson, L., Gilbert, T., Sipos, A., Toma, I., Pennisi, D. J., Peti-Peterdi, J., and Little, M. H. (2009). Loss of renal microvascular integrity in postnatal Crim1 hypomorphic transgenic mice. *Kidney Int.* **76**, 1161–1171.

Winyard, P., and Chitty, L. S. (2008). Dysplastic kidneys. *Semin. Fetal Neonatal Med.* **13**, 142–151.

Woolf, A. S., Kolatsi-Joannou, M., Hardman, P., Andermarcher, E., Moorby, C., Fine, L. G., Jat, P. S., Noble, M. D., and Gherardi, E. (1995). Roles of hepatocyte growth factor/scatter factor and the met receptor in the early development of the metanephros. *J. Cell Biol.* **128**, 171–184.

Xu, P. X., Zheng, W., Huang, L., Maire, P., Laclef, C., and Silvius, D. (2003). Six1 is required for the early organogenesis of mammalian kidney. *Development* **130**, 3085–3094.

Yang, Y., Houle, A. M., Letendre, J., and Richter, A. (2008). RET Gly691Ser mutation is associated with primary vesicoureteral reflux in the French-Canadian population from Quebec. *Hum. Mutat.* **29**, 695–702.

Yosypiv, I. V., Boh, M. K., Spera, M. A., and El-Dahr, S. S. (2008). Downregulation of Spry-1, an inhibitor of GDNF/Ret, causes angiotensin II-induced ureteric bud branching. *Kidney Int.* **74**, 1287–1293.

Yu, J., Carroll, T. J., Rajagopal, J., Kobayashi, A., Ren, Q., and McMahon, A. P. (2009). A Wnt7b-dependent pathway regulates the orientation of epithelial cell division and establishes the cortico-medullary axis of the mammalian kidney. *Development* **136**, 161–171.

Zandi-Nejad, K., Luyckx, V. A., and Brenner, B. M. (2006). Adult hypertension and kidney disease: The role of fetal programming. *Hypertension* **47**, 502–508.

Zhao, H., Kegg, H., Grady, S., Truong, H. T., Robinson, M. L., Baum, M., and Bates, C. M. (2004). Role of fibroblast growth factor receptors 1 and 2 in the ureteric bud. *Dev. Biol.* **276**, 403–415.

CHAPTER SIX

The Game Plan: Cellular and Molecular Mechanisms of Mammalian Testis Development

Elanor N. Wainwright *and* Dagmar Wilhelm

Contents

1. Introduction	234
2. Introducing the Players: Cell Biology and Morphology of the Gonads	235
3. Origin of Sertoli Cells	237
4. Kickoff in Testis Determination: *Sry* and Sertoli Cell Specification	237
5. The Goalkeeper: *Sox9* and Sertoli Cell Differentiation	239
5.1. *Sox9* initiation	240
5.2. *Sox9* maintenance	241
6. Forward Players: Beyond *Sox9*	243
7. The Sweepers: Peritubular Myoid Cells	245
8. Interplay Between PM and Other Testicular Cells	247
9. The Midfielders: Leydig Cells	247
10. The White Lines: Endothelial Cells	251
11. What All the Fuss Is About: Germ Cells	253
12. Concluding Remarks	255
Acknowledgments	256
References	256

Abstract

In mammals, biological differences between males and females, which influence many aspects of their physical, social, and psychological environments, are solely determined genetically. In the presence of a Y chromosome, the gonadal primordium will differentiate into a testis, whereas in the absence of the Y chromosome an ovary will develop. Testis and ovary subsequently direct the differentiation of all secondary sex characteristics down the male and female pathway, respectively. The male-determining factor on the Y chromosome, SRY, was identified some 20 years ago. Since then, significant progress has been made

Division of Molecular Genetics and Development, Institute for Molecular Biosciences, The University of Queensland, Brisbane, Queensland, Australia

Current Topics in Developmental Biology, Volume 90
ISSN 0070-2153, DOI 10.1016/S0070-2153(10)90006-9

toward understanding the molecular and cellular pathways that result in the formation of a testis. Here, we review what is known about testis differentiation in mice and humans, with reference to other species where appropriate.

1. Introduction

Sexual reproduction is arguably the most significant of biological events in that it is essential for life itself. Sexual reproduction involves the fusion of two separate parental nuclei, so the offspring inherit endlessly varied combinations of characteristics that ultimately provide a vast testing ground for new variations that may not only improve the species but also ensure its survival. Therefore, the bifurcation of the developmental pathway into males and females, which is the prerequisite for the union of gametes from the two sexes, is vital to sexual reproduction itself.

In mammals, the first sign of sexual differentiation is when the bipotential gonad starts to develop into either a testis or an ovary in XY and XX individuals, respectively. This decision is made around 11.5 days *post coitum* (dpc) in mice and 6 weeks of gestation in humans, when *Sry* (*SRY* in humans), the male-determining gene on the Y chromosome, is expressed in the supporting cell lineage of the genital ridge (Bullejos and Koopman, 2001; Wilhelm *et al.*, 2005). SRY sets off a chain reaction of gene expression and regulation that controls the differentiation of testis-specific cell types and their migration, association, and organization to form testicular structures. In turn, the testis produces hormones that direct the differentiation of all secondary sexual characteristics such as the regression of the "female" Müllerian duct, development of the "male" Wolffian duct into epididymis, seminal vesicle, and vas deferens, development of the prostate, external genitalia, and sexual dimorphisms of the brain (for review, see Wilhelm and Koopman, 2006). In the absence of correct spatiotemporal SRY function, female-determining genes are expressed and an ovary will develop (see chapter 7).

Sexual differentiation is often described as a black or white decision, resulting in either a boy or a girl. Although this is essentially correct, it fails to recognize the intricacies of the regulatory networks governing these developmental processes and the frailties of many regulatory pathways that may easily get disturbed, potentially leading to disorders of sex development (DSD). Therefore, sexual differentiation is perhaps more correctly viewed as a continuum between the male and the female phenotypes, encompassing a range of intermediate phenotypes with variable severity with regard to reproductive and sexual health.

Sexual differentiation is a multi-tiered process that occurs at many levels, starting at the cellular level and then progressing to tissue, organ, and finally the whole organism. Interestingly, the dichotomy of "boy or girl" is true at the cellular level but is becoming more and more difficult to determine with each higher level: cell, tissue, and organism. Recent studies have shown that, specifically for the supporting cells, each single cell either differentiates into a "male" Sertoli cell, marked by the expression of the transcription factor SOX9, or a "female" granulosa cell, expressing the ovarian marker FOXL2 (Hersmus *et al.*, 2008; Wilhelm *et al.*, 2005). Not yet fully understood inhibitory pathways prevent the expression of ovary- and testis-determining genes, respectively (Kim *et al.*, 2006), so that every cell is committed to either the male or female pathway. At the organ level, testicular and ovarian tissue can exist within the same gonad, so-called ovotestes. Here, testicular tissue generally is found in the center of the gonads and ovarian tissue at the poles (Eicher *et al.*, 1982), although single ovarian cells can be found within the testicular regions (Wilhelm *et al.*, 2009). In most cases where these ovotestes exist during embryonic development, they resolve into either testes or ovaries at later stages. By contrast, the differentiation of secondary sexual characteristics really cannot be described as a dichotomy but rather a continuum, blurring the distinction of what is boy and what is girl. In this chapter we will review the pathways that are important for the formation of a testis, the basis of all other sexual differentiation of the male.

2. Introducing the Players: Cell Biology and Morphology of the Gonads

The development of the gonads is unique in that the same primordium possesses the potential to differentiate into two morphological and functional distinct organs, testis and ovary. These primordia, the bipotential or indifferent genital ridges, are paired structures that develop at the ventro-medial surface of the mesonephros as part of the urogenital system. In mouse, the genital ridges are first visible at around 10 dpc and increase in size due to proliferation of coelomic epithelial cells. The genital ridges consist of several precursor cell lineages including supporting, steroidogenic, and interstitial cell precursors (for review, see Wilhelm *et al.*, 2007b). Primordial germ cells (PGCs), the precursor of sperm and egg, are specified extra-embryonically and then migrate along the hindgut to the developing genital ridges and coalesce with the gonadal somatic cells by 11.5 dpc (for review, see Ewen and Koopman, 2010).

From 10.5 to 12.5 dpc *Sry* is expressed in the supporting precursor cells inducing their differentiation into Sertoli cells (Fig. 6.1A; Bullejos and Koopman, 2001; Wilhelm *et al.*, 2005) as well as proliferation within the coelomic epithelium to generate additional Sertoli cell precursors (Schmahl *et al.*, 2000). The differentiation process is marked by the polarization of

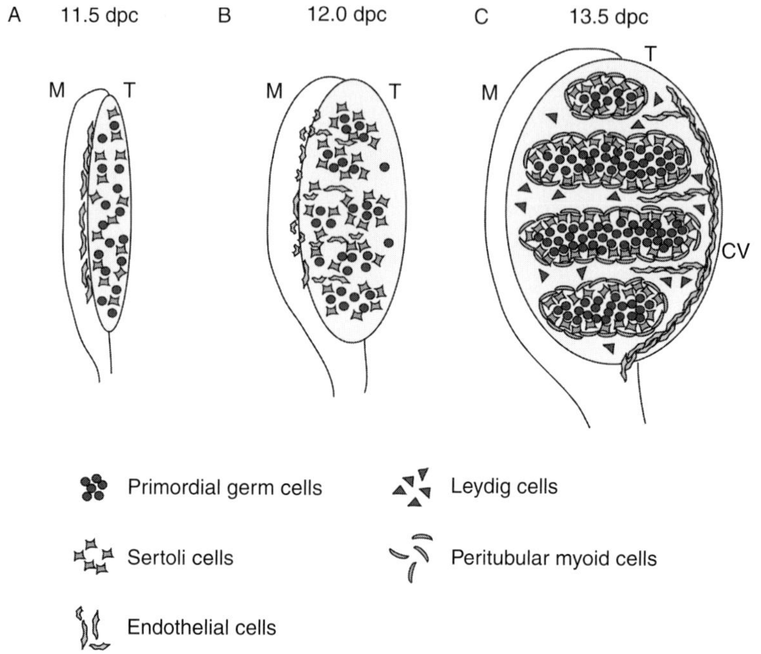

Figure 6.1 Schematic representation of testis development in mouse. (A) At 11.5 dpc SRY is expressed in pre-Sertoli cells (green) that are distributed evenly in between primordial germ cells (red) throughout the genital ridge. A vasculature network within the mesonephros disintegrates and endothelial cells (pink) start to migrate into the gonad. (B) By 12.0 dpc Sertoli cells become polarized and form epithelial aggregates around clusters of germ cells. Endothelial cells partition these clusters into the future testis cords. (C) At 13.5 dpc testis cords are defined with clusters of germ cells surrounded by one layer of Sertoli cells and one layer of peritubular myoid cells (yellow). In the interstitium, Leydig cells (blue) have differentiated to produce testosterone. T, testis; M, mesonephros, CV, coelomic vessel. (See Color Insert.)

Sertoli cells and their assembly around clusters of germ cells. Subsequently, migrating endothelial cells from the mesonephros partition these accumulations of germ and Sertoli cells into testis cords (Fig. 6.1B; Combes *et al.*, 2009b; Coveney *et al.*, 2008). In addition to inducing endothelial cell migration, Sertoli cells also direct the differentiation of other cell types within the developing testis (Fig. 6.1C). In the interstitium, steroidogenic precursor cells differentiate into Leydig cells, and mesenchymal cells into peritubular myoid (PM) cells. PM cells are believed to be the only cell type for which no counterpart exists in the ovary. These long, flattened cells surround Sertoli cells and together they secrete the basal lamina that forms a physical barrier around testis cords. In the following sections, we will describe the different cell types, their origin, the molecular mechanisms underlying their differentiation, and their interaction with other cells.

3. Origin of Sertoli Cells

Sertoli cells are the first somatic cell type to differentiate in the testis. Their origin has been elucidated in a series of elegant experiments. In chicken/quail gonad/mesonephros grafts, ink-labeled cells at the surface epithelium were shown to leave the surface and participate in forming the internal gonadal soma during the indifferent period of gonad development (Rodemer-Lenz, 1989). In addition, 5-bromo-2-deoxyuridine (BrdU) labeling of dividing cells has shown that cells at or near the coelomic epithelium are proliferating and the majority of these proliferating cells are steroidogenic factor 1 (SF1) positive from approximately 11.3 to 11.5 dpc, thereby marking the Sertoli and Leydig cell precursors (Schmahl *et al.*, 2000). To show that these epithelial cells contribute to Sertoli cells, the fate of these coelomic epithelial cells was investigated by labeling them with a fluorescent dye over the time period 11.2–12.5 dpc in mouse XY gonad organ culture. Coelomic epithelial cells that moved into the gonad from 11.2 to 11.4 dpc were found to contribute to Sertoli and interstitial cell populations, whereas cells that migrated from 11.5 to 11.7 dpc solely became interstitial cells. In support of this cell migration, biochemical studies have demonstrated that the subepithelial basement membrane components collagen type I and type III as well as laminin have a discontinuous distribution at 11.5 dpc until 12.5 dpc (Karl and Capel, 1998; Paranko, 1987). Thus, these data suggest that the basement membrane underlying the coelomic epithelium is transiently labile, allowing surface epithelial cells to contribute to somatic cell populations in the interior of the XY gonad. One matter of contention is whether the movement of coelomic epithelial cells into the gonad is a process of active migration or a passive movement from the proliferation of the cells at the coelomic epithelium "pushing" cells into the interior of the gonad.

4. Kickoff in Testis Determination: *Sry* and Sertoli Cell Specification

The male-determining gene *Sry* is expressed in pre-Sertoli cells in a wave starting at 10.5 dpc in the center of XY gonads, spreading to the poles and reaching a maximum at 11.5 dpc before declining in the same center-to-pole like pattern. The last few SRY-positive cells can be detected at 12.5 dpc at the posterior pole (Bullejos and Koopman, 2001; Wilhelm *et al.*, 2005). This dynamic expression pattern suggests a highly controlled transcriptional regulation, although little is known about the molecular mechanism.

Transgenic mice using an approximately 14-kb genomic region containing the *Sry* coding sequences and flanking regions have shown that this

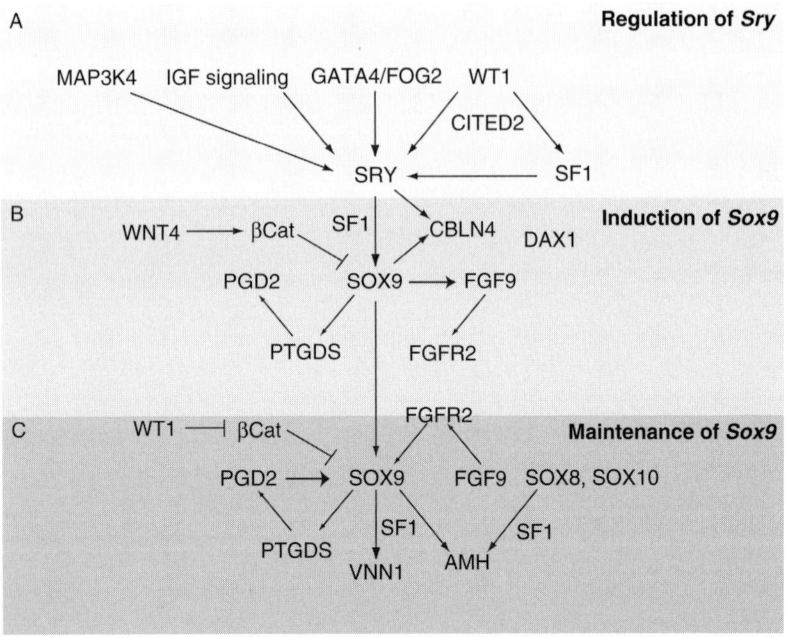

Figure 6.2 Postulated molecular pathways underlying Sertoli cell specification and differentiation. The specification and differentiation of Sertoli cells can be roughly divided into three phases: (A) the regulation of *Sry* expression, (B) the induction of *Sox9* expression, and (C) the maintenance of *Sox9* expression. See text for details. This schematic does not take into account direct or indirect postulated relationships.

region is sufficient to induce XX sex reversal (Koopman *et al.*, 1991). However, comparative genomic approaches for evolutionary conserved elements have proved to be difficult due to very low conservation of the genomic region surrounding *Sry*. Mouse knockout models have implicated a number of factors in the regulation of *Sry* transcription based on the reduction of *Sry* mRNA levels. These factors include the insulin receptor family, comprising IGF1R, IR, and IRR (Nef *et al.*, 2003), the transcription factor GATA4 and its co-factor FOG2 (Tevosian *et al.*, 2002), and the +KTS splice variant of Wilms' tumor suppressor 1 (WT1; Hammes *et al.*, 2001; Fig. 6.2A). However, for most of these models, it is not clear if there is less *Sry* per cell, supporting a direct effect, or only less cells that express *Sry*.

For the +KTS splice variant of WT1, Bradford *et al.* (2009b) used immunofluorescence to show that indeed gonads lacking WT1(+KTS) lower SRY levels per cell as well as less SRY-positive cells. This led to the hypothesis that WT1(+KTS) contributes cell-autonomously to *Sry* activation, but also non-cell-autonomously by increasing the number of pre-Sertoli cells. The role of WT1 in regulating *Sry* expression has been supported by additional *in vitro* and *in vivo* studies. In cell culture, WT1 cooperated with GATA4 to transcriptionally activate the mouse, pig, and human *SRY* promoter. Both WT1 isoforms, –KTS and +KTS, interacted with GATA4; however, the synergism on the *SRY* promoters was stronger with the +KTS form (Miyamoto *et al.*, 2008). Similarly, WT1(–KTS) has been shown in cell culture to bind to and activate the human *SRY* promoter. This activation is mediated by a co-factor of WT1, CITED2 (Cbp/p300-interacting transactivator, with Glu/Asp-rich carboxy-terminal domain, 2), for which genetic analysis demonstrated that it acts with WT1 and SF1 to increase *Sry* levels to a critical threshold for efficient testis differentiation (Buaas *et al.*, 2009). Alternatively, or in addition, CITED2 together with WT1 might upregulate *Sf1* expression (Val *et al.*, 2007; Wilhelm and Englert, 2002), which in turn activates *Sry* (Fig. 6.2A). The latter is supported by the fact that humans with *SF1* mutations exhibit a range of phenotypes including XY sex reversal (Achermann et al., 1999; Correa et al., 2004).

In addition to these transcriptional regulators, a mitogen-activated protein kinase kinase kinase (MAP3K4) has been shown to result in greatly reduced *Sry* levels when mutated in mice (Bogani *et al.*, 2009). MAP3K4 phosphorylates and thereby activates MKK4, which in turn activates p38 MAP kinase through phosphorylation, both of which have been shown to be activated in the developing testis (Bogani *et al.*, 2009). How p38 facilitates the activation of *Sry* transcription is not yet known, but it could be through phosphorylation of a transcriptional activator and/or indirectly through increased proliferation. Similarly, more work needs to be done to identify the upstream activators of MAP3K4, which could potentially identify new candidates for causing DSDs in humans.

5. The Goalkeeper: *Sox9* and Sertoli Cell Differentiation

Shortly after *Sry* expression, the SRY-box containing gene 9 (*Sox9*) is expressed in the pre-Sertoli cells in the same dynamic wave that originates in the center of the gonad and then continues to the rostral and caudal poles. Unlike *Sry, Sox9* expression in mouse Sertoli cells is maintained throughout embryogenesis and into adulthood (Kent *et al.*, 1996; Morais da Silva *et al.*, 1996).

Sox9 is a member of the *Sry*-type HMG box (*Sox*) gene family. It encodes a transcription factor that interacts with DNA through an HMG domain (Sudbeck *et al.*, 1996). Gain of function of *Sox9* in XX transgenic mice induces testis development (Vidal *et al.*, 2001), whereas loss of function results in full XY sex reversal (Barrionuevo *et al.*, 2006; Chaboissier *et al.*, 2004). Therefore, like *Sry, Sox9* is both necessary and sufficient for Sertoli cell differentiation and testis determination. In humans, mutations in *SOX9* result in the disease campomelic dysplasia, which is characterized by skeletal malformation and, in a large proportion of XY patients, male-to-female sex reversal (Foster *et al.*, 1994; Wagner *et al.*, 1994).

The comparable spatial and temporal expression pattern of *Sry* and *Sox9* in testis development has led to the hypothesis that *Sox9* is a direct downstream target of SRY. Recently, Sekido and Lovell-Badge (2008) identified a 3.2-kb testis-specific enhancer in the *Sox9* locus. When this enhancer was linked to a LacZ reporter in transgenic mice, the expression mimicked endogenous *Sox9* expression in both XY and XX gonads. Comparative genome-based approaches narrowed down the enhancer region to a 1.4-kb sequence termed "testis-specific enhancer of *Sox9* core" (TESCO), which had predicted SRY and SF1 binding sites. Chromatin immunoprecipitation (ChIP) assays and mutagenesis of TESCO reporters *in vivo* demonstrated that both SRY and SF1 bind to TESCO and activate it synergistically (Sekido and Lovell-Badge, 2008). Following the initial induction by SRY (Fig. 6.2B), SOX9 binds together with SF1 to the TESCO element to maintain its own transcription in an autoregulatory loop (Fig. 6.2C; Sekido and Lovell-Badge, 2008), providing a simplified model of the molecular mechanisms of sex determination. In the next two sections, we will discuss the initiation (Fig. 6.2B) and the maintenance (Fig. 6.2C) of *Sox9* expression in more detail.

5.1. *Sox9* initiation

In addition to SRY and SF1, the signaling molecule WNT4 (wingless-related MMTV integration site 4), a factor important for ovarian development (see chapter 7), has been implicated in contributing to *Sox9* initiation (Jeays-Ward *et al.*, 2003). In the *Wnt4*-null mice, *Sox9* expression is reduced in XY genital ridges and Sertoli cell differentiation is compromised. However, *Sry* expression is unchanged (Jeays-Ward *et al.*, 2003), suggesting that loss of *Wnt4* weakens the action of SRY. Interestingly, the opposite, active WNT4 signaling, which results in the stabilization of β-catenin, also has been shown to negatively affect *Sox9* expression (Chang *et al.*, 2008). A possible explanation might be that loss of WNT4 signaling reduces initial proliferation of *Sry*-positive cells so that reduced *Sox9* levels is a consequence of a reduced number of Sertoli cells rather than *Sox9* transcription *per se*.

Similarly to *Wnt4, Dax1* (dosage-sensitive sex reversal (DSS), adrenal hypoplasia congenital (AHC)-critical region on the X chromosome gene 1), another gene that had been implicated in ovary development (see chapter 7), was shown to be important for *Sox9* expression and testis cord formation (Bouma *et al.*, 2005; Meeks *et al.*, 2003). *Dax1*-null mutation on a sensitized background, i.e., on a C57Bl/6 background, known to be reactive to disturbance in the early events of testicular development, or in combination with a *Mus domesticus poschiavinus* Y chromosome (Y^{POS}), a "weak" *Sry* allele (Eicher *et al.*, 1982), results in a failure to upregulate *Sox9* expression even though *Sry* is expressed at normal levels (Bouma *et al.*, 2005; Meeks *et al.*, 2003). Interestingly, overexpression of *Sry* is able to rescue this phenotype (Bouma *et al.*, 2005). As with WNT4 signaling, it is not known how DAX1 cooperates with SRY in the upregulation of *Sox9* expression (for review, see Ludbrook and Harley, 2004).

5.2. *Sox9* maintenance

After the initial transcriptional activation by SRY, which itself ceases to be expressed, the maintenance of *Sox9* expression involves pathways independent of SRY. Two positive feedback loops have been described to play a role in *Sox9* maintenance. The first one involves fibroblast growth factor 9 (FGF9) upregulating *Sox9* expression, which in turn activates further *Fgf9* expression. Indeed, *Fgf9*-null mice initially upregulate, but do not maintain, *Sox9* expression and therefore have impaired Sertoli cell differentiation and testis cord formation (Kim *et al.*, 2006), clearly showing that the initiation and the maintenance of *Sox9* expression are controlled by independent pathways. Also, in contrast to SRY, FGF9 is a signaling factor exerting its effect via cellular receptors and not by direct protein-to-DNA interactions, implicating as yet undefined factors in between the receptor and the *Sox9* regulatory region. The receptor responsible for the mediation of FGF9-mediated signaling has been suggested to be FGF receptor 2 (FGFR2; Schmahl *et al.*, 2004). Indeed, *Fgfr2*-null mice on a C57BL/6 background phenocopy *Fgf9*-null mice, with both *Sox9* and *Amh* expression drastically reduced (Bagheri-Fam *et al.*, 2008; Kim *et al.*, 2007b). Furthermore, crossing *Fgfr2* and *Sox9* heterozygous mutants has showed genetic interaction between the two pathways. Mice with either half a dose of *Fgfr2* or *Sox9* display normal testis development, whereas *Fgfr2*/*Sox9* double heterozygous mutants develop ovotestes (Bagheri-Fam *et al.*, 2008).

What is the molecular relationship between *Fgf9* and *Sox9*? Analysis of cell proliferation in *Fgf9*$^{-/-}$ XY gonads found that the number of dividing cells at the coelomic epithelium was reduced at 11.2 dpc and remained at a low level, suggesting that FGF9 regulates the male-specific proliferation that produces pre-Sertoli cells (Schmahl *et al.*, 2004). These pre-Sertoli cells will start to express *Sry* and subsequently *Sox9*. In addition, FGF9 could upregulate *Sox9* expression itself, a possibility that was tested by treating XX gonads with exogenous FGF9. Treatment of whole gonads only resulted in an increased proliferation, but not the induction of *Sox9* expression (Schmahl *et al.*, 2004; Wilhelm *et al.*, 2005). By contrast, using FGF9-coated beads implanted into the gonad proper induced the expression SOX9, albeit only within the local environment of the bead (Kim *et al.*, 2006). However, reducing the dosage of *Wnt4* by using XX *Wnt4* heterozygous gonads,

FGF9 was able to upregulate *Sox9* expression (Kim *et al.*, 2006). Finally, Hiramatsu *et al.* showed that inhibition of FGF9 signaling blocks the pole-ward expansion of *Sox9* expression (Hiramatsu *et al.*, 2010). Taken together, these data support the hypothesis that *Sox9* expression is regulated by FGF9 via FGFR2 (Figs. 6.2B and C), with WNT4 signals functioning antagonistically. What remains unknown is the link between FGFR2 and the *Sox9* regulatory region.

The upregulation of *Fgf9* by SOX9 is less clear. Early null mutation of *Sox9* results in a reduction of *Fgf9* expression (Barrionuevo *et al.*, 2006). However, in two mouse knockout models, conditional deletion of *Wt1* (Gao *et al.*, 2006) and *Sox9*/*Sox8* double knockout (Barrionuevo *et al.*, 2009), in which *Sox9* expression is lost, *Fgf9* expression did not change. In both models *Sox9* was lost after 13.5 dpc, suggesting that at later stages, *Fgf9* expression is independent of *Sox9* (Fig. 6.2C). It still remains to be shown if SOX9 is a *bona fide* transactivator of *Fgf9* during any stage of testis differentiation.

The second positive feedback loop that is integral to Sertoli cell differentiation is prostaglandin signaling (Figs. 6.2B and C). It has been shown that SOX9 activates the expression of the gene encoding prostaglandin D synthase (*Ptgds*; Wilhelm *et al.*, 2007a), which catalyzes the final step in the synthesis of prostaglandin D_2 (PGD_2). Accordingly, *Ptgds* expression is almost completely abolished in *Sox9*-null gonads (Moniot *et al.*, 2009). PGD_2, in turn, induced the upregulation of *Sox9* expression (Adams and McLaren, 2002; Wilhelm *et al.*, 2005) and translocation of SOX9 protein from the cytoplasm to the nucleus via protein kinase A (PKA) phosphorylation (Malki *et al.*, 2005). The importance of this signaling pathway and its independence of the FGF9 signaling pathway has been demonstrated by the null mutation of *Ptgds* in mice, which showed reduced *Sox9* expression and delayed testis cord formation up to 14.5 dpc; however, both recovered by 17.5 dpc (Moniot *et al.*, 2009).

Another critical requirement for *Sox9* maintenance and Sertoli cell differentiation has been shown to be high glucose conditions. Glycogen accumulation early in testis development begins in the pre-Sertoli cells at 11.2 dpc and reaches a peak at 11.5 dpc (Matoba *et al.*, 2008). The accumulation of glycogen might be regulated by insulin growth factor signaling, which also regulates *Sry* expression (Nef *et al.*, 2003). XY genital ridge cultures subjected to glucose starvation showed a disorganized distribution of presumptive Sertoli cells and no testis cord formation. While the initiation of *Sox9* expression was unaffected by glucose starvation, the expression was downregulated thereafter. Quantitative reverse transcriptase polymerase chain reaction (qRT-PCR) and whole mount *in situ* hybridization (ISH) demonstrated that the levels of *Fgf9* and *Ptgds* were normal, suggesting that the mechanism of *Sox9* dependence on glycogen levels is independent of these pathways (Matoba *et al.*, 2008).

The importance of *Sox9* is highlighted by the fact that all reported cases of XY sex reversal show a disturbance in the expression levels or function of SOX9. Therefore, it is surprising that *Sox9* expression seems not to be important for the maintenance of the Sertoli cell fate and testis cord integrity. Null mutation of *Sox9* after 13.5 dpc show normal embryonic testis development and are initially fertile, becoming sterile after about 5 months (Barrionuevo *et al.*, 2009), demonstrating that *Sox9* is dispensable at later stages of testis development. This could be at least in part due to redundancy with the related factors SOX8 and SOX10 (see next section). By contrast, the conditional knockout of *Wt1* after 13.5 dpc resulted in disintegration of testis cords and subsequent loss of Sertoli and germ cells (Gao *et al.*, 2006). *Sox9* expression in these mice was almost immediately lost after *Wt1* excision, demonstrating that first, *Wt1* plays an important role in the regulation of *Sox9* expression and second, *Wt1* has additional roles to *Sox9* in the maintenance of testicular architecture. Interestingly, loss of *Wt1* resulted in the upregulation and stabilization of β-catenin, a hallmark of ovary differentiation (see chapter 7). Accordingly, the genetic stabilization of β-catenin in testes mimicked the *Wt1*-null phenotype, suggesting that WT1 is a negative regulator of β-catenin signaling, which in turn negatively regulates *Sox9* expression (Fig. 6.2C; Chang *et al.*, 2008).

Taken together, these data clearly demonstrate that the regulation and maintenance of the Sertoli cell fate are of key importance and the testis program has a number of additive pathways to ensure correct development. In Fig. 6.2 we have assembled and summarized these pathways into regulatory networks.

6. Forward Players: Beyond *Sox9*

The importance of the two *Sox* genes, *Sry* and *Sox9*, for normal testis development has been well established. However, these two genes belong to a family of 20 members (Schepers *et al.*, 2002) that had been subdivided into 10 subgroups, A to J (Bowles *et al.*, 2000). *Sox9* belongs to group E together with *Sox8* and *Sox10*. Both, *Sox8* and *Sox10* have been shown to be expressed specifically in the developing XY gonads together with *Sox9* (Cory *et al.*, 2007; Polanco *et al.*, 2010; Schepers *et al.*, 2003). Thus, given the importance of *Sox9* in testis differentiation and the close evolutionary relationship between the three *SoxE* members, an important question is if also *Sox8* and *Sox10* have a functional role during gonadogenesis.

With regard to *Sox8*, genetic studies have shown that null mutation results in only idiopathic weight loss, reduced bone density, and a late spermatogenic defect (O'Bryan *et al.*, 2008; Sock *et al.*, 2001), whereas no testis phenotype for *Sox10*-null mutation has been described as yet.

However, *in vitro* and double-knockout analyses of *Sox8* and *Sox9* showed that *Sox8* reinforces *Sox9* function in testis differentiation of mice, i.e., *Sox8* does play a role in testis differentiation, but it is a minor role as compared to *Sox9* (Barrionuevo *et al.*, 2006; Chaboissier *et al.*, 2004). Furthermore, ectopic expression of *Sox10* in XX gonads is sufficient to induce testis differentiation (Polanco *et al.*, 2010), suggesting that, although it might not be necessary for testis formation, *Sox10* still can function as a testis-determining gene. Indeed, masculinized or incompletely feminized XX patients have been described that have a duplication of the region encompassing *SOX10* amongst a number of other genes (Aleck *et al.*, 1999; Cantu *et al.*, 1981; Nicholl *et al.*, 1994; Seeherunvong *et al.*, 2004). There is still the possibility that another gene alone or in combination with *SOX10* is responsible for these human DSD phenotypes, but *SOX10* is certainly a good candidate. Of note is that *Sox9* itself is upregulated in the *Sox10*-transgenic XX gonads. It would be interesting to test if overexpression of *Sox10* and/or *Sox8* would be sufficient to rescue the *Sox9* knockout phenotype.

A number of targets of SOX9 have been identified, including the gene encoding anti-Müllerian hormone (*Amh*; Arango *et al.*, 1999), Vanin-1 (*Vnn1*; Wilson *et al.*, 2005), *Ptgds* (Wilhelm *et al.*, 2007a), and cerebellin 4 precursor (*Cbln4*; Bradford *et al.*, 2009a) (Fig. 6.2C). For some of them, such as *Vnn1* and *Cbln4*, we do not know their role in testis development as yet. The function of others, especially *Amh*, has been described in great detail. SOX9 is essential for *Amh* expression in early Sertoli cells (Arango *et al.*, 1999). SF1 (Arango *et al.*, 1999; Shen and Ingraham, 2002), GATA4 (Tremblay and Viger, 1999), and WT1 (Nachtigal *et al.*, 1998) enhance, whereas DAX1 (Nachtigal *et al.*, 1998) impairs its transcription. AMH is a member of the transforming growth factor-β (TGFβ) superfamily and is responsible for the regression of the Müllerian duct, which in XX animals gives rise to the fallopian tubes, uterus, and upper third of the vagina. It acts through its specific type II receptor together with a nonspecific type I receptor expressed in the mesenchyme of the Müllerian duct. Through yet unknown paracrine signaling pathways, it induces apoptosis in the Müllerian duct epithelial cells (Allard *et al.*, 2000; Baarends *et al.*, 1994; di Clemente *et al.*, 1994; Roberts *et al.*, 1999). Defective AMH action, for example through mutation of the genes encoding AMH and its receptor, results in a condition known as persistent Müllerian duct syndrome, which is often associated with cryptorchidism (undescended testes) and infertility.

Another gene expressed by Sertoli cells that has been shown to play a Sertoli cell-specific role is *Dmrt1* (doublesex and mab-3 related transcription factor 1). Interestingly, *Dmrt1* is one of the few molecular similarities in sex determination found among phyla thus far. *Dmrt1*, the sexual regulators *mab-3* in *Caenorhabditis elegans*, doublesex in *Drosophila*, and *Dmy* in medaka all encode proteins containing a DM domain as DNA-binding motif

(Raymond *et al.*, 1998). In two cases, *Dmrt1* in chicken (Smith *et al.*, 2009) and *Dmy* in medaka (Matsuda *et al.*, 2002; Nanda *et al.*, 2002), these genes are believed to be the sex-determining switch. By contrast, in humans and mice, *Dmrt1* is required for the postnatal differentiation of Sertoli and germ cells. *Dmrt1*-null mice have severely impaired testis development from postnatal day 2 resulting in dysgenic testes (Kim *et al.*, 2007a; Raymond *et al.*, 2000). Similarly, human *DMRT1* hemizygosity is associated with defective testicular development comparable to those of the knockout mouse and consequently 46,XY feminization (Crocker *et al.*, 1988; Raymond *et al.*, 1998; Veitia *et al.*, 1998).

In summary, many genes have been identified to be expressed by Sertoli cells. For some we still need to discover the roles they play and how important they are for Sertoli cell differentiation and testis development. For others, a clear role has been defined. In addition, to the above-mentioned genes, which function mainly cell-autonomously in Sertoli cells, several genes have been identified that mediate the function of Sertoli cells as organizing centers of the testis. These genes will be discussed below in the context of the cell types in which they play a role.

7. The Sweepers: Peritubular Myoid Cells

PM cells are long, flattened cells that surround Sertoli cells within the testis cords (Fig. 6.2C). Postnatally, PM cells have a smooth muscle cell phenotype and are responsible for the contraction of the seminiferous tubules for transportation of spermatozoa (Tripiciano *et al.*, 1998). The signaling agonists prostaglandin F2α, vasopressin, and endothelin are produced locally within the seminiferous tubule and interstitial gonad compartments. These paracrine signals act on PM cells to stimulate actin cytoskeleton rearrangements and subsequently result in contraction of the tubules (Howl *et al.*, 1995; Tripiciano *et al.*, 1996, 1998). In fetal development, PM cells contribute to the formation and architecture of the testis cords. While it is now evident that PM cells are not involved in the initial partitioning of the XY gonad into cord regions of clusters of Sertoli and germ cells (Combes *et al.*, 2009b), PM cells establish cord architecture by interacting with Sertoli cells for the correct formation and organization of the basement membrane (Fig. 6.3; Skinner *et al.*, 1985). Each cell type contributes distinct components to the extracellular matrix with PM cells producing fibronectin, type I and IV collagens, and proteoglycans. The resulting membrane then serves to physically partition the XY gonad into cord and interstitial domains (Skinner *et al.*, 1985; Tung *et al.*, 1984).

The first developmental stage when PM cells can be identified as a single layer of cells localized around the periphery of the testis cords with a

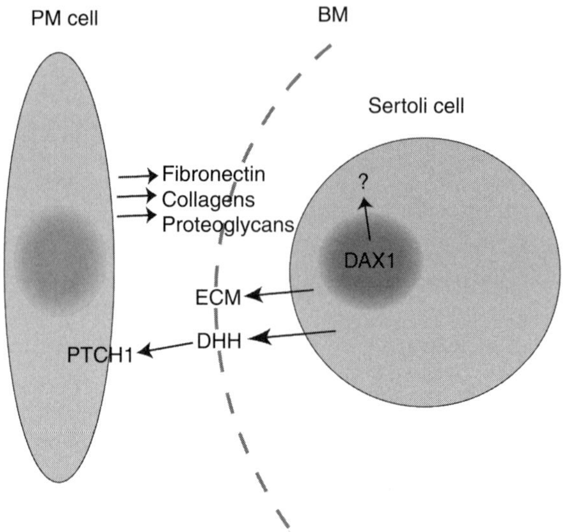

Figure 6.3 Interactions between peritubular myoid (PM) cells and Sertoli cells. The differentiation of PM cells is regulated by desert hedgehog (DHH) signaling from Sertoli cells to the receptor Patched 1 (PTCH1) on PM cells. In addition, *Dax1* expression in Sertoli cells is required for PM cell differentiation; however, downstream signals from *Dax1* are unknown. Interactions between Sertoli cells and differentiating PM cells result in the secretion of extracellular matrix (ECM) molecules by both Sertoli and PM cells for the formation of the basement membrane (BM). Each cell type contributes distinct components to the ECM, with PM cell secreting fibronectin, collagens, and proteoglycans.

characteristic flattened morphology is at around 13.5 dpc. The developmental origin of the PM precursor cells in the XY gonad is still elusive. The lack of a sufficient marker for the PM cell lineage (Jeanes *et al.*, 2005) at early stages of fetal testis development has made cell lineage tracing difficult.

For many years, it was suggested that PM cells were derived from a heterogeneous population of cells that migrated into the testis from the mesonephros at 11.5 dpc (Buehr *et al.*, 1993; Martineau *et al.*, 1997; Merchant-Larios *et al.*, 1993). In 2009, two studies addressed the question of whether PM cells were derived from the mesonephros. The first one made use of a transgenic mouse model that expressed enhanced yellow fluorescent protein (EYFP) under the control of the α-smooth muscle actin (*αSma*) promoter. *αSma*-EYFP was expressed in the mesonephros and in the interstitium including PM cells but not in endothelial cells. Recombination of the *αSma*-EYFP mesonephros with a wild-type gonad demonstrated that none of the *αSma*-EYFP-positive cells migrated from the mesonephros into the gonad, indicating that the migratory population does not give rise to PM cells (Cool *et al.*, 2008). This study was supported by work from Combes *et al.*, in which green fluorescent protein (GFP)-expressing mesonephroi were recombined with wild-type XY gonads in organ culture. Immunofluorescence for a specific endothelial cell marker, VE-cadherin, a marker of interstitial cells, p75, and a vascular associated pericyte marker, NG2, demonstrated that the migrating population were exclusively endothelial cells (Combes *et al.*, 2009b). Together these studies have shown that PM cells are not derived from a migratory subset of cells from the mesonephros. Since, early in fetal development, PM cells express many genes in common with interstitial cells (Jeanes *et al.*, 2005), it is thus hypothesized that PM cells have an interstitial origin either from the initial genital ridge mesenchymal cells or from proliferation of somatic cells at the coelomic epithelium (Combes *et al.*, 2009b; Schmahl *et al.*, 2000).

8. Interplay Between PM and Other Testicular Cells

The differentiation of PM cells from interstitial precursors is influenced by molecular interactions with Sertoli cells. The secreted factor desert hedgehog (DHH) is produced by the Sertoli cells from 11.5 dpc, and its receptor Patched 1 (PTCH1) is present on differentiated PM cells (Bitgood *et al.*, 1996; Clark *et al.*, 2000). In a *Dhh*-null mouse model, PM cells were either absent from tubules or displayed a thickened morphology without the characteristically organized actin cytoskeleton and therefore suggests that signaling from the Sertoli cells stimulates the induction and maturation of the PM cells (Clark *et al.*, 2000). Furthermore, expression of *Dax1* in Sertoli cells also affects PM cell development (Fig. 6.3). Mice with *Dax1*-null mutations have only one-third of the number of PM cells compared to wild-type gonads, and even 12 weeks after birth, the surviving PM cells are relatively undifferentiated (Jeffs *et al.*, 2001; Meeks *et al.*, 2003). It is not yet known how *Dax1* expression in Sertoli cells results in signals to PM cells. Interestingly, both *Dhh*- and *Dax1*-null mouse models display defects in PM cells and Leydig cells (discussed below), suggesting a common precursor mechanism of induction or similar signaling required for maturation.

9. The Midfielders: Leydig Cells

Leydig cells were first described in 1850 by Franz Leydig. There are two populations of Leydig cells, fetal Leydig cells, which are present in the interstitium of embryonic testes from 12.5 dpc until shortly after birth, and adult Leydig cells, which arise at puberty. Based on differences in expression profiles, fetal and adult Leydig cells are thought to arise as two distinct sets of progenitor cells (Lording and De Kretser, 1972; Roosen-Runge and Anderson, 1959; Roosen-Runge and Holstein, 1978). However, these differences might not reflect a distinct origin, but rather differences in the interaction and function of surrounding cells.

Fetal Leydig cells produce hormones to masculinize the embryo, including Wolffian duct differentiation, testicular decent, masculinization of the external genitalia, and sexual dimorphism of the brain. Testosterone production by fetal Leydig cells in mouse testes has been reported to start around 12.5–13.0 dpc and reaches a peak before birth (Gondos *et al.*, 1980). The synthesis of testosterone from cholesterol requires the four enzymes P450 side chain cleavage (CYP11A1/P450SCC/SCC), 3-beta-hydroxy-delta-5-steroid dehydrogenase (3βHSD), cytochrome P450 17-hydroxylase (CYP17), and 17beta-hydroxysteroid dehydrogenase (17βHSDIII). These enzymes are often used as molecular markers of Leydig cells (Fig. 6.4). An important step in testosterone biosynthesis is the acquisition of the precursor cholesterol, which can either be synthesized endogenously by cells or acquired from the blood or surrounding cells (Azhar *et al.*, 2003). Immunohistochemistry has detected expression of the cholesterol synthesis genes lanosterol 14a-demethylase (CYP51) and NADH cytochrome P450 reductase 1 day after the start of testosterone biogenesis, suggesting that, initially, Leydig cells do not synthesize cholesterol and are dependent on exogenous sources (Budefeld *et al.*, 2009). Androgens secreted by Leydig cells induce the embryonic Wolffian ducts to form the internal male genitalia (consisting of the epididymis, vas deferens, and seminal vesicles). Accordingly, in a conditional androgen receptor null mouse model, the vas deferens, epididymis, seminal vesicle, and prostate fail to develop (Yeh *et al.*, 2002).

In addition, Leydig cells play a key role in the regulation of the two phases of testicular decent, transabdominal and inguinoscrotal. Insulin-like factor 3 (INSL3) is a hormone expressed by mature Leydig cells (Fig. 6.4), and *in vitro* studies have suggested that it is regulated transcriptionally by SF1 (Adham *et al.*, 1993; Zimmermann *et al.*, 1998). *Insl3*-null mice exhibit bilateral cryptorchidism suggesting that INSL3 signaling regulates transabdominal testicular decent (Nef *et al.*, 2003). Insights from androgen receptor mutant mice demonstrate failure of the testes to descend from the inner inguinal ring through the inguinal canal to the scrotum, while the initial transabdominal phase was normal (Hutson, 1986). Therefore, different Leydig cell signals, INSL3 and androgens, have distinct functions in mediating testicular decent.

The origin of fetal Leydig cells is unclear. Several sources have been suggested, such as common precursor of adrenal steroidogenic cells (Hatano *et al.*, 1996), neural crest cells (Middendorff *et al.*, 1993), coelomic epithelium (Karl and Capel, 1998), and migrating mesonephric population (Merchant-Larios and Moreno-Mendoza, 1998; Nishino *et al.*, 2001). However, for most of these only circumstantial evidence exists and further research is

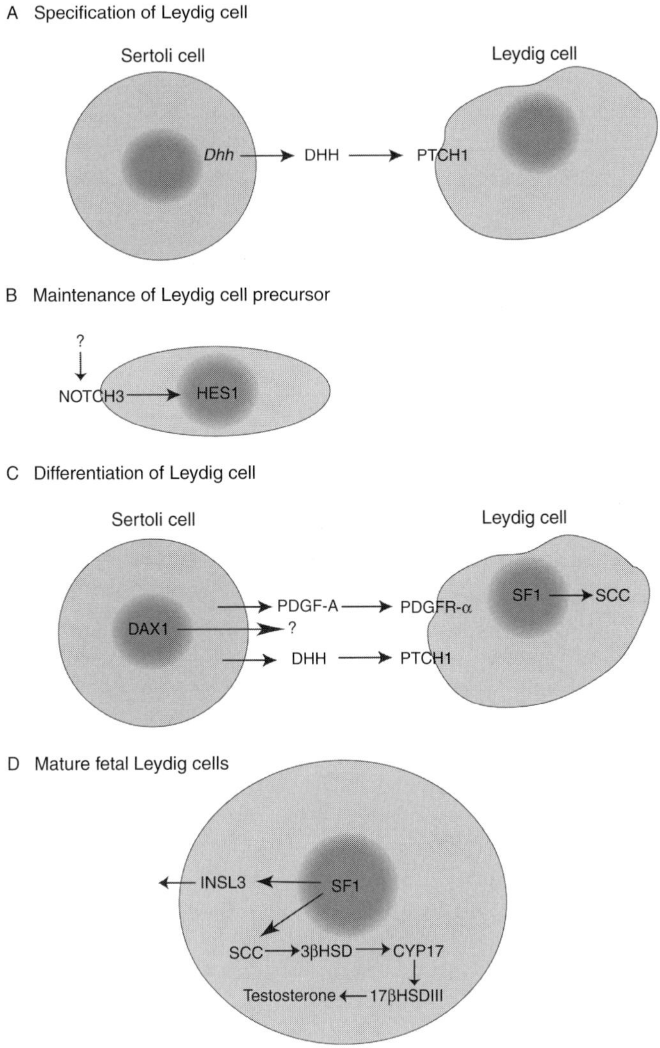

Figure 6.4 The regulation of Leydig cell development. (A) The morphogen DHH expressed by Sertoli cells induces Leydig cell specification through its receptor PTCH1. (B) Notch signaling via the receptor Notch3 and the effector Hes1 is key in the maintenance of the progenitor population and restriction of their differentiation to fetal Leydig cells. (C) Platelet-derived growth factor A (PDGFA) signaling to the PDGFRα receptor plays an integral part in the differentiation of Leydig cells. In addition, *Dax1* expressed by Sertoli cells has been implicated in Leydig cell survival. As Leydig cells differentiate, they start to express genes required for steroid synthesis such as *Scc*. (D) Differentiated Leydig cells synthesize testosterone using the four enzymes P450 side chain cleavage (SCC), 3-beta-hydroxy-delta-5-steroid dehydrogenase (3βHSD), cytochrome P450 17-hydroxylase (CYP17), and 17beta-hydroxysteroid dehydrogenase (17βHSDIII), which masculinize the developing embryo and insulin-like factor 3 (INSL3) for regulation of testicular decent. Both INSL3 and SCC are regulated by SF1 at the transcriptional level.

needed to clarify this issue (for review, see Griswold and Behringer, 2009). By contrast, there is good experimental evidence for the molecular mechanism of specification of fetal Leydig cells (Fig. 6.4A). The morphogen DHH is expressed in Sertoli cells of the XY gonad (Bitgood *et al.*, 1996), and its receptor PTCH1 has been detected by ISH in the majority of the interstitial cells of 12.5 dpc XY gonads including Leydig cells (Yao and Capel, 2002). Null mutations of *Dhh* in mice lead to a greatly diminished Leydig cell number. Furthermore, treatment of *ex vivo* gonad cultures with the global hedgehog inhibitor cyclopamine results in a complete absence of fetal Leydig cells. The more severe phenotype in the cyclopamine-treated gonads compared to $Dhh^{-/-}$ was hypothesized to result from compensation by other hedgehog ligands. To show that hedgehog signaling is sufficient for Leydig cell specification, Barsoum *et al.* (2009) employed a gain-of-function strategy and showed that ectopic expression of DHH in XX gonads induces Leydig cell formation without further masculinization of the fetal ovary.

Following specification, the fetal Leydig cell population at least double in number during embryogenesis; however, quantitative autoradiography has shown that fully differentiated fetal Leydig cells are not mitotically active (Kerr and Knell, 1988; Orth, 1982). This suggests that fetal Leydig cells do not multiply by division of differentiated Leydig cells but rather due to proliferation and differentiation of an undefined population of mesenchymal progenitor or stem cells (Orth, 1982).

The Notch signaling pathway has recently been shown to be a key regulator of the maintenance and differentiation of this progenitor cell population (Fig. 6.4B; Tang *et al.*, 2008). Notch signaling components were detected during testis development with the receptor *Notch3* and the downstream Notch pathway effecter *Hes1* expressed in the interstitial cells from 12.5 dpc. Loss of function of Notch signaling using the γ-secretase inhibitor N-[N-(3,5-difluorophenacetyl)-L-alanyl]-S-phenylglycine-*t*-butyl ester (DABT) *ex vivo* and *Hes1*-null mice resulted in an increase in the number of differentiated Leydig cells as marked by 3βHSD. Furthermore, a gain-of-function approach by overexpressing the Notch intracellular domain in early gonadal somatic cells at 11.5 dpc led to a decrease in Leydig cell number and an increase in the number of somatic progenitor cells as marked by LIM homeobox gene 9 (*Lhx9*). Hence, Notch signaling maintains the progenitor population and restricts their differentiation to fetal Leydig cells (Tang *et al.*, 2008).

In addition to DHH and Notch signaling, which control specification and maintenance, another signaling molecule has been implicated in the differentiation of fetal Leydig cells. Platelet-derived growth factors (PDGF) are homodimers or heterodimers of A, B, C, and D polypeptide chains (PDGF-A, PDGF-B, PDGF-C, PDGF-D). These dimers bind to cell surface tyrosine kinase receptors PDGFRα and PDGFRβ, which homo- or heterodimerize to transduce intracellular signals (reviewed in Betsholtz, 2003; Rosenkranz and Kazlauskas, 1999). Expression of PDGF factors and receptors has been detected in the fetal XY gonad with *Pdgfrα* in the interstitium and *PdgfA* in the Sertoli cells at 12.5 dpc (Brennan *et al.*, 2003; Uzumcu *et al.*, 2002). In *Pdgfrα*-null mice mesonephric cell migration, Sertoli cell proliferation and fetal Leydig cell differentiation are reduced (Brennan *et al.*, 2003). However, the defect in Leydig cell differentiation could be secondary to the Sertoli cell phenotype. In *Dhh*$^{-/-}$-null XY gonads, expression of *Pdgfrα* and *PdgfA* was not perturbed, suggesting that these are independent pathways.

One additional transcription factor plays a role in fetal Leydig cells differentiation. X-linked aristaless-related homeobox gene (*Arx*)-deficient mice have a reduced number of Leydig cells. Intriguingly, immunohistochemistry of ARX in wild-type XY testes at 12.5 dpc and 14.5 dpc detected the protein in PM cells, vascular endothelial cells, and interstitial fibroblast cells but not fetal Leydig cells (Kitamura *et al.*, 2002). It would be interesting to investigate *Arx*-null mice by more in-depth genetic analysis to determine whether diminished Leydig cell differentiation is due to perturbation of cellular interactions with other interstitial cells or a failure of differentiation of an interstitial progenitor pool.

10. The White Lines: Endothelial Cells

Endothelial cells form the vasculature that links the testis to the circulation. The circulation of the fetal testis has been investigated using perfusing embryos with fluorescently labeled lectin, and has been identified that prior to 12.0 dpc, the blood flows from the mesonephros through the vascular plexus at the gonad/mesonephros boundary into the testis. From 12.0 dpc, endothelial cell remodeling redirected the blood flow into the testis via the arterial coelomic vessel (Brennan *et al.*, 2002). The blood flow through the coelomic vessel may facilitate the increasing demand of oxygen and metabolic compounds to the gonad as it differentiates.

In the process of establishing the vasculature, endothelial cells direct the formation of testis cords (Fig. 6.1). Using recombination gonad/mesonephros organ culture experiments, a population of cells was identified that migrates from the mesonephros into the XY gonad beginning at 11.5 dpc (Buehr *et al.*, 1993; Martineau *et al.*, 1997; Merchant-Larios *et al.*, 1993). This migration was found to be downstream of *Sry* expression and critical for testis cord formation (Albrecht *et al.*, 2000; Capel *et al.*, 1999; Tilmann and Capel, 1999). In multiple mouse models, perturbed cord formation is associated with the loss of testicular vascular structures (Albrecht *et al.*, 2000; Brennan *et al.*, 2003). By utilizing genetic markers and specific disruption of endothelial migration with a vascular endothelial cell-blocking antibody (VE-cadherin), it has been demonstrated that the migrating population of cells are exclusively endothelial cells (Combes *et al.*, 2009b; Cool *et al.*, 2008).

Interesting findings regarding the vascularization of the testis have been gained using time-lapse confocal microscopy from 11.5 to 13.5 dpc (Coveney *et al.*, 2008). In these experiments, endothelial cells were visualized in exquisite detail by GFP markers that were generated by crossing endothelial-specific receptor tyrosine kinase (*Tie2*) *Cre* mice or kinase insert domain protein receptor (*Flk1*) *Cre* mice with GFP reporter mice. First, a sex-specific breakdown of a large vessel network within the mesonephric was observed. Endothelial cells underwent a morphological change from an extended squamous cell body to a spherical form. Second, endothelial cells completely detached from the surrounding endothelial cells. Pioneer endothelial cells entered the gonad by extending long filopodia and started to migrate to the coelomic domain on the medial surface of the gonad. Third, additional endothelial cells transverse the gonad following the same paths as the pioneer cells. The coelomic vessel forms under the coelomic epithelium and vessels branch off into the interstitium between the testis cords. Consequently, the gonad is partitioned into approximately 10 avascular domains. Coveney *et al.* (2008) hypothesized that predetermined extracellular matrix scaffolds or boundaries between evenly spaced avascular domains demarcate paths whereby it is permissive for endothelial cells to migrate toward attractive cues secreted by the coelomic epithelium. Conversely, in another study it was proposed that the pioneer migratory endothelial cells would be spaced at random intervals and modify their surroundings such that the subsequent endothelial migration is guided along the same track (Combes *et al.*, 2009a). The next step will be to apply four-dimensional analysis to study known and unidentified factors that influence endothelial migration in transgenic and knockout mouse models and with chemical morphogen/inhibitor treatments. This may allow identification and classification of factors responsible for mesonephric endothelial cell breakdown and remodeling, as well as attraction and guidance of endothelial cells to the coelomic surface. Furthermore, additional cell types could be labeled with other fluorophores to study cellular interactions with endothelial cells.

Many factors have been identified in the genetic regulatory network controlling endothelial cell migration in XY gonads. In the fetal rat testis, vascular endothelial growth factor A (VEGFA) isoforms have been detected in Sertoli cells, and the use of VEGFA signaling inhibitors in organ culture results in reduced vascularization of the testis and impaired cord formation, suggesting that VEGFA secretion by Sertoli cells regulates endothelial cell migration (Bott *et al.*, 2006). VEGF signaling induces endothelial cell migration in other developmental systems (Li *et al.*, 2005), suggesting that this also may be the case during testis differentiation. It has not clear yet how important VEGFA signaling is *in vivo* as compared to other inducer of endothelial cell migration.

For example, the addition of FGF9 to XX mouse gonad organ culture has been shown to induce endothelial cell migration, and conversely, cord formation is impaired in *Fgf9*-null gonads (Colvin *et al.*, 1999). In XX organ culture, AMH induces mesonephric cell migration, male-like vascular formation, and thickening of the coelomic epithelium (Ross *et al.*, 2007). The addition of other factors of the TGFβ family, bone morphogenic protein 2 and 4 (BMP2 and BMP4), also induces similar morphological changes in XX gonads, suggesting that there is redundancy in the signaling pathways. Notably, AMH expression starts after mesonephric cell migration and neither BMP2 nor BMP4 is endogenously expressed in the XY gonad (Behringer *et al.*, 1994; Ross *et al.*, 2007), demonstrating that data obtained with this culturing system have to be treated with care. However, a yet unidentified morphogen may be signaling through the TGFβ pathway to stimulate endothelial migration. PDGF signaling also contributes to testis vasculature development. In *Pdgfr-a*$^{-/-}$-null mice, some XY gonads at 13.5 dpc displayed a disorganized coelomic vessel, fewer vessel branches formed between the cords, and abnormal number and shape of testis cord (Brennan *et al.*, 2003). While these signaling pathways have been shown to be upstream of endothelial cell migration, the exact molecular mechanism is unknown. FGF9 and PDGFR-*a* are expressed and have critical functions in other testis cell types. Hence, the role of these factors in endothelial cell migration may be a secondary rather than direct effect.

In addition to factors that positively induce endothelial cell migration in the XY gonad, ovary-specific factors such as the signaling molecule WNT4 act to antagonize endothelial migration in the XX gonad (see chapter 7). In addition, the co-factor CITED2, which functions to increase *Sry* expression, has been shown to inhibit cell migration in the ovary (Combes *et al.*, 2010). Null mutation in XX gonads induces mesonephric cell migration, a transient upregulation of *Fgf9*, and a delay in *Wnt4* expression (Combes *et al.*, 2010). Whether the transient upregulation of *Fgf9* is the mechanism for this ectopic cell migration, or whether another factor is upregulated in these mice, is not known.

11. What All the Fuss Is About: Germ Cells

The main function of the testis is to provide the necessary environment for the PGCs, the embryonic progenitors of the gametes, to differentiate into functional sperm. PGCs are specified extra-embryonically before gastrulation at the posterior end of the primitive streak (Chiquoine, 1954; Lawson and Hage, 1994). From there they migrate into the adjacent endoderm, along the midline of the hindgut, through the dorsal mesentery to colonize the forming genital ridges between 10.0 and 11.5 dpc (for review, see Richardson and Lehmann, 2010). In the testis, PGC start to form clusters from around 11.5 to 12.5 dpc. These clusters are surrounded by Sertoli cells, which are further divided into testis cords by migrating endothelial cells (Figs. 6.1 and 5). To date, it is not known what factors mediate the clustering of germ cells and Sertoli cells in the early testis.

XX and XY PGCs were always believed to be indistinguishable before they colonize the genital ridges with respect to molecular and morphological properties (Hilscher *et al.*, 1974). Surprisingly, a recent study has shown early differences in the expression of microRNAs (Ciaudo *et al.*, 2009), a class of small RNA molecules involved in gene regulation (for review, see Bartel, 2009; Chekulaeva and Filipowicz, 2009). Ciaudo *et al.* (2009) examined the expression of *miR-302d* in PGCs from 8.5 to 13.5 dpc. Strikingly, *miR-302d* was strongly enriched in XY PGCs at 8.5 dpc and increased through to 9.5 dpc while its expression remained at background levels in XX PGCs at these stages. Thus, it might be that *miR-302d*, and perhaps the *miR-302* cluster, is involved in establishing sex-specific fate in the germline prior to the establishment of the gonad.

Within the gonads, PGCs express pluripotency markers such as OCT4 (pPOU5f1, POU domain, class 5, transcription factor 1). The hallmark of PGC differentiation is their entry into meiosis and therefore downregulation of OCT4 in ovaries (see chapter 7) and mitotic arrest and continuous OCT4 expression in testes (Hilscher *et al.*, 1974; Western *et al.*, 2008). Between 12.5 and 14.5 dpc, PGCs within the testis exit the cell cycle and go into G1/G0 arrest until after birth (Fig. 6.5B), when they re-enter mitotic cycle and start the meiotic cycle at puberty (Fig. 6.5C). Although it has been shown that PGCs in the testis are prevented from entering meiosis by the Sertoli cell-expressed enzyme CYP26B1 (Bowles *et al.*, 2006), what causes their arrest in mitosis is still unexplained. According to their exit of the cell cycle, the expression of key regulators of the G1/S phase is changed. Cell cycle inhibitors such as $p15^{INK4b}$ and $p27^{Kip1}$ are upregulated, whereas promoters of the cell cycle such as cyclin D3 and cyclin E1 and E2 are downregulated (Western *et al.*, 2008). Different mechanisms have been suggested to account for mitotic arrest of PGCs in testes. First, it was often seen as the default pathway that takes place when not entering meiosis. Second, the reduction of retinoic acid by CYP26B1 could not only prevent entry into meiosis but also drive the PGC into mitotic arrest (Trautmann *et al.*, 2008). Third, a HMG box-containing transcription factor, HBP1, which is specifically expressed in XY germ cells at the time of mitotic arrest and has been shown in other systems to induce cell cycle arrest, could be responsible. Fourth, the cell cycle regulator retinoblastoma

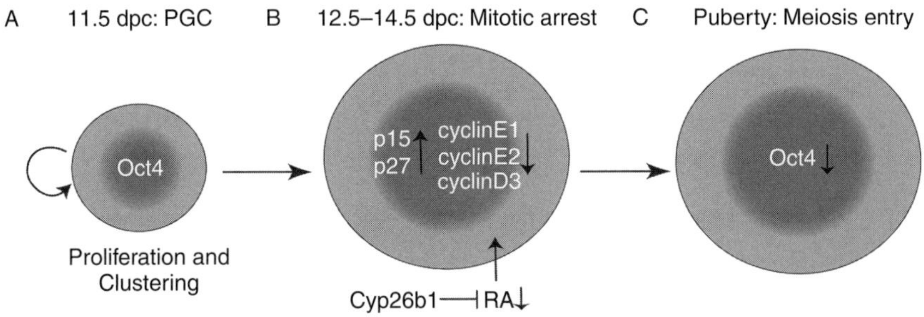

Figure 6.5 Key events in XY germ cell development. (A) By 11.5 dpc, all primordial germ cells (PGCs) have colonized the genital ridges. PGCs proliferate and form clusters. During this phase, PGCs express pluripotency markers such as Oct4. (B) Between 12.5 and 14.5 dpc, PGCs exit the cell cycle and go into mitotic arrest. PGCs in the testis are prevented from entering meiosis by the Sertoli cell-expressed enzyme CYP26B1, which decreases the levels of retinoic acid (RA). Cell cycle inhibitors such as $p15^{INK4b}$ and $p27^{Kip1}$ are upregulated, whereas promoters of the cell cycle such as cyclin D3 and cyclin E1 and E2 are downregulated. (C) At puberty, germ cells resume their journey to becoming spermatozoa. Germ cells re-enter the mitotic cycle, enter meiosis, and therefore downregulate pluripotency genes.

1 (RB1) has recently been shown to be required for correct germ cell entry into mitotic arrest. However, in its absence upregulation of other cell cycle inhibitors induced delayed germ cell arrest (Spiller *et al.*, 2010). And fifth, ectopic expression of the Sertoli cell-expressed transmembrane protein 184a (TMEM184a) in XX genital ridges results in XX germ cell sex reversal, suggesting that signaling through this putative receptor has a role in inhibition of meiosis or promoting mitosis (Svingen *et al.*, 2007, 2009). Clearly, further studies are necessary to identify molecular pathways underlying this male-specific differentiation of PGCs.

12. Concluding Remarks

Since the discovery of the testis- and therefore male-determining gene *Sry* in 1990, exciting progress has been made in unraveling the molecular and cellular mechanisms of testis differentiation. We have highlighted areas of uncertainty throughout this review. In addition to deficiencies in our understanding of the exact relationship between important factors that are involved in testis differentiation, we are faced with the challenge of understanding morphogenesis. Progress has been made due to the development of markers for the different cell types and live-cell imaging techniques. However, for some cell types, such as PM cells, we still do not have a specific marker, and other aspects of morphogenesis such as anterior/posterior and dorsal/ventral patterning have not been studied in detail. In addition, unexplained cases of human DSDs suggest that we are still missing much information about genes involved in these processes, their interaction with each other, and their regulation. Recent genomic approaches are aimed at identifying networks of gene regulation (Munger *et al.*, 2009), but any suggested interaction will need further work for verification and characterization. In addition, the identification of new mechanisms of gene regulation mediated, for example, by noncoding RNAs, including microRNAs, opens a whole new field of possible regulatory pathways that might be important to testis development.

ACKNOWLEDGMENTS

We thank Terje Svingen for helpful comments on the manuscript. We are grateful for research grants from the Australian Research Council (ARC) and National Health and Medical Research Council of Australia (NHMRC). DW is a CDA Fellow of the NHMRC.

REFERENCES

Achermann, J. C., Ito, M., Hindmarsh, P. C., and Jameson, J. L. (1999). A mutation in the gene encoding steroidogenic factor-1 causes XY sex reversal and adrenal failure in humans. *Nat. Genet.* **22**, 125–126.

Adams, I. R., and McLaren, A. (2002). Sexually dimorphic development of mouse primordial germ cells: Switching from oogenesis to spermatogenesis. *Development* **129**, 1155–1164.

Adham, I. M., Burkhardt, E., Benahmed, M., and Engel, W. (1993). Cloning of a cDNA for a novel insulin-like peptide of the testicular leydig cells. *J. Biol. Chem.* **268**, 26668–26672.

Albrecht, K. H., Capel, B., Washburn, L. L., and Eicher, E. M. (2000). Defective mesonephric cell migration is associated with abnormal testis cord development in C57BL/6j XY(Mus domesticus) mice. *Dev. Biol.* **225**, 26–36.

Aleck, K. A., Argueso, L., Stone, J., Hackel, J. G., and Erickson, R. P. (1999). True hermaphroditism with partial duplication of chromosome 22 and without SRY. *Am. J. Med. Genet.* **85**, 2–4.

Allard, S., Adin, P., Gouedard, L., di Clemente, N., Josso, N., Orgebin-Crist, M. C., Picard, J. Y., and Xavier, F. (2000). Molecular mechanisms of hormone-mediated mullerian duct regression: Involvement of beta-catenin. *Development* **127**, 3349–3360.

Arango, N., Lovell-Badge, R., and Behringer, R. (1999). Targeted mutagenesis of the endogenous mouse *mis* gene promoter: In vivo definition of genetic pathways of vertebrate sexual development. *Cell* **99**, 409–419.

Azhar, S., Leers-Sucheta, S., and Reaven, E. (2003). Cholesterol uptake in adrenal and gonadal tissues: The SR-BI and "selective" pathway connection. *Front. Biosci.* **8**, s998–s1029.

Baarends, W. M., van Helmond, M. J., Post, M., van der Schoot, P. J., Hoogerbrugge, J. W., de Winter, J. P., Uilenbroek, J. T., Karels, B., Wilming, L. G., Meijers, J. H., *et al.* (1994). A novel member of the transmembrane serine/threonine kinase receptor family is specifically expressed in the gonads and in mesenchymal cells adjacent to the mullerian duct. *Development* **120**, 189–197.

Bagheri-Fam, S., Sim, H., Bernard, P., Jayakody, I., Taketo, M. M., Scherer, G., and Harley, V. R. (2008). Loss of fgfr2 leads to partial XY sex reversal. *Dev. Biol.* **314**, 71–83.

Barrionuevo, F., Bagheri-Fam, S., Klattig, J., Kist, R., Taketo, M. M., Englert, C., and Scherer, G. (2006). Homozygous inactivation of sox9 causes complete XY sex reversal in mice. *Biol. Reprod.* **74**, 195–201.

Barrionuevo, F., Georg, I., Scherthan, H., Lecureuil, C., Guillou, F., Wegner, M., and Scherer, G. (2009). Testis cord differentiation after the sex determination stage is independent of sox9 but fails in the combined absence of sox9 and sox8. *Dev. Biol.* **327**, 301–312.

Barsoum, I. B., Bingham, N. C., Parker, K. L., Jorgensen, J. S., and Yao, H. H. (2009). Activation of the hedgehog pathway in the mouse fetal ovary leads to ectopic appearance of fetal Leydig cells and female pseudohermaphroditism. *Dev. Biol.* **329**, 96–103.

Bartel, D. P. (2009). MicroRNAs: Target recognition and regulatory functions. *Cell* **136**, 215–233.

Behringer, R. R., Finegold, M. J., and Cate, R. L. (1994). Müllerian-inhibiting substance function during mammalian sexual development. *Cell* **79**, 415–425.

Betsholtz, C. (2003). Biology of platelet-derived growth factors in development. *Birth Defects Res. C Embryo Today* **69**, 272–285.

Bitgood, M. J., Shen, L., and McMahon, A. P. (1996). Sertoli cell signaling by desert hedgehog regulates the male germline. *Curr. Biol.* **6**, 298–304.

Bogani, D., Siggers, P., Brixey, R., Warr, N., Beddow, S., Edwards, J., Williams, D., Wilhelm, D., Koopman, P., Flavell, R. A., Chi, H., Ostrer, H., *et al.* (2009). Loss of mitogen-activated protein kinase kinase kinase 4 (MAP3K4) reveals a requirement for MAPK signalling in mouse sex determination. *PLoS Biol.* **7**, e1000196.

Bott, R. C., McFee, R. M., Clopton, D. T., Toombs, C., and Cupp, A. S. (2006). Vascular endothelial growth factor and kinase domain region receptor are involved in both seminiferous cord formation and vascular development during testis morphogenesis in the rat. *Biol. Reprod.* **75**, 56–67.

Bouma, G. J., Albrecht, K. H., Washburn, L. L., Recknagel, A. K., Churchill, G. A., and Eicher, E. M. (2005). Gonadal sex reversal in mutant dax1 XY mice: A failure to upregulate sox9 in pre-Sertoli cells. *Development* **132**, 3045–3054.

Bowles, J., Knight, D., Smith, C., Wilhelm, D., Richman, J., Mamiya, S., Yashiro, K., Chawengsaksophak, K., Wilson, M. J., Rossant, J., Hamada, H., and Koopman, P. (2006). Retinoid signaling determines germ cell fate in mice. *Science* **312**, 596–600.

Bowles, J., Schepers, G., and Koopman, P. (2000). Phylogeny of the SOX family of developmental transcription factors based on sequence and structural indicators. *Dev. Biol.* **227**, 239–255.

Bradford, S. T., Hiramatsu, R., Maddugoda, M. P., Bernard, P., Chaboissier, M. C., Sinclair, A., Schedl, A., Harley, V., Kanai, Y., Koopman, P., and Wilhelm, D. (2009a). The cerebellin 4 precursor gene is a direct target of SRY and SOX9 in mice. *Biol. Reprod.* **80**, 1178–1188.

Bradford, S. T., Wilhelm, D., Bandiera, R., Vidal, V., Schedl, A., and Koopman, P. (2009b). A cell-autonomous role for WT1 in regulating sry in vivo. *Hum. Mol. Genet.* **18**, 3429–3438.

Brennan, J., Karl, J., and Capel, B. (2002). Divergent vascular mechanisms downstream of *sry* establish the arterial system in the XY gonad. *Dev. Biol.* **244**, 418–428.

Brennan, J., Tilmann, C., and Capel, B. (2003). Pdgfr-alpha mediates testis cord organization and fetal Leydig cell development in the XY gonad. *Genes Dev.* **17**, 800–810.

Buaas, F. W., Val, P., and Swain, A. (2009). The transcription co-factor CITED2 functions during sex determination and early gonad development. *Hum. Mol. Genet.* **18**, 2989–3001.

Budefeld, T., Jezek, D., Rozman, D., and Majdic, G. (2009). Initiation of steroidogenesis precedes expression of cholesterologenic enzymes in the fetal mouse testes. *Anat. Histol. Embryol.* **38**, 461–466.

Buehr, M., Gu, S., and McLaren, A. (1993). Mesonephric contribution to testis differentiation in the fetal mouse. *Development* **117**, 273–281.

Bullejos, M., and Koopman, P. (2001). Spatially dynamic expression of sry in mouse genital ridges. *Dev. Dyn.* **221**, 201–205.

Cantu, J. M., Hernandez, A., Vaca, G., Plascencia, L., Martinez-Basalo, C., Ibarra, B., and Rivera, H. (1981). Trisomy 22q12 leads to qter: "aneusomie de recombinaison" of a pericentric inversion. *Ann. Genet.* **24**, 37–40.

Capel, B., Albrecht, K. H., Washburn, L. L., and Eicher, E. M. (1999). Migration of mesonephric cells into the mammalian gonad depends on *sry*. *Mech. Dev.* **84**, 127–131.

Chaboissier, M. C., Kobayashi, A., Vidal, V. I., Lutzkendorf, S., van de Kant, H. J., Wegner, M., de Rooij, D. G., Behringer, R. R., and Schedl, A. (2004). Functional analysis of sox8 and sox9 during sex determination in the mouse. *Development* **131**, 1891–1901.

Chang, H., Gao, F., Guillou, F., Taketo, M. M., Huff, V., and Behringer, R. R. (2008). Wt1 negatively regulates beta-catenin signaling during testis development. *Development* **135**, 1875–1885.

Chekulaeva, M., and Filipowicz, W. (2009). Mechanisms of miRNA-mediated post-transcriptional regulation in animal cells. *Curr. Opin. Cell Biol.* **21**, 452–460.

Chiquoine, A. D. (1954). The identification, origin, and migration of the primordial germ cells in the mouse embryo. *Anat. Rec.* **118**, 135–146.

Ciaudo, C., Servant, N., Cognat, V., Sarazin, A., Kieffer, E., Viville, S., Colot, V., Barillot, E., Heard, E., and Voinnet, O. (2009). Highly dynamic and sex-specific expression of microRNAs during early ES cell differentiation. *PLoS Genet.* **5**, e1000620.

Clark, A. M., Garland, K. K., and Russell, L. D. (2000). Desert hedgehog (Dhh) gene is required in the mouse testis for formation of adult-type Leydig cells and normal development of peritubular cells and seminiferous tubules. *Biol. Reprod.* **63**, 1825–1838.

Colvin, J. S., Feldman, B., Nadeau, J. H., Goldfarb, M., and Ornitz, D. M. (1999). Genomic organization and embryonic expression of the mouse fibroblast growth factor 9 gene. *Dev. Dyn.* **216**, 72–88.

Combes, A. N., Lesieur, E., Harley, V. R., Sinclair, A. H., Little, M. H., Wilhelm, D., and Koopman, P. (2009a). Three-dimensional visualization of testis cord morphogenesis, a novel tubulogenic mechanism in development. *Dev. Dyn.* **238**, 1033–1041.

Combes, A. N., Spiller, C. M., Harley, V. R., Sinclair, A. H., Dunwoodie, S. L., Wilhelm, D., and Koopman, P. (2010). Gonadal defects in cited2 -mutant mice indicate a role for SF1 in both testis and ovary differentiation. *Int. J. Dev. Biol.* 2010; **54**(4), 683–689.

Combes, A. N., Wilhelm, D., Davidson, T., Dejana, E., Harley, V., Sinclair, A., and Koopman, P. (2009b). Endothelial cell migration directs testis cord formation. *Dev. Biol.* **326**, 112–120.

Cool, J., Carmona, F. D., Szucsik, J. C., and Capel, B. (2008). Peritubular myoid cells are not the migrating population required for testis cord formation in the XY gonad. *Sex Dev.* **2**, 128–133.

Correa, R. V., Domenice, S., Bingham, N. C., Billerbeck, A. E., Rainey, W. E., Parker, K. L., and Mendonca, B. B. (2004). A microdeletion in the ligand binding domain of human steroidogenic factor 1 causes XY sex reversal without adrenal insufficiency. *J. Clin. Endocrinol. Metab.* **89**, 1767–1772.

Cory, A. T., Boyer, A., Pilon, N., Lussier, J. G., and Silversides, D. W. (2007). Presumptive pre-sertoli cells express genes involved in cell proliferation and cell signalling during a critical window in early testis differentiation. *Mol. Reprod. Dev.* **74**, 1491–1504.

Coveney, D., Cool, J., Oliver, T., and Capel, B. (2008). Four-dimensional analysis of vascularization during primary development of an organ, the gonad. *Proc. Natl. Acad. Sci. USA* **105**, 7212–7217.

Crocker, M., Coghill, S. B., and Cortinho, R. (1988). An unbalanced autosomal translocation (7;9) associated with feminization. *Clin. Genet.* **34**, 70–73.

di Clemente, N., Wilson, C., Faure, E., Boussin, L., Carmillo, P., Tizard, R., Picard, J. Y., Vigier, B., Josso, N., and Cate, R. (1994). Cloning, expression, and alternative splicing of the receptor for anti-mullerian hormone. *Mol. Endocrinol.* **8**, 1006–1020.

Eicher, E. M., Washburn, L. L., Whitney, J. B.3rd, and Morrow, K. E. (1982). Mus poschiavinus Y chromosome in the C57BL/6j murine genome causes sex reversal. *Science* **217**, 535–537.

Ewen, K. A., and Koopman, P. (2010). Mouse germ cell development: From specification to sex determination. *Mol. Cell. Endocrinol.* **323**(1), 76–93.

Foster, J. W., Dominguez-Steglich, M. A., Guioli, S., Kwok, C., Weller, P. A., Weissenbach, J., Mansour, S., Young, I. D., Goodfellow, P. N., Brook, J. D., and Schafer, A. J. (1994). Campomelic dysplasia and autosomal sex reversal caused by mutations in an *SRY*-related gene. *Nature* **372**, 525–530.

Gao, F., Maiti, S., Alam, N., Zhang, Z., Deng, J. M., Behringer, R. R., Lecureuil, C., Guillou, F., and Huff, V. (2006). The Wilms tumor gene, wt1, is required for sox9 expression and maintenance of tubular architecture in the developing testis. *Proc. Natl. Acad. Sci. USA* **103**, 11987–11992.

Gondos, B., Rao, A., and Ramachandran, J. (1980). Effects of antiserum to luteinizing hormone on the structure and function of rat leydig cells. *J. Endocrinol.* **87**, 265–270.

Griswold, S. L., and Behringer, R. R. (2009). Fetal leydig cell origin and development. *Sex Dev.* **3**, 1–15.

Hammes, A., Guo, J. K., Lutsch, G., Leheste, J. R., Landrock, D., Ziegler, U., Gubler, M. C., and Schedl, A. (2001). Two splice variants of the Wilms' tumor 1 gene have distinct functions during sex determination and nephron formation. *Cell* **106**, 319–329.

Hatano, O., Takakusu, A., Nomura, M., and Morohashi, K. (1996). Identical origin of adrenal cortex and gonad revealed by expression profiles of Ad4BP/SF-1. *Genes Cells* **1**, 663–671.

Hersmus, R., Kalfa, N., de Leeuw, B., Stoop, H., Oosterhuis, J. W., de Krijger, R., Wolffenbuttel, K. P., Drop, S. L., Veitia, R. A., Fellous, M., Jaubert, F., and Looijenga, L. H. (2008). FOXL2 and SOX9 as parameters of female and male gonadal differentiation in patients with various forms of disorders of sex development (DSD). *J. Pathol.* **215**, 31–38.

Hilscher, B., Hilscher, W., Bulthoff-Ohnolz, B., Kramer, U., Birke, A., Pelzer, H., and Gauss, G. (1974). Kinetics of gametogenesis. I. Comparative histological and autoradiographic studies of oocytes and transitional prospermatogonia during oogenesis and prespermatogenesis. *Cell Tissue Res.* **154**, 443–470.

Hiramatsu, R., Harikae, K., Tsunekawa, N., Kurohmaru, M., Matsuo, I., and Kanai, Y. (2010). FGF signaling directs a center-to-pole expansion of tubulogenesis in mouse testis differentiation. *Development* **137**, 303–312.

Howl, J., Rudge, S. A., Lavis, R. A., Davies, A. R., Parslow, R. A., Hughes, P. J., Kirk, C. J., Michell, R. H., and Wheatley, M. (1995). Rat testicular myoid cells express vasopressin receptors: Receptor structure, signal transduction, and developmental regulation. *Endocrinology* **136**, 2206–2213.

Hutson, J. M. (1986). Testicular feminization: A model for testicular descent in mice and men. *J. Pediatr. Surg.* **21**, 195–198.

Jeanes, A., Wilhelm, D., Wilson, M. J., Bowles, J., McClive, P. J., Sinclair, A. H., and Koopman, P. (2005). Evaluation of candidate markers for the peritubular myoid cell lineage in the developing mouse testis. *Reproduction* **130**, 509–516.

Jeays-Ward, K., Hoyle, C., Brennan, J., Dandonneau, M., Alldus, G., Capel, B., and Swain, A. (2003). Endothelial and steroidogenic cell migration are regulated by WNT4 in the developing mammalian gonad. *Development* **130**, 3663–3670.

Jeffs, B., Ito, M., Yu, R. N., Martinson, F. A., Wang, Z. J., Doglio, L. T., and Jameson, J. L. (2001). Sertoli cell-specific rescue of fertility, but not testicular pathology, in dax1 (ahch)-deficient male mice. *Endocrinology* **142**, 2481–2488.

Karl, J., and Capel, B. (1998). Sertoli cells of the mouse testis originate from the coelomic epithelium. *Dev. Biol.* **203**, 323–333.

Kent, J., Wheatley, S. C., Andrews, J. E., Sinclair, A. H., and Koopman, P. (1996). A male-specific role for SOX9 in vertebrate sex determination. *Development* **122**, 2813–2822.

Kerr, J. B., and Knell, C. M. (1988). The fate of fetal Leydig cells during the development of the fetal and postnatal rat testis. *Development* **103**, 535–544.

Kim, S., Bardwell, V. J., and Zarkower, D. (2007a). Cell type-autonomous and non-autonomous requirements for dmrt1 in postnatal testis differentiation. *Dev. Biol.* **307**, 314–327.

Kim, Y., Bingham, N., Sekido, R., Parker, K. L., Lovell-Badge, R., and Capel, B. (2007b). Fibroblast growth factor receptor 2 regulates proliferation and Sertoli differentiation during male sex determination. *Proc. Natl. Acad. Sci. USA* **104**, 16558–16563.

Kim, Y., Kobayashi, A., Sekido, R., DiNapoli, L., Brennan, J., Chaboissier, M. C., Poulat, F., Behringer, R. R., Lovell-Badge, R., and Capel, B. (2006). Fgf9 and wnt4 act as antagonistic signals to regulate mammalian sex determination. *PLoS Biol.* **4**, e187.

Kitamura, K., Yanazawa, M., Sugiyama, N., Miura, H., Iizuka-Kogo, A., Kusaka, M., Omichi, K., Suzuki, R., Kato-Fukui, Y., Kamiirisa, K., Matsuo, M., Kamijo, S., *et al.* (2002). Mutation of ARX causes abnormal development of forebrain and testes in mice and X-linked lissencephaly with abnormal genitalia in humans. *Nat. Genet.* **32**, 359–369.

Koopman, P., Gubbay, J., Vivian, N., Goodfellow, P., and Lovell-Badge, R. (1991). Male development of chromosomally female mice transgenic for *sry*. *Nature* **351**, 117–121.

Lawson, K. A., and Hage, W. J. (1994). Clonal analysis of the origin of primordial germ cells in the mouse. *Ciba Found. Symp.* **182**, 68–84; discussion 84–91.

Li, Z-D., Bork, J. P., Krueger, B., Patsenker, E., Schulze-Krebs, A., Hahn, E. G., and Schuppan, D. (2005). VEGF induces proliferation, migration and TGF-b1 expression in mouse glomerular endothelial cells via mitogen-activated protein kinase and phosphatidylinositol 3-kinase. *Biochem. Biophys. Res. Commun.* **334**, 1049–1060.

Lording, D. W., and De Kretser, D. M. (1972). Comparative ultrastructural and histochemical studies of the interstitial cells of the rat testis during fetal and postnatal development. *J. Reprod. Fertil.* **29**, 261–269.

Ludbrook, L. M., and Harley, V. R. (2004). Sex determination: A "window" of DAX1 activity. *Trends Endocrinol. Metab.* **15**, 116–121.

Malki, S., Nef, S., Notarnicola, C., Thevenet, L., Gasca, S., Mejean, C., Berta, P., Poulat, F., and Boizet-Bonhoure, B. (2005). Prostaglandin D2 induces nuclear import of the sex-determining factor SOX9 via its cAMP-PKA phosphorylation. *EMBO J.* **24**, 1798–1809.

Martineau, J., Nordqvist, K., Tilmann, C., Lovell-Badge, R., and Capel, B. (1997). Male-specific cell migration into the developing gonad. *Curr. Biol.* **7**, 958–968.

Matoba, S., Hiramatsu, R., Kanai-Azuma, M., Tsunekawa, N., Harikae, K., Kawakami, H., Kurohmaru, M., and Kanai, Y. (2008). Establishment of testis-specific SOX9 activation requires high-glucose metabolism in mouse sex differentiation. *Dev. Biol.* **324**, 76–87.

Matsuda, M., Nagahama, Y., Shinomiya, A., Sato, T., Matsuda, C., Kobayashi, T., Morrey, C. E., Shibata, N., Asakawa, S., Shimizu, N., Hori, H., Hamaguchi, S., *et al.* (2002). DMY is a Y-specific DM-domain gene required for male development in the Medaka fish. *Nature* **417**, 559–563.

Meeks, J. J., Weiss, J., and Jameson, J. L. (2003). Dax1 is required for testis determination. *Nat. Genet.* **34**, 32–3.

Merchant-Larios, H., and Moreno-Mendoza, N. (1998). Mesonephric stromal cells differentiate into Leydig cells in the mouse fetal testis. *Exp. Cell Res.* **244**, 230–238.

Merchant-Larios, H., Moreno-Mendoza, N., and Buehr, M. (1993). The role of the mesonephros in cell differentiation and morphogenesis of the mouse fetal testis. *Int. J. Dev. Biol.* **37**, 407–415.

Middendorff, R., Davidoff, M., and Holstein, A. F. (1993). Neuroendocrine marker substances in human Leydig cells–changes by disturbances of testicular function. *Andrologia* **25**, 257–262.

Miyamoto, Y., Taniguchi, H., Hamel, F., Silversides, D. W., and Viger, R. S. (2008). A GATA4/WT1 cooperation regulates transcription of genes required for mammalian sex determination and differentiation. *BMC Mol. Biol.* **9**, 44.

Moniot, B., Declosmenil, F., Barrionuevo, F., Scherer, G., Aritake, K., Malki, S., Marzi, L., Cohen-Solal, A., Georg, I., Klattig, J., Englert, C., Kim, Y., *et al.* (2009). The PGD2 pathway, independently of FGF9, amplifies SOX9 activity in sertoli cells during male sexual differentiation. *Development* **136**, 1813–1821.

Morais da Silva, S., Hacker, A., Harley, V., Goodfellow, P., Swain, A., and Lovell-Badge, R. (1996). *Sox9* expression during gonadal development implies a conserved role for the gene in testis differentiation in mammals and birds. *Nat. Genet.* **14**, 62–68.

Munger, S. C., Aylor, D. L., Syed, H. A., Magwene, P. M., Threadgill, D. W., and Capel, B. (2009). Elucidation of the transcription network governing mammalian sex determination by exploiting strain-specific susceptibility to sex reversal. *Genes Dev.* **23**, 2521–2536.

Nachtigal, M. W., Hirokawa, Y., Enyeart-van Houten, D. L., Flanagan, J. N., Hammer, G. D., and Ingraham, H. A. (1998). Wilms' tumor 1 and dax1 modulate the orphan nuclear receptor SF1 in sex-specific gene expression. *Cell* **93**, 445–454.

Nanda, I., Kondo, M., Hornung, U., Asakawa, S., Winkler, C., Shimizu, A., Shan, Z., Haaf, T., Shimizu, N., Shima, A., Schmid, M., and Schartl, M. (2002). A duplicated copy of DMRT1 in the sex-determining region of the Y chromosome of the medaka, oryzias latipes. *Proc. Natl. Acad. Sci. USA* **99**, 11778–11783.

Nef, S., Verma-Kurvari, S., Merenmies, J., Vassalli, J. D., Efstratiadis, A., Accili, D., and Parada, L. F. (2003). Testis determination requires insulin receptor family function in mice. *Nature* **426**, 291–295.

Nicholl, R. M., Grimsley, L., Butler, L., Palmer, R. W., Rees, H. C., Savage, M. O., and Costeloe, K. (1994). Trisomy 22 and intersex. *Arch. Dis. Child. Fetal Neonatal. Ed.* **71**, F.

Nishino, K., Yamanouchi, K., Naito, K., and Tojo, H. (2001). Characterization of mesonephric cells that migrate into the XY gonad during testis differentiation. *Exp. Cell Res.* **267**, 225–232.

Orth, J. M. (1982). Proliferation of sertoli cells in fetal and postnatal rats: A quantitative autoradiographic study. *Anat. Rec.* **203**, 485–492.

O'Bryan, M. K., Takada, S., Kennedy, C. L., Scott, G., Harada, S., Ray, M. K., Dai, Q., Wilhelm, D., de Kretser, D. M., Eddy, E. M., Koopman, P., and Mishina, Y. (2008). Sox8 is a critical regulator of adult sertoli cell function and male fertility. *Dev. Biol.* **316**, 359–370.

Paranko, J. (1987). Expression of type I and III collagen during morphogenesis of fetal rat testis and ovary. *Anat. Rec.* **219**, 91–101.

Polanco, J. C., Wilhelm, D., Davidson, T. L., Knight, D., and Koopman, P. (2010). Sox10 gain-of-function causes XX sex reversal in mice: Implications for human 22q-linked disorders of sex development. *Hum. Mol. Genet.* **19**, 506–516.

Raymond, C., Murphy, M., O'Sullivan, M., Bardwell, V., and Zarkower, D. (2000). *Dmrt1*, a gene related to worm and fly sexual regulators, is required for mammalian testis differentiation. *Genes Dev.* **14**, 2587–2595.

Raymond, C., Shamu, C., Shen, M., Seifert, K., Hirsch, B., Hodgkin, J., and Zarkower, D. (1998). Evidence for evolutionary conservation of sex-determining genes. *Nature* **391**, 691–695.

Richardson, B. E., and Lehmann, R. (2010). Mechanisms guiding primordial germ cell migration: Strategies from different organisms. *Nat. Rev. Mol. Cell Biol.* **11**, 37–49.

Roberts, L. M., Hirokawa, Y., Nachtigal, M. W., and Ingraham, H. A. (1999). Paracrine-mediated apoptosis in reproductive tract development. *Dev. Biol.* **208**, 110–122.

Rodemer-Lenz, E. (1989). On cell contribution to gonadal soma formation in quail-chick chimeras during the indifferent stage of gonadal development. *Anat. Embryol. (Berl)* **179**, 237–242.

Roosen-Runge, E. C., and Anderson, D. (1959). The development of the interstitial cells in the testis of the albino rat. *Acta Anat. (Basel)* **37**, 125–137.

Roosen-Runge, E. C., and Holstein, A. F. (1978). The human rete testis. *Cell Tissue Res.* **189**, 409–433.

Rosenkranz, S., and Kazlauskas, A. (1999). Evidence for distinct signaling properties and biological responses induced by the PDGF receptor alpha and beta subtypes. *Growth Factors* **16**, 201–216.

Ross, A., Munger, S., and Capel, B. (2007). Bmp7 regulates germ cell proliferation in mouse fetal gonads. *Sex Dev.* **1**, 127–137.

Schepers, G., Teasdale, R., and Koopman, P. (2002). Twenty pairs of *sox*: Extent, homology, and nomenclature of the mouse and human *sox* transcription factor gene families. *Dev. Cell* **3**, 1–20.

Schepers, G., Wilson, M., Wilhelm, D., and Koopman, P. (2003). SOX8 is expressed during testis differentiation in mice and synergizes with SF1 to activate the Amh promoter in vitro. *J. Biol. Chem.* **278**, 28101–28108.

Schmahl, J., Eicher, E., Washburn, L., and Capel, B. (2000). *Sry* induces cell proliferation in the mouse gonad. *Development* **127**, 65–73.

Schmahl, J., Kim, Y., Colvin, J. S., Ornitz, D. M., and Capel, B. (2004). Fgf9 induces proliferation and nuclear localization of FGFR2 in sertoli precursors during male sex determination. *Development* **131**, 3627–3636.

Seeherunvong, T., Perera, E. M., Bao, Y., Benke, P. J., Benigno, A., Donahue, R. P., and Berkovitz, G. D. (2004). 46,XX sex reversal with partial duplication of chromosome arm 22q. *Am. J. Med. Genet. A* **127A**, 149–151.

Sekido, R., and Lovell-Badge, R. (2008). Sex determination involves synergistic action of SRY and SF1 on a specific sox9 enhancer. *Nature* **453**, 930–934.

Shen, J., and Ingraham, H. (2002). Regulation of the orphan nuclear receptor steroidogenic factor 1 by sox proteins. *Mol. Endocrinol.* **16**, 529–540.

Skinner, M. K., Tung, P. S., and Fritz, I. B. (1985). Cooperativity between sertoli cells and testicular peritubular cells in the production and deposition of extracellular matrix components. *J. Cell Biol.* **100**, 1941–1947.

Smith, C. A., Roeszler, K. N., Ohnesorg, T., Cummins, D. M., Farlie, P. G., Doran, T. J., and Sinclair, A. H. (2009). The avian Z-linked gene DMRT1 is required for male sex determination in the chicken. *Nature* **461**, 267–271.

Sock, E., Schmidt, K., Hermanns-Borgmeyer, I., Bosl, M. R., and Wegner, M. (2001). Idiopathic weight reduction in mice deficient in the high-mobility-group transcription factor sox8. *Mol. Cell. Biol.* **21**, 6951–6959.

Spiller, C. M., Wilhelm, D., and Koopman, P. (2010). Retinoblastoma 1 protein modulates XY germ cell entry into G1/G0 arrest during fetal development in mice. *Biol. Reprod.* **82**, 433–443.

Sudbeck, P., Schmitz, M. L., Baeuerle, P. A., and Scherer, G. (1996). Sex reversal by loss of the C-terminal transctivation domain of human SOX9. *Nat. Genet.* **13**, 230–232.

Svingen, T., Beverdam, A., Bernard, P., McClive, P., Harley, V. R., Sinclair, A. H., and Koopman, P. (2007). Sex-specific expression of a novel gene *tmem184a* during mouse testis differentiaton. *Reproductive* **133**, 983–989.

Svingen, T., Wilhelm, D., Combes, A. N., Hosking, B., Harley, V. R., Sinclair, A. H., and Koopman, P. (2009). Ex vivo magnetofection: A novel strategy for the study of gene function in mouse organogenesis. *Dev. Dyn.* **238**, 956–964.

Tang, H., Brennan, J., Karl, J., Hamada, Y., Raetzman, L., and Capel, B. (2008). Notch signaling maintains Leydig progenitor cells in the mouse testis. *Development* **135**, 3745–3753.

Tevosian, S. G., Albrecht, K. H., Crispino, J. D., Fujiwara, Y., Eicher, E. M., and Orkin, S. H. (2002). Gonadal differentiation, sex determination and normal sry expression in mice require direct interaction between transcription partners GATA4 and FOG2. *Development* **129**, 4627–4634.

Tilmann, C., and Capel, B. (1999). Mesonephric cell migration induces testis cord formation and sertoli cell differentiation in the mammalian gonad. *Development* **126**, 2883–2890.

Trautmann, E., Guerquin, M. J., Duquenne, C., Lahaye, J. B., Habert, R., and Livera, G. (2008). Retinoic acid prevents germ cell mitotic arrest in mouse fetal testes. *Cell Cycle* **7**, 656–664.

Tremblay, J., and Viger, R. (1999). Transcription factor GATA-4 enhances müllerian inhibiting substance gene transcription through a direct interaction with the nuclear receptor SF-1. *Mol. Endocrinol.* **13**, 1388–1401.

Tripiciano, A., Filippini, A., Ballarini, F., and Palombi, F. (1998). Contractile response of peritubular myoid cells to prostaglandin F2alpha. *Mol. Cell. Endocrinol.* **138**, 143–150.

Tripiciano, A., Filippini, A., Giustiniani, Q., and Palombi, F. (1996). Direct visualization of rat peritubular myoid cell contraction in response to endothelin. *Biol. Reprod.* **55**, 25–31.

Tung, P. S., Skinner, M. K., and Fritz, I. B. (1984). Cooperativity between sertoli cells and peritubular myoid cells in the formation of the basal lamina in the seminiferous tubule. *Ann. N. Y. Acad. Sci.* **438**, 435–446.

Uzumcu, M., Dirks, K. A., and Skinner, M. K. (2002). Inhibition of platelet-derived growth factor actions in the embryonic testis influences normal cord development and morphology. *Biol. Reprod.* **66**, 745–753.

Val, P., Martinez-Barbera, J. P., and Swain, A. (2007). Adrenal development is initiated by Cited2 and Wt1 through modulation of Sf-1 dosage. *Development* **134**, 2349–2358.

Veitia, R. A., Nunes, M., Quintana-Murci, L., Rappaport, R., Thibaud, E., Jaubert, F., Fellous, M., McElreavey, K., Goncalves, J., Silva, M., Rodrigues, J. C., Caspurro, M., *et al.* (1998). Swyer syndrome and 46,XY partial gonadal dysgenesis associated with 9p deletions in the absence of monosomy-9p syndrome. *Am. J. Hum. Genet.* **63**, 901–905.

Vidal, V., Chaboissier, M., de Rooij, D., and Schedl, A. (2001). *Sox9* induces testis development in XX transgenic mice. *Nat. Genet.* **28**, 216–217.

Wagner, T., Wirth, J., Meyer, J., Zabel, B., Held, M., Zimmer, J., Pasantes, J., Bricarelli, F. D., Keutel, J., Hustert, E., Wolf, U., Tommerup, N., *et al.* (1994). Autosomal sex reversal and campomelic dysplasia are caused by mutations in and around the *SRY*-related gene *SOX9*. *Cell* **79**, 1111–1120.

Western, P. S., Miles, D. C., van den Bergen, J. A., Burton, M., and Sinclair, A. H. (2008). Dynamic regulation of mitotic arrest in fetal male germ cells. *Stem Cells* **26**, 339–347.

Wilhelm, D., and Englert, C. (2002). The Wilms tumor suppressor WT1 regulates early gonad development by activation of Sf1. *Genes Dev.* **16**, 1839–1851.

Wilhelm, D., Hiramatsu, R., Mizusaki, H., Widjaja, L., Combes, A. N., Kanai, Y., and Koopman, P. (2007a). SOX9 regulates prostaglandin D synthase gene transcription in vivo to ensure testis development. *J. Biol. Chem.* **282**, 10553–10560.

Wilhelm, D., and Koopman, P. (2006). The makings of maleness: Towards an integrated view of male sexual development. *Nat. Rev. Genet.* **7**, 620–631.

Wilhelm, D., Martinson, F., Bradford, S., Wilson, M. J., Combes, A. N., Beverdam, A., Bowles, J., Mizusaki, H., and Koopman, P. (2005). Sertoli cell differentiation is induced both cell-autonomously and through prostaglandin signaling during mammalian sex determination. *Dev. Biol.* **287**, 111–124.

Wilhelm, D., Palmer, S., and Koopman, P. (2007b). Sex determination and gonadal development in mammals. *Physiol. Rev.* **87**, 1–28.

Wilhelm, D., Washburn, L. L., Truong, V., Fellous, M., Eicher, E. M., and Koopman, P. (2009). Antagonism of the testis- and ovary-determining pathways during ovotestis development in mice. *Mech. Dev.* **126**, 324–336.

Wilson, M. J., Jeyasuria, P., Parker, K. L., and Koopman, P. (2005). The transcription factors steroidogenic factor-1 and SOX9 regulate expression of vanin-1 during mouse testis development. *J. Biol. Chem.* **280**, 5917–5923.

Yao, H., and Capel, B. (2002). Disruption of testis cords by cyclopamine or forskolin reveals independent cellular pathways in testis organogenesis. *Dev. Biol.* **246**, 356–365.

Yeh, S., Tsai, M. Y., Xu, Q., Mu, X. M., Lardy, H., Huang, K. E., Lin, H., Yeh, S. D., Altuwaijri, S., Zhou, X., Xing, L., Boyce, B. F., *et al.* (2002). Generation and characterization of androgen receptor knockout (ARKO) mice: An in vivo model for the study of androgen functions in selective tissues. *Proc. Natl. Acad. Sci. USA* **99**, 13498–13503.

Zimmermann, S., Schwarzler, A., Buth, S., Engel, W., and Adham, I. M. (1998). Transcription of the Leydig insulin-like gene is mediated by steroidogenic factor-1. *Mol. Endocrinol.* **12**, 706–713.

CHAPTER SEVEN

Building Pathways for Ovary Organogenesis in the Mouse Embryo

Chia-Feng Liu,[*] Chang Liu,[†] *and* Humphrey H-C Yao[*]

Contents

1. Evolution of the Hypotheses for Ovary Organogenesis in Mammals 264
2. Building the Foundation: Morphogenesis of the Ovary 266
3. Making Eggs: Establishment of the Female Germline 268
4. Determination of the Ovarian Identity: Differentiation of Granulosa Cells 272
5. Emerging Pathways for the Establishment of Female Somatic Environment and Maintenance of Female Germ Cells 276
6. Conclusion and Perspectives 278
References 283

Abstract

Despite its significant role in oocyte generation and hormone production in adulthood, the ovary, with regard to its formation, has received little attention compared to its male counterpart, the testis. With the exception of germ cells, which undergo a female-specific pattern of meiosis, morphological changes in the fetal ovary are subtle. Over the past 40 years, a number of hypotheses have been proposed for the organogenesis of the mammalian ovary. It was not until the turn of the millennium, thanks to the advancement of genetic and genomic approaches, that pathways for ovary organogenesis that consist of positive and negative regulators have started to emerge. Through the action of secreted factors (R-spondin1, WNT4, and follistatin) and transcription regulators (β-catenin and FOXL2), the developmental fate of the somatic cells is directed toward ovarian, while testicular components are suppressed. In this chapter, we review the history of studying ovary organogenesis in mammals and present the most recent discoveries using the mouse as the model organism.

[*] Department of Veterinary Biosciences, University of Illinois at Urbana-Champaign, Illinois, USA
[†] Department of Animal Science, University of Illinois at Urbana-Champaign, Illinois, USA

Current Topics in Developmental Biology, Volume 90
ISSN 0070-2153, DOI 10.1016/S0070-2153(10)90007-0

1. Evolution of the Hypotheses for Ovary Organogenesis in Mammals

The importance of the ovary in mammalian reproduction was not recognized until von Baer's discovery that ovarian follicles are the source of mammalian eggs (von Baer, 1827). Known as "the testicle of the female" before the 16th century, the ovary was considered merely a structure insignificant to generation of species (Harvey, 1653) or a gland that produced female "semen" (Descartes, 1664; Le Grand, 1672; Wharton, 1656). Revelation of the egg-producing ability and endocrine capability of the ovary transformed scientists' view on its role in reproduction. Since then, the ovary has become the centerpiece of the female reproductive system.

The structural and functional foundation of the ovary is established during embryonic development in most eutherian or placental mammals. At the time of fertilization, the sex of the embryo is determined when the sperm carrying either an X or Y chromosome fertilizes the oocyte, which contains one X chromosome. The Y chromosome and its testis-determining element play an indisputable role in testis formation (see Chapter 2 on testis development). However, the number of X chromosomes is irrelevant to the establishment of the ovary, as humans and mice with XO aneuploidy still develop ovaries (Morris, 1968; Ohno and Cattanach, 1962; Singh and Carr, 1966, 1967; Welshons and Russell, 1959).

Alfred Jost, in his groundbreaking work in the early 1950s, revealed the relationship between the sexes of the gonads and phenotypic sexual characteristics. Jost discovered that when the gonads of either sex were removed before the onset of sexual differentiation, female internal and external sexual characteristics arose regardless of the chromosomal sex of the embryo. As a result it was concluded that the female sexual phenotypes appear by default, independent of the presence of gonads (Jost *et al.*, 1953). Jost later extended this paradigm to gonadal differentiation and proposed that a putative "male organizer" prevents the gonadal primordium from developing into an ovary and forces it to become a testis (Jost, 1972) (Fig. 7.1).

The mechanism of ovary organogenesis was explored further by Eicher and Washburn in 1983. Based on the observation of a strain of XY mice where testes were sex-reversed to ovaries (Eicher and Washburn, 1983; Washburn and Eicher, 1983), they proposed that an ovary-determining gene (or *Od*), located on an autosome or the X chromosome, initiates ovary differentiation. In the male embryo, the *Od* gene and subsequent ovary differentiation are inhibited by the testis-determining gene on the Y chromosome (or *Tdy*), which presumably gains its functions preceding the *Od* gene (Fig. 7.1). The discovery of the *SRY* gene (Sex-determining region on the Y chromosome) in the early 1990s confirmed the identity of

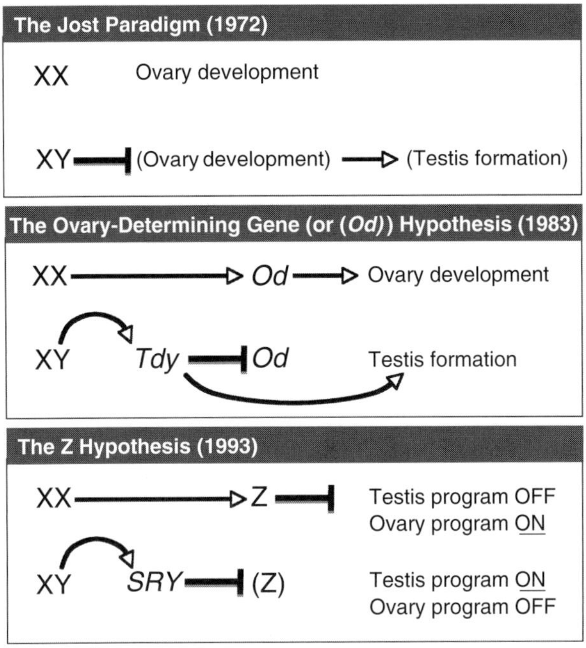

Figure 7.1 Evolution of the hypotheses for sex determination in eutherian mammals. In 1972, Jost extended his paradigm on sexually dimorphic development of the reproductive tracts to gonad differentiation and proposed that a putative "male organizer" prevents the gonadal primordium from developing into an ovary and forces it to become a testis. In 1983, Eicher and Washburn proposed that an ovary-determining gene (*Od*) initiates ovary differentiation in the XX individual. The testis-determining gene on the Y chromosome (or *Tdy*), which becomes active earlier than the *Od* gene, suppresses the *Od* gene in the XY individuals. The Z theory was proposed in 1993 by McElreavey *et al.* after the discovery of *SRY* gene. It was stated that the Z gene in the XX gonad inhibits the testis pathway, therefore leading to the progression of the ovary pathway. In the XY gonad, the *SRY* acts as an inhibitor of the Z gene, allowing the development of the testis.

Tdy and revealed its dominant role in testis determination. However, the puzzling cases of XX male in humans and the polled intersex syndrome (PIS) XX goats, where testes develop in females without *SRY* or any Y chromosome fragments, kindled a rethinking of the mechanism for ovary differentiation. To explain these cases, McElreavey and colleagues proposed the Z hypothesis that in normal XX gonads the Z gene suppresses the emergence of testis program (Fig. 7.1). Loss-of-function of the Z gene in the XX gonad, therefore, results in formation of the testis (McElreavey *et al.*, 1993). In contrast in normal XY gonads, the *SRY* gene antagonizes the functions of the Z gene, allowing the progression of the testis program (Fig. 7.1).

At the turn of the 21st century, mouse genetic models and human clinical cases have made the case that the mechanism for ovary differentiation is beyond X and Y chromosomes and the Z factor. Complicated antagonism and synergism at the levels of cell–cell interaction and transcriptional regulation have already occurred in the undifferentiated ovary. New findings also provide insights into sexually dimorphic regulation of germ cell meiosis and prompt a paradigm shift in our views on how the somatic environment and female germ cells interact. In this chapter, we review the current knowledge of ovary organogenesis using mouse models as the core and make comparisons to humans.

2. Building the Foundation: Morphogenesis of the Ovary

The genital ridge (or gonadal primordium), the structural precursor of both the testis and the ovary, emerges as a thickening of the coelomic epithelium that overlays the ventral aspect of the mesonephros at 10 days *post coitum* (dpc) in the mouse embryo. In both sexes, primordial germ cells (PGCs), which originate from the proximal epiblast, migrate through the wall of the hindgut at ~9 dpc and into the genital ridge at 10.5–11.5 dpc (Anderson *et al.*, 2000; Molyneaux *et al.*, 2001; Tam and Snow, 1981). Once the PGCs colonize the genital ridges, clusters of PGCs coalesce with somatic cells and form the ovigerous cords, which are delineated by the deposition of the basal lamina (Merchant, 1975; Pepling and Spradling, 1998; Ruby *et al.*, 1969). In the developing testis, ovigerous cords differentiate into well-defined tubule structures known as testis cords as a result of the action of Sertoli cells (see Chapter 2 on testis development). In contrast in the fetal ovary, ovigerous cords remain as clusters of female germ cells (or germ cell nests) that are surrounded loosely by somatic cells (Fig. 7.2). Around the time of birth, somatic cells start to break down the germ cell nests by enclosing individual female germ cells or oocytes, leading to the formation of primordial follicles (Merchant-Larios and Chimal-Monroy, 1989; Pepling and Spradling, 1998). Germ cell nests and the forming primordial follicles populate predominantly the outer zone or cortex of the fetal ovary.

In the adult ovary, development of follicles occurs mainly in the ovarian cortex, whereas the medulla is the structure where the vasculature and nerves enter. Ovarian follicles in the cortex consist of oocytes and surrounding somatic cells including granulosa and theca cells. Granulosa cells, an epithelial cell type that forms connections with the oocytes during the fetal stage, support oocyte development and produce hormones responsible for the development and maintenance of the female reproductive system. Theca cells, a mesenchymal cell type that appear only postnatally, are the

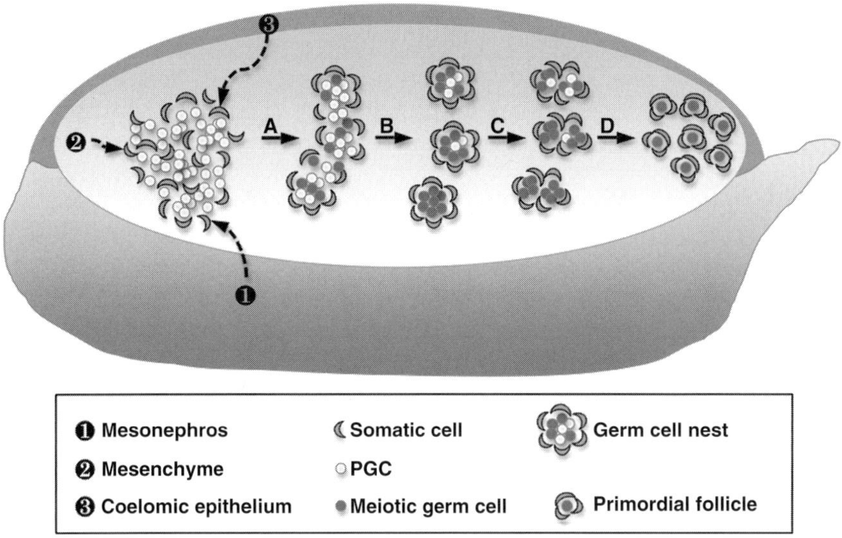

Figure 7.2 Morphogenesis of the mouse fetal ovary. Once the PGCs migrate into and colonize the genital ridge, they coalesce with somatic cells which could be derived from three potential sources: ❶ the neighboring mesonephros, ❷ the existing mesenchyme, and/or ❸ the surrounding coelomic epithelium. (A) Somatic cells and PGCs form ovigerous cords. Female germ cells start to enter meiosis around 13.5 dpc. (B) Germ cell nests that are surrounded loosely by somatic cells begin to form. (C) Around the time of birth, somatic cells start to break down the germ cell nests by enclosing individual oocytes. (D) Breakdown of the germ cell cysts leads to the formation of the primordial follicles.

major source of androgens, which are ultimately converted to estrogens by the granulosa cells (Erickson *et al.*, 1985; Quattropani, 1973).

Granulosa cells in the ovary and Sertoli cells in the testis are derived from common somatic precursor cells in the genital ridge, at least in mice (Albrecht and Eicher, 2001; McLaren, 1991, 2000). Based on morphological and histological observations, granulosa cell precursors could originate from three possible sources (Fig. 7.2): rete ovarii connecting to the neighboring mesonephros (Byskov, 1975, 1978; Byskov and Lintern-Moore, 1973; Byskov and Rasmussen, 1973; Zamboni *et al.*, 1975), the existing mesenchymal cells in the genital ridge (Albrecht and Eicher, 2001; Pinkerton *et al.*, 1961), or ovarian surface epithelium (Gondos, 1975; Motta and Makabe, 1982; Sawyer *et al.*, 2002). Species variation seems evident in regard to the cellular origin(s) of granulosa cell precursors and no definitive sources have been identified in species other than mice. The possible contribution of multiple origins to the granulosa cell lineage cannot be excluded.

Theca cells, the ovarian counterpart of Leydig cells in the testis, begin to appear in the postnatal ovary. Theca cells are thought to be derived from fibroblast-like precursors in the ovarian stroma (Erickson *et al.*, 1985; Hirshfield, 1991; Quattropani, 1973). As the ovarian follicles develop to the secondary stage with multiple layers of granulosa cells surrounding the oocyte, stromal cells adjacent to the basal lamina form a layer of elongated cells. This layer is known as the theca interna, which is a highly vascularized steroidogenic tissue. Outside of the theca interna, a loosely organized band of nonsteroidogenic cells or theca externa is formed. Theca cells are only found in the developing follicle and are adjacent to granulosa cells; therefore, their differentiation is considered to be under the control of granulosa cells (Kotsuji *et al.*, 1990; Orisaka *et al.*, 2006). This concept is supported by the findings that small-molecular-weight proteins enriched from secretions of developing follicles stimulate theca cell differentiation (Magoffin and Magarelli, 1995).

3. Making Eggs: Establishment of the Female Germline

Generation of the female germline (or oogenesis) is the one of the other key functions of the ovary in addition to hormone production. Between 1920 and 1950, the field of oogenesis was dominated by the doctrine that the germinal epithelium, or ovarian surface epithelium encapsulating the ovary, gave rise to oocytes during each estrous or menstrual cycle. This doctrine was later rejected based on evidence that a finite stock of meiotic oocytes is present in the ovary before birth and no new oocytes are generated during adult life (Zuckerman, 1951). However, the discovery of putative germline stem cells in postnatal ovaries led to the resurgence of the controversial idea of "neo-oogenesis" (Johnson *et al.*, 2005, 2004; Zou *et al.*, 2009). In this chapter, we focus only on the establishment of female germline during fetal life.

In the mouse fetal ovary, oocytes start to form around 13.5 dpc when female germ cells (or oogonia) stop proliferating and enter the first meiosis (McLaren, 2000). The oocytes progress through leptonema, zygonema, pachynema, and diplonema and eventually rest at the dictyate stage of meiotic prophase I around the time of birth (Borum, 1961; Speed, 1982). Oocytes do not resume meiosis until ovulation when the female reaches sexual maturity. Once ovulated from the ovary, oocytes complete the first meiotic division, enter the second meiotic division, and arrest again. The second meiotic division is completed after fertilization (Lewis *et al.*, 2006).

Germ cells in the mouse testis behave very differently from their female counterparts. Instead of entering meiosis at 13.5 dpc, male germ cells arrest in mitosis in fetal life and resume mitosis immediately after birth (Hilscher *et al.*, 1974). The male germ cells or spermatogonia then undergo the first meiotic and second meiotic divisions to generate spermatids. The process is repeated many times to constitute a renewing supply of mature sperm (Lewis *et al.*, 2006). In contrast to the finite stock of oocytes at the time of birth, spermatogonia in the testis retain the ability to self-renew throughout the entire reproductive life.

How germ cells make the decision to follow the female or male path has been a central focus of study since 1970s. By creating an XX/XY chimeric embryo, researchers observed that XY germ cells in the ovarian tissue enter meiosis and become functional Y-bearing oocytes (Evans *et al.*, 1977; McLaren *et al.*, 1972; Mystkowska and Tarkowski, 1970). By contrast, XX germ cells avoid entering meiosis if they find themselves in a testicular environment (Palmer and Burgoyne, 1991). The conclusion was, therefore, drawn that all germ cells, regardless of their sex chromosome constitution, are programmed to follow male or female pattern of meiosis according to the surrounding somatic environment (McLaren, 1995, 2003).

Byskov and others proposed that germ cell meiosis is controlled by meiosis-inducing substance or/and a meiosis-preventing substance produced by the somatic cells in the gonads (Byskov, 1974; Byskov *et al.*, 1995, 1998; Byskov *et al.*, 1993; Byskov and Saxen, 1976; Gondos *et al.*, 1996; Grinsted and Byskov, 1981). When an undifferentiated fetal testis is cultured with ovaries containing meiotic germ cells, the male germ cells in the testis are coaxed into entering meiosis (Byskov and Saxen, 1976). On the other hand, when ovaries containing germ cells in meiosis are cultured with fetal testes with well-formed testicular structure, the oocytes are prevented from reaching the diplotene stage of meiotic prophase. Thus it was hypothesized that the fetal ovary secretes a "meiosis-inducing substance", which triggers the meiotic entry of germ cells. The fetal testis instead produces a "meiosis-inhibiting substance" that prevents germ cells from entering meiosis (Byskov and Saxen, 1976).

In contrast to the hypothesis that the ovarian somatic environment secretes factor(s) that induces germ cell meiosis, McLaren and others were in favor of the concept that germ cell entry into meiosis follows a cell-autonomous or intrinsic program (McLaren, 1984; McLaren and Southee, 1997). This concept was based on the observation that when male germ cells lose their way during migration and settle in nongonadal organs such as mesonephros and adrenal, these stray germ cells enter meiosis following the same developmental time frame as their female counterparts in the ovary (Chuma and Nakatsuji, 2001; McLaren, 1983; McLaren and Southee, 1997; Upadhyay and Zamboni, 1982; Zamboni and Upadhyay, 1983). Likewise, XY germ cells enter and progress through meiotic prophase after they are separated from Sertoli cells and then cultured with other nongonadal somatic cells such as embryonic lung cells (McLaren and Southee, 1997). These findings led to the hypothesis that germ cells in the fetal ovary enter meiosis spontaneously, whereas meiosis is inhibited in the testis by factors produced by the somatic cells, probably Sertoli cells (Donovan *et al.*, 1986; McLaren and Southee, 1997).

The debate on how sexually dimorphic pattern of germ cell meiosis is established was resolved by the findings of Bowles *et al.* and Koubova *et al.* in 2006. A mechanism involving retinoic acid (RA) and its degrading enzyme CYP26B1 is in action, with both meiosis-inducing (RA) and meiosis-inhibiting (CYP26B1) properties (Fig. 7.3). The first clue that RA

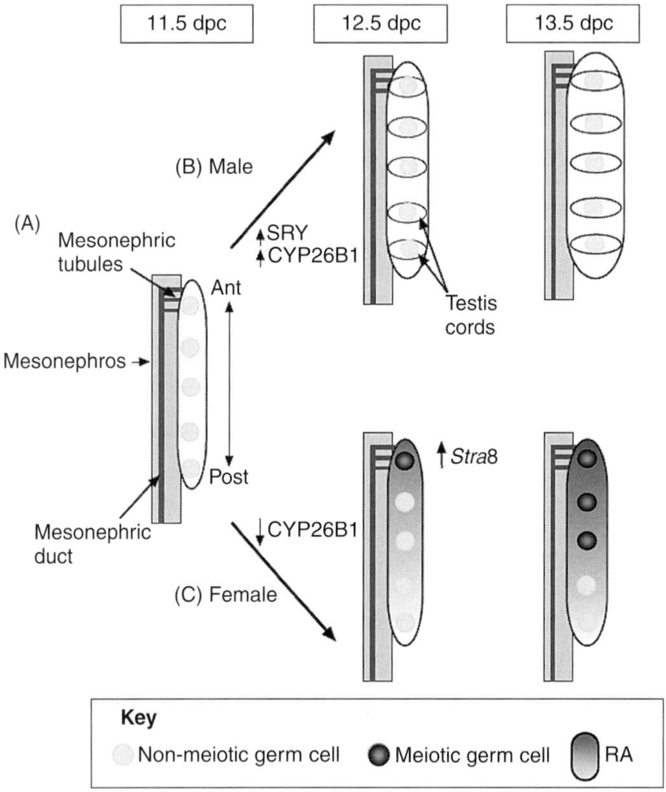

Figure 7.3 Regulation of germ cell entry into meiosis in the developing gonads. (A) At 11.5 dpc, PGCs are present in the genital ridge and RA is produced in the neighboring mesonephric duct and tubules. *Cyp26b1* is expressed at low levels in the gonad of both sexes. The mesonephric tubules, which produce RA, are physically connected with the anterior (Ant) end of the gonad during this time. (B) In the male gonad, once *Sry* is expressed (~11.5 dpc), the RA-degrading enzyme *Cyp26b1* expression is upregulated. The testis cords, which form around germ cell clusters around 12.5 dpc, might concentrate the enzyme in these regions, thereby protecting germ cells from the actions of RA. Germ cells in the male gonad therefore do not enter meiosis at 13.5 dpc. (C) In the female gonad, *Cyp26b1* expression is detectable at 11.5 dpc, but disappears by 12.5 dpc. Germ cells at the anterior end of the gonad begin to express *Stra8* at 12.5 dpc. By 13.5 dpc, female germ cells enter meiosis in an anterior-to-posterior (Post) wave. Germ cells at the anterior end of the gonad might be exposed to RA earlier than those at the posterior end, or the RA concentration might be greater at the anterior end than the posterior end [this figure is modified from Figure 2 in Bowles and Koopman (2007)].

might play a role in meiotic entry of germ cells in the ovary came from an expression screen designed to identify sexually dimorphic genes in mouse fetal gonads. It was found that *Cyp26b1*, the gene encoding a P450 enzyme that degrades RA (Hernandez *et al.*, 2007; Romand *et al.*, 2006; White *et al.*, 2000; Yashiro *et al.*, 2004), shows a testis-specific expression pattern. Initially present in gonads of both sexes, *Cyp26b1* becomes undetectable in female gonads after 11.5 dpc. However in the testis, *Cyp26b1* expression is maintained and reaches its maximum at 13.5 dpc (Bowles *et al.*, 2006). The presence of RA-degrading CYP26b1 in fetal testes is consistent with a significant lower level of RA in the testis compared to the ovary at 13.5 dpc. This evidence suggests that a low RA level is necessary for preventing germ cell meiosis in the testis (Bowles *et al.*, 2006). In other words, a high RA level in the fetal ovary is responsible for inducing germ cell entry into meiosis. Indeed, when exogenous RA is given to fetal testes in culture, XY germ cells enter meiosis (Bowles *et al.*, 2006; Koubova *et al.*, 2006). In addition, treatment of fetal testes with CYP26 inhibitors in culture induces meiotic entry of XY germ cells and an upregulation of stimulated by retinoic acid gene 8 (*Stra8*), which is required for premeiotic DNA replication and the subsequent events of meiotic prophase in germ cells of embryonic ovaries (Baltus *et al.*, 2006; Bowles *et al.*, 2006). Finally, exposure of fetal ovaries to RA antagonists prevents XX germ cells from entering meiosis (Bowles *et al.*, 2006). These *in vitro* results were later substantiated by examination of the *Cyp26b1* knockout mice. In the absence of functional *Cyp26b1* genes, RA levels are increased in embryonic testes and XY germ cells enter meiosis prophase at 13.5 dpc, similar to germ cells in a normal fetal ovary (Bowles *et al.*, 2006; MacLean *et al.*, 2007). Collectively, this evidence supports the concept that a high RA in the fetal ovary due to lack of CYP26B1 is responsible for inducing germ cell meiosis (Fig. 7.3) (Bowles and Koopman, 2007). Presence of CYP26B1 in the fetal testis prevents accumulation of RA and consequent germ cell meiosis.

Intriguingly, fetal testes and ovaries are not the source of RA. Gonads apparently lack the ability to synthesize RA because they do not express *Aldh1a2*, the gene encoding the major enzyme for RA synthesis (Bowles *et al.*, 2006). Instead, RA is secreted by mesonephroi, the mesoderm-derived tissues immediately adjacent to the gonads (Bowles *et al.*, 2006). In 1970s, Byskov proposed that rete ovarii, the extending mesonephric derivative that connects to the ovarian medulla, may be the source of "meiosis-inducing factor" (Byskov, 1974, 1975; Byskov and Lintern-Moore, 1973). In the mouse gonads, mesonephric tubules connect to the anterior portion of the gonads. If indeed the meiosis-inducing factor (or RA) comes from the mesonephric tubules, one would expect that the RA level would be higher at the anterior end than at the posterior end of the gonad. This anterior-to-posterior gradient of RA in the gonads was later confirmed (Fig. 7.3) (Bowles *et al.*, 2006). This phenomenon is also supported by the fact that female germ cells in the anterior part of the fetal ovary enter meiosis earlier than those in the posterior end of the fetal ovary (Bullejos and Koopman, 2004; Menke and Page, 2002; Yao *et al.*, 2003).

Germ cell meiosis in the fetal ovary is controlled by not only the availability of the meiosis-inducing RA, but also the competence of germ cells to respond to RA. Germline specific RNA-binding protein DAZL (deleted in azoospermia-like gene) and NANOS2 have emerged as intrinsic factors in germ cells that define their ability to enter meiosis in response to RA. When *Dazl* became nonfunctional, germ cells in the fetal ovary fail to enter meiosis. In addition, male germ cells in the *Dazl* knockout testis lose their ability to enter meiosis in response to exogenous RA (Lin *et al.*, 2008). This evidence implies that the presence of *Dazl* is a prerequisite for germ cells to gain the ability to respond to RA, therefore, becoming meiosis-competent. *Nanos2*, on the other hand, plays a role in suppressing germ cell meiosis. *Nanos2* is expressed exclusively in germ cells in the fetal testis, whereas it is absent in female germ cells (Tsuda *et al.*, 2003). Loss of *Nanos2* in the fetal testis results in upregulation of *Stra8* and meiosis of male germ cells. Ectopic expression of *Nanos2* in the female germ cells decreases *Stra8* expression and inhibits germ cell meiosis (Suzuki and Saga, 2008). NANOS2 probably inhibits germ cell meiosis by decreasing *Stra8* expression, the downstream target of RA (Suzuki *et al.*, 2010).

In summary, establishment of the female germline, characterized by entry into meiosis in fetal life, requires a synchronized action of both extracellular and intracellular factors. Extrinsic RA, derived from the mesonephros, serves as a meiosis-inducing agent in the fetal ovary. Female germ cells become competent to enter meiosis in response to RA only after they are primed by the presence of intrinsic factor DAZL. The fetal ovary also suppresses production of the meiosis-inhibiting factors including *Cyp26b1* and *Nanos2*, clearing the path for female germ cells to differentiate into oocytes. The observation of putative female germline stem cells in adult ovaries (Johnson *et al.*, 2004, 2005; Zou *et al.*, 2009) raises the question of how these putative germline stem cells escape from the meiosis-inducing RA during fetal development and re-enter the meiosis path later in life. Knowledge of sexually dimorphic regulation of the germline could have implications for solving the controversy around the existence of female germline stem cells in adult ovaries.

4. Determination of the Ovarian Identity: Differentiation of Granulosa Cells

Lessons from regulation of germ cell meiosis highlight the importance of the somatic cell environment in ovary differentiation. In contrast to female germ cells, somatic cells in the fetal ovary have received little attention because their differentiation lacks dramatic elements compared to their counterparts in the fetal testis. The *Sry*-expressing Sertoli cells in the testis orchestrate morphological transformation of the testis (see Chapter 2 on testis development). Granulosa cells, the main somatic cells in the developing ovary, are derived from the same progenitor cells as the Sertoli cells. It was originally thought that the progenitor cells differentiate into granulosa cells by default when the *Sry* is absent, as in XX animals (Fig. 7.1). Influenced by the hypotheses of ovary-determining genes and the Z factor (Fig. 7.1), researchers began to search for genes exclusively expressed in the somatic cells in the ovary. Functional genetic analyses have identified many ovary-specific and somatic cell-derived genes that can be classified into two categories: intracellular factors such as transcription factors (DAX1, FOXL2, and β-catenin) and extracellular factors with paracrine and/or autocrine properties (R-spondin1 and WNT4).

Dax1 (dosage-sensitive sex reversal, adrenal hypoplasia critical region, on chromosome X, gene 1) and *Foxl2* were once the prime candidates for ovary organogenesis and establishment of granulosa cell lineage. *DAX1* was initially considered as the ovary-determining gene because of its X-linked nature and function as a transcriptional regulator (Swain *et al.*, 1998). However, null mutation of *Dax1* in female mouse embryos did not affect formation and development of the ovary (Meeks *et al.*, 2003a–c; Yu *et al.*, 1998). Instead of being an ovary-determining gene, *Dax1* is found to have the anti-testis or Z property in the fetal testis. Duplication of a small piece of the X chromosome that contains *DAX1* in XY humans leads to testis-to-ovary sex reversal (Zanaria *et al.*, 1994). Transgenic mice carrying multiple copies of *Dax1* genes also have testis-to-ovary sex reversal (Swain *et al.*, 1998), suggesting a dose-dependent, anti-testis role of DAX1.

FOXL2, a member of the forkhead transcription factor family, gained attention because of its potential link to the ovary-to-testis sex reversal in the PIS XX goats (Pailhoux *et al.*, 2001, 2002). *FOXL2*, an autosomal gene, shows a granulosa cell-specific pattern conserved among vertebrate species (Loffler *et al.*, 2003; Wang *et al.*, 2004). Originally thought to be the candidate Z factor, *FOXL2* was later found not to be involved in early ovary organogenesis at least in humans and mice. In the absence of functional *FOXL2* genes, human and mouse females develop granulosa cell defects and signs of premature ovarian failure postnatally; however, no signs of ovary-to-testis sex reversal are observed as predicted by the Z hypothesis (Ottolenghi *et al.*, 2005; Schmidt *et al.*, 2004). *FOXL2* and probably *DAX1* may not be critical for early ovary formation, but the possibility of synergy or compensation by other factors cannot be excluded, as discussed below.

In contrast to the uncertain roles of DAX1 and FOXL2, secreted factor R-spondin1 (RSPO1) and WNT4 are doubtlessly critical for establishment of the somatic cell environment in the fetal ovary. Initially expressed in somatic cells in gonads of both sexes, *Rspo1* and *Wnt4* expression become ovary-specific after the time of sex determination (Chassot *et al.*, 2008b; Parma *et al.*, 2006; Tomizuka *et al.*, 2008; Vainio *et al.*, 1999). Female mouse embryos lacking functional *Rspo1* or *Wnt4* develop similar ovarian defects including formation of ectopic testis vasculature, appearance of androgen-producing cells, loss of female germ cells, and appearance of testicular structure at birth (Biason-Lauber *et al.*, 2004; Chassot *et al.*, 2008b; Jeays-Ward *et al.*, 2003; Parma *et al.*, 2006; Tomizuka *et al.*, 2008; Vainio *et al.*, 1999; Yao *et al.*, 2004). The shared phenotypes of the *Rspo1* and *Wnt4* knockout ovaries indicate that these two factors are components of a common pathway in the fetal ovary (see below). Genetic analyses further revealed that RSPO1 is responsible for stimulating the expression of *Wnt4* (Fig. 7.4) (Chassot *et al.*, 2008b; Trautmann *et al.*, 2008; Yao *et al.*, 2004). The RSPO1/WNT4 pathway apparently operates independent of

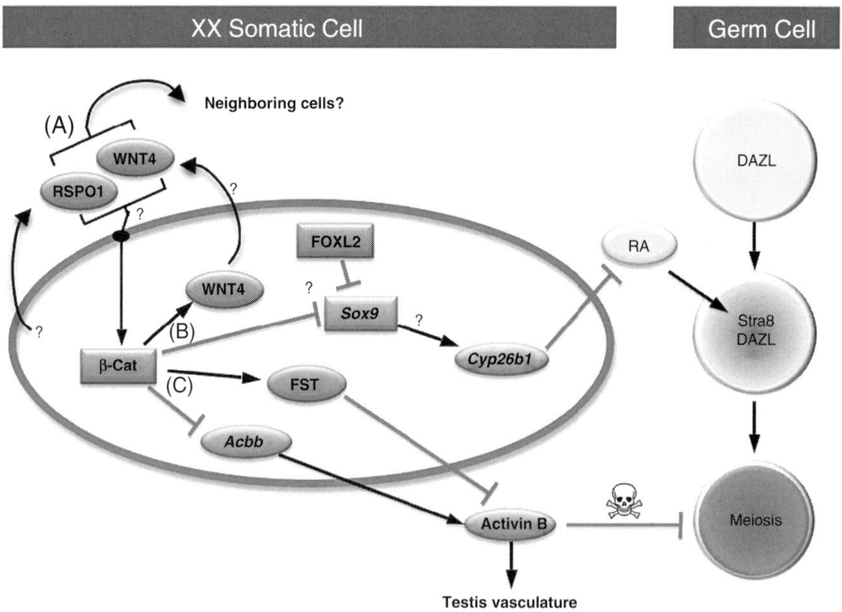

Figure 7.4 Putative pathways for ovary organogenesis in the mouse embryo. (A) Two somatic cell-derived factors, R-spondin1 (RSPO1) and WNT4, activate synergistically or independently the canonical β-catenin (β-cat) pathway in XX somatic cells in an autocrine or paracrine manner. (B) β-catenin then induces expression of *Wnt4* while suppresses expression of *Sox9* (probably through the action of FOXL2) and its putative downstream target *Cyp26b1*. Without the presence of *Cyp26b1*, RA is not degraded and therefore induces *Stra8* expression and meiosis in DAZL-positive germ cells. (C) β-catenin also induces expression of follistatin (*Fst*) and at the same time maintains a low expression of activin βB (*Acbb*). FST antagonizes the action of activin B, which is the protein product of *Acbb*. Lack of activin B ensures that no testis-specific vasculature forms and the survival of female germ cells is maintained. See text for more details. (See Color Insert.)

FOXL2 as the expression of these genes is not affected when either one of these genes become inactive (Chassot *et al.*, 2008b; Ottolenghi *et al.*, 2007).

RSPO1 and WNT4 elicit their actions in ovarian somatic cells via β-catenin, the intracellular regulator of the canonical WNT pathway. Inactivation of β-catenin specifically in the steroidogenic factor 1 (SF1)-positive ovarian somatic cells (putative precursors of granulosa cells) produces ovarian defects similar to those found in *Rspo1* and *Wnt4* knockouts (Liu *et al.*, 2009; Manuylov *et al.*, 2008). The involvement of β-catenin is further confirmed by the gain-of-function experiments where ectopic activation of β-catenin in the absence of *Rspo1* or *Wnt4* restores normal ovarian development (Chassot *et al.*, 2008b; Liu *et al.*, 2010). These experiments also reveal a molecular connection between RSPO1 and WNT4 (Fig. 7.4). In the absence of β-catenin, *Rspo1* expression in the ovary remains unchanged, whereas expression of *Wnt4* is lost, indicating the requirement of RSPO1 and β-catenin for *Wnt4* expression (Liu *et al.*, 2009). RSPO1 is able to induce β-catenin either directly by itself or synergistically in the presence of WNT ligands *in vitro* (Binnerts *et al.*, 2007; Kim *et al.*, 2008, 2006; Lu *et al.*, 2008; Wei *et al.*, 2007). It remains to be determined whether RSPO1 and WNT4 act in a linear fashion or synergistically in activating β-catenin in the somatic cells of the ovary.

Despite their different modes of action (secreted factor versus transcription factor), RSPO1, WNT4, and FOXL2 eventually converge for a common purpose: maintenance of the identity of granulosa cells. In the absence of *Rspo1, Wnt4*, or *Foxl2*, the XX gonadal primoridia develop into ovaries initially but testis components (Sertoli cells and testis cords) start to appear amongst ovarian structure after birth (Chassot *et al.*, 2008b; Ottolenghi *et al.*, 2005; Tomizuka *et al.*, 2008; Vainio *et al.*, 1999). However, when *Wnt4* and *Foxl2* are inactivated together, *Sox9*, the testis-determing gene downstream of SRY (see Chapter 2 on testis development), is significantly upregulated, leading to ovary-to-testis sex reversal (Ottolenghi *et al.*, 2007). These observations support the model that extracellular (RSPO1/WNT4) and intracellular (FOXL2) factors synergistically direct the gonadal somatic cells to follow the ovarian path by antagonizing *Sox9* expression (Figs. 7.4 and 7.5). *Sox9* is expressed in gonads of both sexes before the onset of sex determination. SRY in the testis maintains/stimulates Sox9 expression, whereas in the ovary *Sox9* expression is lost (see Chapter 2 on testis development). Lack of *Sry* and its downstream effectors in the ovary is thought to be responsible for the absence of *Sox9* expression. However, the rise of *Sox9* in the *Wnt4/Foxl2* double knockout ovary without the presence of *Sry* gene argues against this notion. We hypothesize that if neither *Sry* nor RSPO1/WNT4/FOXL2 is present, the default status of gonads is testis, as a result of gradual increase of *Sox9* expression (Fig. 7.5). In the XY individual, SRY in the testis jumpstarts *Sox9* expression, which subsequently suppresses the pro-ovary functions of RSPO1/WNT4/FOXL2. On the other hand, in the absence of *Sry*

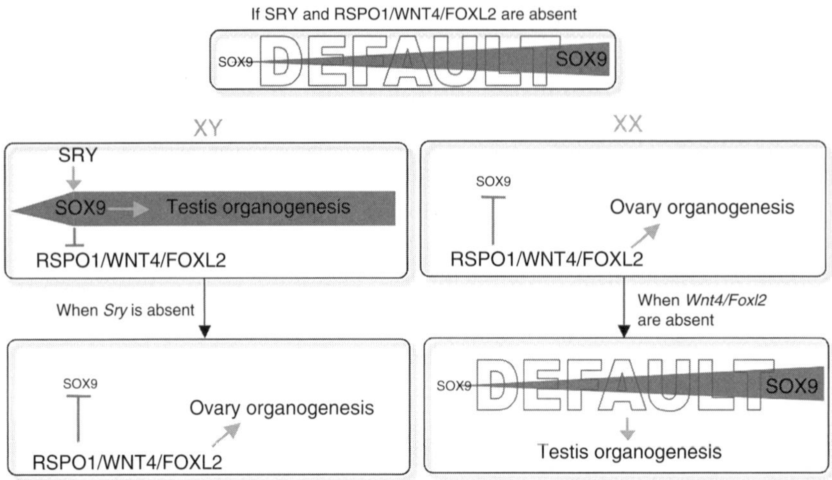

Figure 7.5 The 2010 version of sex determination hypothesis based on mouse genetic models. We propose that if both *SRY* and RSPO1/WNT4/FOXL2 are absent, the default status of gonads is testis, as a result of gradual increase of *Sox9* expression. When both SRY and RSPO1/WNT4/FOXL2 are present as in the XY individual, SRY in the testis jumpstarts *Sox9* expression, which subsequently suppresses the pro-ovary functions of RSPO1/WNT4/FOXL2. On the other hand, in the absence of *Sry* as in the XX individual, RSPO1/WNT4/FOXL2 prevent the rise of *Sox9* and its ability to induce testis differentiation, therefore allowing the gonads to follow the ovarian path. When *Sry* is nonfunctional or lost, RSPO1/WNT4/FOXL2 synergistically suppresses *Sox9* expression and facilitates ovary organogenesis. If the action of RSPO1/WNT4/FOXL2 is silenced in the XX individual, the default testis pathway arises despite the absence of *Sry*.

as in the XX individual, RSPO1/WNT4/FOXL2 prevents the rise of *Sox9* and its ability to induce testis differentiation, therefore, allowing the somatic progenitors to differentiate into granulosa cells and subsequent ovary organogenesis.

5. Emerging Pathways for the Establishment of Female Somatic Environment and Maintenance of Female Germ Cells

In addition to their involvement in ovarian somatic cell differentiation, RSPO1 and WNT4 have unique functions in maintaining proper ovarian environment and survival of female germ cells. One of the earliest morphological differences between fetal testis and ovary is the establishment of an organized vascular network in the testis and the lack of a similar structure in the ovary (Brennan *et al.*, 2002). When either *Rspo1* or *Wnt4* or follistatin (*Fst*) is inactivated, the testis-specific vasculature appears in the fetal ovary at 12.5 dpc (Jeays-Ward *et al.*, 2003; Tomizuka *et al.*, 2008; Yao *et al.*, 2004). In addition to this phenotype, female germ cells undergo apoptosis starting at 15.5 dpc and are lost at the time of birth (Tomizuka *et al.*, 2008a; Vainio *et al.*, 1999; Yao *et al.*, 2004). Similar vasculature and germ cell loss phenotypes are also observed in the fetal ovary that lacks β-catenin in the SF1-positive somatic cells (Liu *et al.*, 2009; Manuylov *et al.*, 2008).

The connection between *Rspo1, Wnt4*, β-catenin, and *Fst* is further confirmed by the genetic models in which the constitutively active form of β-catenin is introduced to the somatic cells of *Rspo1* or *Wnt4* knockout ovary. In the presence of the active β-catenin, normal ovarian characteristics are restored despite a lack of *Rspo1* or *Wnt4* (Chassot *et al.*, 2008b; Liu *et al.*, 2010). In the *Wnt4* knockout ovary, where *Fst* expression is lost, active β-catenin is able to maintain *Fst* expression, placing β-catenin downstream of *Wnt4* (Fig. 7.4). β-catenin probably stimulates *Fst* expression directly via the TCF/LEF consensus sequence in the promoter region of *Fst* (de Groot *et al.*, 2000; Willert *et al.*, 2002).

Fst, a component in the RSPO1/WNT4/β-catenin pathway, encodes a secreted protein that antagonizes the activity of activins. Binding of FST to activins inhibits or limits the ability of activins to interact with their receptors, therefore, silencing the functions of activins (Muttukrishna *et al.*, 2004). Activins consist of either homodimers or heterodimers of activin βA and βB subunits. mRNA for activin βB (*Acbb*), but not activin βA, is present in the fetal mouse ovary although its expression is low (Yao *et al.*, 2006). *Acbb* mRNA expression is suppressed by RSPO1, WNT4, and β-catenin, and when any of these three genes is inactivated, expression of *Acbb* is significantly elevated (Yao *et al.*, 2006). The inhibitory effects of RSPO1/WNT4/β-catenin on *Acbb* expression are further confirmed by the finding that addition of active β-catenin to the *Wnt4* knockout ovary suppresses *Acbb* expression to the low levels seen in the normal ovary (Liu *et al.*, 2010).

The elevated *Acbb* in the *Rspo1, Wnt4*, and β-catenin knockout ovary leads to the hypothesis that *Acbb* could be responsible for appearance of testis-specific vasculature and loss of female germ cells. Two observations support this hypothesis: first, presence of exogenous activin B (protein product of *Acbb*) induces formation of ectopic testis vasculature in normal fetal ovaries in culture (Yao *et al.*, 2006) and second, when the *Acbb* gene is inactivated in the *Wnt4* knockout ovary, normal ovarian development is restored (lack of testis-specific vasculature and maintenance of female germ cells) (Liu *et al.*, 2010; Yao *et al.*, 2006). Intriguingly, inactivation of *Acbb* in the *Fst* knockout ovary also restores normal ovarian development (Yao *et al.*, 2006). These results support the model that in the fetal ovary, *Wnt4* and *Fst* antagonize functions of *Acbb*. WNT4, via the action of β-catenin, suppresses but does not completely abolish the expression of *Acbb*.

The function of FST is to antagonize and inhibit the action of the residual activin B to prevent it from inducing testicular vasculature and demise of female germ cells (Fig. 7.4).

When the ovary-specific functions of *Wnt4* was first discovered in 1999, *Wnt4* was labeled as the inhibitor that prevents the appearance of Leydig cells, the androgen-producing cells present only in the testis (Vainio *et al.*, 1999). In the *Wnt4* knockout ovary, androgen-producing cells appeared ectopically, leading to masculinization of the female. However, the identity of these ectopic androgen-producing cells was later found to be adrenal origin instead of Leydig cells (Heikkila *et al.*, 2002; Jeays-Ward *et al.*, 2003). In addition, these adrenal cells appeared in not only the fetal ovary, but also fetal testis of the *Wnt4* knockout embryos (Heikkila *et al.*, 2002). The appearance of these androgen-producing adrenal cells are also observed in the *Rspo1* and β-catenin knockout ovaries (Chassot *et al.*, 2008a; b; Liu *et al.*, 2009; Manuylov *et al.*, 2008; Tomizuka *et al.*, 2008; Vainio *et al.*, 1999). Adrenals and gonads are derived from a common adrenogonadal primordium, which later separate into two identities. Presence of *Rspo1* and *Wnt4* in the adrenogonadal primordium before the separation of adrenal and gonads suggests that these two genes play a role in proper allocation of the adrenal cell lineage (Heikkila *et al.*, 2002; Jeays-Ward *et al.*, 2003). *Rspo1* and *Wnt4* probably are not involved in suppressing the appearance of Leydig cells in the fetal ovary.

6. Conclusion and Perspectives

Default or not, the pathways that lead to organogenesis of the ovary are far more complicated than what was originally hypothesized (Figs. 7.1 and 7.5). Components of the pathway possess properties of both ovary-determining gene and anti-testis Z factor, which operate at levels of cell-to-cell interaction and transcriptional regulation. Through the action of the secreted factors RSPO1 and WNT4, somatic cells in the fetal mouse ovary are instructed to follow the program for granulosa cell differentiation. Signals from RSPO1 and WNT4 are interpreted intracellularly via β-catenin, which with a synergistic action of FOXL2, silences the SOX9-induced testis differentiation (Figs. 7.4 and 7.5). This antagonism is a critical connection between ovarian somatic cells and germ cells. Lack of SOX9-induced testis differentiation results in absence of CYP26b1 (a putative target of SOX9) that degrades the meiosis-inducing RA. RA, therefore, becomes available only in the fetal ovary and induces meiosis of female germ cells. Once female germ cells have entered meiosis, they can only survive in the ovarian somatic environment that is set up by RSPO1, WNT4, and other components of the pathway (Fig. 7.4).

Despite this progress in our understanding of ovary organogenesis and its regulation, many questions remain. First, in addition to regulating somatic cell development, do *Rspo1* and *Wnt4* have a direct role on female germ cells? Female germ cells are almost completely lost in *Rspo1, Wnt4,* and β-catenin knockout ovaries at the time of birth. However, it is unclear whether the affected germ cells enter meiosis before their demise. Some reports show meiotic defects in female germ cells in the *Rspo1* and *Wnt4* knockout ovaries (Chassot *et al.*, 2008b; Naillat *et al.*, 2010). However, others find proper progression of germ cell meiosis in the absence of *Rspo1, Wnt4,* and β-catenin (Chassot *et al.*, 2008b; Tomizuka *et al.*, 2008; Liu *et al.*, 2009; Liu *et al.*, 2010). It is worth noting that the meiosis-inducing system (RA and Stra8) is not significantly altered in the *Rspo1* knockout ovary, but is compromised in the *Wnt4* knockout ovary (Chassot *et al.*, 2008b; Naillat *et al.*, 2010). More experiments are needed to reconcile these discrepancies and clarify whether RSPO1 and WNT4 intersect with RA in regulating germ cell meiosis.

Second, do fetal and adult ovaries utilize different mechanisms to maintain the differentiated state? Whereas loss of *Foxl2* does not affect formation of the fetal mouse ovary, *Foxl2* is essential for the maintenance of somatic cell identity in the adult ovary. Inactivation of *Foxl2* in the adult mouse ovary leads to upregulation of *Sox9*, transdifferentiation of granulosa cells into Sertoli cells, and appearance of testis structure and cell types (Uhlenhaut *et al.*, 2009). Granulosa-to-Sertoli transdifferentiation is also observed in adult ovaries of estrogen receptor alpha and beta double knockout (ERαβKO) and aromatase knockout mice (Britt *et al.*, 2001; Britt and Findlay, 2003; Britt *et al.*, 2004a, b; Dupont *et al.*, 2003, 2000). FOXL2 is known to stimulate expression of aromatase, the enzyme responsible for estrogen synthesis (Pannetier *et al.*, 2006; Uhlenhaut *et al.*, 2009; Wang *et al.*, 2007). It is therefore proposed that in the adult ovary FOXL2 serves to maintain granulosa cell identity by promoting the synthesis of estrogen, which is the conserved mechanism responsible for ovary formation in species such as reptiles, birds, fish, and even marsupials (Bruggeman *et al.*, 2002; Elf, 2003; Kobayashi and Nagahama, 2009; Mittwoch, 1998; Nakamura, 2009; Pask and Renfree, 2001; Yao, 2005; Yao and Capel, 2005). The maintenance of granulosa cell differentiation apparently shifts from estrogen-independent at the fetal stage to estrogen-dependent in adulthood (Fig. 7.6). Female mammals, particularly eutherian mammals, are constantly exposed to maternal estrogens. The estrogen-insensitive mechanisms for ovary organogenesis make physiological sense; otherwise all the male embryos would be sex-reversed by estrogens. Finding out what roles the RSPO1/WNT4 pathway plays in other vertebrate species will shed light onto the evolution of the mechanism for ovary organogenesis.

Third, where do the theca cell lineage originate from and how is their identity established? Theca cells, the female counterparts of testis Leydig cells, are essential for steroidogenesis and formation of the follicles.

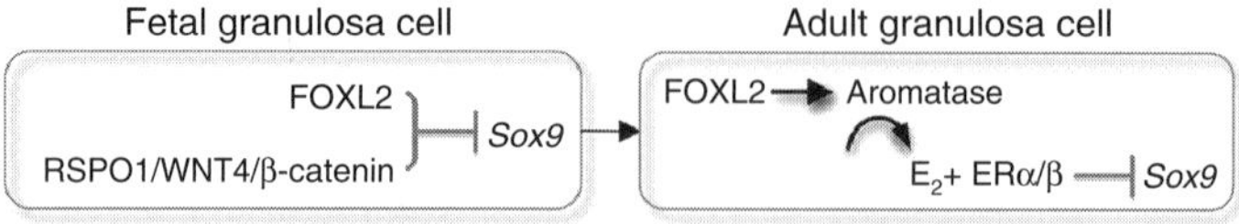

Figure 7.6 Maintenance of granulosa cell identity in fetal and adult ovaries. In the fetal ovary, granulosa cell identity is maintained in an estrogen-independent manner via the RSPO1/WNT4/β-catenin pathway and FOXL2, which together repress the expression of *Sox9*. This mechanism becomes estrogen-dependent after birth and in the adult ovary FOXL2 stimulates estrogen-producing enzyme aromatase and through the action of estrogen (E_2) and their receptors (ERα/β) suppresses *Sox9* and maintain the identity of granulosa cells.

At present, no lineage markers have been identified for theca cells. Based on their similar functions and mesenchymal nature to Leydig cells, we propose that the mechanisms for their specification could also share similarities with Leydig cells. Specification of fetal Leydig cells is under the control of Sertoli cell-derived Desert hedgehog (Barsoum *et al.*, 2009; Barsoum and Yao, 2010; Huang and Yao, 2010; Yao *et al.*, 2002). We are currently investigating whether hedgehog ligands derived from the granulosa cells instruct mesenchymal precursor cells to differentiate into the theca cell lineage in the fetal ovary.

A final, important question is whether knowledge gained from mouse models is applicable to other mammalian species such as human. It is naïve to think that animals utilize identical regulatory pathways for one biological process such as ovary organogenesis. Components of the pathways may be conserved but species variations are expected as the organisms adapt to their unique developmental environment for survival. Comparing the mouse genetic models with human clinical cases collected so far (Table 7.1), *RSPO1* and *WNT4* both are involved in ovary organogenesis with species differences. Loss of function in human *RSPO1* gene leads to complete female-to-male sex reversal (Parma *et al.*, 2006), a phenotype much more severe than that in the mouse *Rspo1* knockout model (Chassot *et al.*, 2008b; Tomizuka *et al.*, 2008). Human patients with defective function of *WNT4* develop various degrees of female-to-male sex reversal (Biason-Lauber *et al.*, 2004; Mandel *et al.*, 2008). In addition, humans seem to be more sensitive to gene dosage than the mouse (see gain-of-function cases in Table 7.1).

Despite these species variations, research using mouse models has identified or confirmed the involvement of components in the pathways toward organogenesis of the ovary. New candidate genes continue to be discovered by mRNA array experiments (Beverdam and Koopman, 2006; Bouma *et al.*, 2009; Cederroth *et al.*, 2007; Coveney *et al.*, 2008; Houmard *et al.*, 2009; Jorgensen and Gao, 2005; Nef *et al.*, 2005), protein screening (Ewen *et al.*, 2009), or regulatory sequence comparison (Lee *et al.*, 2009). It will not be a surprise in the near future if components of ovary organogenesis expand from X, Y, and Z to the entire alphabet.

Table 7.1 Genes involved in fetal ovary development in mice and humans

Gene	Mutation	Sex	Phenotypes in mice	Phenotypes in humans	Reference
*Ctnnb1**	LOF*	F	Appearance of testis-specific vasculature, loss of female germ cells, and appearance of androgen-producing adrenal cells	None reported	Liu *et al.*, 2009; Manuylov *et al.*, 2008.
Ctnnb1	GOF*	M	Testis-to-ovary sex reversal	None reported	Maatouk *et al.*, 2008.
Dax1	LOF	F	No phenotypes in the fetal ovary	X-linked syndrome with adrenal hypoplasia congenital and hypogonadtropic hypogonadism	Meek *et al.*, 2003a; Yu *et al.*, 1998; Zanaria *et al.*,1994.
Dax1	GOF	M	Insertion of multiple copies of *Dax1* gene in the XY mouse delays testis development and causes male-to-female sex reversal in the Y_{Pos} background	Duplication of a portion of the X chromosome containing *DAX1* gene cause male-to-female sex reversal	Barbaro, 2008; Sanlaville *et al.*, 2004; Swain *et al.*, 1998.
Foxl2	LOF	F	Defects in folliculogenesis postnatally and Blepharophimosis Ptosis Epicanthus inversus Syndrome (BPES)-like symptoms	Premature ovarian failure and Blepharophimosis Ptosis Epicanthus inversus Syndrome (BPES)	Crisponi *et al.*, 2001; Ottolenghi *et al.*, 2005; Schmidt *et al.*, 2004.

(Continued)

Table 7.1 *(Continued)*

Gene	Mutation	Sex	Phenotypes in mice	Phenotypes in humans	Reference
Fst	LOF	F	Appearance of testis-specific vasculature and loss of female germ cells	None reported	Yao *et al.*, 2004.
Rspo1	LOF	F	Appearance of testis-specific vasculature, loss of female germ cells, and appearance of androgen-producing adrenal cells	Female-to-male sex reversal; Palmoplantar hyperkeratosis (PPK) and predisposition to squamous,cell carcinoma of the skin (SCC)	Chassot *et al.*, 2008a; Parma *et al.*, 2006; Tomizuka *et al.*, 2008.
Wnt4	LOF	F	Appearance of testis-specific vasculature, loss of female germ cells, and appearance of androgen-producing adrenal cells	Various degree of female-to-male sex reversal and SERKAL syndrome (sex reversal; dysgenesis of kidneys, adrenals, and lungs)	Biason-Lauber *et al.*, 2004; Jeays-Ward *et al.*, 2003; Mandel *et al.*, 2008; Philibert *et al.*, 2008; Sultan *et al.*, 2009; Vainio *et al.*, 1999.
Wnt4	GOF	M	Transgenic mouse carrying human *WNT4* gene disrupts testicular vasculature and decreases testosterone synthesis	Duplication of a DNA fragment containing *WNT4* gene leads to XY female phenotypes	Jordan *et al.*, 2001, 2003.

* *Ctnnb1* = β-catenin; LOF = loss of function; GOF = gain of function

REFERENCES

Albrecht, K. H., and Eicher, E. M. (2001). Evidence that Sry is expressed in pre-Sertoli cells and Sertoli and granulosa cells have a common precursor. *Dev. Biol.* **240**, 92–107.

Anderson, R., Copeland, T. K., Scholer, H., Heasman, J., and Wylie, C. (2000). The onset of germ cell migration in the mouse embryo. *Mech. Dev.* **91**, 61–68.

Baltus, A. E., Menke, D. B., Hu, Y. C., Goodheart, M. L., Carpenter, A. E., de Rooij, D. G., and Page, D. C. (2006). In germ cells of mouse embryonic ovaries, the decision to enter meiosis precedes premeiotic DNA replication. *Nat. Genet.* **38**, 1430–1434.

Barbaro, M., Cicognani, A., Balsamo, A., Lofgren, A., Baldazzi, L., Wedell, A., and Oscarson, M. (2008). Gene dosage imbalances in patients with 46, XY gonadal DSD detected by an in-house-designed synthetic probe set for multiplex ligation-dependent probe amplification analysis. *Clin. Genet.* **73**, 453–464.

Barsoum, I. B., Bingham, N. C., Parker, K. L., Jorgensen, J. S., and Yao, H.H-C. (2009). Activation of the hedgehog pathway in the mouse fetal ovary leads to ectopic appearance of fetal Leydig cells and female pseudohermaphroditism. *Dev. Biol.* **329**, 96–103.

Barsoum, I. B., and Yao, H.H-C. (2010). Fetal Leydig cells: Progenitor cell maintenance and differentiation. *J. Androl.* **31**, 11–15.

Beverdam, A., and Koopman, P. (2006). Expression profiling of purified mouse gonadal somatic cells during the critical time window of sex determination reveals novel candidate genes for human sexual dysgenesis syndromes. *Hum. Mol. Genet.* **15**, 471–431.

Biason-Lauber, A., Konrad, D., Navratil, F., and Schoenle, E. J. (2004). A WNT4 mutation associated with mullerian-duct regression and virilization in a 46,XX woman. *N. Engl. J. Med.* **351**, 792–798.

Binnerts, M. E., Kim, K. A., Bright, J. M., Patel, S. M., Tran, K., Zhou, M., Leung, J. M., Liu, Y., Lomas, W. E. 3rd, Dixon, M., Hazell, S. A., Wagle, M., *et al.* (2007). R-spondin1 regulates Wnt signaling by inhibiting internalization of LRP6. *Proc. Natl. Acad. Sci. USA* **104**, 14700–14705.

Borum, K. (1961). Oogenesis in the mouse. A study of the meiotic prophase. *Exp. Cell Res.* **24**, 495–507.

Bouma, G. J., Hudson, Q. J., Washburn, L. L., and Eicher, E. M. (2009). New candidate genes identified for controlling mouse gonadal sex determination and the early stages of granulosa and Sertoli cell differentiation. *Biol. Reprod.* **82**, 380–389.

Bowles, J., Knight, D., Smith, C., Wilhelm, D., Richman, J., Mamiya, S., Yashiro, K., Chawengsaksophak, K., Wilson, M. J., Rossant, J., Hamada, H., and Koopman, P. (2006). Retinoid signaling determines germ cell fate in mice. *Sciences (New York)* **312**, 596–600.

Bowles, J., and Koopman, P. (2007). Retinoic acid, meiosis and germ cell fate in mammals. *Development (Cambridge, England)* **134**, 3401–3411.

Brennan, J., Karl, J., and Capel, B. (2002). Divergent vascular mechanisms downstream of Sry establish the arterial system in the XY gonad. *Dev. Biol.* **244**, 418–428.

Britt, K. L., Drummond, A. E., Dyson, M., Wreford, N. G., Jones, M. E., Simpson, E. R., and Findlay, J. K. (2001). The ovarian phenotype of the aromatase knockout (ArKO) mouse. *J. Steroid. Biochem. Mol. Biol.* **79**, 181–185.

Britt, K. L., and Findlay, J. K. (2003). Regulation of the phenotype of ovarian somatic cells by estrogen. *Mol. Cell. Endocrinol.* **202**, 11–17.

Britt, K. L., Saunders, P. K., McPherson, S. J., Misso, M. L., Simpson, E. R., and Findlay, J. K. (2004a). Estrogen actions on follicle formation and early follicle development. *Biol. Reprod.* **71**, 1712–1723.

Britt, K. L., Stanton, P. G., Misso, M., Simpson, E. R., and Findlay, J. K. (2004b). The effects of estrogen on the expression of genes underlying the differentiation of somatic cells in the murine gonad. *Endocrinology* **145**, 3950–3960.

Bruggeman, V., Van As, P., and Decuypere, E. (2002). Developmental endocrinology of the reproductive axis in the chicken embryo. *Comp. Biochem. Physiol. A Mol. Integr. Physiol.* **131**, 839–846.

Bullejos, M., and Koopman, P. (2004). Germ cells enter meiosis in a rostro-caudal wave during development of the mouse ovary. *Mol. Reprod. Dev.* **68**, 422–428.

Byskov, A. G. (1974). Does the rete ovarii act as a trigger for the onset of meiosis?. *Nature* **252**, 396–397.

Byskov, A. G. (1975). The role of the rete ovarii in meiosis and follicle formation in the cat, mink and ferret. *J. Reprod. Fertil.* **45**, 201–209.

Byskov, A. G. (1978). The anatomy and ultrastructure of the rete system in the fetal mouse ovary. *Biol. Reprod.* **19**, 720–735.

Byskov, A. G., Andersen, C. Y., Nordholm, L., Thogersen, H., Xia, G., Wassmann, O., Andersen, J. V., Guddal, E., and Roed, T. (1995). Chemical structure of sterols that activate oocyte meiosis. *Nature* **374**, 559–562.

Byskov, A. G., Baltsen, M., and Andersen, C. Y. (1998). Meiosis-activating sterols: Background, discovery, and possible use. *J. Mol. Med.* **76**, 818–823.

Byskov, A. G., Fenger, M., Westergaard, L., and Andersen, C. Y. (1993). Forskolin and the meiosis inducing substance synergistically initiate meiosis in fetal male germ cells. *Mol. Hum. Reprod.* **34**, 47–52.

Byskov, A. G., and Lintern-Moore, S. (1973). Follicle formation in the immature mouse ovary: The role of the rete ovarii. *J. Anat.* **116**, 207–217.

Byskov, A. G., and Rasmussen, G. (1973). Ultrastructural studies of the developing follicle. In the development and maturation of the ovary and its functions. *Excerpta Medica Int. Congr. Ser.* **267**, 55–62.

Byskov, A. G., and Saxen, L. (1976). Induction of meiosis in fetal mouse testis in vitro. *Dev. Biol.* **52**, 193–200.

Cederroth, C. R., Pitetti, J. L., Papaioannou, M. D., and Nef, S. (2007). Genetic programs that regulate testicular and ovarian development. *Mol. Cell. Endocrinol.* **265-266**, 3–9.

Chassot, A. A., Gregoire, E. P., Magliano, M., Lavery, R., and Chaboissier, M. C. (2008a). Genetics of ovarian differentiation: Rspo1, a major player. *Sex Dev.* **2**, 219–227.

Chassot, A. A., Ranc, F., Gregoire, E. P., Roepers-Gajadien, H. L., Taketo, M. M., Camerino, G., de Rooij, D. G., Schedl, A., and Chaboissier, M. C. (2008b). Activation of beta-catenin signaling by Rspo1 controls differentiation of the mammalian ovary. *Hum. Mol. Genet.* **17**, 1264–1277.

Chuma, S., and Nakatsuji, N. (2001). Autonomous transition into meiosis of mouse fetal germ cells in vitro and its inhibition by gp130-mediated signaling. *Dev. Biol.* **229**, 468–479.

Coveney, D., Ross, A. J., Slone, J. D., and Capel, B. (2008). A microarray analysis of the XX Wnt4 mutant gonad targeted at the identification of genes involved in testis vascular differentiation. *Gene Expr. Patterns* **8**, 529–537.

Crisponi, L., Deiana, M., Loi, A., Chiappe, F., Uda, M., Amati, P., Bisceglia, L., Zelante, L., Nagaraja, R., Porcu, S., Ristaldi, M. S., Marzella, R., *et al.*, (2001). The putative forkhead transcription factor FOXL2 is mutated in blepharophimosis/ptosis/epicanthus inversus syndrome. *Nat. Genet.* **27**, 159–166.

de Groot, E., Veltmaat, J., Caricasole, A., Defize, L., and van den Eijnden-van Raaij, A. (2000). Cloning and analysis of the mouse follistatin promoter. *Mol. Biol. Rep.* **27**, 129–139.

Descartes, R., L'Homme, et un traitt'e la formation du foetus du mesme autheur. C. Angot, Paris., 1664.

Donovan, P. J., Stott, D., Cairns, L. A., Heasman, J., and Wylie, C. C. (1986). Migratory and postmigratory mouse primordial germ cells behave differently in culture. *Cell* **44**, 831–838.

Dupont, S., Dennefeld, C., Krust, A., Chambon, P., and Mark, M. (2003). Expression of Sox9 in granulosa cells lacking the estrogen receptors, ERalpha and ERbeta. *Dev. Dyn.* **226**, 103–106.

Dupont, S., Krust, A., Gansmuller, A., Dierich, A., Chambon, P., and Mark, M. (2000). Effect of single and compound knockouts of estrogen receptors alpha (ERalpha) and beta (ERbeta) on mouse reproductive phenotypes. *Development* **127**, 4277–4291.

Eicher, E. M., and Washburn, L. L. (1983). Inherited sex reversal in mice: Identification of a new primary sex-determining gene. *J. Exp. Zool.* **228**, 297–304.

Elf, P. K. (2003). Yolk steroid hormones and sex determination in reptiles with TSD. *Gen. Comp. Endocrinol.* **132**, 349–355.

Erickson, G. F., Magoffin, D. A., Dyer, C. A., and Hofeditz, C. (1985). The ovarian androgen producing cells: A review of structure/function relationships. *Endocr. Rev.* **6**, 371–399.

Evans, E. P., Ford, C. E., and Lyon, M. F. (1977). Direct evidence of the capacity of the XY germ cell in the mouse to become an oocyte. *Nature* **267**, 430–431.

Ewen, K., Baker, M., Wilhelm, D., Aitken, R. J., and Koopman, P. (2009). Global survey of protein expression during gonadal sex determination in mice. *Mol. Cell Proteomics* **8**, 2624–2641.

Gondos, B. (1975). Surface epithelium of the developing ovary. Possible correlation with ovarian neoplasia. *Am. J. Pathol.* **81**, 303–321.

Gondos, B., Byskov, A. G., and Hansen, J. L. (1996). Regulation of the onset of meiosis in the developing testis. *Ann. Clin. Lab. Sci.* **26**, 421–425.

Grinsted, J., and Byskov, A. G. (1981). Meiosis-inducing and meiosis-preventing substances in human male reproductive organs. *Fertil. Steril.* **35**, 199–204.

Harvey, W. (1653). *Anatomical Exercitations Concerning the Generation of Living Creatures.* London: Printed by J. Young for O. Pulleyn, 1653.

Heikkila, M., Peltoketo, H., Leppaluoto, J., Ilves, M., Vuolteenaho, O., and Vainio, S. (2002). Wnt-4 deficiency alters mouse adrenal cortex function, reducing aldosterone production. *Endocrinology* **143**, 4358–4365.

Hernandez, R. E., Putzke, A. P., Myers, J. P., Margaretha, L., and Moens, C. B. (2007). Cyp26 enzymes generate the retinoic acid response pattern necessary for hindbrain development. *Development* **134**, 177–187.

Hilscher, B., Hilscher, W., Bulthoff-Ohnolz, B., Kramer, U., Birke, A., Pelzer, H., and Gauss, G. (1974). Kinetics of gametogenesis. I. Comparative histological and autoradiographic studies of oocytes and transitional prospermatogonia during oogenesis and prespermatogenesis. *Cell Tissue Res.* **154**, 443–470.

Hirshfield, A. N. (1991). Theca cells may be present at the outset of follicular growth. *Biol. Reprod.* **44**, 1157–1162.

Houmard, B., Small, C., Yang, L., Naluai-Cecchini, T., Cheng, E., Hassold, T., and Griswold, M. (2009). Global gene expression in the human fetal testis and ovary. *Biol. Reprod.* **81**, 438–443.

Huang, C.C., and H-C, Yao, H. H. C. (2010). Diverse functions of the hedgehog signaling in formation and physiology of steroidogenic organs. *Mol. Reprod. Dev.* **77** 489–496.

Jeays-Ward, K., Hoyle, C., Brennan, J., Dandonneau, M., Alldus, G., Capel, B., and Swain, A. (2003). Endothelial and steroidogenic cell migration are regulated by WNT4 in the developing mammalian gonad. *Development* **130**, 3663–3670.

Johnson, J., Bagley, J., Skaznik-Wikiel, M., Lee, H. J., Adams, G. B., Niikura, Y., Tschudy, K. S., Tilly, J. C., Cortes, M. L., Forkert, R., Spitzer, T., Iacomini, J., *et al.* (2005). Oocyte generation in adult mammalian ovaries by putative germ cells in bone marrow and peripheral blood. *Cell* **122**, 303–315.

Johnson, J., Canning, J., Kaneko, T., Pru, J. K., and Tilly, J. L. (2004). Germline stem cells and follicular renewal in the postnatal mammalian ovary. *Nature* **428**, 145–150.

Jordan, B. K., Mohammed, M., Ching, S. T., Delot, E., Chen, X. N., Dewing, P., Swain, A., Rao, P. N., Elejalde, B. R., and Vilain, E. (2001). Up-regulation of WNT-4 signaling and dosage-sensitive sex reversal in humans. *Am. J. Hum. Genet.* **68**, 1102–1109.

Jordan, B. K., Shen, J. H., Olaso, R., Ingraham, H. A., and Vilain, E. (2003). Wnt4 overexpression disrupts normal testicular vasculature and inhibits testosterone synthesis by repressing steroidogenic factor 1/beta-catenin synergy. *Proc. Natl. Acad. Sci. USA* **100**, 10866–10871.

Jorgensen, J. S., and Gao, L. (2005). Irx3 is differentially up-regulated in female gonads during sex determination. *Gene Expr. Patterns* **5**, 756–762.

Jost, A. (1972). A new look at the mechanisms controlling sex differentiation in mammals. *Johns Hopkins Med. J.* **130**, 38–53.

Jost, A., Gonse-Danysz, P., and Jacquot, R. (1953). Studies on physiology of fetal hypophysis in rabbits and its relation to testicular function.. *J. Physiol. (Paris)* **45**, 134–136.

Kim, K. A., Wagle, M., Tran, K., Zhan, X., Dixon, M. A., Liu, S., Gros, D., Korver, W., Yonkovich, S., Tomasevic, N., Binnerts, M., and Abo, A. (2008). R-spondin family members regulate the Wnt pathway by a common mechanism. *Mol. Biol. Cell* **19**, 2588–2596.

Kim, K. A., Zhao, J., Andarmani, S., Kakitani, M., Oshima, T., Binnerts, M. E., Abo, A., Tomizuka, K., and Funk, W. D. (2006). R-spondin proteins: A novel link to beta-catenin activation. *Cell Cycle* **5**, 23–26.

Kobayashi, T., and Nagahama, Y. (2009). Molecular aspects of gonadal differentiation in a teleost fish, the nile tilapia. *Sex Dev.* **3**, 108–117.

Kotsuji, F., Kamitani, N., Goto, K., and Tominaga, T. (1990). Bovine theca and granulosa cell interactions modulate their growth, morphology, and function. *Biol. Reprod.* **43**, 726–732.

Koubova, J., Menke, D. B., Zhou, Q., Capel, B., Griswold, M. D., and Page, D. C. (2006). Retinoic acid regulates sex-specific timing of meiotic initiation in mice. *Proc. Natl. Acad. Sci. USA* **103**, 2474–2479.

Le Grand, A. (1672). Institutio Philosophiae, Secundum Principia Renati Descartes Londini.

Lee, T. L., Li, Y., Cheung, H. H., Claus, J., Singh, S., Sastry, C., Rennert, O. M., Lau, Y. F., and Chan, W. Y. (2009). GonadSAGE: A comprehensive SAGE database for transcript discovery on male embryonic gonad development. *Bioinformatics* **26**(4), 585–586.

Lewis, W., Rosa, B., Jeremy, B., and Thomas, J. (2006). *Principles of Development.* Oxford University Press, USA.

Lin, Y., Gill, M. E., Koubova, J., and Page, D. C. (2008). Germ cell-intrinsic and -extrinsic factors govern meiotic initiation in mouse embryos. *Science (New York)* **322**, 1685–1687.

Liu, C. F., Bingham, N., Parker, K., and Yao, H. H. (2009). Sex-specific roles of beta-catenin in mouse gonadal development. *Hum. Mol. Genet.* **18**, 405–417.

Liu, C. F., Parker, K., and Yao, H.H.-C. (2010). Wnt4/beta-catenin pathway maintains female germ cell survival by inhibiting activin beta B in the mouse fetal ovary. *PLoS One* **5**, e10382.

Loffler, K. A., Zarkower, D., and Koopman, P. (2003). Etiology of ovarian failure in blepharophimosis ptosis epicanthus inversus syndrome: FOXL2 is a conserved, early-acting gene in vertebrate ovarian development. *Endocrinology* **144**, 3237–3243.

Lu, W., Kim, K. A., Liu, J., Abo, A., Feng, X., Cao, X., and Li, Y. (2008). R-spondin1 synergizes with Wnt3a in inducing osteoblast differentiation and osteoprotegerin expression. *FEBS Lett.* **582**, 643–650.

Maatouk, D. M., DiNapoli, L., Alvers, A., Parker, K. L., Taketo, M. M., and Capel, B. (2008). Stabilization of beta-catenin in XY gonads causes male-to-female sex-reversal. *Hum. Mol. Genet.* **17**, 2949–2955.

MacLean, G., Li, H., Metzger, D., Chambon, P., and Petkovich, M. (2007). Apoptotic extinction of germ cells in testes of cyp26b1 knockout mice. *Endocrinology* **148**, 4560–4567.

Magoffin, D. A., and Magarelli, P. C. (1995). Preantral follicles stimulate luteinizing hormone independent differentiation of ovarian theca-interstitial cells by an intrafollicular paracrine mechanism. *Endocrine* **3**, 107–112.

Mandel, H., Shemer, R., Borochowitz, Z. U., Okopnik, M., Knopf, C., Indelman, M., Drugan, A., Tiosano, D., Gershoni-Baruch, R., Choder, M., and Sprecher, E. (2008). SERKAL syndrome: An autosomal-recessive disorder caused by a loss-of-function mutation in WNT4. *Am. J. Hum. Genet.* **82**, 39–47.

Manuylov, N. L., Smagulova, F. O., Leach, L., and Tevosian, S. G. (2008). Ovarian development in mice requires the GATA4-FOG2 transcription complex. *Development* **135**, 3731–3743.

McElreavey, K., Vilain, E., Abbas, N., Herskowitz, I., and Fellous, M. (1993). A regulatory cascade hypothesis for mammalian sex determination: SRY represses a negative regulator of male development. *Proc. Natl. Acad. Sci. USA* **90**, 3368–3372.

McLaren, A. (1983). Studies on mouse germ cells inside and outside the gonad. *J. Exp. Zool.* **228**, 167–171.

McLaren, A. (1984). Meiosis and differentiation of mouse germ cells. *Symp. Soc. Exp. Biol.* **38**, 7–23.

McLaren, A. (1991). Development of the mammalian gonad: The fate of the supporting cell lineage. *BioEssays* **13**, 151–156.

McLaren, A. (1995). Germ cells and germ cell sex. *Philos. Trans. R. Soc. Lond, Ser. B Biol. Sci.* **350**, 229–233.

McLaren, A. (2000). Germ and somatic cell lineages in the developing gonad. *Mol. Cell. Endocrinol.* **163**, 3–9.

McLaren, A. (2003). Primordial germ cells in the mouse. *Dev. Biol.* **262**, 1–15.

McLaren, A., Chandley, A. C., and Kofman-Alfaro, S. (1972). A study of meiotic germ cells in the gonads of foetal mouse chimaeras. *J. Embryol. Exp. Morphol.* **27**, 515–524.

McLaren, A., and Southee, D. (1997). Entry of mouse embryonic germ cells into meiosis. *Dev. Biol.* **187**, 107–113.

Meeks, J. J., Crawford, S. E., Russell, T. A., Morohashi, K., Weiss, J., and Jameson, J. L. (2003a). Dax1 regulates testis cord organization during gonadal differentiation. *Development* **130**, 1029–1036.

Meeks, J. J., Russell, T. A., Jeffs, B., Huhtaniemi, I., Weiss, J., and Jameson, J. L. (2003b). Leydig cell-specific expression of DAX1 improves fertility of the dax1-deficient mouse. *Biol. Reprod.* **69**, 154–160.

Meeks, J. J., Weiss, J., and Jameson, J. L. (2003c). Dax1 is required for testis determination. *Nat. Genet.* **34**, 32–33.

Menke, D. B., and Page, D. C. (2002). Sexually dimorphic gene expression in the developing mouse gonad. *Gene Expr. Patterns* **2**, 359–367.

Merchant, H. (1975). Rat gonadal and ovarioan organogenesis with and without germ cells. An ultrastructural study. *Dev. Biol.* **44**, 1–21.

Merchant-Larios, H., and Chimal-Monroy, J. (1989). The ontogeny of primordial follicles in the mouse ovary. *Prog. Clin. Biol. Res.* **296**, 55–63.

Mittwoch, U. (1998). Phenotypic manifestations during the development of the dominant and default gonads in mammals and birds. *J. Exp. Zool.* **281**, 466–471.

Molyneaux, K. A., Stallock, J., Schaible, K., and Wylie, C. (2001). Time-lapse analysis of living mouse germ cell migration. *Dev. Biol.* **240**, 488–498.

Morris, T. (1968). The XO and OY chromosome constitutions in the mouse. *Genet. Res.* **12**, 125–137.

Motta, P. M., and Makabe, S. (1982). Development of the ovarian surface and associated germ cells in the human fetus. A correlated study by scanning and transmission electron microscopy. *Cell Tissue Res.* **226**, 493–510.

Muttukrishna, S., Tannetta, D., Groome, N., and Sargent, I. (2004). Activin and follistatin in female reproduction. *Mol. Cell. Endocrinol.* **225**, 45–56.

Mystkowska, E. T., and Tarkowski, A. K. (1970). Behaviour of germ cells and sexual differentiation in late embryonic and early postnatal mouse chimeras. *J. Embryol. Exp. Morphol.* **23**, 395–405.

Naillat, F., Prunskaite-Hyyrylainen, R., Pietila, I., Sormunen, R., Jokela, T., Shan, J., and Vainio, S. J. (2010). Wnt4/5a signalling coordinates cell adhesion and entry into meiosis during presumptive ovarian follicle development. *Hum. Mol. Genet.* **19**(8), 1539–1550.

Nakamura, M. (2009). Sex determination in amphibians. *Semin. Cell Dev. Biol.* **20**, 271–282.

Nef, S., Schaad, O., Stallings, N. R., Cederroth, C. R., Pitetti, J. L., Schaer, G., Malki, S., Dubois-Dauphin, M., Boizet-Bonhoure, B., Descombes, P., Parker, K. L., and Vassalli, J. D.

(2005). Gene expression during sex determination reveals a robust female genetic program at the onset of ovarian development. *Dev. Biol.* **287**, 361–377.

Ohno, S., and Cattanach, B. M. (1962). Cytological study of an X-autosome translocation in mus musculus. *Cytogenetics* **1**, 129–140.

Orisaka, M., Tajima, K., Mizutani, T., Miyamoto, K., Tsang, B. K., Fukuda, S., Yoshida, Y., and Kotsuji, F. (2006). Granulosa cells promote differentiation of cortical stromal cells into theca cells in the bovine ovary. *Biol. Reprod.* **75**, 734–740.

Ottolenghi, C., Omari, S., Garcia-Ortiz, J. E., Uda, M., Crisponi, L., Forabosco, A., Pilia, G., and Schlessinger, D. (2005). Foxl2 is required for commitment to ovary differentiation. *Hum. Mol. Genet.* **14**, 2053–2062.

Ottolenghi, C., Pelosi, E., Tran, J., Colombino, M., Douglass, E., Nedorezov, T., Cao, A., Forabosco, A., and Schlessinger, D. (2007). Loss of Wnt4 and Foxl2 leads to female-to-male sex reversal extending to germ cells. *Hum. Mol. Genet.* **16**, 2795–2804.

Pailhoux, E., Vigier, B., Chaffaux, S., Servel, N., Taourit, S., Furet, J. P., Fellous, M., Grosclaude, F., Cribiu, E. P., Cotinot, C., and Vaiman, D. (2001). A 11.7-Kb deletion triggers intersexuality and polledness in goats. *Nat. Genet.* **29**, 453–458.

Pailhoux, E., Vigier, B., Vaiman, D., Servel, N., Chaffaux, S., Cribiu, E. P., and Cotinot, C. (2002). Ontogenesis of female-to-male sex-reversal in XX polled goats. *Dev. Dyn.* **224**, 39–50.

Palmer, S. J., and Burgoyne, P. S. (1991). In situ analysis of fetal, prepuberal and adult XX–XY chimaeric mouse testes: Sertoli cells are predominantly, but not exclusively, XY. *Development (Cambridge, England)* **112**, 265–268.

Pannetier, M., Fabre, S., Batista, F., Kocer, A., Renault, L., Jolivet, G., Mandon-Pepin, B., Cotinot, C., Veitia, R., and Pailhoux, E. (2006). FOXL2 activates P450 aromatase gene transcription: Towards a better characterization of the early steps of mammalian ovarian development. *J. Mol. Genet. Med.* **36**, 399–413.

Parma, P., Radi, O., Vidal, V., Chaboissier, M. C., Dellambra, E., Valentini, S., Guerra, L., Schedl, A., and Camerino, G. (2006). R-spondin1 is essential in sex determination, skin differentiation and malignancy. *Nat. Genet.* **38**, 1304–1309.

Pask, A., and Renfree, M. B. (2001). Sex determining genes and sexual differentiation in a marsupial. *J. Exp. Zool.* **290**, 586–596.

Pepling, M. E., and Spradling, A. C. (1998). Female mouse germ cells form synchronously dividing cysts. *Development* **125**, 3323–3328.

Philibert, P., Biason-Lauber, A., Rouzier, R., Pienkowski, C., Paris, F., Konrad, D., Schoenle, E., and Sultan, C. (2008). Identification and functional analysis of a new WNT4 gene mutation among 28 adolescent girls with primary amenorrhea and müllerian duct abnormalities: a French collaborative study. *J. Clin. Endocrinol. Metab.* **93**, 895–900.

Pinkerton, J. H., Mc, K. D., Adams, E. C., and Hertig, A. T. (1961). Development of the human ovary—a study using histochemical technics. *Obstet. Gynecol.* **18**, 152–181.

Quattropani, S. L. (1973). Morphogenesis of the ovarian interstitial tissue in the neonatal mouse. *Anat. Rec.* **177**, 569–583.

Romand, R., Dolle, P., and Hashino, E. (2006). Retinoid signaling in inner ear development. *J. Neuroendocrinol.* **66**, 687–704.

Ruby, J. R., Dyer, R. F., and Skalko, R. G. (1969). The occurrence of intercellular bridges during oogenesis in the mouse. *J. Morphol.* **127**, 307–339.

Sanlaville, D., Vialard, F., Thepot, F., Vue-Droy, L., Ardalan, A., Nizard, P., Corre, A., Devauchelle, B., Martin-Denavit, T., Nouchy, M., Malan, V., Taillemite, J. L., and Portnoi, M. F. (2004). Functional disomy of Xp including duplication of DAX1 gene with sex reversal due to t(X;Y)(p21.2;p11.3). *Am. J. Med. Genet. A.* **128A**, 325–330.

Sawyer, H. R., Smith, P., Heath, D. A., Juengel, J. L., Wakefield, S. J., and McNatty, K. P. (2002). Formation of ovarian follicles during fetal development in sheep. *Biol. Reprod.* **66**, 1134–1150.

Schmidt, D., Ovitt, C. E., Anlag, K., Fehsenfeld, S., Gredsted, L., Treier, A. C., and Treier, M. (2004). The murine winged-helix transcription factor Foxl2 is required for granulosa cell differentiation and ovary maintenance. *Development* **131**, 933–942.

Singh, R. P., and Carr, D. H. (1966). The anatomy and histology of XO human embryos and fetuses. *Anat. Rec.* **155**, 369–383.

Singh, R. P., and Carr, D. H. (1967). Anatomic findings in human abortions of known chromosomal constitution. *Obstet. Gynecol.* **29**, 806–818.

Speed, R. M. (1982). Meiosis in the foetal mouse ovary. I. An analysis at the light microscope level using surface-spreading. *Chromosoma* **85**, 427–437.

Sultan, C., Biason-Lauber, A., and Philibert, P. (2009). Mayer-Rokitansky-Kuster-Hauser syndrome: recent clinical and genetic findings. *Gynecol. Endocrinol.* **25**, 8–11.

Suzuki, A., Igarashi, K., Aisaki, K., Kanno, J., and Saga, Y. (2010). NANOS2 interacts with the CCR4-NOT deadenylation complex and leads to suppression of specific RNAs. *Proc. Natl. Acad. Sci. USA* **107**, 3594–3599.

Suzuki, A., and Saga, Y. (2008). Nanos2 suppresses meiosis and promotes male germ cell differentiation. *Genes Dev.* **22**, 430–435.

Swain, A., Narvaez, V., Burgoyne, P., Camerino, G., and Lovell-Badge, R. (1998). Dax1 antagonizes Sry action in mammalian sex determination. *Nature* **391**, 761–767.

Tam, P. P., and Snow, M. H. (1981). Proliferation and migration of primordial germ cells during compensatory growth in mouse embryos. *J. Embryol. Exp. Morphol.* **64**, 133–147.

Tomizuka, K., Horikoshi, K., Kitada, R., Sugawara, Y., Iba, Y., Kojima, A., Yoshitome, A., Yamawaki, K., Amagai, M., Inoue, A., Oshima, T., and Kakitani, M. (2008). R-spondin1 plays an essential role in ovarian development through positively regulating Wnt-4 signaling. *Hum. Mol. Genet.* **17**, 1278–1291.

Trautmann, E., Guerquin, M. J., Duquenne, C., Lahaye, J. B., Habert, R., and Livera, G. (2008). Retinoic acid prevents germ cell mitotic arrest in mouse fetal testes. *Cell Cycle (Georgetown Tex.).* **7**, 656–664.

Tsuda, M., Sasaoka, Y., Kiso, M., Abe, K., Haraguchi, S., Kobayashi, S., and Saga, Y. (2003). Conserved role of nanos proteins in germ cell development. *Sciences (New York)* **301**, 1239–1241.

Uhlenhaut, N. H., Jakob, S., Anlag, K., Eisenberger, T., Sekido, R., Kress, J., Treier, A. C., Klugmann, C., Klasen, C., Holter, N. I., Riethmacher, D., Schutz, G., *et al.* (2009). Somatic sex reprogramming of adult ovaries to testes by FOXL2 ablation. *Cell* **139**, 1130–1142.

Upadhyay, S., and Zamboni, L. (1982). Ectopic germ cells: Natural model for the study of germ cell sexual differentiation. *Proc. Natl. Acad. Sci. USA* **79**, 6584–6588.

Vainio, S., Heikkila, M., Kispert, A., Chin, N., and McMahon, A. P. (1999). Female development in mammals is regulated by Wnt-4 signalling. *Nature* **397**, 405–409.

vonBaer (1827). Epistola de Ovo Mammalium et Hominis Genesi (On the Origin of the Mammalian and Human Ovum).

Wang, D., Kobayashi, T., Zhou, L., and Nagahama, Y. (2004). Molecular cloning and gene expression of Foxl2 in the nile tilapia, oreochromis niloticus. *Biochem. Biophys. Res. Commun.* **320**, 83–89.

Wang, D. S., Kobayashi, T., Zhou, L. Y., Paul-Prasanth, B., Ijiri, S., Sakai, F., Okubo, K., Morohashi, K., and Nagahama, Y. (2007). Foxl2 up-regulates aromatase gene transcription in a female-specific manner by binding to the promoter as well as interacting with ad4 binding protein/steroidogenic factor 1. *Mol. Endocrinol.* **21**, 712–725.

Washburn, L. L., and Eicher, E. M. (1983). Sex reversal in XY mice caused by dominant mutation on chromosome 17. *Nature* **303**, 338–340.

Wei, Q., Yokota, C., Semenov, M. V., Doble, B., Woodgett, J., and He, X. (2007). R-spondin1 is a high affinity ligand for LRP6 and induces LRP6 phosphorylation and beta-catenin signaling. *J. Biol. Chem.* **282**, 15903–15911.

Welshons, W. J., and Russell, L. B. (1959). The Y-chromosome as the bearer of male determining factors in the mouse. *Proc. Natl. Acad. Sci. USA* **45**, 560–566.

Wharton, T. (1656), Adenographia: Sive glandularum totius corporis descriptio. Londini, typ. J. G. Impens. Authoris.

White, J. A., Ramshaw, H., Taimi, M., Stangle, W., Zhang, A., Everingham, S., Creighton, S., Tam, S. P., Jones, G., and Petkovich, M. (2000). Identification of the human cytochrome P450, P450RAI-2, which is predominantly expressed in the adult cerebellum and is responsible for all-trans-retinoic acid metabolism. *Proc. Natl. Acad. Sci. USA* **97**, 6403–6408.

Willert, J., Epping, M., Pollack, J. R., Brown, P. O., and Nusse, R. (2002). A transcriptional response to Wnt protein in human embryonic carcinoma cells. *BMC Dev. Biol.* **2**, 8.

Yao, H. H. (2005). The pathway to femaleness: Current knowledge on embryonic development of the ovary. *Mol. Cell. Endocrinol.* **230**, 87–93.

Yao, H. H., Aardema, J., and Holthusen, K. (2006). Sexually dimorphic regulation of inhibin beta B in establishing gonadal vasculature in mice. *Biol. Reprod.* **74**, 978–983.

Yao, H. H., and Capel, B. (2005). Temperature, genes, and sex: A comparative view of sex determination in trachemys scripta and mus musculus. *J. Biochem.* **138**, 5–12.

Yao, H. H., DiNapoli, L., and Capel, B. (2003). Meiotic germ cells antagonize mesonephric cell migration and testis cord formation in mouse gonads. *Development* **130**, 5895–5902.

Yao, H. H., Matzuk, M. M., Jorgez, C. J., Menke, D. B., Page, D. C., Swain, A., and Capel, B. (2004). Follistatin operates downstream of Wnt4 in mammalian ovary organogenesis. *Dev. Dyn.* **230**, 210–215.

Yao, H.H.C., Whoriskey, W., and Capel, B. (2002). Desert hedgehog/patched-1 signaling specifies early Leydig cell fate in testis organogenesis. *Genes Dev.* **16**, 1433–1440.

Yashiro, K., Zhao, X., Uehara, M., Yamashita, K., Nishijima, M., Nishino, J., Saijoh, Y., Sakai, Y., and Hamada, H. (2004). Regulation of retinoic acid distribution is required for proximodistal patterning and outgrowth of the developing mouse limb. *Dev. Cell* **6**, 411–422.

Yu, R. N., Ito, M., Saunders, T. L., Camper, S. A., and Jameson, J. L. (1998). Role of Ahch in gonadal development and gametogenesis. *Nat. Genet.* **20**, 353–357.

Zamboni, L., Bezard, J., and Mauleon, P. (1975). The role of the mesonephros in the development of the sheep fetal ovary. *Ann. Biol. Anim. Biochem. Biophys.* **19**, 1153–1178.

Zamboni, L., and Upadhyay, S. (1983). Germ cell differentiation in mouse adrenal glands. *J. Exp. Zool.* **228**, 173–193.

Zanaria, E., Muscatelli, F., Bardoni, B., Strom, T. M., Guioli, S., Guo, W., Lalli, E., Moser, C., Walker, A. P., McCabe, E. R., *et al.*, (1994). An unusual member of the nuclear hormone receptor superfamily responsible for X-linked adrenal hypoplasia congenita. *Nature* **372**, 635–641.

Zou, K., Yuan, Z., Yang, Z., Luo, H., Sun, K., Zhou, L., Xiang, J., Shi, L., Yu, Q., Zhang, Y., Hou, R., and Wu, J. (2009). Production of offspring from a germline stem cell line derived from neonatal ovaries. *Nat. Cell Biol.* **11**, 631–636.

Zuckerman, S. (1951). *Rec. Prog. Horm. Res.* **6**, 63–108.

CHAPTER EIGHT

Vertebrate Skeletogenesis

Véronique Lefebvre *and* Pallavi Bhattaram

Contents

1. Introduction 292
2. Structural Organization and Advantages of the Vertebrate Skeleton 293
3. Development of Skeletogenic Cells 294
4. Development of Cartilage Anlagen 298
5. Development of Cartilage Growth Plates 300
6. Bone Development 302
7. Synovial Joint Formation 306
8. Skeleton Variation 309
9. Perspectives 310
Acknowledgments 311
References 311

Abstract

Vertebrate skeletogenesis consists in elaborating an edifice of more than 200 pieces of bone and cartilage. Each skeletal piece is crafted at a distinct location in the body, is articulated with others, and reaches a specific size, shape, and tissue composition according to both species instructions and individual determinants. This complex, customized body frame fulfills multiple essential tasks. It confers morphological features, allows controlled postures and movements, protects vital organs, houses hematopoiesis, stores minerals, and adsorbs toxins. This review provides an overview of the multiple facets of this ingenious process for experts as well as nonexperts of skeletogenesis. We explain how the developing vertebrate uses both specific and ubiquitously expressed genes to generate multipotent mesenchymal cells, specify them to a skeletogenic fate, control their survival and proliferation, and direct their differentiation into cartilage, bone, and joint cells. We review milestone discoveries made toward uncovering the intricate networks of regulatory factors that are involved in these processes, with an emphasis on signaling pathways and transcription factors. We describe numerous skeletal malformation and degeneration diseases that occur in humans as a result of mutations in regulatory genes, and

Department of Cell Biology, and Orthopaedic and Rheumatologic Research Center, Lerner Research Institute, Cleveland Clinic, Euclid Avenue, Cleveland, Ohio, USA

Current Topics in Developmental Biology, Volume 90
ISSN 0070-2153, DOI 10.1016/S0070-2153(10)90008-2

explain how these diseases both help and motivate us to further decipher skeletogenic processes. Upon discussing current knowledge and gaps in knowledge in the control of skeletogenesis, we highlight ultimate research goals and propose research priorities and approaches for future endeavors.

1. Introduction

The development of a skeleton made of cartilage, bone, and joints is a novel process that has critically contributed to the emergence of vertebrates (Ota *et al.*, 2009). Its importance is reflected in the word "vertebrate," which means "having a vertebral column" or "having joints." Both meanings are justified. The vertebral column is indeed the skeletal feature shared by all vertebrates, as other skeletal elements were acquired later, were never acquired, or were lost in some vertebrate species. Joints are also characteristic features of the vertebrate skeleton, and the vertebral column uses the notochord, a primitive skeletal structure in vertebrate ancestors and embryos, to form the core part (nucleus pulposus) of its joints intervertebral discs (IVD). The rigid, articulated elements of the vertebrate skeleton permit vertebrates deliberate postures and movements. The subsequent acquisition of a skull, jaw, and appendicular skeleton allowed vertebrates to develop a defined brain, face, and limbs. The thoracic cage and marrow space evolved to protect the brain, hematopoietic tissue, and other organs. Furthermore, bones became mineral reserves and toxin clearance centers. Interestingly, being aware that the skeleton has key physical roles, ancient cultures even endowed it with spiritual meaning and thought that it housed the soul. Skeletogenesis is thus an essential process in the development of vertebrates.

Skeletogenesis starts in the vertebrate embryo once multipotent mesenchymal cells arise from ectoderm and mesoderm, migrate to specific locations in the body, and commit to a skeletal fate. Most skeletogenic cells later develop into cartilage cells (chondrocytes), bone cells (osteoblasts), or joint cells (mainly articular chondrocytes and synovial cells), while some may persist as mesenchymal stem cells throughout life. The primary skeleton is entirely cartilaginous. It grows quickly and most of it is progressively replaced by bone throughout fetal and postnatal growth. The process is called endochondral ossification. Concomitantly, joints and additional bones form. The latter develop upon a mesenchymal template, without cartilage intermediate, through a process called intramembranous ossification. Bone, cartilage, and joints differ in composition and regulation, but their associated developments are tightly coordinated. Our knowledge of the cellular and molecular events that govern skeletogenesis has greatly increased over the last two decades thanks to the identification of disease-causing mutations, gene manipulations in animals, and novel molecular and cellular approaches. It is now clear that an amazingly large number of factors are involved in skeletogenesis. In fact, no other process, except perhaps brain development, may recruit as many factors. These factors are hormones, growth factors, receptors, signaling mediators, transcription factors, extracellular matrix components, and enzymes. Factors determining the identity of skeletal cells are called differentiation factors, and factors specifying the number, size, and shape of skeletal elements are called patterning factors. The latter greatly outnumber the former and contribute to the incredible skeletal variations that exist between individuals both within and between species.

A corollary of the importance and complexity of the vertebrate skeleton is a high frequency and diversity of severe skeleton malformation diseases in humans (Spranger, 2006). The incidence of these diseases is estimated at 1 in 4000 births, with half of them being early lethal. Their true incidence, however, may be twice as high because many develop only years after birth. Osteochondrodysplasias are generalized skeleton malformation diseases, whereas dysostoses affect only a specific subset of skeletal elements. As genes controlling skeletogenesis are often involved in several processes, their alterations can cause complex syndromes, of which skeletal disease is only one component. In addition, it must be noted that skeletal variations at the edge of normalcy are often inconsequential at a young age, but constitute high-risk factors for skeleton degeneration diseases later in life, such as osteoarthritis [articular cartilage (AC) degeneration] and osteoporosis (bone loss disease).

Vertebrate skeletogenesis is thus a fascinating process to study for developmental and evolutionary biologists who want to understand how the skeleton and its variations are generated in developing vertebrates. Furthermore, it is also a mandatory process to study for geneticists and clinician scientists who want to decipher the molecular basis of skeletal diseases in humans and develop greatly needed therapies for these diseases. We review here key aspects of the vertebrate skeleton composition and development. We analyze milestone discoveries made toward understanding mechanisms underlying skeletogenic cell fate determination, chondrogenesis, osteogenesis, joint formation, and individual skeletal variations. We discuss current knowledge and gaps in knowledge, and end with suggestions for important goals and approaches for future research.

2. Structural Organization and Advantages of the Vertebrate Skeleton

Invertebrates have no skeleton or a skeleton made exclusively of minerals (e.g., calcium carbonate or silica) or carbohydrates (e.g., chitin). The vertebrate skeleton, in contrast, is made of specialized connective tissues, i.e., structures composed of cells embedded in an abundant extracellular matrix. The invertebrate and vertebrate skeletons are thus very different from each other and this can be explained by independent evolutions (Ota *et al.*, 2009).

Cartilage and bone are built and populated by chondrocytes and osteoblasts, respectively. Cartilage is capable of growing rapidly and is, therefore, profuse in embryos and youngsters. It is also highly resilient and thus maintained in adult skeletal sites subject to load, such as airways and synovial joints. Bone is very strong mechanically and thus preponderant in the adult skeleton. Both tissues are constructed upon a collagenous fibrillar network that confers structure and tensile strength. Collagen-2 is a very abundant and specific protein in cartilage, and collagen-1 is the major protein in bone. The cartilage collagen network entraps a highly charged gel of aggrecan and other proteoglycans, which confer tissue resiliency and control the diffusion of growth factors. Bone has a low content of proteoglycan, but contains specific proteins, such as osteocalcin and bone sialoprotein. These proteins control the deposition of hydroxyapatite, the main calcium phosphate mineral of the bone matrix. Its high collagen and mineral content confer on bone its unique mechanical and mineral homeostatic functions.

As they invented bone and cartilage elements, vertebrates also devised joint structures to interconnect these elements. They created synovial joints between limb elements to be able to display great ranges of motions; IVD to confer pliability to the vertebral column; and fibrous joints to minimize movements between skull bones. These different types of joints are made of distinct tissues. Synovial joints, for instance, consist of AC coating bone surfaces, a fibrous capsule surrounding the joint and sealing its cavity, a synovial membrane lining the capsule internally and producing a lubricating fluid, tendons transmitting muscle force to bones, and ligaments stabilizing the joints by connecting bones. The vertebrate skeleton is thus a complex and highly advantageous edifice.

3. Development of Skeletogenic Cells

The first step in skeletogenesis consists in generating skeletogenic cells. The origin of these cells can be tracked back to the onset of organogenesis, when the vertebrate embryo is composed of three germ layers: ectoderm, mesoderm, and endoderm (Fig. 8.1A). These layers transform themselves into multiple early derivatives. These include the ectoderm-derived neural tube, mesoderm-derived notochord, paraxial mesoderm, and lateral plate mesoderm, which give rise to the skeleton as well as other organs. The neural crest is a population of cells that delaminates from the neural tube, undergoes epithelial-to-mesenchymal transformation, and migrates to

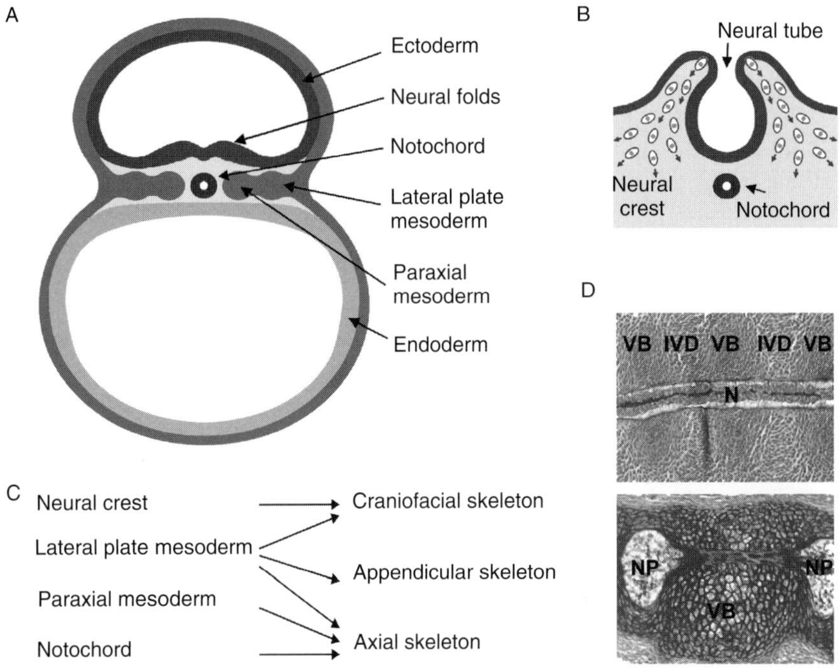

Figure 8.1 Origin of skeletal cells in the vertebrate embryo. (A) Schematic of a cross-section through a mouse embryo soon after gastrulation at day 8 of development (equivalent to gestation day 17 in humans). The three germ layers are seen: ectoderm, endoderm, and mesoderm. Ectoderm-derived neural folds are rising. The mesoderm has formed the notochord and is starting to form lateral plate and paraxial derivatives on either sides of the midline. (B) Schematic showing the delamination of neural crest cells from the neural folds at the time of neural tube closure. These cells are starting to migrate inside the embryo (blue arrows), where they will participate in the formation of various structures. (C) Schematic showing the contribution of the neural crest, lateral plate mesoderm, paraxial mesoderm, and notochord to the three major parts of the skeleton. (D) Mid-sagittal sections through the notochord of mouse embryos at the gestation days 12.5 (E12.5, top) and E15.5 (bottom). The E12.5 notochord is a rod-like structure that becomes surrounded by the mesenchymal cell condensations of the prospective vertebral bodies (VB) and IVD. E15.5 VB are cartilaginous and notochord cells have migrated into the intervertebral disc spaces, where they have formed NP. Sections are stained with nuclear fast red and with Alcian blue, which is specific of the notochord and cartilage extracellular matrix. (See Color Insert.)

numerous locations in the embryo (Fig. 8.1B). Once there, the cells develop into various types, such as neuronal cells, melanocytes, and skeletogenic cells. The latter give rise to several throat and craniofacial skeletal elements (Fig. 8.1C). The lateral plate mesoderm gives rise to the other craniofacial skeletal structures, the limb skeletal elements (appendicular skeleton), the sternum (part of the axial skeleton), and the non-skeletal structures.

The paraxial mesoderm gives rise to somites, which develop into dermomyotomes and sclerotomes. The latter form most of the axial skeleton, i.e., the ribs and vertebrae. As they develop around the notochord, the vertebrae force the notochord cells to change phenotype, migrate to the intervertebral spaces, and develop the nuclei pulposi (NP) of IVD (Fig. 8.1D). Many factors control neural crest and mesoderm cells before they reach skeletal sites. We will not review them, but suggest excellent recent reviews on this topic (Limura *et al.*, 2009; Sauka-Spengler and Bronner-Fraser, 2008).

Upon settling in skeletal sites, neural crest- and mesoderm-derived cells produce a matrix rich in collagen-1, fibronectin, and hyaluronan, and they proliferate or die in a tightly controlled spatial and temporal manner (Li *et al.*, 2007; reviewed by Shum *et al.*, 2003). They thereby establish mesenchymal structures that prefigure the future skeletal elements. They are often called osteochondroprogenitors because most of them give rise to osteoblasts and chondrocytes, and are able to switch fate under specific conditions (Fig. 8.2). Some, however, give rise to synovial cells, tenocytes, bone marrow stromal cells, endothelial cells, and presumably mesenchymal stem cells. We, therefore, prefer to call them skeletogenic cells.

As of today, it is unknown whether master transcription factors control the identity and maintenance of these cells in all sites, but Sox4, Sox11, and Sox12, which form the group C of Sry-related HMG box transcription factors, are strong candidates, as they are expressed in all embryonic mesenchymal cells and are required for the survival of these cells (Bhattaram *et al.*, 2010; Dy *et al.*, 2008). Numerous factors are known, in contrast, that participate in the formation of skeletogenic templates in a site- and time-specific manner. These patterning regulators include transcription factors from many families and most were identified through dysostosis-causing mutations (reviewed by Hermanns *et al.*, 2001). Our knowledge of their exact roles and targets remains incomplete, as studies are complicated by a high degree of redundancy existing between co-expressed factors. Patterning factors are too numerous to be reviewed in detail here. Instead, we will focus on the Hox proteins, which were among the first ones to be discovered and which spectacularly illustrate the importance of patterning factors (reviewed by Wellik, 2009).

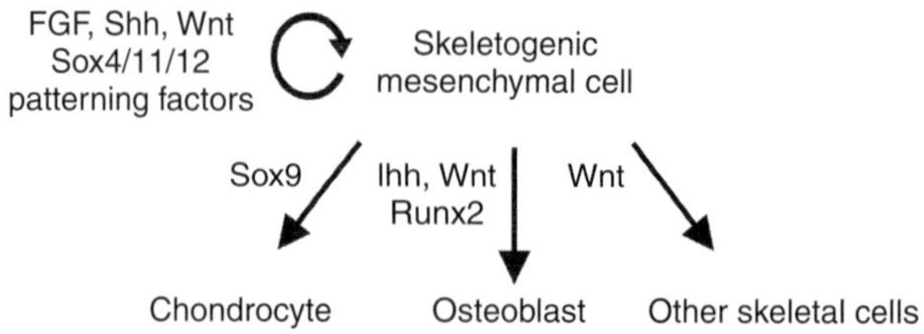

Figure 8.2 Fate and molecular control of skeletogenic mesenchymal cells.

Hox proteins are encoded by a family of 39 genes in humans and most other vertebrates, and these genes are distributed in four paralogous clusters. The position of each gene in a cluster correlates with its expression pattern along the anteroposterior axis of the trunk, limb, and head. Hox proteins feature a DNA-binding domain called the homeodomain, in reference to the fact that *Hox* mutations often cause homeotic transformations. For instance, inactivation of the mouse *Hox9* genes results in transformations of the posterior thoracic vertebrae such that up to 14 pairs of full-length ribs form and attach to the sternum instead of 7 (McIntyre *et al.*, 2007). The fact that all four *Hox9* genes (*a, b, c,* and *d*) need to be inactivated to generate this phenotype demonstrates the high degree of redundancy existing between *Hox* paralogues. Despite redundancy, single *HOX* gene mutations can cause severe skeletal malformations in humans. For instance, amplification of an alanine stretch in *HOXD13* causes syndactyly (fusion of digits), polydactyly (extranumerary digits), brachydactyly (short digits), and transformation of metacarpals and metatarsals to carpals and tarsals, respectively (reviewed by Zhao *et al.*, 2007). A similar phenotype was observed in mice with an equivalent mutation and was proposed to result from a dominant-negative action of the mutant protein. These examples thus highlight how locally but yet powerfully patterning transcription factors participate in the design of skeletal structures.

Complex networks of morphogens control the expression and activity of patterning transcription factors. These morphogens include but are not limited to Sonic hedgehog (Shh), fibroblast growth factors (FGFs), Wnt ligands, bone morphogenetic proteins (BMPs), and retinoic acid. Their roles and interactions are reviewed in an accompanying paper with a special emphasis on the developing limb (reviewed by Butterfield *et al.*, 2010).

The lineage potential of skeletogenic cells is established early, as revealed by expression of master transcription factors, such as the chondrogenic factor Sox9 and the osteogenic Runt-domain transcription factor Runx2 (Ducy *et al.*, 1997; Eames *et al.*, 2004; Ng *et al.*, 1997). By definition, however, skeletogenic cells are undifferentiated. Mechanisms must, therefore, be in place to keep them as such. Several have been identified. For instance, Wnt/beta-catenin signaling is able to block the activity of Sox9 (Akiyama *et al.*, 2004; Day *et al.*, 2005; Hill *et al.*, 2005), and both Sox9 and the Twist1/2 homeodomain transcription factor can physically interact with Runx2 to block its activity (Bialek *et al.*, 2004; Zhou *et al.*, 2006). Moreover, and as further described later, lack of expression of Sox5 and Sox6, two potent partners of Sox9 in chondrogenesis, helps keep chondrogenic cells undifferentiated, while lack of expression of Osx, a zinc-finger transcription factor required for osteoblast differentiation, keeps osteogenic cells undifferentiated. The generation of skeletogenic cells and the proper control of their fate are thus critical early steps in the formation of the vertebrate skeleton.

4. Development of Cartilage Anlagen

Chondrogenesis, the fate of most skeletogenic cells, results in the construction of a multitude of cartilage anlagen, which altogether constitute the primary skeleton of the vertebrate embryo (Fig. 8.3A). This process occurs in two steps: precartilaginous condensation and chondrocyte early differentiation (Fig. 8.3B).

Precartilaginous condensation requires that skeletogenic cells stop proliferating and expressing collagen-1 and hyaluronan and that they start expressing N-cadherin, tenascin-C, and other adhesion proteins that allow them to tightly aggregate (reviewed by Hall *et al.*, 2000; Shum *et al.*, 2003). *In vitro* studies have suggested that transforming growth factor-beta (Tgf-beta) and Wnt/beta-catenin signaling govern this process, but *in vivo* validation is still lacking (Tuli *et al.*, 2003). *In vivo* data have indicated that Sox9 is somehow required for precartilaginous cell condensation and survival, but exact roles in this step, probably beyond TGF-beta and Wnt/beta-catenin signaling, remain unknown (Akiyama *et al.*, 2002; Bi *et al.*, 1999). Our knowledge of the regulation of precartilaginous condensation is thus still scarce.

Condensed cells undergoing chondrocyte early differentiation stop expressing adhesion molecules, resume proliferation, and start producing

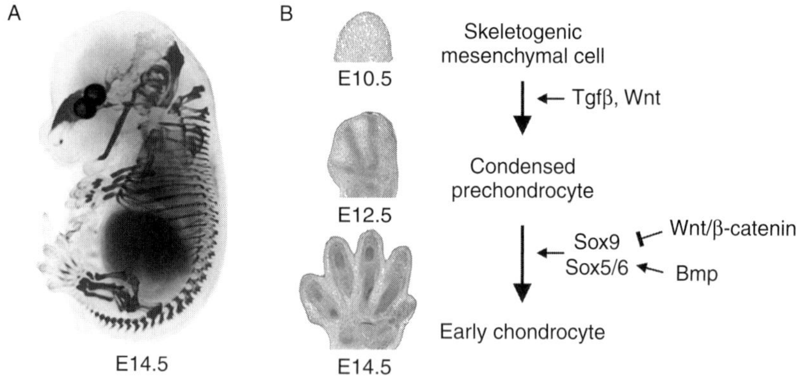

Figure 8.3 Chondrocyte early differentiation and development of cartilage primordia. (A) Alcian blue staining of a mouse embryo at E14.5 demonstrates that chondrocyte differentiation of skeletogenic cells leads to the formation of a primary skeleton that is entirely cartilaginous. (B) Sections through the developing paws of mouse embryos illustrate the major steps of early chondrogenesis. At E10.5, the limb bud is filled with skeletogenic cells. By E12.5, some of these cells have formed precartilaginous condensations that prefigure the future digits. By E14.5, condensed prechondrocytes have undergone chondrocyte early differentiation. The sections are stained with Alcian blue and nuclear fast red. (See Color Insert.)

profuse amounts of cartilage extracellular matrix. This event first kicks off in the center of precartilaginous condensations, but quickly spreads towards the periphery. A few layers of peripheral cells, however, remain skeletogenic and form a so-called perichondrium. They express patterning factors and concomitantly give rise to new chondrocytes for appositional growth of cartilage primordia. Chondrocyte early differentiation is driven at the transcriptional level by Sox5, Sox6, and Sox9 (reviewed by Lefebvre *et al.*, 2005; and Akiyama, 2008). The Sox5 and Sox6 proteins are very similar to each other and act in redundancy, but they differ from Sox9 in DNA-binding specificity and lack of intrinsic transactivation function. The importance of Sox9 in chondrogenesis was revealed when heterozygous mutations within and around *SOX9* were found to cause Campomelic Dysplasia, a severe form of human chondrodysplasia (Foster *et al.*, 1994; Wagner *et al.*, 1994). *Sox9* is expressed in skeletogenic cells prior to *Sox5* and *Sox6*, and the three genes are co-expressed in all precartilaginous condensations and cartilage elements. Homozygous inactivation of mouse *Sox9* in early chondrocytes blocks cell differentiation (Akiyama *et al.*, 2002; Bi *et al.*, 1999), whereas inactivation of *Sox5* and *Sox6* severely impairs, but does not block chondrocyte early differentiation (Smits *et al.*, 2001). Gain-of-function experiments *in vivo* and *in vitro* have revealed that the three proteins are capable together of inducing chondrocyte differentiation of mesenchymal cells (Ikeda *et al.*, 2004). Molecular studies have demonstrated that they cooperatively bind and activate the genes for many cartilage-specific extracellular matrix components (Lefebvre *et al.*, 1997, 1998; Han *et al.*, 2008). Sox5/Sox6 and Sox9 thus constitute a trio of transcription factors that is both needed and sufficient for chondrocyte early differentiation.

Since this trio controls cell differentiation in other lineages besides chondrocytes, it is likely that specific factors determine its chondrogenic activity. Such factors, however, have not been identified yet. The control of the expression of the Sox trio in chondrocytes remains scantily known. *Sox9* is expressed upstream of *Sox5* and *Sox6* (Akiyama *et al.*, 2002; Smits *et al.*, 2001), and the Sox9 protein was proposed *in vitro* to be able to control its own gene expression *via* a positive feedback loop (Kumar and Lassar, 2009). BMP signaling is required for expression of the three genes in skeletal sites *in vivo* (Yoon *et al.*, 2005), and Shh signaling was shown *in vitro* to confer competence in somitic tissue for subsequent BMP signals to induce overt chondrogenesis (Zeng *et al.*, 2002). FGF signaling increases *Sox9* expression in cultured mesenchymal cells (Murakami *et al.*, 2000). Shh, BMP, and FGF signaling are active in most skeletogenic sites and may thus cooperatively activate *Sox9* expression. In contrast, Wnt/beta-catenin signaling blocks the chondrogenic activity of Sox9 and its gene expression *in vivo* and may thus help restrict *Sox9* expression to early chondrocytes (Akiyama *et al.*, 2004; Hartmann and Tabin, 2001; Hill *et al.*, 2005). Major progress has thus been made in recent years by identifying the Sox5/6/9 trio, but major progress remains to be accomplished to identify the factors that control precartilaginous condensation and to better understand the mechanisms underlying the expression and specific action of the chondrogenic trio.

5. Development of Cartilage Growth Plates

Soon after they form, most cartilage primordia embark into rapid growth and shaping into a long shaft (diaphysis) flanked by globular ends (epiphyses). The growth of the shafts is achieved by so-called growth plates (GPs), i.e., layers of chondrocytes proceeding in a staggered manner through a series of maturation steps (Fig. 8.4A). Proliferating chondrocytes precede prehypertrophic, hypertrophic, and matrix-mineralizing terminal chondrocytes. Chondrocyte proliferation and hypertrophy are the two drivers of tissue elongation. Chondrocytes express an early phenotype through prehypertrophy, but also express stage-specific markers, such as matrilin-1 (*Matn1*) and the FGF receptor 3 (*Fgfr3*) at the proliferative stage, and the receptor for the parathyroid hormone and *Ppr, Ihh*, and collagen-10 (*Col10a1*) at the prehypertrophic stage. *Col10a1* is the only one of these markers that is still expressed at the hypertrophic stage. Terminal

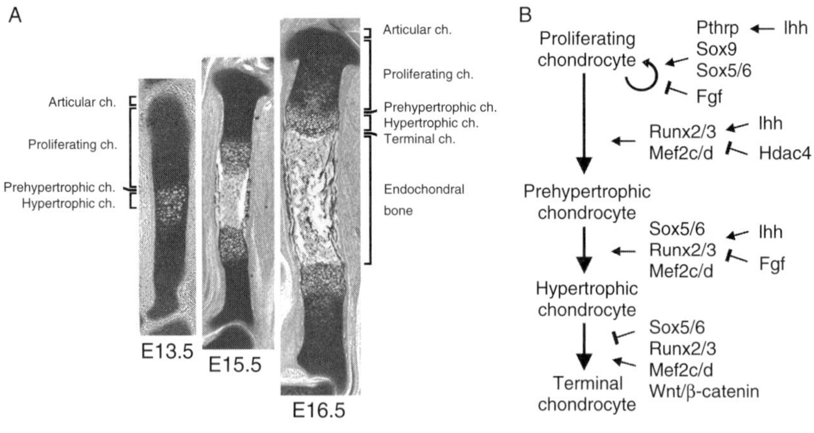

Figure 8.4 Chondrocyte maturation and development of cartilage growth plates. (A) Sections through a mouse embryo tibia (T) illustrate the development of growth plates and endochondral bone. At E13.5, early chondrocytes in the center of cartilage primordia undergo prehypertrophic and hypertrophic maturation. They reach terminal maturation and are replaced by endochondral bone by E15.5. Later on, growth plates maintain themselves and elongate developing bones. Chondrocytes keep proliferating and give rise, layer by layer, to maturing chondrocytes. These cells, which eventually die, are replaced by bone. The sections are stained with Alcian blue and nuclear fast red. (B) Schematic of the molecular control of GP chondrocytes. (See Color Insert.)

chondrocytes no longer express *Col10a1*, but express osteoblast markers, including the gene for the matrix metalloproteinase 13 (*Mmp13*). Like osteoblasts, they also mineralize their surrounding extracellular matrix and many are found positive for cell death by the TUNEL assay. It is thus believed that all terminal chondrocytes die. To our knowledge, however, conclusive lineage tracing experiments have not been published yet to definitively prove that no cells survive and directly participate as osteoblasts to endochondral ossification. We also like to stress that many studies on growth plate chondrocytes do not distinguish prehypertrophic from hypertrophic chondrocytes. In reviewing these studies, we paid attention to the actual phenotype of maturing chondrocytes and, therefore, sometimes reached a conclusion different from that of the authors.

The Sox trio is still required for the development of proliferative and prehypertrophic chondrocytes (Fig. 8.4B). It continues to generate expression of early cartilage markers in these cells, and it also activates expression of *Matn1* (Rentsendorj *et al.*, 2005; Smits *et al.*, 2001). Its role in chondrocyte proliferation is unclear, as chondrocyte proliferation is blocked in the absence of *Sox5/6* or *Sox9*, but also slowed down when *Sox9* is overexpressed (Akiyama *et al.*, 2002, 2004; Smits *et al.*, 2001, 2004). *Sox5/6* are needed to delay chondrocyte prehypertrophy, i.e., *Ppr* and *Ihh* expression, and to allow the cells to activate *Col10a1* and to undergo hypertrophy instead of immediately undergoing terminal maturation (Smits *et al.*, 2004). Sox9 may delay chondrocyte prehypertrophy downstream of Pthrp signaling (Huang *et al.*, 2001), but *in vivo* validation of this property is still missing.

Chondrocyte maturation is driven from prehypertrophy to the terminal stage by the Runt-domain transcription factors Runx2 and Runx3 (also known as Cbfa1 and Cbfa3, respectively) and by the MADS-box transcription factors Mef2c and Mef2d (myocyte enhancer factor 2c and 2d, respectively). The activity of these factors is inhibited by the histone deacetylase Hdac4. Runx2 and Runx3 were shown in knockout mice to be needed for chondrocyte maturation in a largely redundant manner (Inada *et al.*, 1999; Kim *et al.*, 1999; Yoshida *et al.*, 2004), and gain-of-function experiments in transgenic mice demonstrated that Runx2 is sufficient to cause ectopic and precocious maturation of chondrocytes (Takeda *et al.*, 2001). Runx2 directly binds and activates *Ihh, Col10a1*, and *Mmp13* (Selvamurugan *et al.*, 2004; Yoshida *et al.*, 2004; Zheng *et al.*, 2003). Mef2c and Mef2d were originally identified as essential regulators of muscle and cardiovascular development. Deletion of *Mef2c* and *Mef2d* in the mouse results in severe impairment of chondrocyte maturation, including downregulation of *Runx2* expression, while forced expression of a superactivating form of MEF2C causes precocious maturation (Arnold *et al.*, 2007). Mef2c directly binds and activates *Col10a1*, but it has not been shown whether it cooperates with Runx2. Mice lacking the Hdac4 histone deacetylase precociously mature chondrocytes, whereas mice overexpressing Hdac4 in proliferating chondrocytes exhibit a delay of chondrocyte maturation

(Vega *et al.*, 2004). Hdac4 can physically interact with Runx2 and Mef2c and thereby block *Col10a1* transactivation (Arnold *et al.*, 2007; Kozhemyakina *et al.*, 2009; Vega *et al.*, 2004). Runx2/3 and Mef2c/d are thus master switches for chondrocyte maturation, with Mef2c/d acting upstream and possibly together with Runx2, and with Hdac4 blocking the activity of both types of factors.

Many signaling pathways interact with each other to control chondrocyte proliferation and maturation in the GP. Their roles are so critical that mutations in any one of them cause severe skeleton malformation diseases (reviewed by Rimoin *et al.*, 2007). Ihh signaling has a pivotal role. Secreted by prehypertrophic chondrocytes, Ihh stimulates chondrocyte proliferation on one hand, and chondrocyte maturation on the other hand (St-Jacques *et al.*, 1999). Furthermore, it activates the gene for the *Pthrp* in subarticular chondrocytes. Pthrp signals through its Ppr receptor expressed at a low but functional level in proliferating chondrocytes and at a high level in prehypertrophic chondrocytes. It does not affect chondrocyte proliferation directly, but establishes a negative feedback loop with Ihh to delay prehypertrophy. Counteracting Ihh signaling, FGF signaling inhibits chondrocyte proliferation and maturation. The Fgf18 and Fgf9 ligands produced by the perichondrium bind to the FGF receptor-3 (Ffgr3) abundantly expressed on proliferating chondrocytes (reviewed by Ornitz, 2005). They activate the MAPK pathway to inhibit chondrocyte maturation and the Stat1 transcription factor to inhibit cell proliferation (Murakami *et al.*, 2004). Several Wnt ligands and pathways control GP development. Best known is the canonical pathway mediated by beta-catenin. Its main action appears to lead hypertrophic chondrocytes to the terminal stage (Guo *et al.*, 2009). Other pathways with key roles in the GP during gestation and postnatally are initiated by BMPs, TGF-beta, growth hormone, insulin-like growth factors, thyroid hormone (TH), retinoic acid, connective tissue growth factor, C-type natriuretic peptide, and estrogen. We refer the readers to recent reviews for details on their actions (Mackie *et al.*, 2008; Olney, 2009; Pogue *et al.*, 2006). Cartilage GPs are thus essential drivers of skeleton growth and the multistep maturation program of chondrocytes in these plates is tightly controlled by numerous signaling pathways and transcription factors. While this aspect of skeletogenesis appears to be one of the best known, additional studies are still warranted to identify additional regulatory players and fully understand how all pathways and factors interact with each other.

6. Bone Development

Bones form through two processes: endochondral and intramembranous ossification. The former consists in replacing GP cartilage by bone (Fig. 8.5A, B). As soon as terminal chondrocytes die at the end of GPs, their

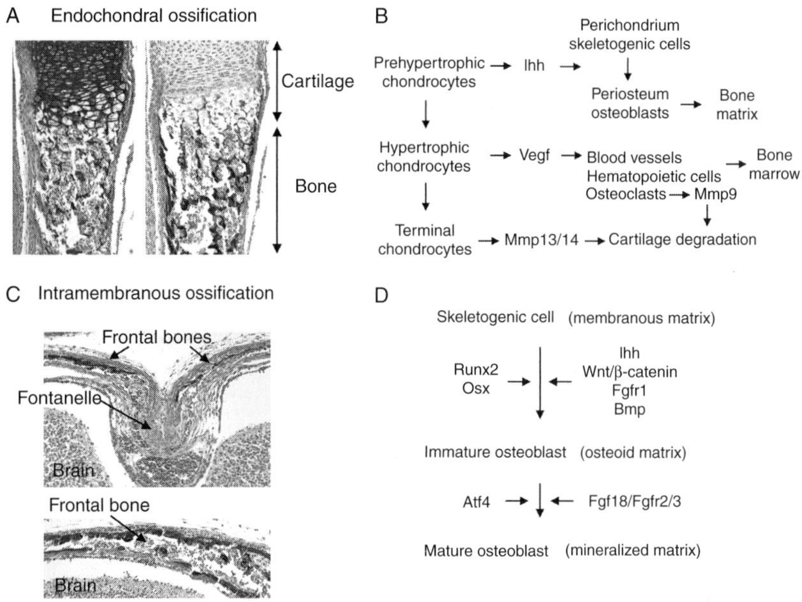

Figure 8.5 Osteoblast differentiation and intramembranous and endochondral ossification. (A) Sections through an endochondral bone in a newborn mouse show the replacement of cartilage by bone. The left section is stained with Alcian blue and the right one with the von Kossa reagent, which leaves a brown precipitate on the mineralized bone matrix. (B) Schematic showing how GP chondrocytes and bone-forming cells interact with each other to achieve endochondral ossification. (C) Coronal sections of a newborn mouse head. In the suture linking the two frontal bones (top panel), osteoblast precursors are surrounded by an abundant collagenous matrix. Further away (bottom panel), osteoblasts mature and deposit a mineralized bone matrix. This matrix is stained with the von Kossa reagent. (D) Schematic of the molecular control of osteoblast differentiation. (See Color Insert.)

lacunae are invaded by a team of bone-forming cells: osteoclasts help remove the cartilage matrix; osteoblasts lay down bone matrix; endothelial cells vascularize the newly formed tissue; and hematopoietic and stromal cells generate bone marrow. Chondrocytes actively contribute to this process. At prehypertrophy, their production of Ihh induces osteoblast differentiation in the perichondrium, then renamed periosteum or bone collar (St-Jacques *et al.*, 1999). At hypertrophy, they produce the vascular endothelial growth factor to trigger blood vessel formation in developing bone, and the matrix metalloproteinases Mmp13 and Mmp14 to help osteoclast-produced Mmp9 degrade the cartilage matrix (Stickens *et al.*, 2004; Zelzer *et al.*, 2004). The exact mechanisms that cause chondrocyte death remain unknown, but it does not occur if bone formation is impaired, further demonstrating that close interactions occur between cartilage and bone cells during endochondral ossification. The bone tissue forming in the diaphysis of fetal long bones is called the primary ossification center. Secondary centers similarly develop in the epiphyses postnatally, and the two types of centers fuse when GPs close in early adulthood.

Intramembranous ossification occurs when skeletogenic cells condense into compact nodules in the sites of future skull bones and in the perichondrium of GPs at the level of prehypertrophic chondrocytes (reviewed by Opperman, 2000) (Fig. 8.5C). While some cells develop capillaries, most become osteoblastic. They first produce an organic matrix and later calcify it through radial production of bony spicules from an ossification center located in the middle of each future bone. Developing skull bones are connected by fibrous, elastic tissues called sutures and whose primary role is not to serve as joints, but as bone growth centers. They maintain a pool of undifferentiated, osteogenic cells and simultaneously produce new bone cells that are recruited into ossification fronts. They disappear at about two years of age in humans, resulting in fusion of cranial bones. Insufficient bone growth results in suture agenesis, whereas too much bone growth results in craniosynostosis, i.e., precocious obliteration of sutures. Both types of defects result in severe skull malformations.

Both endochondral and intramembranous bones are built by osteoblasts, while osteoclasts degrade the tissue and have a key role in bone homeostasis postnatally. We will restrict our review of bone cells to osteoblasts, and will refer readers interested in osteoclast development to several recent reviews (reviewed by Karsenty *et al.*, 2009; Yavropoulou *et al.*, 2008). Osteoblasts differentiate from skeletogenic cells in two main steps (reviewed by Hartmann, 2009; Jensen *et al.*, 2010; Karsenty *et al.*, 2009) (Fig. 8.5D). As mentioned earlier, they first build an organic non-mineralized (osteoid) matrix. They strongly upregulate collagen-1 expression and express alkaline phosphatase, an important enzyme in matrix mineralization. As they mature, they produce bone-specific proteins, such as osteocalcin (*Bgp*) and bone sialoprotein (*Bsp*), and mineralize the osteoid matrix. These differentiation steps are governed by three specific transcription factors: Runx2, Osx, and ATF4.

Runx2 is a master osteogenic factor. While both *Runx2* and *Runx3* are expressed in GP chondrocytes, *Runx2* but not *Runx3* is expressed in skeletogenic cells and it remains expressed in differentiating osteoblasts until maturation. The importance of *Runx2* in this lineage was uncovered in 1997 when it was found that *RUNX2* haploinsufficiency causes cleidocranial dysplasia, a human disorder characterized by clavicle hypoplasia or aplasia (cleido-) and persistently open skull sutures (-cranial) (Mundlos *et al.*, 1997). At the same time, inactivation of *Runx2* in mice was shown to result in a block of osteoblast differentiation at the skeletogenic stage and thus in complete absence of bone (Komori *et al.*, 1997; Otto *et al.*, 1997). Runx2 directly binds and activates *Col1a1* (collagen-1), *Bsp*, and *Bgp*, and is sufficient to activate these genes in mesenchymal cells *in vitro* (Ducy *et al.*,

1997). Interestingly, however, forced expression of *Runx2* in chondrocytes in transgenic mice results in precocious and ectopic maturation of the cells, but not in their osteoblastic transformation (Takeda *et al.*, 2001). *Runx2* is thus needed, but may not be sufficient for osteoblast differentiation.

Osterix (Osx or Sp7) is a Krüppel-like zinc finger domain-containing transcription factor expressed exclusively in osteoblasts (Nakashima *et al.*, 2002). It was identified through potent induction of its expression by Bmp2 in mesenchymal cells *in vitro*. Its inactivation in the mouse revealed that it is required for the differentiation of *Runx2*-expressing skeletogenic cells into osteoblasts in both endochondral and intramembranous bones, including expression of *Col1a1, Bgp*, and *Bsp*. Osx is not only needed, but also sufficient to activate *Col1a1, Bgp*, and *Bsp* in mesenchymal cells *in vitro* and evidence exists that it directly activates these genes (Nakashima *et al.*, 2002; Sinha *et al.*, 2010). It is unknown, however, whether Osx cooperates with Runx2 in this function. Interestingly, *Osx*-null osteogenic cells exhibit a prehypertrophic chondrocyte phenotype in endochondral and intramembranous bone primordia (Nakashima *et al.*, 2002). *Osx* is thus needed to repress the chondrocytic fate of skeletogenic cells and it is directly needed to ensure osteoblast differentiation of *Runx2*-positive osteogenic cells.

Atf4 (activating transcription factor-4, or cyclic AMP-responsive element-binding protein 2, Creb2) controls osteoblast maturation. Its gene is widely expressed, but the protein is detected almost exclusively in osteoblasts. *Atf4* deficiency results in delayed bone formation during mouse embryonic development (Yang *et al.*, 2004). Atf4 is needed for amino acid import and thereby for synthesis of collagen, the main protein in bone. It is also directly involved in activating terminal markers in a Runx2-dependent manner. Each step of osteoblast differentiation is thus under the control of a specific master transcription factor.

Besides Runx2, Osx, and Atf4, many widely expressed transcription factors also have important roles in osteoblast development. They control cell proliferation or modulate the expression and activity of the masters. They include Twist1/2, already mentioned earlier, AP1 family members (reviewed by Wagner, 2002), and the forkhead factor FoxO1. The latter was very recently shown to stimulate osteoblast proliferation through interaction with Atf4, as well as through the regulation of a stress-dependent pathway influencing p53 signaling (Rached *et al.*, 2010).

Many signaling pathways control osteoblastogenesis. We have already mentioned the key role of Ihh in the induction of osteoblast differentiation (Day *et al.*, 2008). Downstream of Ihh signaling, the Wnt/beta-catenin pathway plays an important role in the decision of skeletogenic cells to undergo osteogenesis rather than chondrogenesis, and in boosting osteoblast differentiation (Rodda *et al.*, 2006; reviewed by Karsenty *et al.*, 2009). It is believed to work by downregulating *Sox9* expression and upregulating *Runx2* expression. FGF signaling plays important roles in endochondral and intramembranous ossification, as demonstrated for instance by the fact that gain-of-function mutations in the human FGFR1, FGFR2, and FGFR3 receptors cause craniosynostosis (reviewed by Ornitz, 2005; Su *et al.*, 2008). FGF signaling influences various stages of osteoblast development. FGFR1 promotes the differentiation of early osteoblasts, but prevents their maturation, whereas FGFR2 and FGFR3 promote osteoblast maturation. While Fgf2, 4, and 8 are essential in skeletogenic mesenchyme, Fgf18 promotes osteoblast progenitor proliferation as well as osteoblast maturation, mainly through Fgfr2. Bmps were the first molecules identified as possessing bone-inducing properties *in vivo* (reviewed by Rosen, 2006). Their osteogenic activities *in vitro* have been profusely documented. For instance, Bmp2 was shown to upregulate *Runx2* and *Osx* expression in osteogenic cells, and in turn, Bmp2 expression was shown to regulate *Runx2* expression. However, as of today, our knowledge of the specific roles of Bmps in osteoblasts *in vivo* remains scanty. Like many other cell developmental pathways, osteoblast generation thus involves a few cell type-specific master transcription factors that are controlled by and act together with many other widely expressed factors and signaling pathways. As for the chondrocyte lineage, most osteogenic regulatory factors were identified in the last two decades, and important progress remains to be done to elucidate the precise actions and mutual interactions of many known factors and likely to identify additional key factors.

7. Synovial Joint Formation

As mentioned earlier, several types of joints link skeletal elements together. Their composition and roles are very different from one type to another. Sutures are fibrous joints allowing minimal movement between skull bones and ensuring bone growth. Intervertebral and other cartilaginous joints allow more movement than sutures, but much less movement than synovial joints. We will focus here on the latter because they have key roles and are subject to highly prevalent of degeneration diseases in humans, but also because our knowledge of the mechanisms underlying their development remains very modest compared with that of other skeletogenic processes. We thus owe to draw attention to this process and stimulate research efforts in this area.

Synovial joints develop concomitantly with the skeletal elements that they articulate (reviewed by Khan *et al.*, 2007; Pacifici *et al.*, 2005; Pitsillides *et al.*, 2008). They do so in two main steps. In the first step, called joint specification, skeletogenic cells are directed to the articular fate (Fig. 8.6A, B). In the second step, called joint morphogenesis, articular cells differentiate and develop the various joint structures.

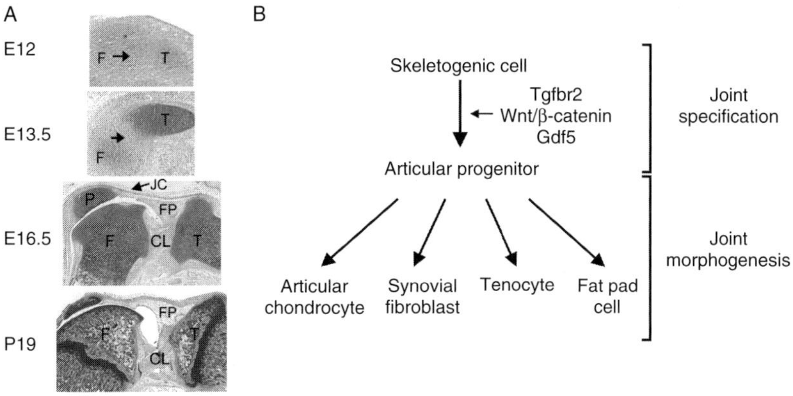

Figure 8.6 Synovial joint development. (A) Sections through the mouse knee joint at various stages of development. At E12, the presumptive joint region (arrow) is not distinguishable from the femur (F) and T precartilaginous condensations. At E13.5, this region becomes distinguishable as surrounding cartilage primordia are overtly developing. At E16.5, joint morphogenesis is well advanced. The joint cavity has formed between the patella (P) and F. Cruciate ligaments and FP, lined with synovial tissue, are developed. At the postnatal day 19, the knee joint is mature. The AC is separated from the epiphyseal GP by a secondary center of ossification. The sections are stained with Alcian blue and nuclear fast red. (B) Schematic of the molecular control of synovial joint cell differentiation. (See Color Insert.)

As mentioned earlier, articular progenitor cells arise in the first step of joint formation from the same pool of skeletogenic cells as chondrocytes and osteoblasts (Koyama *et al.*, 2008; Rountree *et al.*, 2004). They express *Sox5/6/9* as do adjacent chondrogenic cells, but specifically express the genes for the Tgf-beta receptor 2 (*Tgfbr2*), Wnt canonical ligands (*Wnt4, Wnt9a/14*, and *Wnt16*), and the growth differentiation factor-5 (*Gdf5*) (Dy *et al.*, 2010; Guo *et al.*, 2004; Hartmann and Tabin, 2001; Seo and Serra, 2007; Spagnoli *et al.*, 2007). A Tgfbr2/Wnt/Gdf5 signaling cascade is required and sufficient to specify progenitors to the articular fate. Its action results in the formation of presumptive joint regions, called interzones. These zones feature highly condensed cells, among which the precursors of non-cartilaginous tissues downregulate expression of *Sox5/6/9*.

An important aspect of synovial joint morphogenesis is the formation of an internal cavity, a process referred to as joint cavitation. Cell death was originally thought to trigger this process, but this notion is now contested by evidence that cell death is detected in presumptive phalangeal joints, but not in other sites (Ito and Kida, 2000). Instead, hyaluronan secretion and shifts in extracellular matrix composition are now believed to promote joint cavitation (Matsumoto *et al.*, 2009), and skeletal movement and cartilage primordia development were recently demonstrated to be required for this event (Dy *et al.*, 2010; Kahn *et al.*, 2009). The underlying mechanisms, however, remain unknown.

Another important aspect of joint morphogenesis is the differentiation of articular chondrocytes. These cells overtly differentiate in the mouse only postnatally and, in contrast to GP chondrocytes, they do never terminally mature. Instead, they maintain a chondrocyte early phenotype throughout life. Not unexpectedly, their differentiation requires expression of *Sox5/6* (Dy *et al.*, 2010). It is likely that it requires *Sox9* expression as well, but this has not been demonstrated yet. The Ets-domain transcription factor Erg has been proposed to control the specific fate of articular chondrocytes based on the fact that *Erg* was reported to be highly expressed in articular progenitors, but to be absent in GP chondrocytes, and that its forced expression in mouse cartilage primordia appeared to delay GP development (Iwamoto *et al.*, 2007). However, recent evidence that *Erg* is actually highly expressed in both GP and articular chondrocyte progenitors and that it is down-regulated upon differentiation of both chondrocyte types, strongly suggests that *Erg* might help maintain chondrogenic cells at a precursor stage rather than help them acquire an articular fate (Dy *et al.*, 2010). There is presently no evidence that transcription factors other than the Sox trio control articular chondrocyte differentiation. It is thus possible that the maintenance of an early phenotype in articular chondrocytes depends on continued expression of the Sox trio rather than on expression of additional lineage-specific transcription factors. Knowledge of the roles of signaling pathways that control articular chondrocytes remains scanty. Notch signaling is receiving increasing attention, as it is believed to have key roles in the maintenance of articular progenitors (reviewed by Karlsson *et al.*, 2009). Pthrp, Ihh, Bmp, and Tgf-beta signaling are required to maintain healthy AC postnatally, and Wnt/beta-catenin signaling may have a key role in preventing chondrocyte differentiation of articular progenitors not destined to form AC (Chen *et al.*, 2008; Rountree *et al.*, 2004; Serra *et al.*, 1997; Yang *et al.*, 2001; Yuasa *et al.*, 2009; Zhu *et al.*, 2008). The direct involvement of these pathways in the development of AC remains, however, unclear. This is mostly due to the lack of suitable mouse models to assess their roles in articular chondrocytes independently of their roles in GP chondrocytes and other cell types.

Finally, synovial joint morphogenesis is also characterized by the development of synovial fibroblasts, fat pad (FP) cells, and tenocytes that form the non-cartilaginous structures of synovial joints. The mechanisms underlying the differentiation of these cells remain virtually unknown. This brief review thus highlights that our current understanding of synovial joint development lags way behind that of other skeletogenic processes. While several pathways have been identified that are required for joint specification, major gaps in knowledge remain regarding the transcription factors and signaling pathways involved in the differentiation of all types of articular cells and in the coordination of their development into the multiple joint tissues.

8. Skeleton Variation

The skeleton, probably more than any other system, exhibits an astonishingly great variability. Variability exists across species in the number of skeletal elements making up the skeleton, and variability also exists across and within species in the size and shape of each element. A fascinating field of research for skeleton biologists, geneticists, and evolutionary developmental biologists is to elucidate the mechanisms underlying this variability. New mechanisms were recently proposed and will be discussed here.

Variability across species could be due to differences in the number of genes involved in skeletogenesis. The *SHOX* (short stature homeobox) genes, for instance, are candidates for this mechanism. Two of them exist in humans (*SHOX* and *SHOX2*), one in mice (*Shox2*), and none in invertebrates. Mutations in the X-linked *SHOX* gene occur at a very high frequency, about 1 in 1000, in humans. Haploinsufficiency causes short stature and malformation of limb extremities (Ellison *et al.*, 1997; Gahunia *et al.*, 2009). Homozygous mutations cause Langer's mesomelic dysplasia, characterized by extremely short and bowed limbs. The human genes and the mouse gene have similar expression patterns (Clement-Jones *et al.*, 2000) and *Shox2* inactivation in the mouse results in a phenotype overlapping with that caused by *SHOX* mutations in humans (Cobb *et al.*, 2006). The *Shox* genes thus encode essential patterning factors, whose alterations result in striking skeletal differences between human individuals. It can, therefore, be hypothesized that the duplication of the *SHOX* genes in humans compared to other species might contribute to some major skeletal differences existing between them.

It is evident that most of the variability existing between and within species cannot be explained by differences in gene numbers, as all humans have the same gene set and largely share this set with other vertebrates. How do divergent skeletal morphologies then arise? Studies on identical twins and siblings have demonstrated that environmental and hormonal conditions affect skeletal growth, but to an extent that is less than 10% (Macgregor *et al.*, 2006). Heritability must thus account for the rest and could have impacts at several levels. For instance, differences in regulatory DNA sequences can cause differences in gene expression patterns and levels. This was recently demonstrated for a *Prx1* enhancer variation that may contribute to the spectacular difference in forelimb length between mice and bats (Cretekos *et al.*, 2008). The replacement of a mouse *Prx1* enhancer with an orthologous bat sequence resulted in elevated transcript levels in developing forelimbs and in abnormally long limb skeletal elements. Interestingly, deletion of the mouse *Prx1* enhancer resulted in normal forelimb length and *Prx1* expression, suggesting regulatory redundancy. These findings suggest that mutations in regulatory sequences can result in morphological differences between and within species and that *cis*-regulatory redundancy may facilitate accumulation of such mutations.

Protein sequence differences also can contribute to variations. This was recently demonstrated in genome-wide studies that associated 47 loci with height variation in humans (Lettre, 2009). In particular, a variant allele of *HMGA2* (high-mobility-group DNA-binding protein A2) was estimated to explain approximately 0.3% of the human population variation in height. The effect of each single nucleotide polymorphism was small, but in aggregate these polymorphisms could correctly assign individuals to the lower or upper tail of the human population height distribution. Interestingly, *Hmga2* mutations cause dwarfism in pigmy mice (Zhou *et al.*, 1995) and truncations of human *HMGA2* and mouse *Hmga2* are associated with extreme overgrowth (Ligon *et al.*, 2005). *Hmga2* is expressed in mesenchymal cells and its roles are unknown. Several of the other 47 loci also included genes previously implicated in tall or short stature syndromes (Lettre, 2009). These studies thus indicate that genes mutated in severe syndromes can also harbor common alleles with a weaker effect on stature. They also demonstrate that skeletogenesis is under the control of a few master genes as well as numerous patterning genes whose allele variants determine skeletal variability within and among species.

9. Perspectives

We have reviewed here many regulatory factors and pathways that were unknown 5, 10, or 20 years ago and that have become the focus of attention of skeletal biologists throughout the world since milestone discoveries were made on their roles in skeletogenesis or skeleton malformation diseases. Despite tremendous progress, it is nevertheless striking that our knowledge of the genes involved in human skeletal diseases and in the control of skeletogenesis remains incomplete. Indeed, we often do not fully understand how factors are regulated and participate in the complex molecular networks that govern skeletogenic programs. It also appears likely in some cases and obvious in others that additional factors remain to be uncovered. Having reviewed our current knowledge and identified gaps in knowledge, it is now time to make long-term and short-term plans for future research efforts.

Identifying mechanisms to prevent and cure skeletal malformation diseases must continue to be our chief ultimate goal. To reach it, we must pursue our searches for disease-causing mutations. Positional cloning and candidate gene approaches remain fully appropriate and although laborious they are becoming easier with each technological improvement and the progressive deciphering of human and other genomes. Once mutations are identified, the most logical strategy is to try and fix these mutations using gene therapy or tissue engineering approaches. Major scientific, technological, and even ethical challenges, however, have still to be overcome before such a strategy becomes successful. It is, therefore, wise to consider alternative strategies, which would ameliorate skeletogenesis by bypassing the impact of gene mutations. The proper design of such strategies requires that we understand the roles of affected genes, their modes of regulation, and the impact of the gene mutations. For these purposes, gene manipulations in cultured cells, vertebrates, and even invertebrates, such as the fly and sea urchin, are instrumental, as are various types of cellular and molecular approaches *in vitro*. Major technological discoveries made in recent years are opening doors to new and more efficient approaches. For instance, comparative genomics has become a powerful approach to identify important coding and noncoding regulatory elements. Mass spectrometry has become highly proficient to identify protein partners and modifications. DNA, RNA, and protein delivery methods in cells using viruses and other reagents have also become very effective to address myriads of questions *in vivo* or *in vitro*. One can thus be very optimistic that the landmark discoveries made in recent years have set the stage for many more chief discoveries such that 5, 10, or 20 years from now we will not only understand the molecular basis of most skeletal dysplasias, but we may also be able to provide preventive and therapeutic strategies to alleviate the distress of many affected children.

ACKNOWLEDGMENTS

Work in the Lefebvre laboratory described in this review was supported by grants from the National Institutes of Health (AR46249 and AR54153 to V.L.). We apologize to all authors whose important work could not be cited due to space constraints.

REFERENCES

Akiyama, H., (2008). Control of chondrogenesis by the transcription factor Sox9. *Mod. Rheumatol.* **18**, 213–219.

Akiyama, H., Chaboissier, M. C., Martin, J. F., Schedl, A., and de Crombrugghe, B. (2002). The transcription factor Sox9 has essential roles in successive steps of the chondrocyte differentiation pathway and is required for expression of Sox5 and Sox6. *Genes Dev.* **16**, 2813–2828.

Akiyama, H., Lyons, J. P., Mori-Akiyama, Y., Yang, X., Zhang, R., Zhang, Z., Deng, J. M., Taketo, M. M., Nakamura, T., Behringer, R. R., McCrea, P. D., and de Crombrugghe,

B. (2004). Interactions between Sox9 and beta-catenin control chondrocyte differentiation. *Genes Dev.* **18**, 1072–1087.

Arnold, M. A., Kim, Y., Czubryt, M. P., Phan, D., McAnally, J., Qi, X., Shelton, J. M., Richardson, J. A., Bassel-Duby, R., and Olson, E. N. (2007). MEF2C transcription factor controls chondrocyte hypertrophy and bone development. *Dev. Cell* **12**, 377–389.

Bhattaram, P., Penzo-Méndez, A., Sock, E., Colmenares, C., Kaneko, K. J., Vassilev, A., DePamphilis, M. L., Wegner, M., and Lefebvre, V. (2010). Organogenesis relies on Soxc transcription factors for the survival of neural and mesenchymal progenitors. *Nat. Commun.* 1:9 doi: 10.1038/ncomms 1008.

Bi, W., Deng, J. M., Zhang, Z., Behringer, R. R., and de Crombrugghe, B. (1999). Sox9 is required for cartilage formation. *Nat. Genet.* **22**, 85–89.

Bialek, P., Kern, B., Yang, X., Schrock, M., Sosic, D., Hong, N., Wu, H., Yu, K., Ornitz, D. M., Olson, E. N., Justice, M. J., and Karsenty, G. (2004). A twist code determines the onset of osteoblast differentiation. *Dev. Cell* **6**, 423–435.

Butterfield, N. C., McGlinn, E., and Wicking, C. (2010). The molecular regulation of vertebrate limb patterning. *Curr. Top. Dev. Biol.* (submitted).

Chen, X., Macica, C. M., Nasiri, A., and Broadus, A. E. (2008). Regulation of articular chondrocyte proliferation and differentiation by Indian hedgehog and parathyroid hormone-related protein in mice. *Arthritis Rheum.* **58**, 3788–3797.

Clement-Jones, M., Schiller, S., Rao, E., Blaschke, R. J., Zuniga, A., Zeller, R., Robson, S. C., Binder, G., Glass, I., Strachan, T., Lindsay, S., and Rappold, G. A. (2000). The short stature homeobox gene SHOX is involved in skeletal abnormalities in turner syndrome. *Hum. Mol. Genet.* **9**, 695–702.

Cobb, J., Dierich, A., Huss-Garcia, Y., and Duboule, D. (2006). A mouse model for human short-stature syndromes identifies shox2 as an upstream regulator of runx2 during long-bone development. *Proc. Natl. Acad. Sci. USA* **103**, 4511–4515.

Cretekos, C. J., Wang, Y., Green, E. D., NISC Comparative Sequencing Program, Martin, J. F., Rasweiler, J. J., IV, and Behringer, R. R. (2008). Regulatory divergence modifies limb length between mammals. *Genes Dev.* 22, 141–151.

Day, T. F., Guo, X., Garrett-Beal, L., and Yang, Y. (2005). Wnt/beta-catenin signaling in mesenchymal progenitors controls osteoblast and chondrocyte differentiation during vertebrate skeletogenesis. *Dev. Cell* **8**, 739–750.**

Day, T. F., and Yang, Y. (2008). Wnt and hedgehog signaling pathways in bone development. *J. Bone Joint Surg. Am.* **90**(Suppl. 1), 19–24.

Ducy, P., Zhang, R., Geoffroy, V., Ridall, A. L., and Karsenty, G. (1997). Osf2/cbfa1: a transcriptional activator of osteoblast differentiation. *Cell.* **89**, 747–754.

Dy, P., Penzo-Méndez, A., Wang, H., Pedraza, C. E., Macklin, W. B., and Lefebvre, V. (2008). The three SoxC proteins—Sox4, Sox11 and Sox12—exhibit overlapping expression patterns and molecular properties. *Nucl. Acids Res.* **36**, 3101–3117.

Dy, P., Smits, P., Silvester, A., Penzo-Méndez, A., Dumitriu, B., Han, Y., de la Motte, C. A., Kingsley, D. M., and Lefebvre, V. (2010). Synovial joint morphogenesis requires the chondrogenic action of Sox5 and Sox6 in growth plate and articular cartilage. *Dev. Biol.* **341**, 346–359.

Eames, B. F., Sharpe, P. T., and Helms, J. A. (2004). Hierarchy revealed in the specification of three skeletal fates by Sox9 and Runx2. *Dev. Biol.* **274**, 188–200.

Ellison, J. W., Wardak, Z., Young, M. F., Robey, P. G., Webster, M., and Chiong, W. (1997). PHOG, a candidate gene for involvement in the short stature of turner syndrome. *Hum. Mol. Genet.* **6**, 1341–1347.

Foster, J. W., Dominguez-Steglich, M. A., Guioli, S., Kwok, C., Weller, P. A., Stevanović, M., Weissenbach, J., Mansour, S., Young, I. D., Goodfellow, P. N., Brook, J. D., and Schafer, A. J. (1994). Campomelic dysplasia and autosomal sex reversal caused by mutations in an SRY-related gene. *Nature* **372**, 525–530.

Gahunia, H. K., Babyn, P. S., Kirsch, S., and Mendoza-Londono, R. (2009). Imaging of SHOX-associated anomalies. *Semin. Musculoskelet. Radiol.* **13**, 236–254.

Guo, X., Day, T. F., Jiang, X., Garrett-Beal, L., Topol, L., and Yang, Y. (2004). Wnt/beta-catenin signaling is sufficient and necessary for synovial joint formation. *Genes Dev.* **18**, 2404–2417.

Guo, X., Mak, K. K., Taketo, M. M., and Yang, Y. (2009). The wnt/beta-catenin pathway interacts differentially with PTHrP signaling to control chondrocyte hypertrophy and final maturation. *PLoS One* **4**, e6067.

Hall, B. K., and Miyake, T. (2000). All for one and one for all: condensations and the initiation of skeletal development. *BioEssays* **22**, 138–147.

Han, Y., and Lefebvre, V. (2008). L-Sox5/Sox6 drive expression of the aggrecan gene in cartilage by securing binding of Sox9 to a far-upstream enhancer. *Mol. Cell. Biol.* **28**, 4999–5013.

Hartmann, C. (2009). Transcriptional networks controlling skeletal development. *Curr. Opin. Genet. Dev.* **19**, 437–443.

Hartmann, C., and Tabin, C. J. (2001). Wnt-14 plays a pivotal role in inducing synovial joint formation in the developing appendicular skeleton. *Cell* **104**, 341–351.

Hermanns, P., and Lee, B. (2001). Transcriptional dysregulation in skeletal malformation syndromes. *Am. J. Med. Genet.* **106**, 258–271.

Hill, T. P., Später, D., Taketo, M. M., Birchmeier, W., and Hartmann, C. (2005). Canonical wnt/beta-catenin signaling prevents osteoblasts from differentiating into chondrocytes. *Dev. Cell* **8**, 727–738.

Huang, W., Chung, U. I., Kronenberg, H. M., and de Crombrugghe, B. (2001). The chondrogenic transcription factor Sox9 is a target of signaling by the parathyroid hormone-related peptide in the growth plate of endochondral bones. *Proc. Natl. Acad. Sci. USA* **98**, 160–165.

Ikeda, T., Kamekura, S., Mabuchi, A., Kou, I., Seki, S., Takato, T., Nakamura, K., Kawaguchi, H., Ikegawa, S., and Chung, U. I. (2004). The combination of SOX5, SOX6, and SOX9 (the SOX trio) provides signal sufficient for induction of permanent cartilage. *Arthritis Rheum.* **50**, 3561–3573.

Inada, M., Yasui, T., Nomura, S., Miyake, S., Deguchi, K., Himeno, M., Sato, M., Yamagiwa, H., Kimura, T., Yasui, N., Ochi, T., Endo, N., *et al.* (1999). Maturational disturbance of chondrocytes in cbfa1-deficient mice. *Dev. Dyn.* **214**, 279–290.

Ito, M. M., and Kida, M.Y. (2000). Morphological and biochemical re-evaluation of the process of cavitation in the rat knee joint: cellular and cell strata alterations in the interzone. *J. Anat.* **4**, 659–679.

Iwamoto, M., Tamamura, Y., Koyama, E., Komori, T., Takeshita, N., Williams, J. A., Nakamura, T., Enomoto-Iwamoto, M., and Pacifici, M. (2007). Transcription factor ERG and joint and articular cartilage formation during mouse limb and spine skeletogenesis. *Dev. Biol.* **305**, 40–51.

Jensen, E. D., Gopalakrishnan, R., and Westendorf, J. J. (2010). Regulation of gene expression in osteoblasts. *Biofactors* **36**, 25–32.

Kahn, J., Shwartz, Y., Blitz, E., Krief, S., Sharir, A., Breitel, D. A., Rattenbach, R., Relaix, F., Maire, P., Rountree, R. B., Kingsley, D. M., and Zelzer, E. (2009). Muscle contraction is necessary to maintain joint progenitor cell fate. *Dev. Cell* **16**, 734–743.

Karlsson, C., and Lindahl, A. (2009). Notch signaling in chondrogenesis. *Int. Rev. Cell. Mol. Biol.* **275**, 65–88.

Karsenty, G., Kronenberg, H. M., and Settembre, C. (2009). Genetic control of bone formation. *Annu. Rev. Cell. Dev. Biol.* **25**, 629–648.

Khan, I. M., Redman, S. N., Williams, R., Dowthwaite, G. P., Oldfield, S. F., and Archer, C. W. (2007). The development of synovial joints. *Curr. Top. Dev. Biol.* **79**, 1–36.

Kim, I. S., Otto, F., Zabel, B., and Mundlos, S. (1999). Regulation of chondrocyte differentiation by cbfa1. *Mech. Dev.* **80**, 159–170.

Komori, T., Yagi, H., Nomura, S., Yamaguchi, A., Sasaki, K., Deguchi, K., Shimizu, Y., Bronson, R. T., Gao, Y. H., Inada, M., Sato, M., Okamoto, R., *et al.* (1997). Targeted disruption of cbfa1 results in a complete lack of bone formation owing to maturational arrest of osteoblasts. *Cell* **89**, 755–764.

Koyama, E., Shibukawa, Y., Nagayama, M., Sugito, H., Young, B., Yuasa, T., Okabe, T., Ochiai, T., Kamiya, N., Rountree, R. B., Kingsley, D. M., Iwamoto, M., *et al.* (2008). A distinct cohort of progenitor cells participates in synovial joint and articular cartilage formation during mouse limb skeletogenesis. *Dev. Biol.* **316**, 62–73.

Kozhemyakina, E., Cohen, T., Yao, T. P., and Lassar, A. B. (2009). Parathyroid hormone-related peptide represses chondrocyte hypertrophy through a protein phosphatase 2a/histone deacetylase 4/MEF2 pathway. *Mol. Cell. Biol.* **29**, 5751–5762.

Kumar, D., and Lassar, A. B. (2009). The transcriptional activity of Sox9 in chondrocytes is regulated by RhoA signaling and actin polymerization. *Mol. Cell. Biol.* **29**, 4262–4273.

Lefebvre, V., Huang, W., Harley, V. R., Goodfellow, P. N., and de Crombrugghe, B. (1997). SOX9 is a potent activator of the chondrocyte-specific enhancer of the pro alpha1(II) collagen gene. *Mol. Cell. Biol.* **17**, 2336–2346.

Lefebvre, V., Li, P., and de Crombrugghe, B. (1998). A new long form of Sox5 (L-Sox5), Sox6 and Sox9 are co-expressed in chondrogenesis and cooperatively activate the type II collagen gene. *EMBO J.* **17**, 5718–5733.

Lefebvre, V., and Smits, P. (2005). Transcriptional control of chondrocyte fate and differentiation. *Birth Defects Res. C Embryo Today* **75**, 200–212.

Lettre, G. (2009). Genetic regulation of adult stature. *Curr. Opin. Pediatr.* **21**, 515–522.

Li, Y., Toole, B. P., Dealy, C. N., and Kosher, R. A. (2007). Hyaluronan in limb morphogenesis. *Dev. Biol.* **305**, 411–420.

Ligon, A. H., Moore, S.D.P., Parisi, M. A., Mealiffe, M. E., Harris, D. J., Ferguson, H. L., Quade, B. J., and Morton, C. C. (2005). Constitutional rearrangement of the architectural factor HMGA2: a novel human phenotype including overgrowth and lipomas. *Am. J. Hum. Genet.* **76**, 340–348.

Limura, T., Denans, N., and Pourquié, O. (2009). Establishment of hox vertebral identities in the embryonic spine precursors. *Curr. Top. Dev. Biol.* **88**, 201–234.

Macgregor, S., Cornes, B. K., Martin, N. G., and Visscher, P. M. (2006). Bias, precision and heritability of self-reported and clinically measured height in Australian twins. *Hum. Genet.* **120**, 571–580.

Mackie, E. J., Ahmed, Y. A., Tatarczuch, L., Chen, K. S., and Mirams, M. (2008). Endochondral ossification: how cartilage is converted into bone in the developing skeleton. *Int. J. Biochem. Cell Biol.* **40**, 46–62.

Matsumoto, K., Li, Y., Jakuba, C., Sugiyama, Y., Sayo, T., Okuno, M., Dealy, C. N., Toole, B. P., Takeda, J., Yamaguchi, Y., and Kosher, R. A. (2009). Conditional inactivation of has2 reveals a crucial role for hyaluronan in skeletal growth, patterning, chondrocyte maturation and joint formation in the developing limb. *Development* **136**, 2825–2835.

McIntyre, D. C., Rakshit, S., Yallowitz, A. R., Loken, L., Jeannotte, L., Capecchi, M. R., and Wellik, D. M. (2007). Hox patterning of the vertebrate rib cage. *Development* **134**, 2981–2989.

Mundlos, S., Otto, F., Mundlos, C., Mulliken, J. B., Aylsworth, A. S., Albright, S., Lindhout, D., Cole, W. G., Henn, W., Knoll, J. H., Owen, M. J., Mertelsmann, R., *et al.* (1997). Mutations involving the transcription factor CBFA1 cause cleidocranial dysplasia. *Cell* **89**, 773–779.

Murakami, S., Balmes, G., McKinney, S., Zhang, Z., Givol, D., and de Crombrugghe, B. (2004). Constitutive activation of MEK1 in chondrocytes causes stat1-independent achondroplasia-like dwarfism and rescues the fgfr3-deficient mouse phenotype. *Genes Dev.* **18**, 290–305.

Murakami, S., Kan, M., McKeehan, W. L., and de Crombrugghe, B. (2000). Up-regulation of the chondrogenic Sox9 gene by fibroblast growth factors is mediated by the mitogen-activated protein kinase pathway. *Proc. Natl. Acad. Sci. USA* **97**, 1113–1118.

Nakashima, K., Zhou, X., Kunkel, G., Zhang, Z., Deng, J. M., Behringer, R. R., and de Crombrugghe, B. (2002). The novel zinc finger-containing transcription factor osterix is required for osteoblast differentiation and bone formation. *Cell* **108**, 17–29.

Ng, L. J., Wheatley, S., Muscat, G. E., Conway-Campbell, J., Bowles, J., Wright, E., Bell, D. M., Tam, P. P., Cheah, K. S., and Koopman, P. (1997). SOX9 binds DNA, activates transcription, and coexpresses with type II collagen during chondrogenesis in the mouse. *Dev. Biol.* **183**, 108–121.

Olney, R. C. (2009). Mechanisms of impaired growth: effect of steroids on bone and cartilage. *Horm. Res.* **72**(Suppl 1), 30–35.

Opperman, L. A. (2000). Cranial sutures as intramembranous bone growth sites. *Dev. Dyn.* **219**, 472–485.

Ornitz, D. M. (2005). FGF signaling in the developing endochondral skeleton. *Cytokine Growth Factor Rev.* **16**, 205–213.

Ota, K. G., and Kuratani, S. (2009). Evolutionary origin of bone and cartilage in vertebrates. *In* The Skeletal System (O. Pourquié, Ed.) Cold Spring Harbor Laboratory Press, Cold Spring Harbor, New York, pp. 1–18.

Otto, F., Thornell, A. P., Crompton, T., Denzel, A., Gilmour, K. C., Rosewell, I. R., Stamp, G. W., Beddington, R. S., Mundlos, S., Olsen, B. R., Selby, P. B., and Owen, M. J. (1997). Cbfa1, a candidate gene for cleidocranial dysplasia syndrome, is essential for osteoblast differentiation and bone development. *Cell* **89**, 765–771.

Pacifici, M., Koyama, E., and Iwamoto, M. (2005). Mechanisms of synovial joint and articular cartilage formation: recent advances, but many lingering mysteries. *Birth Defects Res. C Embryo Today* **75**, 237–248.

Pitsillides, A. A., and Ashhurst, D. E. (2008). A critical evaluation of specific aspects of joint development. *Dev. Dyn.* **237**, 2284–2294.

Pogue, R., and Lyons, K. (2006). BMP signaling in the cartilage growth plate. *Curr. Top. Dev. Biol.* **76**, 1–48.

Rached, M. T., Kode, A., Xu, L., Yoshikawa, Y., Paik, J. H., Depinho, R. A., and Kousteni, S. (2010). FoxO1 is a positive regulator of bone formation by favoring protein synthesis and resistance to oxidative stress in osteoblasts. *Cell Metab.* **11**, 147–160.

Rentsendorj, O., Nagy, A., Sinkó, I., Daraba, A., Barta, E., and Kiss, I. (2005). Highly conserved proximal promoter element harbouring paired Sox9-binding sites contributes to the tissue- and developmental stage-specific activity of the matrilin-1 gene. *Biochem. J.* **389**, 705–716.

Rimoin, D. L., Cohn, D., Krakow, D., Wilcox, W., Lachman, R. S., and Alanay, Y. (2007). The skeletal dysplasias: clinical-molecular correlations. *Ann. NY Acad. Sci.* **1117**, 302–309.

Rodda, S. J., and McMahon, A. P. (2006). Distinct roles for hedgehog and canonical wnt signaling in specification, differentiation and maintenance of osteoblast progenitors. *Development* **133**, 3231–3244.

Rosen, V. (2006). BMP and BMP inhibitors in bone. *Ann. NY Acad. Sci.* **1068**, 19–25.

Rountree, R. B., Schoor, M., Chen, H., Marks, M. E., Harley, V., Mishina, Y., and Kingsley, D. M. (2004). BMP receptor signaling is required for postnatal maintenance of articular cartilage. *PLoS Biol.* **2**, e355.

Sauka-Spengler, T., and Bronner-Fraser, M. (2008). A gene regulatory network orchestrates neural crest formation. *Nat. Rev. Mol. Cell. Biol.* **9**, 557–568.

Selvamurugan, N., Kwok, S., Alliston, T., Reiss, M., and Partridge, N. C. (2004). Transforming growth factor-beta 1 regulation of collagenase-3 expression in osteoblastic cells by cross-talk between the smad and MAPK signaling pathways and their components, Smad2 and Runx2. *J. Biol. Chem.* **279**, 19327–19334.

Seo, H. S., and Serra, R. (2007). Deletion of Tgfbr2 in Prx1-cre expressing mesenchyme results in defects in development of the long bones and joints. *Dev. Biol.* **310**, 304–316.

Serra, R., Johnson, M., Filvaroff, E. H., LaBorde, J., Sheehan, D. M., Derynck, R., and Moses, H. L. (1997). Expression of a truncated, kinase-defective TGF-beta type II receptor in mouse skeletal tissue promotes terminal chondrocyte differentiation and osteoarthritis. *J. Cell Biol.* **139**, 541–552.

Shum, L., Coleman, C. M., Hatakeyama, Y., and Tuan, R. S. (2003). Morphogenesis and dysmorphogenesis of the appendicular skeleton. *Birth Defects Res. C Embryo Today* **69**, 102–122.

Sinha, K. M., Yasuda, H., Coombes, M. M., Dent, S. Y., and de Crombrugghe, B. (2010). Regulation of the osteoblast-specific transcription factor osterix by NO66, a jumonji family histone demethylase. *EMBO J.* **29**, 68–79.

Smits, P., Dy, P., Mitra, S., and Lefebvre, V. (2004). Sox5 and Sox6 are needed to develop and maintain source, columnar and hypertrophic chondrocytes in the cartilage growth plate. *J. Cell Biol.* **164**, 747–758.

Smits, P., Li, P., Mandel, J., Zhang, Z., Deng, J. M., Behringer, R. R., de Crombrugghe, B., and Lefebvre, V. (2001). The transcription factors L-Sox5 and Sox6 are essential for cartilage formation. *Dev. Cell* **1**, 277–290.

Spagnoli, A., O'Rear, L., Chandler, R. L., Granero-Molto, F., Mortlock, D. P., Gorska, A. E., Weis, J. A., Longobardi, L., Chytil, A., Shimer, K., and Moses, H. L. (2007). TGF-beta signaling is essential for joint morphogenesis. *J. Cell Biol.* **177**, 1105–1117.

Spranger, J. (2006). Skeletal dysplasias. *In* Human Malformations and Related Anomalies, second edition (R. E.Stevenson and J. G. Hall, Ed.) Oxford University Press, Oxford, New York Chapter 22.

St-Jacques, B., Hammerschmidt, M., and McMahon, A. P. (1999). Indian hedgehog signaling regulates proliferation and differentiation of chondrocytes and is essential for bone formation. *Genes Dev.* **13**, 2072–2086.

Stickens, D., Behonick, D. J., Ortega, N., Heyer, B., Hartenstein, B., Yu, Y., Fosang, A. J., Schorpp-Kistner, M., Angel, P., and Werb, Z. (2004). Altered endochondral bone development in matrix metalloproteinase 13-deficient mice. *Development* **131**, 5883–5895.

Su, N., Du, X., and Chen, L. (2008). FGF signaling: its role in bone development and human skeleton diseases. *Front. Biosci.* **13**, 2842–2865.

Takeda, S., Bonnamy, J. P., Owen, M. J., Ducy, P., and Karsenty, G. (2001). Continuous expression of Cbfa1 in nonhypertrophic chondrocytes uncovers its ability to induce hypertrophic chondrocyte differentiation and partially rescues Cbfa1-deficient mice. *Genes Dev.* **15**, 467–481.

Tuli, R., Tuli, S., Nandi, S., Huang, X., Manner, P. A., Hozack, W. J., Danielson, K. G., Hall, D. J., and Tuan, R. S. (2003). Transforming growth factor-beta-mediated chondrogenesis of human mesenchymal progenitor cells involves N-cadherin and mitogen-activated protein kinase and wnt signaling cross-talk. *J. Biol. Chem.* **278**, 41227–41236.

Vega, R. B., Matsuda, K., Oh, J., Barbosa, A. C., Yang, X., Meadows, E., McAnally, J., Pomajzl, C., Shelton, J. M., Richardson, J. A., Karsenty, G., and Olson, E. N. (2004). Histone deacetylase 4 controls chondrocyte hypertrophy during skeletogenesis. *Cell* **119**, 555–566.

Wagner, E. F. (2002). Functions of AP1 (fos/Jun) in bone development. *Ann. Rheum. Dis.* **61(**Suppl 2:ii), 40–42.

Wagner, T., Wirth, J., Meyer, J., Zabel, B., Held, M., Zimmer, J., Pasantes, J., Bricarelli, F. D., Keutel, J., Hustert, E., Wolf, U., Tommerup, N., *et al.* (1994). Autosomal sex reversal and campomelic dysplasia are caused by mutations in and around the SRY-related gene. *Cell* **79**, 1111–1120.

Wellik, D. M. (2009). Hox genes and vertebrate axial pattern. *Curr. Top. Dev. Biol.* **88**, 257–278.

Yang, X., Chen, L., Xu, X., Li, C., Huang, C., and Deng, C. X. (2001). TGF-beta/Smad3 signals repress chondrocyte hypertrophic differentiation and are required for maintaining articular cartilage. *J. Cell Biol.* **153**, 35–46.

Yang, X., Matsuda, K., Bialek, P., Jacquot, S., Masuoka, H. C., Schinke, T. Li, L., Brancorsini, S., Sassone-Corsi, P., Townes, T. M., Hanauer, A., and Karsenty, G. (2004). ATF4 is a substrate of RSK2 and an essential regulator of osteoblast biology; implication for Coffin-Lowry Syndrome. *Cell* **117**, 387–398.

Yavropoulou, M. P., and Yovos, J. G. (2008). Osteoclastogenesis—current knowledge and future perspectives. *J. Musculoskelet. Neuronal Interact.* **8**, 204–216.

Yoon, B. S., Ovchinnikov, D. A., Yoshii, I., Mishina, Y., Behringer, R. R., and Lyons, K. M. (2005). Bmpr1a and bmpr1b have overlapping functions and are essential for chondrogenesis in vivo. *Proc. Natl. Acad. Sci. USA* **102**, 5062–5067.

Yoshida, C. A., Yamamoto, H., Fujita, T., Furuichi, T., Ito, K., Inoue, K., Yamana, K., Zanma, A., Takada, K., Ito, Y., and Komori, T. (2004). Runx2 and Runx3 are essential for chondrocyte maturation, and Runx2 regulates limb growth through induction of indian hedgehog. *Genes Dev.* **18**, 952–963.

Yuasa, T., Kondo, N., Yasuhara, R., Shimono, K., Mackem, S., Pacifici, M., Iwamoto, M., and Enomoto-Iwamoto, M. (2009). Transient activation of wnt/{beta}-catenin signaling induces abnormal growth plate closure and articular cartilage thickening in postnatal mice. *Am. J. Pathol.* **175**, 1993–2003.

Zelzer, E., Mamluk, R., Ferrara, N., Johnson, R. S., Schipani, E., and Olsen, B. R. (2004). VEGFA is necessary for chondrocyte survival during bone development. *Development* **131**, 2161–2171.

Zeng, L., Kempf, H., Murtaugh, L. C., Sato, M. E., and Lassar, A. B. (2002). Shh establishes an nkx3.2/Sox9 autoregulatory loop that is maintained by BMP signals to induce somitic chondrogenesis. *Genes Dev.* **16**, 1990–2005.

Zhao, X., Sun, M., Zhao, J., Leyva, J. A., Zhu, H., Yang, W., Zeng, X., Ao, Y., Liu, Q., Liu, G., Lo, W.H.Y., Jabs, E. W., *et al.* (2007). Mutations in HOXD13 underlie syndactyly type V and a novel brachydactyly–syndactyly syndrome. *Am. J. Hum. Genet.* **80**, 361–371.

Zheng, Q., Zhou, G., Morello, R., Chen, Y., Garcia-Rojas, X., and Lee, B. (2003). Type X collagen gene regulation by Runx2 contributes directly to its hypertrophic chondrocyte-specific expression in vivo. *J. Cell Biol.* **162**, 833–842.

Zhou, X., Benson, K. F., Ashar, H. R., and Chada, K. (1995). Mutation responsible for the mouse pygmy phenotype in the developmentally regulated factor HMGI-C. *Nature* **377**, 771–774.

Zhou, G., Zheng, Q., Engin, F., Munivez, E., Chen, Y., Sebald, E., Krakow, D., and Lee, B. (2006). Dominance of SOX9 function over RUNX2 during skeletogenesis. *Proc. Natl. Acad. Sci. USA* **103**, 19004–19009.

Zhu, M., Chen, M., Zuscik, M., Wu, Q., Wang, Y. J., Rosier, R. N., O'Keefe, R. J., and Chen, D. (2008). Inhibition of beta-catenin signaling in articular chondrocytes results in articular cartilage destruction. *Arthritis Rheum.* **58**, 2053–2064.

CHAPTER NINE

The Molecular Regulation of Vertebrate Limb Patterning

Natalie C. Butterfield,[*,‡] Edwina McGlinn,[†] *and* Carol Wicking[*]

Contents

1. Introduction	320
2. The Early Limb	322
2.1. Limb induction	322
2.2. Pre-patterning, positioning of the ZPA, and establishing asymmetry	323
3. Patterning Along the AP Axis	325
3.1. Interactions between SHH and GLI3	325
3.2. The primary cilium and GLI3 function in the limb	326
3.3. SHH-mediated digit specification and growth	327
4. Proximal–Distal Patterning and Outgrowth	328
4.1. FGF signaling from the AER	328
4.2. Models of PD patterning	329
5. Interaction Between the ZPA and the AER	332
5.1. The SHH–GREM1–FGF feedback loop	332
5.2. Termination of limb outgrowth	333
6. Towards a Systems Biology Approach to Limb Patterning	334
Acknowledgments	335
References	335

Abstract

The limb has long been considered a paradigm for organogenesis because of its simplicity and ease of manipulation. However, it has become increasingly clear that the processes required to produce a perfectly formed limb involve complex molecular interactions across all three axes of limb development. Old models have evolved with acquisition of molecular knowledge, and in more recent times mathematical modeling approaches have been invoked to explain the precise spatio-temporal regulation of gene networks that coordinate limb

* Institute for Molecular Bioscience, The University of Queensland, Brisbane, Queensland, Australia
† Department of Genetics, Harvard Medical School, Boston, Massachusetts, USA
‡ Present address: Division of Developmental Biology, MRC-National Institute for Medical Research, Mill Hill London, United Kingdom

Current Topics in Developmental Biology, Volume 90
ISSN 0070-2153, DOI 10.1016/S0070-2153(10)90009-4

patterning and outgrowth. This review focuses on recent advances in our understanding of vertebrate limb development, highlighting the signaling interactions required to lay down the pattern on which the processes of differentiation will act to ultimately produce the final limb.

1. Introduction

Decades of classical analysis of the vertebrate limb have shaped our understanding of the underlying processes involved in organogenesis. With the advent of modern molecular tools, new bioinformatic capabilities and sophisticated imaging technologies, old models are being revisited and new molecular circuitry is being uncovered. In recent times appreciation of the complexities of the signaling interactions that pattern the limb has reached new levels, and the limb is once again emerging as a dominant model for organogenesis.

The limb initially protrudes from the lateral plate mesoderm (LPM) as a bud of undifferentiated mesenchymal cells encased in an outer epithelial layer. The ensuing period of organogenesis is characterized by a coordinated program of cell proliferation, adhesion, differentiation and, in some cases cell death, which ultimately forms the mature limb. These physical changes are underpinned by complex inter-connected gene regulatory networks which provide the pattern on which the final limb morphology is based. While the structure of the mature limb varies across species, and indeed between fore- and hindlimbs, the underlying molecular cues are well conserved. The usefulness of complementary analyses in two model systems, the chick and mouse, is well illustrated in the limb. Elegant *in ovo* manipulations in the chick have dovetailed with mouse molecular genetics to shape our understanding of limb development. In addition, data generated in other models such as the bat (Cretekos *et al.*, 2008; Hockman *et al.*, 2008) have provided further insight into the evolution of molecular patterning in the limb.

The vertebrate limb develops along three axes: anterior–posterior (AP; from thumb to little finger), proximal–distal (PD; from shoulder to finger tip), and dorsal–ventral (DV; from back of hand to palm). Unlike a number of more complex organs, a defined signaling center has been identified as the main player in each axis. The major center responsible for AP patterning is a group of posterior mesenchymal cells that constitute the zone of polarizing activity (ZPA), and secrete the morphogen Sonic hedgehog (SHH; Fig. 9.1A). Patterning across the AP axis determines the number and identity of digits as well as skeletal elements in the mid-region of the limb. The PD growth is controlled primarily by the apical ectodermal ridge (AER), a distal strip of thickened epithelium which secretes fibroblast

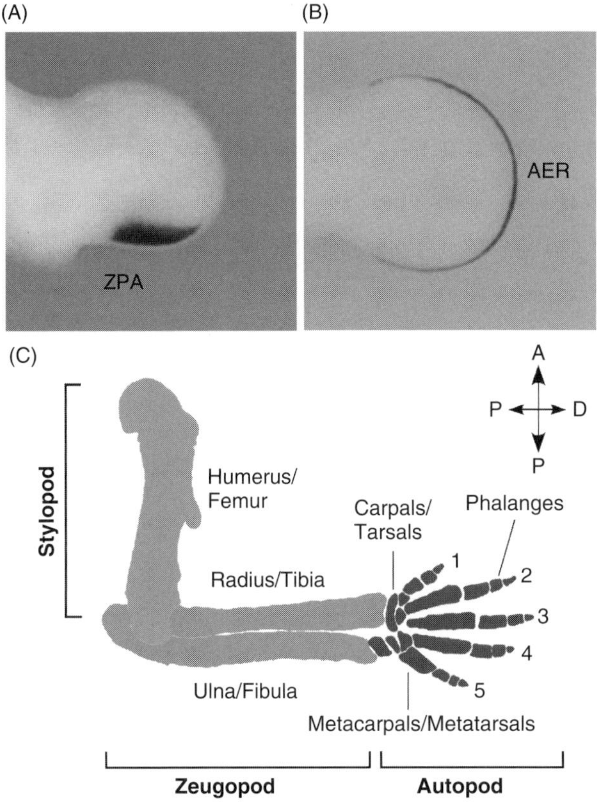

Figure 9.1 Major signaling centers in the 11.5 dpc mouse limb. Whole mount *in situ* hybridization analysis reveals (A) the ZPA characterized by *Shh* expression, and (B) the apical ectodermal ridge marked by expression of *Fgf8*. (C) The mature limb showing the autopod (handplate, consisting of phalanges, metacarpals, and carpals; and footplate consisting of phalanges, metatarsals, and tarsals), the zeugopod (radius–ulna in forelimb; tibia–fibula in hindlimb), and stylopod (humerus in forelimb; femur in hindlimb). A–P, anterior–posterior; P–D, proximal–distal.

growth factors (FGFs; Fig. 9.1B). Growth and patterning along this axis ultimately determines the shape and length of the mature limb, incorporating the stylopod (upper arm–thigh), zeugopod (forearm–lower leg), and autopod (hand–foot; Fig. 9.1C). During the early stages of limb budding in the mouse, DV polarity is established by the demarcation between those surface ectoderm cells expressing the dorsal lineage marker *Lmx1b* through the influence of WNT7a, and those in which the dorsal lineage is inhibited by the ventrally expressed gene *engrailed-1* (Arques *et al.*, 2007; Chen and Johnson, 2002; Davis and Joyner, 1988; Pearse *et al.*, 2007). Interactions between the signaling centers controlling patterning in all three axes of the limb are pivotal to the formation of a correctly patterned limb. Given the complexities of these interactions, it is not surprising that limb anomalies are a common feature of human congenital dysmorphology syndromes.

Here we will focus on the molecular cues involved in limb induction, establishment of the early pre-pattern, and subsequent interactions between the AP and PD axes. Patterning along the DV axis has been reviewed elsewhere (Chen and Johnson, 1999). While our particular focus is on data gleaned from the mouse limb, distinctions between the mouse and chick are discussed where appropriate.

2. The Early Limb

2.1. Limb induction

The paired appendages of the vertebrate embryo emerge from the body wall at precisely defined positions along the rostro-caudal axis. Implicated in determining this position is a nested collinear pattern of *Hox* gene expression along the body axis (Burke *et al.*, 1995; Cohn and Tickle, 1999; Cohn *et al.*, 1997; Krumlauf, 1994; Rancourt *et al.*, 1995). As demonstrated in the chick, rapid differential proliferation of the LPM at the level of the presumptive forelimb and hindlimb subsequently results in limb budding (Searls and Janners, 1971; Summerbell and Wolpert, 1973). Initiation of this process in the limbs of both mouse and chick is triggered by *Fgf10* expression in the LPM, which induces *Fgf8* in the overlying ectoderm (Min *et al.*, 1998; Ohuchi *et al.*, 1997; Sekine *et al.*, 1999). FGF8 in turn maintains *Fgf10* expression in the underlying mesenchyme to ensure limb outgrowth. This mesenchyme–ectoderm signaling loop is mediated by the two isoforms of FGFR2, as studies in the mouse have shown that ablation of either the ectodermal (FGFR2b) or mesenchymal (FGFR2c) receptor effectively abolishes limb initiation (Xu *et al.*, 1998). Mice lacking *Fgf10* show some evidence of limb budding but subsequent outgrowth is severely impaired (Sekine *et al.*, 1999), suggesting that there are factors upstream of FGF10 which are also required for the initiation of limb budding.

The T-box transcription factors TBX5 and TBX4 have been shown to act upstream of *Fgf10* in the forelimb and hindlimb fields, respectively, during limb initiation (Agarwal *et al.*, 2003; Ahn *et al.*, 2002; Naiche and Papaioannou, 2003; Ng *et al.*, 2002; Rallis *et al.*, 2003). However, while inactivation of *Tbx5* in the mouse completely abolishes forelimb formation and *Fgf10* expression (Agarwal *et al.*, 2003; Rallis *et al.*, 2003), *Tbx4*-ablated hindlimbs display a milder phenotype, with early weak expression of *Fgf10* (Naiche and Papaioannou, 2007). Thus, while TBX5 directly activates *Fgf10* expression during forelimb bud initiation, an additional factor(s) may act co-operatively with TBX4 in the initiation of *Fgf10* expression and hindlimb budding. In addition, some commonality of role for TBX4 and TBX5 during limb initiation has been demonstrated, since ectopic expression of *Tbx4* is able to completely rescue forelimb outgrowth in the absence of *Tbx5* (Minguillon *et al.*, 2005).

A role for retinoic acid (RA) produced by the LPM in forelimb initiation has also been postulated in the mouse, as embryos lacking the RA synthesis enzyme RALDH2 show no evidence of forelimb budding (Niederreither *et al.*, 1999, 2002). Treatment of chick embryos with the RA synthesis inhibitor disulfiram also abolishes limb initiation (Stratford *et al.*, 1996), and zebrafish lacking *raldh2* do not develop pectoral fins (Gibert *et al.*, 2006; Grandel *et al.*, 2002). However, maternal RA depletion in the rat embryo does not affect limb initiation (Power *et al.*, 1999). While RA undoubtedly plays an important role in limb development, the precise nature of the overall contribution of this molecule remains to be elucidated.

2.2. Pre-patterning, positioning of the ZPA, and establishing asymmetry

Once limb budding has been initiated at the correct sites, a molecular hierarchy is established to polarize the limb across the AP axis. Asymmetry is a fundamental characteristic of the vertebrate limb, and is ultimately determined by the polarizing activities of the ZPA. Correct positioning of the ZPA in the posterior limb is a critical factor in controlling these events, and a tightly regulated pre-pattern is established in the early limb to ensure appropriate establishment of the ZPA. This pre-patterning stage polarizes the nascent limb bud into anterior and posterior domains, and is mediated by several key interactions.

Reciprocal genetic interactions between the transcription factors GLI3 and HAND2 play a key role in generating the pre-pattern (Fig. 9.2A). The glioma-associated oncogene (GLI) proteins are the primary transcriptional mediators of the hedgehog signal, and knockout studies in the limb suggest that GLI3 is the major transcriptional regulator of hedgehog signaling during limb patterning (Bai *et al.*, 2002; Park *et al.*, 2000). GLI3 is a bipotential transcription factor that is proteolytically cleaved to a truncated transcriptional repressor [GLI3R; (Wang *et al.*, 2000)]. Even prior to establishment of *Shh* expression in the ZPA, GLI3 restricts expression of *Hand2* to the posterior limb (Charite *et al.*, 2000; Fernandez-Teran *et al.*, 2000; te Welscher *et al.*, 2002a). In turn, HAND2 restricts the expression of *Gli3* and *Alx4* to the anterior limb (te Welscher *et al.*, 2002a). This mutual genetic antagonism between GLI3 and HAND2 pre-patterns the limb bud, and contributes to the restriction of *Shh* expression and hence the ZPA to the posterior margin of the limb (Zuniga and Zeller, 1999).

In addition to GLI3 and HAND2, a number of other molecules play a role in determining the site of the ZPA. Most notably the early collinear

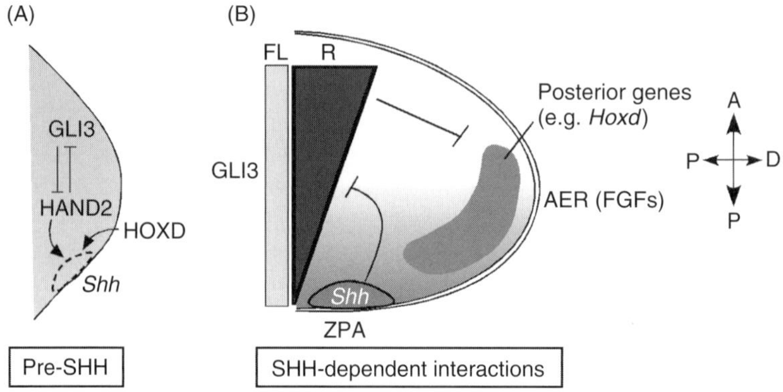

Figure 9.2 SHH-independent and dependent genetic interactions in the vertebrate limb. (A) Prior to *Shh* expression (9.0–9.5 dpc in the mouse), the limb is polarized along the A–P axis by reciprocal antagonism between GLI3 and HAND2. Along with expression of *Hoxd11–13* genes, this pre-pattern serves to restrict expression of *Shh* expression to the ZPA. (B) After *Shh* expression is initiated in the ZPA it now regulates expression of a number of these genes including the *Hoxd* genes. SHH now controls the processing of the full-length GLI3 protein into a truncated transcriptional repressor (GLI3R) and thus establishes a gradient of GLI3R across the A–P axis of the limb. FL, full-length form of GLI3; R, truncated repressor form of GLI3.

activation of the *Hoxd* genes appears to be required for establishment of the ZPA and its restriction to the posterior margin (Kmita *et al.*, 2005; Zakany *et al.*, 2004). Prior to establishment of the ZPA, *Hoxd* expression is controlled temporally by the 3′ chromosomal regulator termed the early limb control region, and spatially by the posterior restriction (POST) region 5′ of the cluster (Tarchini and Duboule, 2006). Due to these regulators, 5′ members of the cluster (e.g., *Hoxd11–13*) are activated later, and in a more posteriorly restricted pattern than 3′ *Hoxd* genes (e.g., *Hoxd1*) (Tarchini and Duboule, 2006). It is the posterior expression of the 5′ *Hoxd* genes that is pivotal to restricting *Shh* expression to the posterior margin of the limb bud (Kmita *et al.*, 2005; Zakany *et al.*, 2004).

Disruption of the pre-pattern by ectopic expression of an overactive form of *Hand2,* or of the *5′Hoxd11–13* genes, results in an anterior redistribution of *Shh* expression and posteriorization of anterior skeletal elements, reinforcing the role of these genes in defining limb asymmetry (McFadden *et al.*, 2002; Zakany *et al.*, 2004). A similar phenotype is observed specifically in the forelimbs of mice characterized by conditional inactivation of the hedgehog receptor patched1 in the limb mesenchyme (Butterfield *et al.*, 2009). In the latter case, the phenotype was attributed to high level ligand-independent activation of the hedgehog pathway across the limb paddle prior to normal establishment of the ZPA, highlighting the importance of correct spatial and temporal positioning of *Shh* expression for limb asymmetry.

In addition to the early pre-patterning events described so far, a number of other molecular interactions work in the limb to further confine *Shh* expression to the posterior margin. These include *Tbx2* and *Tbx3* which are expressed in the anterior and posterior margins of the limb mesoderm and are postulated to act upstream of *Shh* in the posterior limb (Nissim *et al.*, 2007; Rallis *et al.*, 2005). Elegant grafting experiments in the chick have identified the non-AER (DV) border ectoderm as a new signaling center in the limb responsible for maintaining adjacent marginal expression of *Tbx2* (Nissim *et al.*, 2007). These authors propose that while *Tbx2* is expressed at both margins, it induces *Shh* only in the posterior due to the action of pre-patterning factors such as HAND2 in defining a posterior domain competent to express *Shh* (Nissim *et al.*, 2007). It has also been demonstrated recently that the transcription factors ETV4 and ETV5 (Mao *et al.*, 2009; Zhang *et al.*, 2009), and BMP signaling (Bastida *et al.*, 2009) are involved in maintaining posterior restriction of *Shh*. These interactions are discussed in more detail below.

3. Patterning Along the AP Axis

3.1. Interactions between SHH and GLI3

Once *Shh* expression is established in the ZPA, many of the molecules involved in positioning the ZPA now come under the control of SHH. At this stage, expression of the *Hoxd* genes is controlled by a global control region 5′ of the *Hoxd* complex (Tarchini and Duboule, 2006), and is modulated by SHH from the ZPA (Fig. 9.2B). SHH also regulates expression of the gene encoding the BMP antagonist Gremlin (GREM1) (Capdevila *et al.*, 1999; Khokha *et al.*, 2003; Zuniga *et al.*, 1999), although this regulation is thought to be indirect, and paradoxically requires the intermediate action of BMP signaling in a concentration-dependent manner (Nissim *et al.*, 2006). However, the most significant interaction in defining AP patterning is the interaction between SHH and GLI3 (Fig. 9.2B). Cleavage of the full-length GLI3 protein (GLI3FL) to a truncated transcriptional repressor (GLI3R) is inhibited by SHH (Wang *et al.*, 2000), and seminal studies in SHH–GLI3 double mutant embryos showed that this is the major role for SHH in limb patterning (Litingtung *et al.*, 2002; te Welscher *et al.*, 2002b). These studies further established SHH and GLI3 as the primary determinants of digit number and identity in the vertebrate limb, with GLI3 believed to restrain, and SHH to enhance, digit number. Support for this comes from the phenotypes of relevant mouse models. *Shh* null mice display a single digit in the hindlimb (Chiang *et al.*, 1996), while the *Gli3* mutant *extra-toes* is characterized by extra digits [polydactyly; (Hui and Joyner, 1993)]. Furthermore, mutations in the human *GLI3* gene cause a range of human syndromes associated with polydactyly (reviewed by Nieuwenhuis and Hui, 2005).

At this stage, expression of *Gli3* extends across the AP axis of the limb, although with a marked anterior bias. As a result of SHH-dependent inhibition of cleavage, likely combined with SHH-independent regulation of *Gli3* expression (Hill *et al.*, 2009), a gradient is established whereby GLI3R predominates in the anterior limb. The gradient of GLI3R across the limb was initially thought to mediate AP limb patterning (Wang *et al.*, 2000), although more recent studies suggest that the ratio of GLI3FL to GLI3R across the limb paddle is the major determinant of GLI3 action in the limb (Wang *et al.*, 2007). However, although GLI3FL acts as a potent transcriptional activator in some contexts, whether this isoform plays a role in early limb patterning is still the subject of some debate (Hill *et al.*, 2009; Wang *et al.*, 2007). Studies in which the relative concentration of the GLI3R and FL isoforms have been manipulated in the mouse have led to conflicting conclusions (Hill *et al.*, 2009; Wang *et al.*, 2007), with the most recent studies suggesting that GLI3FL plays no appreciable role in AP limb patterning (Hill *et al.*, 2009).

3.2. The primary cilium and GLI3 function in the limb

It is possible that a resolution to the discrepancies discussed above will come at least in part from analysis of mutants in which the formation or function of the primary cilium is altered. The primary cilium is a single microtubule-based organelle that extends from the surface of virtually all vertebrate cells and serves both a mechanosensory and signaling function. Specifically, mouse genetic studies have uncovered an indispensable role for the primary cilium in regulating hedgehog signaling downstream of ligand reception and upstream of the GLI transcription factors (Huangfu and Anderson, 2005; Huangfu *et al.*, 2003). Virtually all cilia-related mouse mutants are characterized by polydactyly, and most show altered ratios of GLI3FL to GLI3R. Moreover, polydactyly is considered one of the hallmark features of ciliopathies, a group of pleiotropic human disorders that have been attributed to ciliary dysfunction (reviewed by Baker and Beales, 2009).

Assembly and disassembly of cilia rely on transport of essential cargo to (anterograde) and from (retrograde) the cilium tip *via* a polarized trafficking system known as intraflagellar transport (IFT). The IFT particle is a multi-protein complex, a detailed description of which is beyond the scope of this review but is covered in a number of recent review articles (Pedersen and Rosenbaum, 2008; Pedersen *et al.*, 2008).

A number of mouse mutants with defective primary cilia due to mutation in genes encoding components of the IFT machinery, show evidence of both reduced GLI repressor and activator levels in the limbs (Liu *et al.*, 2005; May *et al.*, 2005). The reduction in GLI activators, as evidenced by decreased expression of the downstream pathway markers *Ptch1* and *Gli1*, is often accompanied by an accumulation of GLI3FL levels in mutant embryos (Liu *et al.*, 2005; May *et al.*, 2005). These observations suggest that in this context Gli3FL is not acting as a transcriptional activator, possibly because an additional conversion step is required. While levels of GLI3FL may not necessarily impact on limb patterning, they nevertheless provide a readout of GLI3 status. Analysis of many of the cilia-related mutants to date suggests that the formation of both GLI repressors and activators is dependent on the primary cilium. These studies also support the notion that in these mutants polydactyly, generally considered a hallmark of Shh gain-of-function mutants, results primarily from de-repression due to loss of GLI3R. However, it should be noted that at least one mutant with a defect in the retrograde IFT protein (THM1) displays enhanced hedgehog signaling in the limb (Tran *et al.*, 2008). Likewise, mice harboring mutations in the Meckel syndrome gene (*Mks1*), which is implicated in ciliogenesis, show some evidence of expanded hedgehog signaling in the limb (Weatherbee *et al.*, 2009). These data imply that the regulation of GLI function by the primary cilium is complex, and the full elucidation of the mechanism by which this occurs is likely to come from detailed trafficking studies of the type currently underway in a number of laboratories (Kim *et al.*, 2009; Rohatgi *et al.*, 2007). The limb provides an ideal model to support these studies, since both GLI3 repressor and activator readouts can be monitored using a battery of well-characterized markers.

3.3. SHH-mediated digit specification and growth

While the GLI3 gradient is undoubtedly crucial to how the SHH signal is interpreted in the limb, the exquisitely localized expression and graded action of *Shh* itself lies at the heart of the complex interactions that define the AP axis. While *Shh* RNA is confined to the ZPA, SHH protein levels decrease in a graded fashion from the posterior, consistent with a spatial morphogen gradient (Lewis *et al.*, 2001). However, when all cells that had ever expressed *Shh* were marked in a reporter mouse expressing Cre recombinase from the endogenous *Shh* promoter, a role for temporal exposure to SHH in digit patterning was revealed (Harfe *et al.*, 2004). Specifically, it was shown that those posterior cells expressing *Shh* expand anteriorly through proliferation, stop expressing *Shh*, and contribute to digits 5, 4 and part of 3. These individual digits are thought to be specified by a temporal gradient of exposure to high-level autocrine SHH signaling, with those *Shh* descendants contributing to digit 3 being the first to move beyond the ZPA. Digit 2 and part of digit 3 are not formed from *Shh* expressing cells, but are instead subject to paracrine *Shh* signaling in a concentration-dependent manner (Harfe *et al.*, 2004). Digit 1 specification is thought to be independent of SHH, as supported by the presence of a single digit 1 in the hindlimb of *Shh* null mice (Chiang *et al.*, 1996). Fate mapping studies based on *Gli* expressing cells to mark SHH responding cells support these findings by showing that digits 2–5 all develop from cells that have responded to SHH (Ahn and Joyner, 2004). Together these studies suggest that digit 2 is the only digit dependent on a classical concentration gradient of SHH. How this gradient is established is beyond the scope of this review, but several studies primarily in *Drosophila* implicate a role for the interaction of lipid-modified SHH with lipoproteins during secretion (reviewed by Eaton, 2008; Willnow *et al.*, 2007).

Recent data suggest that the early SHH-mediated specification of digit identity described above is just one aspect of the action of SHH in limb patterning. Subsequent to these events, which occur from 9.5 to 10.5 days post coitum (dpc) in the mouse, SHH participates in a later, separable phase to control expansion of the limb field to a size that accommodates the formation of all five digits (Zhu *et al.*, 2008). These authors showed that formation of the digits in this growth phase occurs in a particular sequence. Use of an inducible Cre-recombinase to remove *Shh* at various times during limb patterning revealed that SHH directs digit formation, the identity of which is pre-specified in the patterning phase, in the order 4,2,5,3, with digit 1 being independent of SHH, as mentioned previously. However, it should be noted that the order of digit specification revealed using optical projection tomography analysis of cultured mouse limbs was inconsistent with these data (Boot *et al.*, 2008). These variations may be due to differences in experimental systems, but without reliable markers for each digit this is difficult to determine unequivocally. In the chick, a similar growth-morphogen model for SHH action was also uncovered, but in this case the specification and growth roles for SHH are not clearly separable (Towers *et al.*, 2008). The conclusion from these studies was that these roles are intimately linked, with control of proliferation by SHH determining time of exposure and hence mediating digit identity. While SHH was shown to mediate cell proliferation in the limb through control of known cell cycle regulators (Towers *et al.*, 2008), it is possible that less direct effects of SHH on the cell cycle may also be at play in this system. Whether the differences observed in these two studies represent true species-specific variations in the role of SHH in mediating both growth and patterning is not currently clear.

4. Proximal–Distal Patterning and Outgrowth

4.1. FGF signaling from the AER

When the AER is removed, distal limb structures fail to form, with the degree of severity in limb truncation correlating with the timing of AER removal (Saunders, 1948; Summerbell, 1974). These early studies established the AER as the major signaling center determining PD outgrowth.

Furthermore, limb outgrowth defects following removal of the AER can be rescued by application of FGFs, suggesting that the function of the AER is defined by secreted FGFs (Fallon *et al.*, 1994; Niswander *et al.*, 1993). While this has been known for some time, the relative roles of individual FGFs in mediating the function of the AER have only recently been elucidated (Mariani *et al.*, 2008).

Fgf4, Fgf8, Fgf9, and *Fgf17* are expressed specifically in the AER, and are known as the AER–*Fgfs*. *Fgf8* is expressed prior to establishment of a morphological ridge and subsequently across the AER until it eventually regresses (Crossley and Martin, 1995). Expression of the remaining AER–FGF genes commences after AER formation, ceases prior to its regression, and is posteriorly biased (reviewed by Fernandez-Teran and Ros, 2008). When the AER–*Fgfs* are inactivated individually in mice, only loss of *Fgf8* generates a substantial limb patterning phenotype (Colvin *et al.*, 2001; Lewandoski *et al.*, 2000; Moon and Capecchi, 2000; Moon *et al.*, 2000; Sun *et al.*, 2000; Xu *et al.*, 2000). These data imply that while *Fgf4, 9*, and *17* are likely to play a role in limb patterning, this role is redundant when *Fgf8* is present. These findings were further extended with recent data based on combined elimination of the AER–*Fgf* genes in mice (Mariani *et al.*, 2008), showing that in the presence of *Fgf8*, individual or combined removal of *Fgf4–9–17* does not affect limb development. However, combinatorial removal of *Fgf4–9–17* in the absence of *Fgf8* produces more severe defects, suggesting that there is a threshold level of AER–FGF signaling below which limb development is severely compromised.

4.2. Models of PD patterning

While the morphology and molecular nature of the AER is well established, the mechanism by which the AER controls PD outgrowth and patterning of the limb has come under a considerable degree of controversy in recent years. The classical progress zone (PZ) model of limb specification is based on the existence of a region of mesenchyme directly under the AER where cells are maintained in a proliferative state through the action of the AER. The cells in this so-called PZ are assigned proximal or distal identity based on the amount of time spent there (Summerbell *et al.*, 1973). The longer a cell spends in the PZ under the influence of the AER the more distal it is fated to become, in a process regulated by an internal clock-like mechanism (Fig. 9.3A). Once a cell at the proximal edge of the PZ is pushed out of range of AER-induced proliferation, it undergoes differentiation and its fate along the PD axis is set. Under this model those cells that exit the PZ first will contribute to the proximal limb elements, while those cells exposed to the AER for longer will become the digits (Summerbell *et al.*, 1973).

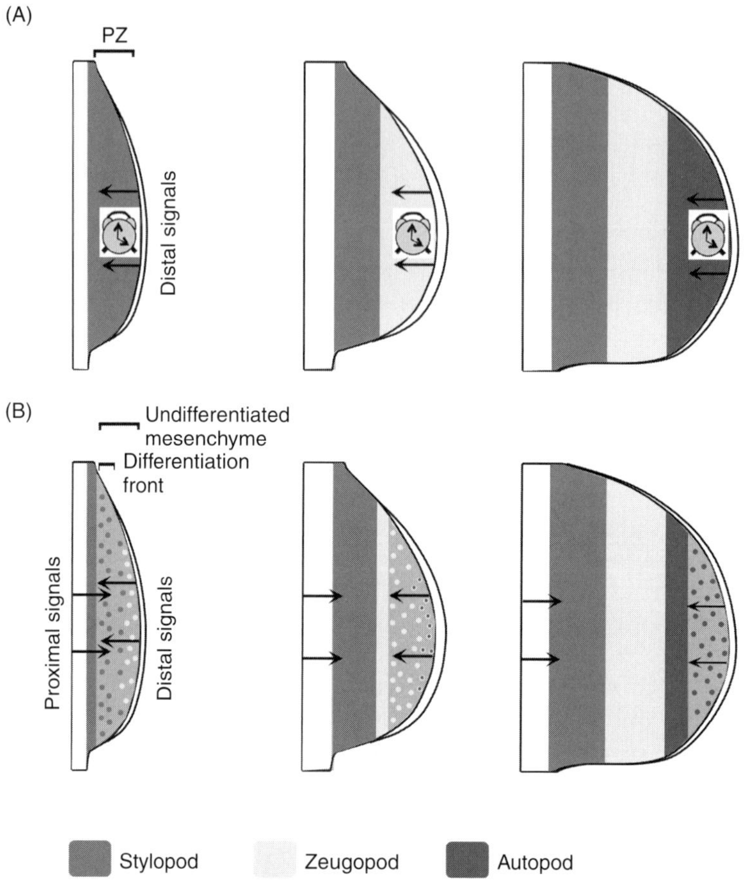

Figure 9.3 Evolving models of PD limb patterning. (A) The PZ model of PD limb specification proposed that cells are maintained in a proliferative state in the PZ region immediately under the AER. Cells that are maintained in the PZ for longer acquire more distal information, in a process regulated by an internal clock mechanism. This model is thought to be inconsistent with available data, prompting the development of new theories to explain PD specification. (B) The two-signal or differentiation front theory suggests that PD specification is controlled by opposing gradients of proximalizing (possibly RA) and distalizing (likely AER–FGFs) factors. The proximal boundary of the zone of FGF influence is called the differentiation front, and once cells pass this front their fate is specified. (See Color Insert.)

The PZ model is consistent with the early studies of AER removal, but falls short in accommodating more recent data. In particular, the PZ model predicts that FGF signals from the AER are permissive for PD patterning but do not play an instructive role in this process. Findings from the combinatorial knockout of mouse FGFs described above contradict this prediction by showing that AER–FGF signaling regulates expression of the PD specification gene *Meis1*, thus establishing the instructive nature of the FGF signal (Mariani *et al.*, 2008). In addition, the fact that a number of these mutants show loss or reduction of the zeugopod while retaining more distal elements, is inconsistent with the PZ model prediction that specification of distal elements occurs only after proximal (Lewandoski *et al.*, 2000; Mariani *et al.*, 2008; Sun *et al.*, 2002). Furthermore, the PZ model does not fit well with expression data based on available markers (discussed by Tabin and Wolpert, 2007), or with recent studies of X-irradiation-induced phocomelia in chick (Galloway *et al.*, 2009). This condition mimics the effects of thalidomide exposure in humans and was previously assumed to be a PD patterning defect that could be rationalized with the PZ model of limb outgrowth. However, lineage tracing and marker analysis in X-irradiated chick limbs have now revealed that this condition is due primarily to a loss of skeletal progenitors (Galloway *et al.*, 2009). A reassessment of all available data has led to the formulation of a new understanding of how PD patterning and outgrowth is regulated (Tabin and Wolpert, 2007). This has been termed the two-signal or differentiation front model.

The underlying basis of the two-signal model of PD specification was first described by Mercader and colleagues (2000). In this scenario, PD specification in the limb is dictated by opposing morphogen gradients of proximalizing factor(s) (proposed to be RA) from the flank and distalizing factor(s) (FGFs) from the AER. The distalizing FGFs keep distal cells in a proliferating state in the "undifferentiated zone" (previously the PZ; Dealy *et al.*, 1997; Globus and Vethamany-Globus, 1976; Niswander and Martin, 1993; Niswander *et al.*, 1993). The proximal boundary of this zone of FGF influence has been termed the "differentiation front" (Tabin and Wolpert, 2007). Under this model, cells commit to differentiation only once they pass beyond this front. The proximal and distal compartments of the limb express a specific genetic profile based on the combination of proximalizing and distalizing factors they receive. At earlier stages these genetic domains are influenced by both gradients and hence overlap, but, during PD outgrowth these domains become separated along the PD axis. The point at which each domain becomes entirely separated appears to be the time at which each segment progenitor pool becomes specified (Fig. 9.3A). The PD fate of a cell is based on the set of genes it expresses as it crosses the differentiation front and, as the limb expands in the PD direction, progenitor pools specific for each limb segment are generated. This conceptualization of PD growth is the most plausible based on current data, although a recent study based on the analysis of *Raldh2–Raldh3* mutant mice rescued with exogenous RA supplementation, proposes that RA is unlikely to be the proximal signal responsible for PD patterning (Zhao *et al.*, 2009). It is likely that current discrepancies will be resolved and a more definitive model of PD outgrowth will evolve over time as more molecular markers become available.

5. Interaction Between the ZPA and the AER

5.1. The SHH–GREM1–FGF feedback loop

Coordinated interaction between the ZPA and AER are fundamental for correct growth and patterning of the limb, and a positive feedback loop between these signaling centers is crucial in regulating this interaction (Fig. 9.4; Laufer *et al.*, 1994; Niswander *et al.*, 1994). A key intermediate in this interaction is the diffusible BMP antagonist GREM1 which is initially activated in the mesenchyme by BMP4 in a regulatory loop that will be discussed in more detail below (Fig. 9.4A; Benazet *et al.*, 2009). Once the ZPA is established, SHH induces *Grem1* expression (Capdevila *et al.*, 1999; Khokha *et al.*, 2003; Zuniga *et al.*, 1999), likely acting through BMP signaling (Nissim *et al.*, 2006). Upregulated GREM1 in turn antagonizes BMP-mediated inhibition of *Fgf* expression in the AER (Khokha *et al.*, 2003; Michos *et al.*, 2004). In the final step of the loop, FGFs from the AER maintain *Shh* expression in the ZPA (Fig. 9.4B; Laufer *et al.*, 1994; Niswander *et al.*, 1994).

More recently, novel interactions between the ZPA and AER have been uncovered. BMP signaling was shown to negatively regulate *Shh* transcription in both chick and mouse limbs (Bastida *et al.*, 2009). Because

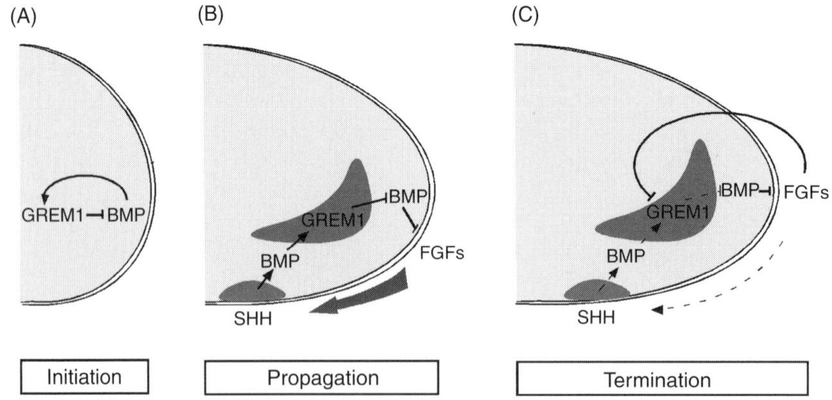

Figure 9.4 Regulatory circuits controlling limb patterning and outgrowth rely on interactions between SHH in the ZPA and FGFs in the AER. (A) The early phase is SHH-independent and involves high levels of BMP4 signaling inducing *Grem1* expression which in turn inhibits BMP4 in a fast-acting regulatory loop. (B) Decreasing levels of BMP4 allow establishment of the slower SHH–GREM1–FGF loop. SHH now controls *Grem1* expression which, through inhibition of BMP, allows *Fgf* expression in the AER. FGF in turn maintains *Shh* expression. (C) When FGF levels reach a threshold they feedback to inhibit *Grem1* expression (FGF–GREM inhibitory loop) thus allowing an increase in BMP action, and subsequent termination of the signaling circuit.

SHH can also regulate expression of *Bmp* genes in the limb (Nissim *et al.*, 2006), a BMP–*Shh* negative feedback loop is thought to exist. Moreover, the downregulation of *Shh* transcription by BMP signaling involves antagonism of FGF signaling (Bastida *et al.*, 2009). These data identify a novel role for BMP signaling in regulating *Shh* transcription and thus controlling the ZPA. They also reveal that in addition to maintaining *Shh* expression, FGFs from the AER independently induce BMP signaling, which subsequently inhibits *Shh* expression. This link with FGF signaling in the negative control of *Shh* expression is reminiscent of the role of ETV4 and ETV5, two members of the E twenty-six (ETS) family of transcription factors that act downstream of FGF signaling to restrict *Shh* expression from the anterior limb (Mao *et al.*, 2009; Zhang *et al.*, 2009). These studies highlight the complexity of the interaction between FGF signaling from the AER and SHH from the ZPA, and attempts to understand these interactions based on mathematical modeling will be discussed below.

5.2. Termination of limb outgrowth

There are two current models for how the growth of the limb bud is terminated, and it is likely that aspects of both models operate *in vivo*. The earlier model is based on the observation that, although SHH induces *Grem1* expression, *Shh* expressing cells and their descendants are unable to express *Grem1* (Scherz *et al.*, 2004). The mechanism for this refractoriness is yet to be elucidated, but it has been proposed that the high levels of autocrine signaling achieved by *Shh* expressing cells may contribute to this phenomenon (Scherz *et al.*, 2004). As *Shh* expressing cells and their descendants proliferate and expand anteriorly, the *Grem1*-expressing zone of cells eventually moves beyond the influence of SHH from the ZPA. As a result, SHH can no longer induce *Grem1*, BMP activity inhibits expression of *Fgfs* in the AER, and the SHH–GREM1–FGF signaling loop is broken. The correct termination of this signaling loop is essential for regulation of limb size (Scherz *et al.*, 2004).

While the above model is consistent with the sequence of signal termination observed in chick, analysis of gene expression in mouse limb buds revealed that *Fgf4* expression ceases first, then *Shh* and lastly *Grem1* (Verheyden and Sun, 2008). Careful analysis of gene expression in a number of relevant mouse mutants, combined with bead implantation studies in chick, led to the formulation of a potentially complementary model for termination of limb outgrowth which is consistent with the sequence of signal cessation in both the mouse and chick (Fig. 9.4C; Verheyden and Sun, 2008). This model is based on the demonstration that AER–FGF signaling represses *Grem1* expression, and proposes that the SHH–GREM1–FGF loop is based on relatively low levels of AER–FGF signals. As AER–FGF signaling increases over time, it eventually reaches a threshold level above which it inhibits *Grem1* expression. This results in the downregulation of *Grem1* levels below the AER, establishing a zone of mesenchyme immediately abutting the AER that does not express *Grem1*. Over time, the *Grem1*-negative domain expands until, in the mouse limb, *Grem1* expressing cells are separated from the AER by too great a distance to allow maintenance of *Fgf* gene expression, subsequently leading to decreased *Shh* expression. In the chick, it is proposed that the *Grem1*-negative domain first prevents *Shh* from maintaining *Grem1* expression, thus inhibiting AER–FGFs and subsequently SHH (Verheyden and Sun, 2008). As a result of this FGF–GREM1 inhibitory loop, the signaling interactions between the ZPA and AER subside, and limb outgrowth is terminated.

6. Towards a Systems Biology Approach to Limb Patterning

Our current knowledge of the molecular basis of limb patterning has to a large extent been derived from gene-by-gene analyses in model organisms. However, the advent of more systems biology based approaches is now moving this analysis to a new level. Genomics-based screens have already begun to elucidate new players in limb patterning (McGlinn *et al.*, 2005; Vokes *et al.*, 2008). The challenge now is to integrate new and existing players into a comprehensive roadmap of molecular limb patterning in time and space. A recent study has begun this task by adopting a computational approach to modeling the genetic hierarchy of regulatory loops operating in concert to pattern the mouse limb (Benazet *et al.*, 2009). Using a hypomorphic *Bmp4* allele to successively reduce *Bmp4* dosage in a *Grem1* null background, Benazet and co-workers established the existence of an early BMP4–GREM1 regulatory loop. Kinetic analyses based on experimental manipulations in mouse limb culture were combined with mathematical modeling to simulate the timing of this loop relative to the SHH–GREM1–FGF loop (Fig. 9.4). This showed that in an early initiation phase around 9.0 dpc in the mouse, BMP4 induces *Grem1* expression, which subsequently antagonizes BMP4 in a fast-acting regulatory loop (approximately 2 h loop time). As a result of this interaction, BMP4 signaling is reduced to a level that allows establishment of the slower SHH–GREM1–FGF loop, which operates with a loop time of approximately 12 h. At this stage, SHH takes over from BMP4 in controlling *Grem1* expression, and the initiation phase is replaced by SHH-dependent propagation of limb patterning. Theoretical simulations failed in the absence of BMP4, predicting that BMP4 is the key factor required for initiation of limb patterning, and that GREM1 is the essential hub linking initiation with

SHH-dependent propagation, and ultimate termination mediated by the FGF–GREM1 inhibitory loop (Verheyden and Sun, 2008). The termination phase is marked by an increase in BMP4 signaling, highlighting the biphasic nature of BMP4 function in the limb. The experiments of Benazet and co-workers (2009) also revealed that this inter-connected series of regulatory loops acts to buffer the system against genetic and environmental insult, creating a robustness distinct from that achieved by genetic redundancy.

These studies undoubtedly represent the tip of the iceberg with respect to what can be learnt by applying systems biology approaches to the analysis of limb patterning. As more molecular players are uncovered and incorporated into existing hierarchies, our understanding of the complexities of limb patterning will increase. Appreciation of the detailed signaling interactions in the developing limb will also further illuminate the mechanisms employed to integrate the processes of cell adhesion, differentiation, and death in shaping the mature limb. How these processes are coordinated in time and space remains a major focus of current research efforts. In particular, the specification of the skeletal elements in the limb is intricately linked to patterning, and plays a major role in determining final limb form. The molecular regulation of skeletogenesis is beyond the scope of this review, but is covered in detail elsewhere in this edition.

Lessons learnt from decades of experimental analysis in the vertebrate limb have not only shaped our understanding of limb patterning, but have greatly enhanced our appreciation of organogenesis in general. There is no doubt that the coming years will be an exciting time as we move further towards a detailed systems biology approach, complemented by more classical analyses, to reveal in full the complex regulatory circuits underpinning vertebrate limb patterning.

ACKNOWLEDGMENTS

The authors would like to thank Malcolm Logan, Veronique Duboc, Peleg Hasson and Jimmy Kuang-Hsien Hu for critical reading of the manuscript and helpful comments. We apologise to those of our colleagues whose work we could not cite due to space constraints. Work on limb development in the Wicking lab is supported by the Australian National Health and Medical Research Council (NHMRC). CW is an NHMRC Senior Research Fellow.

REFERENCES

Agarwal, P., Wylie, J. N., Galceran, J., Arkhitko, O., Li, C., Deng, C., Grosschedl, R., and Bruneau, B. G. (2003). Tbx5 is essential for forelimb bud initiation following patterning of the limb field in the mouse embryo. *Development* **130**, 623–633.

Ahn, S., and Joyner, A. L. (2004). Dynamic changes in the response of cells to positive hedgehog signaling during mouse limb patterning. *Cell* **118**, 505–516.

Ahn, D. G., Kourakis, M. J., Rohde, L. A., Silver, L. M., and Ho, R. K. (2002). T-box gene tbx5 is essential for formation of the pectoral limb bud. *Nature* **417**, 754–758.

Arques, C. G., Doohan, R., Sharpe, J., and Torres, M. (2007). Cell tracing reveals a dorsoventral lineage restriction plane in the mouse limb bud mesenchyme. *Development* **134**, 3713–3722.

Bai, C. B., Auerbach, W., Lee, J. S., Stephen, D., and Joyner, A. L. (2002). Gli2, but not gli1, is required for initial Shh signaling and ectopic activation of the Shh pathway. *Development* **129**, 4753–4761.

Baker, K., and Beales, P. L. (2009). Making sense of cilia in disease: the human ciliopathies. *Am. J. Med. Genet. C Semin. Med. Genet.* **151C**, 281–295.

Bastida, M. F., Sheth, R., and Ros, M. A. (2009). A BMP–Shh negative-feedback loop restricts Shh expression during limb development. *Development* **136**, 3779–3789.

Benazet, J. D., Bischofberger, M., Tiecke, E., Goncalves, A., Martin, J. F., Zuniga, A., Naef, F., and Zeller, R. (2009). A self-regulatory system of interlinked signaling feedback loops controls mouse limb patterning. *Science* **323**, 1050–1053.

Boot, M. J., Westerberg, C. H., Sanz-Ezquerro, J., Cotterell, J., Schweitzer, R., Torres, M., and Sharpe, J. (2008). In vitro whole-organ imaging: 4d quantification of growing mouse limb buds. *Nat. Methods* **5**, 609–612.

Burke, A. C., Nelson, C. E., Morgan, B. A., and Tabin, C. (1995). Hox genes and the evolution of vertebrate axial morphology. *Development* **121**, 333–346.

Butterfield, N. C., Metzis, V., McGlinn, E., Bruce, S. J., Wainwright, B. J., and Wicking, C. (2009). Patched 1 is a crucial determinant of asymmetry and digit number in the vertebrate limb. *Development* **136**, 3515–3524.

Capdevila, J., Tsukui, T., Rodriquez Esteban, C., Zappavigna, V., and Izpisua Belmonte, J. C. (1999). Control of vertebrate limb outgrowth by the proximal factor meis2 and distal antagonism of BMPs by gremlin. *Mol. Cell* **4**, 839–849.

Charite, J., McFadden, D. G., and Olson, E. N. (2000). The bHLH transcription factor dHAND controls sonic hedgehog expression and establishment of the zone of polarizing activity during limb development. *Development* **127**, 2461–2470.

Chen, H., and Johnson, R. L. (1999). Dorsoventral patterning of the vertebrate limb: a process governed by multiple events. *Cell Tissue Res.* **296**, 67–73.

Chen, H., and Johnson, R. L. (2002). Interactions between dorsal–ventral patterning genes lmx1b, engrailed-1 and wnt-7a in the vertebrate limb. *Int. J. Dev. Biol.* **46**, 937–941.

Chiang, C., Litingtung, Y., Lee, E., Young, K. E., Corden, J. L., Westphal, H., and Beachy, P. A. (1996). Cyclopia and defective axial patterning in mice lacking sonic hedgehog gene function. *Nature* **383**, 407–413.

Cohn, M. J., Patel, K., Krumlauf, R., Wilkinson, D. G., Clarke, J. D., and Tickle, C. (1997). Hox9 genes and vertebrate limb specification. *Nature* **387**, 97–101.

Cohn, M. J., and Tickle, C. (1999). Developmental basis of limblessness and axial patterning in snakes. *Nature* **399**, 474–479.

Colvin, J. S., White, A. C., Pratt, S. J., and Ornitz, D. M. (2001). Lung hypoplasia and neonatal death in fgf9-null mice identify this gene as an essential regulator of lung mesenchyme. *Development* **128**, 2095–2106.

Cretekos, C. J., Wang, Y., Green, E. D., Martin, J. F., Rasweiler, J. Jr., and Behringer, R. R. (2008). Regulatory divergence modifies limb length between mammals. *Genes Dev.* **22**, 141–151.

Crossley, P. H., and Martin, G. R. (1995). The mouse fgf8 gene encodes a family of polypeptides and is expressed in regions that direct outgrowth and patterning in the developing embryo. *Development* **121**, 439–451.

Davis, C. A., and Joyner, A. L. (1988). Expression patterns of the homeo box-containing genes en-1 and en-2 and the proto-oncogene int-1 diverge during mouse development. *Genes Dev.* **2**, 1736–1744.

Dealy, C. N., Seghatoleslami, M. R., Ferrari, D., and Kosher, R. A. (1997). FGF-stimulated outgrowth and proliferation of limb mesoderm is dependent on syndecan-3. *Dev. Biol.* **184**, 343–350.

Eaton, S. (2008). Multiple roles for lipids in the hedgehog signalling pathway. *Nat. Rev. Mol. Cell Biol.* **9**, 437–445.

Fallon, J. F., Lopez, A., Ros, M. A., Savage, M. P., Olwin, B. B., and Simandl, B. K. (1994). FGF-2: apical ectodermal ridge growth signal for chick limb development. *Science* **264**, 104–107.

Fernandez-Teran, M., Piedra, M. E., Kathiriya, I. S., Srivastava, D., Rodriguez-Rey, J. C., and Ros, M. A. (2000). Role of dHAND in the anterior-posterior polarization of the limb bud: implications for the sonic hedgehog pathway. *Development* **127**, 2133–2142.

Fernandez-Teran, M., and Ros, M. A. (2008). The apical ectodermal ridge: morphological aspects and signaling pathways. *Int. J. Dev. Biol.* **52**, 857–871.

Galloway, J. L., Delgado, I., Ros, M. A., and Tabin, C. J. (2009). A reevaluation of X-irradiation-induced phocomelia and proximodistal limb patterning. *Nature* **460**, 400–404.

Gibert, Y., Gajewski, A., Meyer, A., and Begemann, G. (2006). Induction and prepatterning of the zebrafish pectoral fin bud requires axial retinoic acid signaling. *Development* **133**, 2649–2659.

Globus, M., and Vethamany-Globus, S. (1976). An in vitro analogue of early chick limb bud outgrowth. *Differentiation* **6**, 91–96.

Grandel, H., Lun, K., Rauch, G. J., Rhinn, M., Piotrowski, T., Houart, C., Sordino, P., Kuchler, A. M., Schulte-Merker, S., Geisler, R., Holder, N., Wilson, S. W., and Brand, M. (2002). Retinoic acid signalling in the zebrafish embryo is necessary during pre-segmentation stages to pattern the anterior–posterior axis of the CNS and to induce a pectoral fin bud. *Development* **129**, 2851–2865.

Harfe, B. D., Scherz, P. J., Nissim, S., Tian, H., McMahon, A. P., and Tabin, C. J. (2004). Evidence for an expansion-based temporal Shh gradient in specifying vertebrate digit identities. *Cell* **118**, 517–528.

Hill, P., Gotz, K., and Ruther, U. (2009). A SHH-independent regulation of gli3 is a significant determinant of anteroposterior patterning of the limb bud. *Dev. Biol.* **328**, 506–516.

Hockman, D., Cretekos, C. J., Mason, M. K., Behringer, R. R., Jacobs, D. S., and Illing, N. (2008). A second wave of sonic hedgehog expression during the development of the bat limb. *Proc. Natl. Acad. Sci. USA* **105**, 16982–16987.

Huangfu, D., and Anderson, K. V. (2005). Cilia and hedgehog responsiveness in the mouse. *Proc. Natl. Acad. Sci. USA* **102**, 11325–11330.

Huangfu, D., Liu, A., Rakeman, A. S., Murcia, N. S., Niswander, L., and Anderson, K. V. (2003). Hedgehog signalling in the mouse requires intraflagellar transport proteins. *Nature* **426**, 83–87.

Hui, C. C., and Joyner, A. L. (1993). A mouse model of greig cephalopolysyndactyly syndrome: the extra-toesJ mutation contains an intragenic deletion of the gli3 gene. *Nat. Genet.* **3**, 241–246.

Khokha, M. K., Hsu, D., Brunet, L. J., Dionne, M. S., and Harland, R. M. (2003). Gremlin is the BMP antagonist required for maintenance of Shh and Fgf signals during limb patterning. *Nat. Genet* **34**, 303–307.

Kim, J., Kato, M., and Beachy, P. A., 2009. Gli2 trafficking links Hedgehog-dependent activation of smoothened in the primary cilium to transcriptional activation in the nucleus. *Proc. Natl. Acad. Sci. USA* **106**, 21666–21671.

Kmita, M., Tarchini, B., Zakany, J., Logan, M., Tabin, C. J., and Duboule, D. (2005). Early developmental arrest of mammalian limbs lacking HoxA/HoxD gene function. *Nature* **435**, 1113–1116.

Krumlauf, R. (1994). Hox genes in vertebrate development. *Cell* **78**, 191–201.

Laufer, E., Nelson, C. E., Johnson, R. L., Morgan, B. A., and Tabin, C. (1994). Sonic hedgehog and fgf-4 act through a signaling cascade and feedback loop to integrate growth and patterning of the developing limb bud. *Cell* **79**, 993–1003.

Lewandoski, M., Sun, X., and Martin, G. R. (2000). Fgf8 signalling from the AER is essential for normal limb development. *Nat. Genet.* **26**, 460–463.

Lewis, P. M., Dunn, M. P., McMahon, J. A., Logan, M., Martin, J. F., St-Jacques, B., and McMahon, A. P. (2001). Cholesterol modification of sonic hedgehog is required for long-range signaling activity and effective modulation of signaling by ptc1. *Cell* **105**, 599–612.

Litingtung, Y., Dahn, R. D., Li, Y., Fallon, J. F., and Chiang, C. (2002). Shh and gli3 are dispensable for limb skeleton formation but regulate digit number and identity. *Nature* **418**, 979–983.

Liu, A., Wang, B., and Niswander, L. A. (2005). Mouse intraflagellar transport proteins regulate both the activator and repressor functions of gli transcription factors. *Development* **132**, 3103–3111.

Mao, J., McGlinn, E., Huang, P., Tabin, C. J., and McMahon, A. P. (2009). Fgf-dependent ETV4/5 activity is required for posterior restriction of sonic hedgehog and promoting outgrowth of the vertebrate limb. *Dev. Cell* **16**, 600–606.

Mariani, F. V., Ahn, C. P., and Martin, G. R. (2008). Genetic evidence that FGFs have an instructive role in limb proximal–distal patterning. *Nature* **453**, 401–405.

May, S. R., Ashique, A. M., Karlen, M., Wang, B., Shen, Y., Zarbalis, K., Reiter, J., Ericson, J., and Peterson, A. S. (2005). Loss of the retrograde motor for IFT disrupts localization of smo to cilia and prevents the expression of both activator and repressor functions of gli. *Dev. Biol.* **287**, 378–389.

McFadden, D. G., McAnally, J., Richardson, J. A., Charite, J., and Olson, E. N. (2002). Misexpression of dHAND induces ectopic digits in the developing limb bud in the absence of direct DNA binding. *Development* **129**, 3077–3088.

McGlinn, E., van Bueren, K. L., Fiorenza, S., Mo, R., Poh, A. M., Forrest, A., Soares, M. B., Bonaldo Mde, F., Grimmond, S., Hui, C. C., Wainwright, B., and Wicking, C. (2005). Pax9 and jagged1 act downstream of gli3 in vertebrate limb development. *Mech. Dev.* **122**, 1218–1233.

Mercader, N., Leonardo, E., Piedra, M. E., Martinez, A. C., Ros, M. A., and Torres, M. (2000). Opposing RA and FGF signals control proximodistal vertebrate limb development through regulation of meis genes. *Development* **127**, 3961–3970.

Michos, O., Panman, L., Vintersten, K., Beier, K., Zeller, R., and Zuniga, A. (2004). Gremlin-mediated BMP antagonism induces the epithelial–mesenchymal feedback signaling controlling metanephric kidney and limb organogenesis. *Development* **131**, 3401–3410.

Min, H., Danilenko, D. M., Scully, S. A., Bolon, B., Ring, B. D., Tarpley, J. E., DeRose, M., and Simonet, W. S. (1998). Fgf-10 is required for both limb and lung development and exhibits striking functional similarity to drosophila branchless. *Genes Dev.* **12**, 3156–3161.

Minguillon, C., Del Buono, J., and Logan, M. P. (2005). Tbx5 and tbx4 are not sufficient to determine limb-specific morphologies but have common roles in initiating limb outgrowth. *Dev. Cell* **8**, 75–84.

Moon, A. M., Boulet, A. M., and Capecchi, M. R. (2000). Normal limb development in conditional mutants of fgf4. *Development* **127**, 989–996.

Moon, A. M., and Capecchi, M. R. (2000). Fgf8 is required for outgrowth and patterning of the limbs. *Nat. Genet.* **26**, 455–459.

Naiche, L. A., and Papaioannou, V. E. (2003). Loss of tbx4 blocks hindlimb development and affects vascularization and fusion of the allantois. *Development* **130**, 2681–2693.

Naiche, L. A., and Papaioannou, V. E. (2007). Tbx4 is not required for hindlimb identity or post-bud hindlimb outgrowth. *Development* **134**, 93–103.

Ng, J. K., Kawakami, Y., Buscher, D., Raya, A., Itoh, T., Koth, C. M., Rodriguez Esteban, C., Rodriguez-Leon, J., Garrity, D. M., Fishman, M. C., and Izpisua Belmonte, J. C. (2002). The limb identity gene tbx5 promotes limb initiation by interacting with wnt2b and fgf10. *Development* **129**, 5161–5170.

Niederreither, K., Subbarayan, V., Dolle, P., and Chambon, P. (1999). Embryonic retinoic acid synthesis is essential for early mouse post-implantation development. *Nat. Genet.* **21**, 444–448.

Niederreither, K., Vermot, J., Schuhbaur, B., Chambon, P., and Dolle, P. (2002). Embryonic retinoic acid synthesis is required for forelimb growth and anteroposterior patterning in the mouse. *Development* **129**, 3563–3574.

Nieuwenhuis, E., and Hui, C. C. (2005). Hedgehog signaling and congenital malformations. *Clin. Genet.* **67**, 193–208.

Nissim, S., Allard, P., Bandyopadhyay, A., Harfe, B. D., and Tabin, C. J. (2007). Characterization of a novel ectodermal signaling center regulating tbx2 and Shh in the vertebrate limb. *Dev. Biol.* **304**, 9–21.

Nissim, S., Hasso, S. M., Fallon, J. F., and Tabin, C. J. (2006). Regulation of gremlin expression in the posterior limb bud. *Dev. Biol.* **299**, 12–21.

Niswander, L., Jeffrey, S., Martin, G. R., and Tickle, C. (1994). A positive feedback loop coordinates growth and patterning in the vertebrate limb. *Nature* **371**, 609–612.

Niswander, L., and Martin, G. R. (1993). FGF-4 and BMP-2 have opposite effects on limb growth. *Nature* **361**, 68–71.

Niswander, L., Tickle, C., Vogel, A., Booth, I., and Martin, G. R. (1993). FGF-4 replaces the apical ectodermal ridge and directs outgrowth and patterning of the limb. *Cell* **75**, 579–587.

Ohuchi, H., Nakagawa, T., Yamamoto, A., Araga, A., Ohata, T., Ishimaru, Y., Yoshioka, H., Kuwana, T., Nohno, T., Yamasaki, M., Itoh, N., and Noji, S. (1997). The mesenchymal factor, FGF10, initiates and maintains the outgrowth of the chick limb bud through interaction with FGF8, an apical ectodermal factor. *Development* **124**, 2235–2244.

Park, H. L., Bai, C., Platt, K. A., Matise, M. P., Beeghly, A., Hui, C. C., Nakashima, M., and Joyner, A. L. (2000). Mouse gli1 mutants are viable but have defects in SHH signaling in combination with a gli2 mutation. *Development* **127**, 1593–1605.

Pearse, R. V. 2nd, Scherz, P. J., Campbell, J. K., and Tabin, C. J. (2007). A cellular lineage analysis of the chick limb bud. *Dev. Biol.* **310**, 388–400.

Pedersen, L. B., and Rosenbaum, J. L. (2008). Intraflagellar transport (IFT) role in ciliary assembly, resorption and signalling. *Curr. Top. Dev. Biol.* **85**, 23–61.

Pedersen, L. B., Veland, I. R., Schroder, J. M., and Christensen, S. T. (2008). Assembly of primary cilia. *Dev. Dyn.* **237**, 1993–2006.

Power, S. C., Lancman, J., and Smith, S. M. (1999). Retinoic acid is essential for shh/hoxd signaling during rat limb outgrowth but not for limb initiation. *Dev. Dyn.* **216**, 469–480.

Rallis, C., Bruneau, B. G., Del Buono, J., Seidman, C. E., Seidman, J. G., Nissim, S., Tabin, C. J., and Logan, M. P. (2003). Tbx5 is required for forelimb bud formation and continued outgrowth. *Development* **130**, 2741–2751.

Rallis, C., Del Buono, J., and Logan, M. P. (2005). Tbx3 can alter limb position along the rostrocaudal axis of the developing embryo. *Development* **132**, 1961–1970.

Rancourt, D. E., Tsuzuki, T., and Capecchi, M. R. (1995). Genetic interaction between hoxb-5 and hoxb-6 is revealed by nonallelic noncomplementation. *Genes Dev.* **9**, 108–122.

Rohatgi, R., Milenkovic, L., and Scott, M. P. (2007). Patched1 regulates hedgehog signaling at the primary cilium. *Science* **317**, 372–376.

Saunders, J. J. (1948). The proximo-distal sequence of origin of the parts of the chick wing and the role of the ectoderm. *J. Exp. Zool.* **108**, 363–403.

Scherz, P. J., Harfe, B. D., McMahon, A. P., and Tabin, C. J. (2004). The limb bud Shh–Fgf feedback loop is terminated by expansion of former ZPA cells. *Science* **305**, 396–399.

Searls, R. L., and Janners, M. Y. (1971). The initiation of limb bud outgrowth in the embryonic chick. *Dev. Biol.* **24**, 198–213.

Sekine, K., Ohuchi, H., Fujiwara, M., Yamasaki, M., Yoshizawa, T., Sato, T., Yagishita, N., Matsui, D., Koga, Y., Itoh, N., and Kato, S. (1999). Fgf10 is essential for limb and lung formation. *Nat. Genet.* **21**, 138–141.

Stratford, T., Horton, C., and Maden, M. (1996). Retinoic acid is required for the initiation of outgrowth in the chick limb bud. *Curr. Biol.* **6**, 1124–1133.

Summerbell, D. (1974). Interaction between the proximo-distal and antero-posterior co-ordinates of positional value during the specification of positional information in the early development of the chick limb-bud. *J. Embryol. Exp. Morphol.* **32**, 227–237.

Summerbell, D., Lewis, J. H., and Wolpert, L. (1973). Positional information in chick limb morphogenesis. *Nature* **244**, 492–496.

Summerbell, D., and Wolpert, L. (1973). Precision of development in chick limb morphogenesis. *Nature* **244**, 228–230.

Sun, X., Lewandoski, M., Meyers, E. N., Liu, Y. H., Maxson, R. E. Jr., and Martin, G. R. (2000). Conditional inactivation of fgf4 reveals complexity of signalling during limb bud development. *Nat. Genet.* **25**, 83–86.

Sun, X., Mariani, F. V., and Martin, G. R. (2002). Functions of FGF signalling from the apical ectodermal ridge in limb development. *Nature* **418**, 501–508.

Tabin, C., and Wolpert, L. (2007). Rethinking the proximodistal axis of the vertebrate limb in the molecular era. *Genes Dev.* **21**, 1433–1442.

Tarchini, B., and Duboule, D. (2006). Control of hoxd genes' collinearity during early limb development. *Dev. Cell* **10**, 93–103.

te Welscher, P., Fernandez-Teran, M., Ros, M. A., and Zeller, R. (2002a). Mutual genetic antagonism involving GLI3 and dHAND prepatterns the vertebrate limb bud mesenchyme prior to SHH signaling. *Genes Dev.* **16**, 421–426.

te Welscher, P., Zuniga, A., Kuijper, S., Drenth, T., Goedemans, H. J., Meijlink, F., and Zeller, R. (2002b). Progression of vertebrate limb development through SHH-mediated counteraction of GLI3. *Science* **298**, 827–830.

Towers, M., Mahood, R., Yin, Y., and Tickle, C. (2008). Integration of growth and specification in chick wing digit-patterning. *Nature* **452**, 882–886.

Tran, P. V., Haycraft, C. J., Besschetnova, T. Y., Turbe-Doan, A., Stottmann, R. W., Herron, B. J., Chesebro, A. L., Qiu, H., Scherz, P. J., Shah, J. V., Yoder, B. K., and Beier, D. R. (2008). THM1 negatively modulates mouse sonic hedgehog signal transduction and affects retrograde intraflagellar transport in cilia. *Nat. Genet.* **40**, 403–410.

Verheyden, J. M., and Sun, X. (2008). An Fgf/Gremlin inhibitory feedback loop triggers termination of limb bud outgrowth. *Nature* **454**, 638–641.

Vokes, S. A., Ji, H., Wong, W. H., and McMahon, A. P. (2008). A genome-scale analysis of the cis-regulatory circuitry underlying sonic hedgehog-mediated patterning of the mammalian limb. *Genes Dev.* **22**, 2651–2663.

Wang, B., Fallon, J. F., and Beachy, P. A. (2000). Hedgehog-regulated processing of gli3 produces an anterior/posterior repressor gradient in the developing vertebrate limb. *Cell* **100**, 423–434.

Wang, C., Ruther, U., and Wang, B. (2007). The Shh-independent activator function of the full-length gli3 protein and its role in vertebrate limb digit patterning. *Dev. Biol.* **305**, 460–469.

Weatherbee, S. D., Niswander, L. A., and Anderson, K. V. (2009). A mouse model for Meckel syndrome reveals mks1 is required for ciliogenesis and hedgehog signaling. *Hum. Mol. Genet.* **18**, 4565–4575.

Willnow, T. E., Hammes, A., and Eaton, S. (2007). Lipoproteins and their receptors in embryonic development: more than cholesterol clearance. *Development* **134**, 3239–3249.

Xu, J., Liu, Z., and Ornitz, D. M. (2000). Temporal and spatial gradients of fgf8 and fgf17 regulate proliferation and differentiation of midline cerebellar structures. *Development* **127**, 1833–1843.

Xu, X., Weinstein, M., Li, C., Naski, M., Cohen, R. I., Ornitz, D. M., Leder, P., and Deng, C. (1998). Fibroblast growth factor receptor 2 (FGFR2)-mediated reciprocal regulation loop between FGF8 and FGF10 is essential for limb induction. *Development* **125**, 753–765.

Zakany, J., Kmita, M., and Duboule, D. (2004). A dual role for hox genes in limb anterior–posterior asymmetry. *Science* **304**, 1669–1672.

Zhang, Z., Verheyden, J. M., Hassell, J. A., and Sun, X. (2009). FGF-regulated ETV genes are essential for repressing Shh expression in mouse limb buds. *Dev. Cell* **16**, 607–613.

Zhao, X., Sirbu, I. O., Mic, F. A., Molotkova, N., Molotkov, A., Kumar, S., and Duester, G. (2009). Retinoic acid promotes limb induction through effects on body axis extension but is unnecessary for limb patterning. *Curr. Biol.* **19**, 1050–1057.

Zhu, J., Nakamura, E., Nguyen, M. T., Bao, X., Akiyama, H., and Mackem, S. (2008). Uncoupling sonic hedgehog control of pattern and expansion of the developing limb bud. *Dev. Cell* **14**, 624–632.

Zuniga, A., Haramis, A. P., McMahon, A. P., and Zeller, R. (1999). Signal relay by BMP antagonism controls the SHH/FGF4 feedback loop in vertebrate limb buds. *Nature* **401**, 598–602.

Zuniga, A., and Zeller, R. (1999). Gli3 (xt) and formin (ld) participate in the positioning of the polarising region and control of posterior limb-bud identity. *Development* **126**, 13–21.

CHAPTER TEN

Eye Development

Jochen Graw

Contents

1. Introduction 344
2. Overview of Eye Development 346
3. Early Stage: The Eye Field 346
   3.1. Formation of the eye field 348
   3.2. Patterning of the eye field 349
   3.3. Splitting the eye field 351
   3.4. From optic vesicle to optic cup 352
4. Lens Development 354
   4.1. Formation of the lens placode and *Pax6* as its master control gene 354
   4.2. Signaling cascades in early lens development 357
   4.3. From lens vesicle to the mature lens 358
5. The Cornea 363
6. The Iris and the Ciliary Body 366
7. The Retina 367
   7.1. The retinal pigmented epithelium 367
   7.2. The neural retina 368
   7.3. Development of the hyaloid and retinal vasculature 373
8. The Optic Nerve 374
   8.1. Outlook: the visual system 376
9. Conclusion and Perspectives 377
Acknowledgments 378
Databases used 378
References 378

Abstract

The vertebrate eye comprises tissues from different embryonic origins: the lens and the cornea are derived from the surface ectoderm, but the retina and the epithelial layers of the iris and ciliary body are from the anterior neural plate. The timely action of transcription factors and inductive signals ensure the

Helmholtz Center Munich—German Research Center for Environmental Health, Institute of Developmental Genetics, Neuherberg, Germany

Current Topics in Developmental Biology, Volume 90
ISSN 0070-2153, DOI 10.1016/S0070-2153(10)90010-0

correct development of the different eye components. Establishing the genetic basis of eye defects in zebrafishes, mouse, and human has been an important tool for the detailed analysis of this complex process. A single eye field forms centrally within the anterior neural plate during gastrulation; it is characterized on the molecular level by the expression of "eye-field transcription factors." The single eye field is separated into two, forming the optic vesicle and later (under influence of the lens placode) the optic cup. The lens develops from the lens placode (surface ectoderm) under influence of the underlying optic vesicle. *Pax6* acts in this phase as master control gene, and genes encoding cytoskeletal proteins, structural proteins, or membrane proteins become activated. The cornea forms from the surface ectoderm, and cells from the periocular mesenchyme migrate into the cornea giving rise for the future cornea stroma. Similarly, the iris and ciliary body form from the optic cup. The outer layer of the optic cup becomes the retinal pigmented epithelium, and the main part of the inner layer of the optic cup forms later the neural retina with six different types of cells including the photoreceptors. The retinal ganglion cells grow toward the optic stalk forming the optic nerve.This review describes the major molecular players and cellular processes during eye development as they are known from frogs, zebrafish, chick, and mice—showing also differences among species and missing links for future research. The relevance to human disorders is one of the major aspects covered throughout the review.

1. Introduction

The vertebrate eye is a very complex organ (Fig. 10.1) built up by the three major tissues, the cornea, the lens, and the retina. It is obvious that its formation depends on highly organized processes that take place during embryonic development, and mutations in key genes lead to severe congenital disorders. In many vertebrates, the eye is a very prominent organ in the head, and major alterations can be recognized easily. In particular for humans, the eye is one of the most important sensory systems and loss of its function causes many social handicaps and changes in personality. Therefore, the eye has provided a fascinating topic for research since decades.

There are two highlights in eye research, which changed our view fundamentally:

- First of all, at the beginning of the last century Hans Spemann made a careful analysis of eye development: his finding of the dependence of lens induction from the underlying optic cup leads to the discovery of the basic concept of "organizers" in development biology, which became a prototype for tissue interactions in embryonic development (Spemann, 1924).

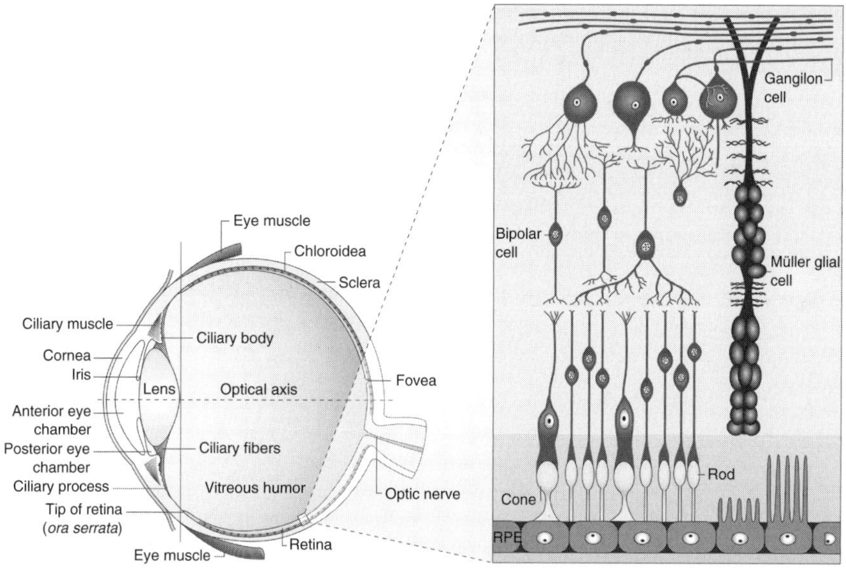

Figure 10.1 Schematic section through an adult human eye. This illustration shows the main tissues of the human eye; note that the optical axis does not coincide with the exit of the optic nerve or with the fovea, which is the area of the highest resolution at the retina. The vertical line divides the anterior segment of the eye (the cornea, lens, iris, and ciliary body) from the posterior segment (consisting mainly of the vitreous humor, retina, and the choroidea). Light enters the eye through the cornea, the anterior chamber, and the lens. Before it meets the retina, light has to pass through the vitreous humor. The panel on the right shows a close-up view of the components of the retina, which are (from the outside to the inside of the retina) the retinal pigmented epithelium (RPE), photoreceptor cells (rod and cone), Müller glial cells, bipolar cells, and retinal ganglion cells (RGC) (Graw, 2003; with permission from the Nature Publishing Group).

- Later, thanks to modern genetics, the characterization of causative mutations in congenital human disorders and their comparison to mutations in various model organisms changed a central dogma in zoology, namely the independent evolution of eyes in flies and vertebrates: since the transcription factor Pax6 can induce both, rhabdomeric eyes in *Drosophila* as well as complex eyes in mammals (Halder *et al.*, 1995), a common genetic network of eye development was suggested first in flies, mice, and humans and included later also frogs and fishes.

However, ongoing research activities revealed also some diversity in eye development among these organisms. This review will focus mainly on eye development in mammals and its key steps, but in some cases it will be compared also to other vertebrates such as *Xenopus*, zebrafish, or chick.

2. Overview of Eye Development

During gastrulation, the developing eye is organized as a single field located centrally in the developing forebrain (Adelmann, 1929) (Fig. 10.2). During establishment of the midline, the single "eye field" is separated. During neurulation, two lateral **optic pits** become apparent as the first signs of the developing eyes. They form when the lateral walls of the diencephalon begin to bulge out (in the mouse at embryonic day 8.5; stage with 11–13 pairs of somites). They enlarge to form half a day later (12–14 pairs of somites) the **optic vesicles**, which terminate very close to the overlying surface ectoderm. The neuroectodermal optic vesicle is connected to the lumen of the primitive forebrain by the **optic stalk** (giving rise to the later optic nerve).

In parallel, in the overlying surface ectoderm the **lens placode** is formed as a thickened surface ectoderm. In the mouse it is first evident at embryonic day 8 (E8), a stage with 5–7 pairs of somites; in humans, this happens at ~28 days of gestation (Carnegie stage 13; ~30 pairs of somites). The lens placode comes into close contact with the underlying optic vesicle (in the mouse at the stage of 20–25 pairs of somites, E9.5). As a result of this interaction, the neuroectoderm folds inward and forms the **optic cup** (which Hans Spemann referred to as the *organizer of the lens*; Spemann, 1924). In the mouse this process begins at E10 (a stage with 25–30 pairs of somites). For morphological details, refer to Kaufman (1992) and Hinrichsen (1993). Later, the inner layer of the optic cup gives rise to the **retina**, whereas its outer layer will form the retinal pigment epithelium. By contrast, the **lens** is formed in a similar invagination process from the lens placode, and the **cornea** is formed after the detachment of the lens from the surface epithelium. The corneal stroma is made from cells invading from the periocular mesenchyme, a derivative of the neural crest.

In summary, the major ocular structures are made from three major sources: neural ectoderm forms the retina, surface ectoderm gives rise to the lens and part of the cornea, and neural crest cells form the central part of the cornea.

3. Early Stage: The Eye Field

As mentioned above, the developing eye is organized during gastrulation as a single field located centrally in the developing forebrain. Failure in its formation leads to eyeless phenotypes (anophthalmia). However, if the eye field is formed, but not split into two hemispheres, it results in one single eye, a structure which is named "cyclopia"—a reference to the

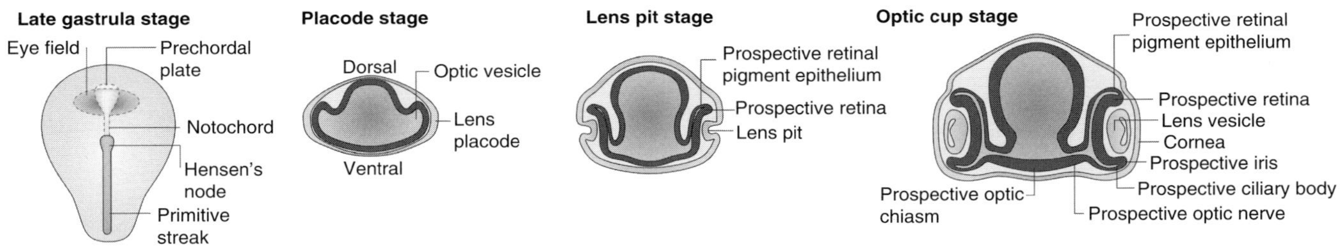

Figure 10.2 Schematic view of a developing vertebrate eye: from late gastrula to the optic cup. The most important stages from the late gastrula to the optic cup stage are shown. The first main step occurs when the single central eye field splits into two lateral parts to form the optic vesicle. At the same time (at embryonic day (E) 9.5 in the mouse and 28 days of gestation in the human) the lens placode forms (placode stage). The invagination of the lens placode occurs at E10.5 in the mouse (lens pit stage). By the optic cup stage, at E11.5 in the mouse and 31–35 days of gestation in the human, the lens pit closes to form the lens vesicle, the future cornea becomes visible, and the retina begins to differentiate (modified according to Graw, 2003; with permission from the Nature Publishing Group).

Cyclops people mentioned by Homer in his epic "Odyssey"; these giants had just one eye in the middle of the forehead.

3.1. Formation of the eye field

One of the early mouse mutants affecting eye development is the spontaneous mutant *eyeless*, which was found in the 1940s (Chase, 1944). Molecular analysis showed almost 60 years later that the eyeless phenotype is caused by a point mutation affecting an alternative translation initiation codon of the homeobox gene *Rax* (retina and anterior neural fold homeobox; Tucker *et al.*, 2001). Mouse embryos carrying a null allele of this gene do not form optic cups and so do not develop eyes (Mathers *et al.*, 1997). Actually, seven *Rax* alleles have been reported; homozygous null mutants die neonatally with severe brain defects including absence of forebrain/midbrain structures, and fail to form eye structures. Homozygous hypomorph mutants are viable, but lack eyes and optic tracts and have hypothalamic defects (MGI database, March 2010).

In zebrafish, the ortholog to the *Rax* gene in mammals is *rx3* (retinal homeobox gene 3), which is mutated in the *chockh* mutants (gene symbol: *chk*) lacking eyes from the earliest stages in development. The phenotype is caused by a nonsense mutation in the homeodomain of the *rx3* gene. In this mutant, retinal pigmented epithelium (RPE) and neuroretina are missing; a lens forms, but it is markedly reduced in size (Loosli *et al.*, 2003). This zebrafish phenotype is similar to the spontaneous, recessive *eyeless* mutant (gene symbol: *el*) in the Japanese medakafish: the *el* mutant embryos do not develop morphologically visible eye structures and die at early larval stages (Winkler *et al.*, 2000). In total, the MGI database contains 74 genes responsible for "anophthalmia" if mutated (March 2010).

These examples indicate a similar function of *Rax/rx3* in early eye development of mouse and fish. The mouse *Rax* gene is expressed in the anterior neural plate (E8.5), and restricted to a region including the optic sulci and a narrow band of cells in the ventral forebrain (Mathers *et al.*, 1997). Recent studies in zebrafish indicated that Rx3 controls the segregation of telencephalic and eye-field identities inside the zebrafish forebrain territory in a cell-autonomous manner (Stigloher *et al.*, 2006). It should be mentioned that the *Drosophila* ortholog of *Rx, Drx*, is not expressed in the embryonic eye primordia or in the larval eye imaginal discs (Eggert *et al.*, 1998).

Similar to the examples given above for zebrafish and mouse, mutations in the human *RAX* gene lead also to an eyeless phenotype (anophthalmia). In case of a 2-year-old girl suffering from bilateral anophthalmia, magnetic resonance imaging scan showed the absence of ocular structures being replaced by fibrous tissue. In this patient, anophthalmia is caused by compound heterozygous *RAX* mutations; both mutations are located in exon 3: the first, c.664delT, is a frame-shift deletion leading downstream to a premature stop codon, and the second, c.909C→G, is a nonsense mutation (Tyr303Stop). The heterozygous parents are healthy indicating a classical recessive mode of inheritance (Lequeux *et al.*, 2008).

Recently, Danno *et al.* (2008) characterized a noncoding element (*CNS1*), located ~2 kb upstream of the *Rax* promoter. This element is highly conserved among mammals and even in frogs. It can bind to Otx2 and Sox2 transcription factors activating *Rax* expression synergistically, and Otx2 and Sox2 proteins physically interact with each other when binding to *CNS1*. In humans, it was shown by Regge *et al.* (2005) that *SOX2* mutations frequently lead to anophthalmia; similar phenotypes, however, have not yet been reported for mouse or zebrafish mutations affecting *Sox2*.

Studies on *Xenopus* and zebrafish have established the role of Wnt-signaling in the formation of the eye field. The mutants *masterblind* (affected gene *axin*) and *headless* (affected gene: *Tcf3*) pointed out that canonical Wnt signaling inhibits eye formation. By contrast, overexpression of *Fzd3* receptors in pregastrulation *Xenopus* embryos resulted in ectopic expression of top-hierarchical genes for eye development, i.e., *Pax6, Rx*, and *Otx2*, and finally to ectopic eye formation. Finally, it turned out that specification of the eye field in the anterior neural plate of the zebrafish arises through mutual antagonism of the canonical and noncanonical Wnt pathways (de Iongh *et al.*, 2006). However, it remains at the moment an open question whether these mechanisms also operate in mammals, particularly mice. For the corresponding genes (*axin, Tcf3*), several mutated alleles are described for mice; however, there is no observation reported on eye phenotypes in these mouse mutants (MGI database, March 2010).

A new and unexploited topic has been touched recently, the role of miRNAs in eye development. Qiu *et al.* (2009) demonstrated that microinjection of a synthetic miRNA precursor molecule for mammalian *miR-196a* into *Xenopus* embryo is sufficient for miR-196a overexpression during early eye development. The misexpression of *miR-196a* in the anterior embryo led to dose-dependent eye anomalies and downregulation of most members of the eye-field transcription factors (EFTFs) (*ET, Rx1, Six3, Pax6, Lhx2, Optx2*, and *Ath5*) in the eye field (or later in the optic cup). These results indicate that *miR-196a* can target gene(s) in the genetic network involved in eye formation, providing a potential tool for studying the mechanisms of eye development and diseases (Qiu *et al.*, 2009).

3.2. Patterning of the eye field

The eye field (in *Xenopus*) is specified at the neural plate stage not only by Otx2, Six3, and Rax, but also by some other transcription factors, like ET (also known as transcription repression factor Tbx3), Pax6, Lhx2, Tll (also known as Nr2e1—encoding a nuclear receptor subfamily 2, group E, member 1), and Optx2 (also known as Six6); they can be summarized under the heading of "eye field transcription factors" (EFTFs). Genetic evidence demonstrated the importance of these EFTFs in eye formation of several organisms, mainly in lower vertebrates. They are expressed in a dynamic, overlapping pattern in the presumptive eye field. Expression of an EFTF cocktail together with *Otx2* in developing *Xenopus* embryos was shown to induce ectopic eyes outside the nervous system at high frequency. Detailed analysis of the interactions of the participating genes led finally to a model very close to that described for *Drosophila* (Zuber *et al.*, 2003; Fig. 10.3). Recently, it was shown by Massé *et al.* (2007) that the activities of the EFTFs can be induced in *Xenopus* by ADP, which is released by ectonucleoside triphosphate diphosphohydrolase 2 (E-NTPDase2) activity offering a new mechanism for induction of eye development.

However, in mammals, only mutations affecting *Pax6* and *Lhx2* show effects in early eye development. In mouse knockout mutants of *Lhx2* (encoding LIM homeobox protein 2), eye development is arrested at the optic vesicle stage and will be discussed there. Similarly, *Pax6* mutations in the mouse affect the formation of the optic vesicle and the lens placode and will be discussed later.

There are even some more differences among the species concerning the role of some genes in eye development. Two examples are *dachshund* (gene symbol: *dac*) and *eyes absent* (gene symbol: *eya*). *Dac* was shown in *Drosophila* to lead to ectopic eye development (Shen and Mardon, 1997), and

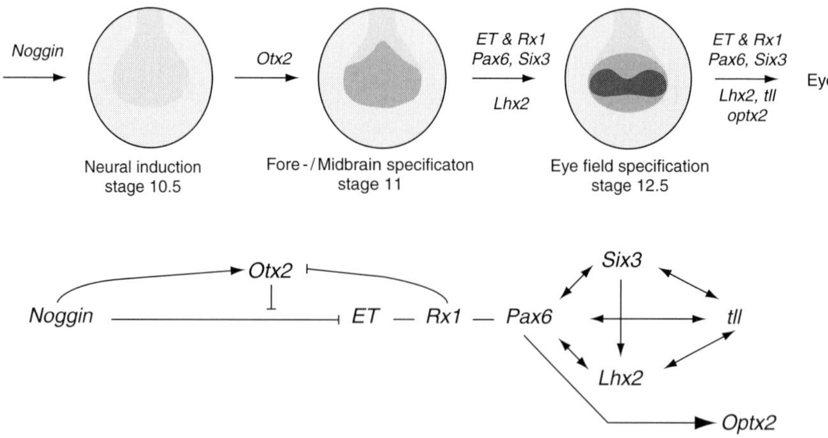

Figure 10.3 Genetic factors forming the single eye field. A simplified model of eye field induction in the anterior neural plate of vertebrates is given including the major transcription factors. Light blue indicates the neural plate, blue shows the area of *Otx2* expression, and dark blue represents the eye field (Zuber *et al.*, 2003; with permission from the Company of Biologists Ltd.). (See Color Insert.)

therefore, it was suggested to be involved also in early eye development in vertebrates. In mice, two dachshund genes are known (gene symbols: *Dach1* and *Dach2*); however, there is no eye defect reported in the corresponding knockout mutants (MGI database, December 2009). Similarly, a homozygous knockout of *eya* function in *D. melanogaster* results in severe embryonic defects and absence of compound eyes due to eye progenitor cell death (Bonini *et al.*, 1993, 1998). In vertebrates, four *Eya* genes are present (*Eya1* → *Eya4*); however, only in a few cases congenital cataracts and ocular anterior segment anomalies could be associated with *EYA1* mutations in humans (Azuma *et al.*, 2000). In case of *Eya3*, a striking difference among vertebrates was reported: studies in Xenopus revealed a strong influence of the *Eya3* on survival and proliferation of neural progenitor cells in the anterior neural plate of *Xenopus* embryos (Kriebel *et al.*, 2007). There are, however, no indications for a similar expression pattern, neither in zebrafish nor in mice, and neither a pathological eye nor brain phenotype could be observed in the mouse *Eya3* knockout mutant (Söker *et al.*, 2008).

3.3. Splitting the eye field

The splitting of the eye field occurs in parallel with the introduction of the midline, which is the main function of Sonic hedgehog (*Shh*): it is important for the formation of the midline and therefore also for the separation of the single eye field into two. The other gene important for very early eye development in mammals is *Six3*. Both, *Shh* and *Six3*, are expressed in the single eye field. Functional studies in mice revealed that Six3 protects the anterior neural ectoderm from the posteriorizing activity of Wnt1 via repression of *Wnt1* transcription; the absence of Six3 leads to an expansion of *Wnt1* expression and causes the absence of rostral diencephalon (Lagutin *et al.*, 2003). Moreover, Geng *et al.* (2008) showed that SIX3 regulates *SHH* expression in the rostral diencephalon ventral midline. Therefore, it is not surprising that loss of either *SIX3* or *SHH* gene activity leads to holoprosencephaly with cyclopia. In humans, loss of activity of these genes is causative for ~5% of all cases of holoprosencephaly. In the most severe form (alobar holoprosencephaly), only a single ventricle without an interhemispheric fissure can be observed; typically, the olfactory bulbs and tracts and the *corpus callosum* are absent.

In zebrafish, also cyclopic mutants have been reported; one of these is (*cyclops*, gene symbol: *cyc*). It is characterized by a loss of medial floorplate and severe deficits in ventral forebrain development including cyclopia caused by incomplete splitting of the eye field. The *cyc* gene encodes the nodal-related protein Ndr2, a member of the transforming growth factor β (TGFβ) superfamily. The corresponding point mutation in the *cyclops* mutants affects the initial start codon (ATG→ATA); it is very likely that the next in-frame ATG is used as start codon for translation, leading to the loss of a 38 amino-acid signal sequence (Rebagliati *et al.*, 1998). The missing Ndr2 function causes a failure in the anterior movement of median cells. Cell fate map analysis in wild-type zebrafishes showed that this movement separates the eye field and forms ventral anterior diencephalon and the primitive hypothalamus (Varga *et al.*, 1999). In zebrafish, two other nodal signaling proteins are known, Squint (gene symbol: *Sqt*) and Southpaw (gene symbol: *Spaw*); however, only mutations in *Sqt* lead to cyclopia in a similar manner as described above for *cyc* mutants (Pei and Feldmann, 2009). In the mouse, homozygous *nodal* null mutants fail to form a primitive streak, show placental defects, and die at gastrulation. Hypomorphic mutants are defective in anterior–posterior, anterior–midline, and left–right body patterning, resulting in multiple organ defects (MGI database, March 2010).

Another zebrafish mutant resulting in cyclopia is *one-eyed pinhead* (gene symbol: *oep*). Molecular analysis showed that the *oep* gene encodes an EGF-related protein, which has similarity to the proteins cripto, cryptic, and FRL-1 (Zhang *et al.*, 1998). Later on, it was shown that the oep protein acts as a co-receptor in the TGFβ/nodal pathway (Pézeron *et al.*, 2008) linking the three cyclopic mutations in a common pathway.

3.4. From optic vesicle to optic cup

When the split eye field further evaginates forming the optic vesicle, it comes in close contact with the surface ectoderm. These two tissues form a highly interactive system with several mutual interactions leading finally to the lens placode as a thickening of the surface ectoderm, which is considered as the first step into lens development (see next section). The optic vesicle invaginates and forms the optic cup, which is considered as a crucial step for the formation of iris and ciliary body (Section 7), the retina (see Section 8), and the optic stalk, the presumptive optic nerve (Section 9). It is obvious that the different developmental outcomes of distinct regions in the optic vesicle and even more in the optic cup lead to different expression patterns of signaling molecules making the analysis (and its description) difficult and complex.

One of the key players in the transformation of the optic vesicle to the optic cup might be the retinoic acid (RA) signaling system (for recent reviews, see Adler and Canto-Soler, 2007; Cvekl and Wang, 2009; and references therein). Even if RA is also important for other developmental processes in the eye and other tissues, there is increasing experimental evidence that paracrine RA signaling generated by Raldh2 (retinaldehyde dehydrogenase; actual gene symbol: *Aldh1a2*, aldehyde dehydrogenase family 1, subfamily A2; MGI database, January 2010) from the temporal mesenchyme reaches the optic vescle and is required for both, the lens pit and optic cup invagination. In particular, mouse *Raldh2*$^{-/-}$ embryos lacking RA synthesis in the optic vesicle exhibit a failure in optic vesicle invagination, which is the first step in optic cup development (Mic *et al.*, 2004). Molotkov *et al.* (2006) showed that RA signaling is not required for the establishment or maintenance of dorsoventral patterning in the future retina (there was normal expression of *Tbx5* and ephrin B2 [*Efnb2*] dorsally, plus *Vax2* and *Ephb2* ventrally). In summary, *Raldh2* is first expressed transiently in periocular mesenchyme. Later, *Raldh1* (actual genetic symbol: *Aldh1a1*) and *Raldh3* (actual genetic symbol: *Aldh1a3*) expression begins in the dorsal and ventral retina, respectively, and these sources of RA are maintained in the fetus. RA is required for morphogenetic movements that form the optic cup, ventral retina, cornea, and eyelids (Duester, 2009).

Another important player in the transition from optic vesicle to optic cup is the transcription factor Lhx2 (LIM homeobox protein 2). In *Lhx2*$^{-/-}$ mouse embryos, eye field specification and optic vesicle morphogenesis occur, but development arrests prior to optic cup formation in both the optic neuroepithelium and lens ectoderm. This is accompanied by failure to maintain or initiate the characteristic expression patterns of the optic

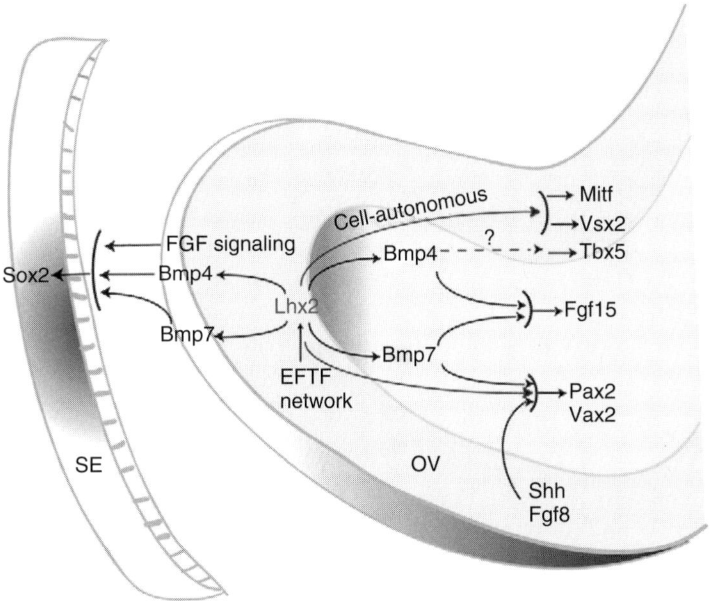

Figure 10.4 Model of *Lhx2* function during mouse early eye organogenesis. *Lhx2*, under the control of the EFTF network, links lens specification and optic vesicle patterning through the regulation of BMP signaling (black arrows). *Lhx2* also promotes optic vesicle patterning by cell-autonomous mechanisms (red arrows). Why Bmp4 fails to upregulate *Tbx5* expression is not resolved (dashed line). The timing of action and influence of *Lhx2* on several pathways suggest that it acts to coordinate the multiple patterning events necessary for optic cup formation (Yun *et al.*, 2009; with permission from the Company of Biologists Ltd.). (See Color Insert.)

vesicle and lens-inducing determinants. These data reported by Yun *et al.* (2009) indicate that *Lhx2* is required for optic vesicle patterning and lens formation in part by regulating bone morphogenetic protein (BMP) signaling in an autocrine manner in the optic neuroepithelium, and in a paracrine manner in the lens ectoderm. The authors propose a model in which *Lhx2* is a central link in a genetic network that coordinates the multiple pathways (formed by the EFTFs) leading to optic cup formation (Fig. 10.4). There is no corresponding human disease reported up to now (OMIM March 2010).

4. Lens Development

4.1. Formation of the lens placode and *Pax6* as its master control gene

Besides the formation of the eye field within the anterior neural plate and its splitting into the future bilateral optic vesicles, the second important step in early eye development is the formation of the lens placode in the surface ectoderm. It starts when the preplacodal region develops in the ectoderm—a transient bilateral structure exhibiting placodal competence leading finally to the anterior pituitary, olfactory neurons, the lens, inner ear, and the trigeminal and epibranchial cranial placodes (Streit, 2007). Within the preplacodal region, the lens placode is induced at its contact region with the underlying optic vesicle as a protuberance of the diencephalon. Morphologically, lens placode formation is characterized by a thickening of the surface ectoderm and further invagination forming the **lens cup** (also referred to as lens pit) and subsequently the **lens vesicle** (Fig. 10.2). In a reciprocal process, the optic vesicle forms the optic cup. In human embryos, this event takes place approximately at day 33 of gestation and in the mouse at day 9.5 of embryonic development (E9.5; Fig. 10.3). The rebuilt surface ectoderm above the lens vesicle gives rise to the future **cornea epithelium**.

The current literature concerning genes involved in the development and differentiation of the lens lineage covers not only several homeobox-transcription factors (predominantly Pax6), but also Fgf- and Wnt-signaling cascades.

Pax6 is the paradigm for a master control gene in eye development. It belongs to the familiy of genes that encode transcription factors with a homeo- and a paired-domain. Loss of *Pax6* function leads to the *eyeless* phenotype in *Drosophila* (Quiring *et al.*, 1994). Pioneering work of Walter Gehring's group in 1995 (Halder *et al.*, 1995) showed that ectopic expression of the mouse *Pax6* induces functional ommatidal eyes in *Drosophila* in antennae or legs. This result suggested that at least from a genetic point of view, there is one way to make an eye—the corresponding cascade of signals leads to the development of the ommatidal eye in insects and a lens eye in mammals. A corresponding experiment was done in the frog: injection of *Pax6* mRNA leads to the formation of ectopical, but differentiated eyes (Chow *et al.*, 1999).

The first mouse mutation described in *Pax6* leads in heterozygous mutants to small eyes, but homozygous mutants have only remnants of ocular tissues and die shortly after birth because of nasal dysfunction (Hill *et al.*, 1991). Actually, in mice 39 distinct alleles have been described with different consequences for eye development (MGI database, December 2009). The most severe group has no (or almost no) Pax6 activity and includes the homozygous mutants $Pax6^{3Neu}$ and $Pax6^{7Neu}$. The $Pax6^{3Neu}$ mutation is an insertion of an A after nt position 598 resulting in a truncation after the paired domain, whereas $Pax6^{7Neu}$ affects splicing of exon 3 and is suggested to lead to a reduced translation. In both mutants, eye development terminates very early: neither the lens placode nor the optic vesicle invaginate. On the other side of the allelic series are hypomorphic alleles like $Pax6^{132\text{-}14Neu}$: in these mutants (even if they are homozygous for the underlying Phe272Ile mutation), all major eye tissues (cornea, lens, and retina) develop. However, eye size is reduced and there is a large plug of persistent epithelial cells remaining attached between the lens and the cornea (Favor *et al.*, 2001, 2008).

In the mouse, a series of experiments using either tissue recombination experiments or conditional knockout mutants for *Pax6* made clear that *Pax6* is essential for the formation of the lens placode and later for the lens, but not for the formation of the optic cup. *Pax6* acts in a cell-autonomous manner; i.e., mutated *Pax6* in the optic cup has no influence on wild-type *Pax6* in the lens placode—on the other hand, wild-type *Pax6* in the optic vesicle cannot rescue a defect of *Pax6* in the lens placode (Lang, 2004; and references therein).

To understand the different steps in lens placode formation and the role of *Pax6* during this process, it is necessary to address the regulation of Pax6 expression itself. In the mouse, *Pax6* is expressed during the preplacode period in the entire preplacodal region and becomes inactivated later in most of the non-lens placodes. Major problems in the analysis of mouse mutants which carry mutations in genes being important in very early stages of development are their pleiotropic effects, which lead frequently to death of the embryos. Therefore, conditional mutations have been designed and developed which shut off the activity of a gene of interest in a tissue-specific manner using either Cre- or Flp-mediated recombinations (for a general review, see Wirth *et al.*, 2007). Using this approach, the effect of *Six3* and *Sox2* has been tested for their role in placode formation of the mouse. As discussed above, both genes have been shown previously to be expressed also in the single eye field.

Six3 was shown by Oliver *et al.* (1996) to induce a lens if ectopically expressed in the otic placode of Medaka fish. In the mouse, *Six3* expression precedes that of *Pax6* in the presumptive lens ectoderm, and Cre-mediated loss of *Six3* expression in the presumptive lens ectoderm causes abnormal lenses of different severity (Liu *et al.*, 2006). The severity of the defect correlates with an earlier onset of *Six3* inactivation (*Six3* expression in the optic cup was not affected in these conditional mutants). In the most severe phenotypes, the surface ectoderm did not thicken to form a lens placode, and therefore no lens cup and no lens vesicle were made. The authors also could show that Six3 binds to *Pax6* and *Sox2* lens enhancers suggesting that *Pax6* is downstream of *Six3* in the preplacodal phase. This could be confirmed in the analysis of Six3 expression pattern in the homozygous $Pax6^{Sey}$ mutant mice.

As mentioned above, *Sox2* is on the one hand a direct target of the transcription factor Six3; however, its expression in the lens placode is also dependent on the optic vesicle and responsive to inductive signals (Kamachi *et al.*, 1998). In particular, *Sox2* is upregulated by BMP4, an important signaling ligand from the optic cup to be implicated in lens induction (Furuta & Hogan, 1998). Another BMP, BMP7, is involved in *Pax6* induction (Wawersik *et al.*, 1999). On the other hand, there is increasing evidence that *Pax6* and *Sox2* are cross-regulated during development. To test this hypothesis, Smith *et al.* (2009) developed a conditional double-knockout consisting of Sox2 and Pax6. They could demonstrate that in the preplacodal stage, *Pax6* and *Sox2* expression is not interdependent, but the two transcription factors cooperate functionally. However, after the formation of the lens placode, *Sox2* expression becomes dependent on Pax6. Further Pax6-regulated genes encode transcription factors (such as FoxE3, Maf, Mitf, Prox1, Lhx2, Pitx3) and are involved in the formation of the lens and the cornea; however, there are also others (such as Pax2, Chx10, Eya1) being involved in the retina and the optic nerve development (Cvekl *et al.*, 2004).

In humans, *PAX6* mutations cause mainly aniridia, a panocular disorder, and less commonly isolated cataracts, macular hypoplasia, keratitis, and Peters anomaly. As in the mouse, homozygous loss of *PAX6* function in human affects all expressing tissues and is neonatal lethal. It might be of medical interest that *PAX6* is not only expressed in the optic field and in the lens, but also in several brain regions and in the pancreas. Therefore, it is not surprising that there is a growing body of evidence that *PAX6* mutations cause in addition to the ocular diseases behavioral and neurodevelopmental phenotypes as well as disorders of the pancreas (Davis *et al.*, 2008; Sander *et al.*, 1997; Tsonis and Fuentes, 2006). The PAX6 database contains more than 300 entries of human mutations (http://lsdb.hgu.mrc.ac.uk/home.php?select_db=PAX6).

4.2. Signaling cascades in early lens development

Fgf signaling is widely accepted to play an important role for the lens placode formation and later for lens development and differentiation. The developing lens expresses all four genes encoding the Fgf receptors (gene symbols: *Fgfr1–4*). Conditional inactivation of *Fgfr2* (using a *Cre* transgene under the control of a Pax6 promoter) shows that *Fgfr2* signaling is needed to drive lens fiber cells out of the cell cycle during their terminal differentiation and contributes to the normal elongation of primary lens fiber cell (Garcia *et al.*, 2005). By contrast, deletion of three Fgf-receptor genes (*Fgfr1–3*) early in lens development demonstrated that expression of only a single allele of *Fgfr2* or *Fgfr3* was sufficient for grossly regular lens development, whereas mice possessing only a single *Fgfr1* allele developed cataracts and microphthalmia. Severe defects occurred in lenses lacking all three Fgf receptor genes such as lack of lens fiber cell elongation, abnormal proliferation in prospective lens fiber cells, reduced expression of the cell cycle inhibitors $p27^{kip1}$ and $p57^{kip2}$, increased apoptosis, and aberrant or reduced expression of *Prox1, Pax6, c-Maf, E-cadherin*, and genes coding for α-, β-, and γ-crystallins (Zhao *et al.*, 2008).

However, there is a lack of evidence demonstrating an essential Fgf ligand in lens induction, which might be due to its functional redundancy—there are more than 20 Fgf genes listed in the MGI database. Some of these have been shown to be expressed in the surface ectoderm and presumptive lens, e.g., *Fgf1* and *Fgf2* (de Iongh and McAvoy, 1993), *Fgf8* (Kurose *et al.*, 2005), and *Fgf19* (also referred to as *Fgf15*; Kurose *et al.*, 2004, 2005). However, even the double mutation for *Fgf1* and *Fgf2* does not lead to a pathological eye phenotype (Miller *et al.*, 2000). Therefore, it is still unknown which Fgf ligands are involved in lens induction (Smith *et al.*, 2010).

Similarly, **Wnt signaling** may play also a role during early lens development and morphogenesis, since *Wnt2b* expression was found in the presumptive lens ectoderm of chicken. At the lens vesicle stage, it is restricted to the lens epithelium. For a long time, there was no mutation available affecting the Wnt pathway demonstrating pathological alterations in early lens development. However, recent papers showed that lens morphogenesis is dependent on the inhibition of the canonical Wnt/β-catenin signaling: since there is no Wnt signaling going on in lens development, inactivation of β-catenin in mice does not perturb the regular appearance of lens fate markers, but results in a failure of coordinated epithelial cell behavior and in abnormal morphogenesis of the lens most likely due to the missing cytoskeletal function of β-catenin. By contrast, when Wnt signaling is active (like in the developing nose or in the periocular ectoderm), inactivation of β-catenin results in the formation of crystallin-positive ectopic lentoid bodies (Kreslova *et al.*, 2007; Smith *et al.*, 2005). Moreover, ectopic Wnt activation in the retina and lens abrogates lens formation (Machon *et al.*, 2010). Furthermore, these authors demonstrated that inhibition of Wnt signaling during eye development is dependent on Pax6.

One of the activators of *Pax6* in the mouse is pygopus2 (gene symbol: *Pygo2*). *Pygo2* was shown recently to have a crucial role in mouse lens induction. Null mutants of *Pygo2* result in microphthalmia with severe, but variable lens defects ranging from small lenses to no lenses. The first morphological difference in these mutants is a thinner lens placode and subsequently a smaller lens pit. Even if Pygo2 was identified originally as part of the Wnt pathway in *Drosophila* (Belenkaya *et al.*, 2002), Song *et al.* (2007) could demonstrate that Pygo2 acts during mouse lens development independent of the Wnt pathway.

BMP-mediated signaling in early lens development is also essential for lens formation, particularly involving two members of this protein family of morphogens, BMP4 and BMP7. Similarly, also the corresponding receptors (type I) are present in the lens placode (Acvr1 and Bmpr1a). In the mouse, *Bmp4* is expressed initially in both, the surface ectoderm and the underlying optic vesicle, but it becomes restricted to the optic vesicle during lens placode formation. However, germline deletion prevents lens formation. There was no change in *Pax6* expression in *Bmp4* knockout mutant eyes, and *Bmp4* expression appears unaffected in eyes of homozygous $Pax6^{Sey\text{-}1Neu}$ mice, suggesting that *Pax6* and *Bmp4* function independently (Furuta and Hogan, 1998). *Bmp7* is expressed in the lens placode; it is suggested to function predominantly to regulate lens induction. Germline mutations in *Bmp7* prevent lens formation in most cases (Wawersik *et al.*, 1999). Recent conditional knockouts of the genes coding for the BMP receptors Acvr1 and Bmpr1a demonstrated that only the deletion of both genes reduced lens thickening and prevented lens invagination, leading to eyes without lenses (Rajagopal *et al.*, 2009).

Among other signaling events, Murato and Hashimoto (2009) reported interesting effects of morpholino-dependent silencing of *hairy2* in *Xenopus* during the gastrula stage. The consequence was the reduced expression of lens marker genes at every step of lens development, eventually resulting in lens malformation. By contrast, retina marker genes expression remained normal. The effect of *hairy2* silencing could be rescued at least in part by simultaneous knockdown of *p27*, a gene encoding a cell cycle inhibitor. In this context it might be interesting that rat *p27* mutants suffer first from cataract, but later in life, they develop a broad variety of neuroendocrine tumors (Fritz *et al.*, 2002; Pellegata *et al.*, 2006). For the mouse, no *hairy2* (=*hes2*) mutant hase been reported (MGI database, March 2010).

4.3. From lens vesicle to the mature lens

The lens vesicle forms by closing the lens cup (also known as lens pit) and detaching from the surface ectoderm. An intermediate step is the

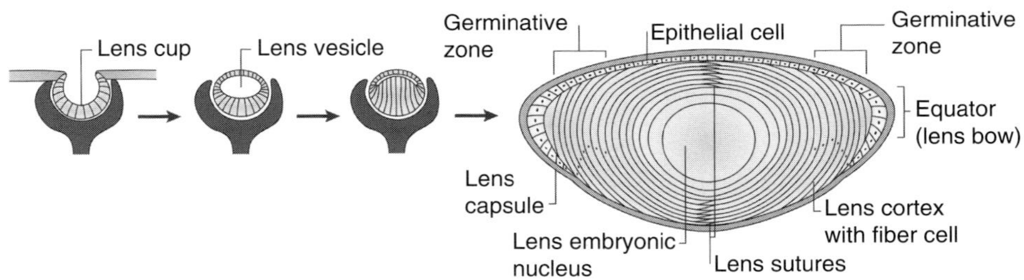

Figure 10.5 Formation of the lens. Once the lens vesicle has formed, the primary lens fibers elongate from the posterior epithelium of the lens vesicle and fill its entire lumen. The secondary fiber cells start to elongate at the lens bow region; the fibers from opposite sides meet at the anterior and posterior pole, and give rise to the lens sutures (which are Y-shaped in the three-dimensional view). The final step in lens differentiation is the degradation of the cell nuclei and mitochondria, which takes place around the time of birth in the mouse (modified according to Graw, 2003; with permission from the Nature Publishing Group).

development of a lens stalk keeping the closed vesicle and the surface ectoderm together for a few hours (in the mouse). The lens vesicle is nearly spherical with a large central cavity; the cells from its posterior pole elongate till they reach the anterior epithelial cells and fill the entire lens vesicle; these elongated cells are referred to as **primary lens fiber cells**. This step occurs around day 44 of gestation in human embryos and at E11.5 in the mouse (Fig. 10.5). The cells at the anterior pole of the lens vesicle remain as epithelial cells. Mitotically active cells surrounding the central region of the lens epithelium move into the equatorial region (or lens bow region), where they elongate and differentiate into **secondary lens fibers**. The midline, where secondary lens fibers from opposite points of the equator join, is referred to as the anterior and posterior lens suture. The secondary lens fibers form concentric layers around the primary fibers of the lens nucleus (in the mouse at day E15.5; Fig. 10.5). With this arrangement, the lens fibers toward the periphery are successively younger in developmental and differentiation terms. As long as the lens grows, new secondary fibers move in from the equator onto the outer cortex of the lens.

Both the primary and secondary fiber cells lose their mitochondria and cell nuclei during the final differentiation process: for the primary fibers, it takes place in mice at E17/E18 and is finalized 2 weeks after birth, when the mice open their eyelids (Vrensen *et al.*, 1991). The secondary fiber cells, which encircle the primary fiber cells, lose their organelles, when they move from the outer to the inner cortex (Kuwabara and Imaizumi, 1974).

The anterior epithelial cells, however, remain mitotically active as a stem cell niche producing secondary fiber cells. These secondary lens fiber cells are terminally differentiated cells and lose their organelles also, when they are pressed deeper into the lens by the successive fiber cells.

In the zebrafish, however, several differences in lens development and differentiation occur. In particular, primary fiber cell elongation occurs in a circular fashion resulting in an embryonic lens nucleus with concentric shells of fibers. The very close spacing of the nuclei of the differentiating secondary fibers in a narrow zone close to the equatorial epithelium, however, suggests that secondary fiber cell differentiation deviates from that described for mammalian or avian lenses. Because of these differences, one should be cautious when extrapolating findings on the zebrafish to mouse or human lens development or function (Dahm *et al.*, 2007).

In mice, at least two genes, *Pitx3* and *Foxe3*, characterize the importance of the transient nature of the lens stalk stage. In mouse embryos, *Pitx3* is expressed in the developing lens starting at E11, first in the lens vesicle, and later in the anterior epithelium and the lens equator. Mutations in the regulatory or coding regions of the *Pitx3* gene have been shown to cause the phenotype of *aphakia* (*ak*) or *eyeless* (*eyl*) mouse mutants, which lack lenses and pupils (Rieger *et al.*, 2001; Rosemann *et al.*, 2010; Semina *et al.*, 2000). In these mice, the lens stalk persists for several days leading finally to a degradation of the rudimentary lens vesicle, and retinal tissue fills the entire eye globe. Since *Pitx3* is also expressed in dopaminergic neurons of the *substantia nigra*, these mice are also excellent models for Parkinson's disease (Rosemann *et al.*, 2010). In contrast to the mouse, mutations in the human *PITX3* cause anterior segment mesenchymal dysgenesis (ASMD; Semina *et al.*, 1998).

The *ak/ak* mice have an ocular phenotype that is very similar to the *dyl* (dysgenic lens) mice, indicating that both genes are involved in the same biological process. Blixt *et al.* (2000) showed that the *dyl* phenotype is mediated by a mutation in the *Foxe3* gene. In the mouse, *FoxE3* is expressed in the developing eye around E9.5, at the start of lens placode induction (Fig. 10.2). As the lens placode forms, the expression of *FoxE3* increases and becomes confined to the lens vesicle as it detaches from the surface ectoderm. Two mutations within the DNA-binding domain of *FoxE3* were identified in *dyl* mice. In humans, mutations in *FOXE3* are responsible for anterior segment optical dysgenesis (ASOD). Because of the expression pattern of *FOXE3* and the variable phenotype of the heterozygous *dyl* mice, a small cohort of patients with Peters anomaly in whom no *PAX6* mutations could be detected were screened for *FOXE3* mutations. One of the patients turned out to be heterozygous for an Arg90Leu substitution affecting the DNA-binding domain of *FOXE3* (Ormestad *et al.*, 2002).

The second important step is the elongation of the cells at the posterior half of the lens vesicle filling it with **primary fiber cells**. In the mouse mutant "opaque flecks in the lens," a point mutation affects the basic region of Maf (encoded by an oncogene, responsible for musculoaponeurotic fibrosarcoma) and prevents correct formation of the primary lens fibers leading to a phenotype that is similar to the pulverulent cataract in a human family (Lyon *et al.*, 2003). Mammalian *MAF* is expressed in the lens placode and lens vesicle, and later in the primary lens fibers.

Similarly, Puk *et al.* (2008) recently characterized a new ethyl nitrosourea (ENU)-induced mouse mutant with a small-eye phenotype and an empty lens vesicle in the homozygous state. In this case, a mutation in the gene *Gjf1* (also referred to as *Gje1*) was identified. In the mouse, the gene *Gjf1* encodes a connexin-like protein of 23.8 kDa, which is expressed in the posterior part of the lens vesicle, where the primary fiber elongation starts. In the mutants, the expression pattern of *Pax6, Prox1, Six3*, and *Crygd* are modified, but not the pattern of *Pax2*. The gene *Gjf1* is thought to be essential for the formation of the primary lens fibers (Puk *et al.*, 2008) and might be considered a downstream target of the transcription factor c-Maf; mutations in the corresponding *Maf* gene lead to a similar phenotype in the mouse (Lyon *et al.*, 2003; Perveen *et al.*, 2007). At present, it is not clear whether there is a functional human counterpart of the mouse *Gjf1* gene.

A third phenotype without elongation of the primary lens fibers is caused by the knockout of the *Pparbp* gene (coding for the peroxisome proliferator activator receptor binding protein; Crawford *et al.*, 2002). The relationship between these three functionally distinct proteins for the formation of the primary lens fiber cells is not yet clear.

In addition to these three genes, Wnt signaling might also play a role in the elongation of the primary fiber cells. Faber *et al.* reported in 2002 a dominant-negative form of the Bmp-receptor 1b (gene symbol: *Bmpr1b*) in transgenic mice. These transgenic mouse mutants show an inhibition of the primary fiber cell development, however, in an asymmetric fashion: it appeared only on the nasal side of the lens in the ventral half. The authors concluded that distinct differentiation stimuli might be active in different quadrants.

On the anterior side, the lens epithelial cells remain the only mitotically active cells in the lens. They are characterized by an ongoing expression of several *Wnt* genes: however, the detailed expression data reported are not only different between chick and mice, but vary also among different strains of mice (for details, see a review by de Iongh *et al.*, 2006). Nevertheless, it remains clear that Wnt signaling pathway genes are expressed predominantly in the lens epithelial cells. Consistently, Fzd receptors (gene symbols: *Fzd1–8*) and co-receptors *Lrp5* and *Lrp6, Sfrp1–3* and *Dkk1–3* genes have also been demonstrated to be expressed during lens development. They are mainly present in the epithelial cells; the only exception is *Fzd6* being increasingly expressed in differentiating fiber cells (de Iongh *et al.*, 2006). As an example, *lrp6* null mutants have been analyzed showing (besides some other defects; see MGI database) small eyes and aberrant lenses characterized by an incompletely formed anterior epithelium resulting in extrusion of lens fibers into the overlying corneal stroma (Stump *et al.*, 2003).

However, the key trigger for lens fiber cell differentiation is Fgf signaling. One of the most significant findings demonstrated in rat lens explants that different concentrations of Fgf2 (previously known as "basic Fgf" or "bFGF") are responsible for lens cell proliferation, migration, and lens fiber cell differentiation (McAvoy and Chamberlain, 1989). Since it is still unknown which of the several Fgfs are involved in lens induction (Smith *et al.*, 2010), research had focused on the Fgf receptors. As mentioned above, severe defects in lens fiber cell elongation occurred in lenses lacking three Fgf receptor genes (*Fgfr1–3*; Zhao *et al.*, 2008). Fgf signaling is also necessary for priming the noncanonical Wnt pathway (i.e., independent of β-catenin) in lens epithelial cells; in lens explants, it leads to accumulation of β-crystallin, a marker for fiber cell differentiation (Lyo and Joo, 2004).

The **mature lens** contains several classes of structural proteins: the crystallins (α-, β-, γ-, δ-, μ-, ζ-crystallins), transmembrane proteins (such as MP19 and MIP26, and the connexins 43, 46, and 50), some collagens, and cytoskeletal and intermediate filament proteins. Mutations in the corresponding genes (or specific transcription factors) lead to functional imbalances and lens opacities (cataract). The age of onset of the cataracts and their mode of inheritance depend on the expression of the corresponding genes and by the domain which is affected by the underlying mutation. In total, ~60 different genes are known to be responsible for cataract formation in mice and humans. A detailed discussion of the corresponding mutations and their functional consequences is beyond the scope of this chapter; reviews corresponding to this particular topic were published recently by the author (Graw, 2009a,b).

5. The Cornea

Cornea forms as a result of the last series of major inductive events in eye development with the lens vesicle interacting with the overlying surface ectoderm (Fig. 10.6); finally, the cornea consists of an anterior epithelium and a posterior endothelium with the corneal stroma within between. The basis of corneal embryonic development in chick was summarized in great detail by Hay (1979).

During detaching of the lens vesicle from the surface ectoderm, the two tissues remain transiently connected via the lens stalk. When the lens vesicle and the surface ectoderm have been completely separated, the space between is filled by invading cells from the perinuclear mesenchyme (which are of neural crest origin). This wave of neural crest cells leads in the mouse at E12 to a layer four to seven cells thick (Cvekl and Tamm, 2004). Interestingly, this cell migration appears to be species specific: in reptiles, birds, and primates, two waves are observed—first, endothelial cells and then keratocytes. However, in rodents, cats, rabbits, and cattle, only a single migration of cells is observed resulting in both cell types (Zieske, 2004). Later, the mesenchyme condenses and forms several layers being separated from each other by a loose extracellular matrix. The posterior cells closest to the lens form the **corneal endothelium**.

The surface ectoderm at the anterior side becomes the **corneal epithelium**. After opening of the eye lid (which takes place in the mouse ~2 weeks after birth), the thickness of the corneal epithelium grows from two cells up to six to seven cells—depending on EGF. During this process, the basal cells change their shape from ovoid to columnar. In contrast to the lens, they do not loose their cell nuclei during their terminal differentiation from the basal cells to the epithelial cells (Cvekl and Tamm, 2004). However, similar to the lens, they contain high proportions of taxon-specific, multifunctional proteins, e.g., aldehyde dehydrogenase 3 in most mammals but gelsolin in the zebrafish (for a detailed review, see Piatigorsky, 2001).

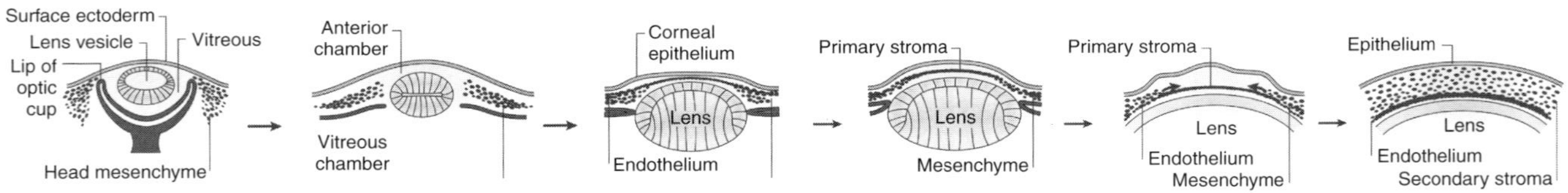

Figure 10.6 Formation of the cornea. The cornea begins to develop when the surface ectoderm closes after the formation of the lens vesicle and its detachment from the surface ectoderm. Mesenchymal cells (neural crest cells) invade the cornea and form the corneal stroma after condensation (modified according to Graw, 2003; with permission from the Nature Publishing Group).

The first layers of the **corneal stroma** are formed by aligned bundles of collagen ("lamellae"), which are made by the epithelial cells and stacked in an orthogonal pattern; each new layer is deposited next to the epithelium (this primary stroma is not observed in rabbits, mice, or primates). The layers of collagen fibrils, which are deposited after the invasion of fibroblasts, are not strictly orthogonal (Hay, 1979). In addition to collagen, the corneal stroma contains a variety of other extracellular matrix proteins like fibronectin, laminin, and vitronection, which can bind to integrin receptor subunits and reorganize the actin cytoskeleton (Svoboda *et al.*, 2008).

Under the influence of thyroxine from the developing thyroid gland, the corneal stroma is dehydrated, and the collagen-rich matrix of epithelial and mesenchymal tissues becomes the transparent cornea. In the mouse, the mesenchymal cells of the stroma differentiate to keratocytes representing the major cell type of the cornea. The keratocytes derived from invaded neural crest cells do not terminally differentiate in the corneal stroma, but remain in a G0 stage, which allows further proliferation during corneal wound healing if necessary (Cvekl and Tamm, 2004).

A major function of the corneal endothelium is to keep the cornea in a dehydrated state. Therefore, a barrier is formed by focal tight junctions, which prevent fluid flow, and on a "pumping" action provided by Na^+/K^+-ATPase and Mg^{2+}-dependent bicarbonate enzymes present in the lateral membranes of the endothelial cells. The corneal endothelial cells are arrested in the G1 cell cycle stage. For more details of the late cornea development, see Zieske (2004).

One of the major questions during cornea development is "how is the migration of the neural crest cells from the periocular mesenchyme to the cornea regulated?" Recently, an interesting observation was published by Lwigale and Bronner-Fraser (2009) demonstrating that the secreted and well-known axon-guidance protein semaphorin-3A (mouse gene symbol: *Sema3a*) is expressed in the lens placode, in the lens vesicle, and later in the anterior epithelium of the chicken lens. The corresponding neuroreceptor complex contains neuropilin-1 (mouse gene symbol: *Nrp1*), a transmembrane protein which is present in periocular neural crest cells, but downregulated when these cells migrate between the ectoderm and lens to form the cornea. Since Sema3a acts as a chemorepellant, it inhibits the migration of the periocular neural crest cells via the Nrp1-mediated receptor complex. When the receptor is switched off, the cells can migrate attracted by a still unknown factor. However, in chicken embryos with ablated lenses the migration starts earlier, since the repulsive function of Sema3a is missing, and the corneal cells differentiate abnormally. Later in corneal development, the same system is responsible for the correct innervation by trigeminal sensory afferents (Lwigale and Bronner-Fraser, 2007).

6. The Iris and the Ciliary Body

Whilst the lens and the cornea are being formed, profound changes also occur in the optic cup. The two layers of the optic cup begin to differentiate in distinct directions. The cells of the outer layer produce pigment and eventually form the **pigmented layer of the retina**, and the inner layer will further differentiate to the **neural retina** (see below/ next section). The area where the developing neural and pigmental retinas meet, the outer lips of the optic cup or the margin of the optic cup, is very close to the lens vesicle and undergoes a different transformation into the **iris** and **ciliary body**.

In the mouse, iris development starts at the end of embryonic development (~E17) by extending the outer lips of the optic cup, which did not take part in the differentiation process of the retina, which started earlier. When the margin of the optic cup extends to form the iris, the periocular mesenchymal cells proliferate and migrate along the iris epithelial layers and differentiate into the iris stroma. The origin of the periocular mesenchymal cells is still a matter of debate; most likely they are derivatives of neural crest and mesoderm. Between the iris stroma and the (pigmented) iris epithelium, the iris muscle forms as indicated by the expression of smooth muscle-specific markers. At the root of the iris, the cells differentiate further to form the ciliary body (Davis-Silberman and Ashery-Padan, 2008).

From earlier studies it is well known that the lens is required for proper development of the iris and ciliary body. In a classic experiment it was shown that ablation of the lens, either mechanical or by lens-specific expression of the cytotoxic diphtheria toxin A, disrupted the development of the iris and ciliary body (and also the cornea; Beebe and Coats, 2000; Harrington *et al.*, 1991). The identification of the underlying molecular mechanisms is a subject of ongoing research.

Actually, several signaling pathways have been identified to be involved in iris and ciliary body development. First of all, Bmp signaling has been demonstrated to have an important role in this process. The ciliary body was completely absent in transgenic mice engineered to overexpress Noggin, a Bmp antagonist, using a lens-specific promoter (Zhao *et al.*, 2002). Dias da Silva *et al.* (2007) proposed for chicken a model of anti-parallel gradients of Bmp and Fgf signals to define the region where the ciliary body can further develop. Moreover, Wnt2b signaling in the developing optic vesicle and optic cup of chick induces ciliary body and iris by losing retinal identity upon Wnt signal activation. This conclusion is supported by the fact that *in vivo* activation of Wnt signaling in the retina interferes with the maintenance of retinal progenitor identity and leads to the conversion of retinal cells into the peripheral fates of ciliary body and iris. The same conclusion can be drawn from loss-of-function studies of Wnt signaling: these set of experiments demonstrated an inhibition of peripheral marker expression and iris hypoplasia without affecting retinal tissue (Cho and Cepko, 2006).

In humans, inherited disorders are known affecting the iris (**iridogoniodysgenesis**). Iridogoniodysgenesis is characterized by hypoplasia of the iris stroma, gonadal dysgenesis, and juvenile glaucoma caused by additional abnormalities of the trabecular meshwork (Pearce *et al.*, 1999); we can distinguish type I and type II. Type I is caused by mutations in *FOXC1*; however, mutations in this gene can lead also to Axenfeld–Rieger syndrome. The penetrance of the clinical phenotype varies with the genetic background, indicating the influence of modifier(s). For example, *CYP1B1*, a human gene associated with congenital/infantile glaucoma may have such a modulating effect in the development of anterior segment anomalies such as Peters anomaly. A nonsense mutation in the mouse homolog of *FOXC1* leads to multiple and severe developmental defects and finally to lethality. Heterozygotes that suffer from a glaucoma-related eye disease, including multiple anterior segment defects, resemble Axenfeld–Rieger anomaly patients (Graw, 2003; and references therein; for a recent case report, see Weisschuh *et al.*, 2008). *FoxC1* knockout mice also have anterior segment abnormalities that are similar to those reported in humans (Smith *et al.*, 2000). Iridogoniodysgenesis type II is caused by mutations in *PITX2*; mutations in this gene have also been identified in patients suffering from Axenfeld–Rieger syndrome (OMIM 137600, March 2010). Homozygous mouse knockout mutants show among other phenotypes absent ocular muscles and optic nerve defects suggesting different function of *Pitx2* in mice and humans (MGI database, December 2009). Gage *et al.* (2008) demonstrated recently *Dkk2* as a downstream target gene of Pitx2 providing a mechanism to suppress locally the Wnt pathway; the authors further propose a model placing Pitx2 as an essential integration node between RA and canonical Wnt signaling during eye development.

7. The Retina

7.1. The retinal pigmented epithelium

The two layers of the optic cup differentiate (Fig. 10.2): the cells of the outer layer produce pigment and eventually form the RPE. This outer layer is in close contact with the periocular mesenchyme, and signals coming from these cells seem to be important for the further steps to the formation of the RPE. In zebrafish, one of these signaling molecules was identified as activin A, a member of the TGFβ family (Fuhrmann *et al.*, 2000). By contrast, Fgf signaling from the surface ectoderm was shown in chicken to inhibit RPE formation. Transgenic mice that ectopically express *Fgf9* in the presumptive RPE do not develop an RPE, but rather develop a second neural retina, while *Fgf9*-null mice tend to expand the RPE into the neural retina (for an overview, see Martinez-Morales *et al.*, 2004; and references therein). Furthermore, a few genes coding for transcription factors have been proven to be essential for RPE specification: *Mitf, Otx1/Otx2*, and *Pax6*. Pax6 and Otx1/Otx2 also have several functions at the top of the hierarchy (see previous sections), but *Mitf* (encoding the microphthalmia-associated transcription factor) is important for the regulation of the genes coding for the melanogenic enzyme Tyrosinase (gene symbol: *Tyr*) and Tyrosinase-related proteins (*Trp1* and *Trp2*). *Mitf* is a member of the basic helix–loop–helix leucine zipper family of transcription factors.

Microphthalmia (*Mi*) was one of the first mouse mutants in which development of the retina was affected. Since its first discovery almost 70 years ago (Hertwig, 1942), an interesting allelic series ranging from weak recessive to severe dominant phenotypes has been compiled and genetically characterized. The eyes of the mutants develop poorly because of the affected retinal pigment epithelium. Mutations in *Mitf* have been identified underlying these pathological events; in particular, they are responsible for the phenotypes of *mibA* rat mutant, *anophthalmic white* and Wh^{V203} hamster mutants, and *nacre; nac*^{W2} zebrafish mutants. Mutations in the human homolog, *MITF*, cause 20% of Waardenburg syndrome type 2 (for an overview, see Graw, 2003; and references therein).

7.2. The neural retina

The cells of the inner layer of the optic cup constitute the neural retina. At these stage, the pool of retinal progenitor cells expands by proliferation and will subsequently generate the six types of neurons, i.e., **retinal ganglion cells** (RGCs), **amacrine cells, horizontal cells, bipolar cells**, and light-sensitive **photoreceptor cells** (rods and cones). In contrast to these neuronal cells, the **Müller cells** are glial cells (Fig. 10.7). In humans, retinal differentiation begins around day 47 of gestation, and cone and rod photoreceptors can first be distinguished in week 15 of gestation. Full development continues until the 8th month, and the *fovea centralis* (point of maximum optical resolution) becomes fully functional only after birth (Hinrichsen, 1993). In the mouse, the corresponding process starts at E12 and is finished 2 weeks after birth when the eye lids are open (Ohsawa and Kageyama, 2008). Three major layers can be recognized: the ganglion cell layer (GCL) contains the RGCs and displaced amacrine cells; the inner nuclear layer (INL) consists of amacrine, Müller glia, and bipolar and horizontal cells; and the outer nuclear layer (ONL) is formed by the photoreceptor cells. Besides vision, the RGCs expressing melanopsin as photopigment are responsible for several other responses to light including phototrainment to the circadian oscillator and constriction of the pupil (Panda *et al.*, 2003).

Retinal development has been extensively studied in several model systems in addition to the mouse such as zebrafish, *Xenopus laevis*, and chick and rat (for reviews, see Adler, 2000; Andreazzoli, 2009; Fadool and Dowling, 2008; Glass and Dahm, 2004). The investigations have discovered many similarities, but also important differences among the various species; to avoid confusion and misunderstanding, the present section will therefore focus almost exclusively on studies carried out with mice.

As mentioned above, the inner layer of the optic cup will give rise to the different layers of the mature retina. The retinal progenitor cells differentiate to the neuronal and glial cells in a temporal order, which is conserved among many species. In this process, the ganglion cells are the first ones, followed by the amacrine cells, cone photoreceptors, and horizontal cells. The bipolar cells and rod photoreceptors are formed at later stages. However, the ongoing differentiation process needs also a continuous support of retinal precursor cells, which is regulated by the basic helix–loop–helix transcription factors Hes1 and Hes5. The corresponding *Hes* genes are homologs of the genes encoding "hairy and enhancer of split" in *Drosophila*, acting as transcriptional repressors. In *Hes1* mutant embryos, cell proliferation of the retina is severely impaired; in *Hes5* mutant mice, ~1/3 of the Müller glial cells are not formed. *Hes1* and *Hes5* are downstream targets of the Notch signaling pathway (for review see, Ohsawa and Kageyama 2008; and references therein).

A severely affected mouse mutant—the recessive *ocular retardation* (*or*)—is characterized (when homozygous) by blindness with obvious microphthalmia, a cataractous lens, a thin retina that is morphologically poorly differentiated, and a lack of optic nerve. The *or* phenotype is caused by a mutation in *Chx10*—a gene that encodes a homeobox transcription factor (Burmeister *et al.*, 1996; it is also referred to as *Vsx2*, MGI database). A similar phenotype (microphthalmia, cataracts, and severe abnormalities of the iris) has been reported recently in two families suffering from recessive mutations in the *CHX10* gene. Both mutations affect the Arg residue at position 200 leading to a severe disruption of CHX10 transcription factor activity (Percin *et al.*, 2000). Further functional analysis indicated that Chx10 is required for the repression of *Mitf* and therefore for the maintenance of mammalian neuroretinal identity (Horsford *et al.*, 2005).

In the following, the genesis of the different retinal cell types is discussed briefly. A generalized overview of the genes involved in the regulation of retinal patterning is given in Fig. 10.8.

The RGCs are the first being formed during retinal differentiation (around E10.5 in the mouse embryo). They are largely lost in *Math5* mutants; *Math5* is a member of the mouse gene family homologous to the *Drosophila* proneural gene *atonal*; therefore, the current nomenclature in the mouse is *Atoh7* (atonal homolog 7; MGI database). It encodes a murine basic helix–loop–helix transcription factor being activated by Pax6

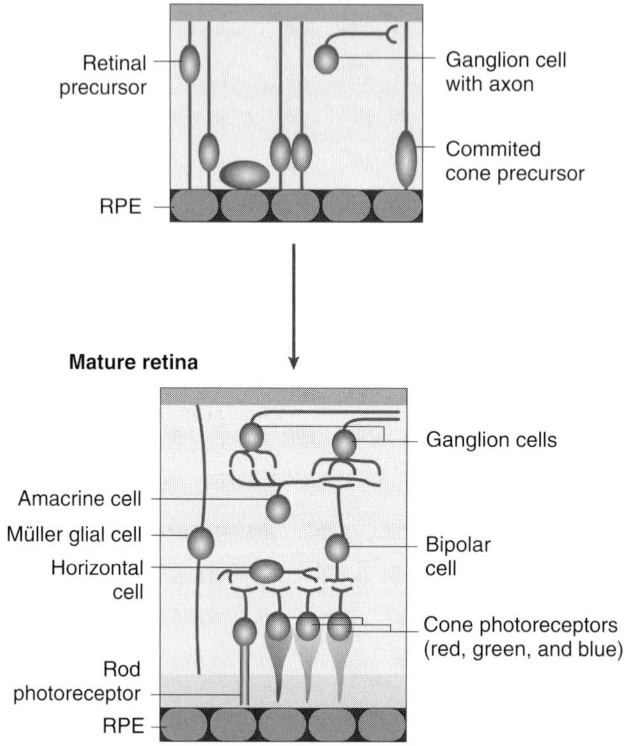

Figure 10.7 Formation of the retina. The outer layer of the optic cup gives rise to the retinal pigmented epithelium (RPE), whereas the inner layer differentiates into the neural retina (beginning at E10.5 in the mouse and 4–5 weeks after gestation in the human; pigmented epithelium is visible in 6-week-old human embryos). The first rods and cones appear in human embryos at 10–15 weeks, and the horizontal, bipolar, and ganglion cell layers appear in the middle of embryonic development. The retina is fully developed in the mouse a few days after birth and several months after birth in humans (modified according to Graw, 2003; with permission from the Nature Publishing Group).

(Riesenberg *et al.*, 2009). If the RGCs are almost all lost as in *Math5/brn3b* double mutants, it is accompanied by a drastic loss in the number of all other retinal cell types (Moshiri *et al.*, 2008).

The **amacrine cells** require different classes of transcription factors for their differentiation. Misexpression of *Pax6* together with *Neurod1* (also referred to as *NeuroD*) or *Math3* leads to amacrine cell formation, but the forkhead box gene *Foxn4* alone efficiently generates amacrine cells in retinal explants. The knockouts of *Foxn4* and of *Ptf1a* cause a total loss of horizontal cells and decreased amacrine cell number; the double-knockout of *Math3/NeuroD* shows a selective loss of amacrine cells. The situation is even more complex, since amacrine cells can be subdivided

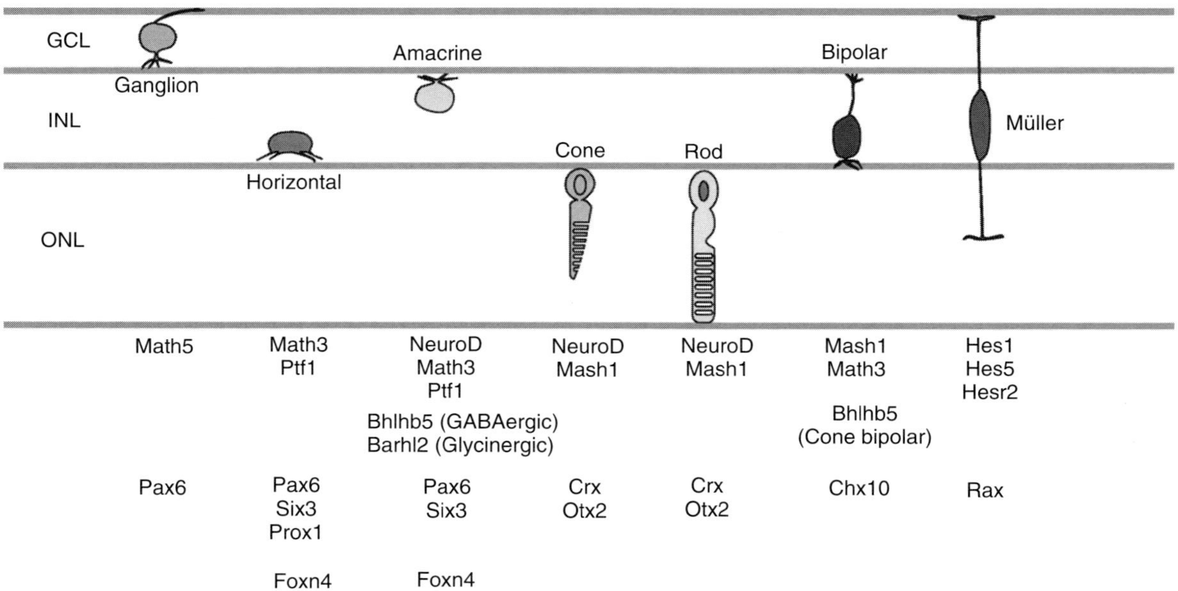

Figure 10.8 Regulation of retinal cell fate specification by transcription factors. Combinations of multiple transcription factors, such as bHLH-type and homeobox-type factors, are required for proper specification of retinal cell types. GCL, ganglion cell layer; INL, inner nuclear layer; ONL, outer nuclear layer (Ohsawa and Kageyama, 2008; with permission from Elsevier).

further into GABAergic and gylcinergic cells. There are experimental hints that the basic helix–loop–helix transcription factor Barhl2 plays an important role in the formation of glycinergic amacrine cells and Bhlhb5 for the GABAergic amacrine cells (for review, see Ohsawa and Kageyama, 2008; and references therein).

As indicated above, the **horizontal cells** have some regulators in common with the amacrine cells. Additionally, Prox1 is discussed as a further transcription factor involved in the formation of horizontal cells (Dyer *et al.*, 2003).

The differentiation of **bipolar cells** is obviously dependent on Mash1 and Math3, since in the corresponding double mutants bipolar cells are completely lost (Tomita *et al.*, 2000). However, the regulation of bipolar cell differentiation is more complex because of the existence of ON and OFF subtypes (for an extensive review on this topic, see Westheimer, 2007).

The differentiation of **photoreceptor cells** is mainly driven by *Otx2* and *Crx*, since deletion of either gene leads to a conversion of photoreceptor cells to amacrine cells or defects in their genesis. Obviously, the default pathway leads to cone photoreceptors, but the expression of *Nrl* changes this default pathway for the differentiation to rod photoreceptors; knockout mutation of *Nrl* leads to the loss of rod cells (for review, see Ohsawa and Kageyama, 2008; and references therein).

Understanding the basics of embryonic eye development in different model organisms is a prerequisitive to make a better diagnosis for congenital (retinal) disorders and on a long term to improve therapy. Till spring 2010, 160 retinal disease-causing genes have been identified; 42 additional loci or genes are mapped but not yet characterized in detail (http://www.sph.uth.tmc.edu/retnet/home.htm).

Among these disorders, major progress has been made during recent years to understand the molecular basis of the **Bardet–Biedl syndrome** (BBS; OMIM 209900). It is a pleiotropic disorder characterized by several symptoms including retinal degeneration, obesity, and cystic kidneys. To date, 14 loci are reported by OMIM (March 2010), and there might be some more; there are three additional publications in early 2010 (Hjortshøj *et al.*, 2010; Muller *et al.*, 2010; Pereiro *et al.*, 2010). BBS is understood as an example of the rapidly growing numbers of ciliopathies. In several organisms like worms, mice, zebrafish, and humans, the affected genes and their encoded proteins mediate and regulate microtubule-based intracellular transport processes (Blacque and Leroux, 2006). The most frequently mutated genes are *BBS1, BBS10*, and *BBS2*; further examples for affected genes are *ARL6* (ADP-ribosylation factor 6), *TTC8* (tetratricopeptide repeat domain-containing gene), *CEP290* (centrosomal protein 290), and *TRIM32* (tripartite motif-containing protein 32; Muller *et al.*, 2010).

One example of congenital disorders, which are characterized for a longer time, is **Leber congenital amourosis** (LCA). It is a group of autosomal recessive retinal dystrophies that represent the most common genetic cause of congenital retinal disorders in infants and children. Its incidence is 2–3 per 100,000 births and it accounts for 10–18% of cases of congenital blindness. At least 14 genes contribute to this disorder explaining together approximately 70% of the cases. Among them, *CEP290* (centrosomal protein 290 kDa; 15%), *GUCY2D* (guanylate cyclase 2D; 12%), and *CRB1* (*crumbs* homolog-1; 10%) are the most frequently mutated genes causing LCA; one intronic *CEP290* mutation is found in ~20% of all LCA patients from northwestern Europe (for a recent review, see den Hollander *et al.*, 2008). However, the genetic heterogeneity is not only due to the number of genes that have been implicated in LCA, but also due to the consequences of the different mutations within these genes. At least for two genes (*RPE65* and *CRX*) it is well established that mutations do not only lead to LCA, but also to other and later-onset retinal dystrophies like retinitis pigmentosa; similarly, *CEP290* is mutated in LCA as well as in BBS.

Disorders like LCA are a paradigm to study somatic gene therapy. Since the affected area, the retina, is relatively small (as compared to other tissues in the body), and since it can easily be treated by subretinal injections, it was one of the first targets of gene therapy. Therefore, it is not surprising that also positive results can be found in the literature. One example is the successful treatment of LCA, caused by mutations in *RPE65*. Simonelli *et al.* reported in 2010 that the safety and the efficacy noted at early time points persist through at least 1.5 years after injection in the three LCA2 patients enrolled in the low-dose cohort of their trial. Results like this let us suggest that further improvement of the therapeutic schedule will lead also to a more significant improvement of vision in patients.

7.3. Development of the hyaloid and retinal vasculature

During embryogenesis, the development and differentiation of the eye requires support of nutrients by a dense vascular network. Therefore, the hyaloid vasculature is established as a transient embryonic vascular system which is complete at birth in mammals and regresses contemporaneously with the formation of the retinal vasculature. The process of intraocular vascularization begins with the entry of the hyaloid artery into the optic cup through the fetal fissure. Rapidly, the hyaloid artery extends and reaches the posterior pole of the developing lens. The main vessels are branching intensely over the lens surface forming a dense capillary network (*tunica vasculosa lentis*; TVL). Since all vessels of the hyaloid system are arteries, it needs to be connected to a venous system—here to the choroidal vasculature at the anterior border of the optic cup. During lens development, the TVL expands and reaches the anterior part of the lens forming the pupillary membrane. When the development of the hyaloid vascular system is completed, it provides all nutrients to the intraocular components of the developing eye. When the development of the retinal vasculature starts, the regression of the hyaloid vessel system is initiated (for a review, see Saint-Geniez and D'Amore, 2004).

In many mammals, the retina remains avascular because of its thinness permitting an efficient diffusion of oxygen from outside. However, in mice and rat as the major mammalian model systems, retinal blood vessels are present. They are restricted to the inner layers of the retina and organized in two planar layers. The retinal vascular network is characterized by a blood–barrier status similar to that in the central nervous system. In the human retina, the primitive vessels emerge from the optic disc at the base of the hyaloid artery during the fourth month of gestation. From that time, the initial network extends and spreads to the periphery of the retina, but its formation is completed only after birth (for a review, see Saint-Geniez and D'Amore, 2004).

Major players in the formation of the retinal vessels are Vegf (vascular endothelial growth factor) and its receptor Vegfr2. Recently, Alvarez *et al.* (2007) characterized genetic determinants of hyaloid and retinal vasculature in zebrafish and identified nine genes (including those coding for Plexin D1, Synapsin II, and Laminin α1) with cell membrane or extracellular matrix identity that is necessary for zebrafish hyaloid and retinal vasculature development. Anti-VEGF antibodies are widely used in the therapy for neovascular ocular diseases including the age-related macular degeneration (Ciulla and Rosenfeld, 2009).

8. The Optic Nerve

At approximately 47/48 days of gestation (around E11.5 in the mouse), the **optic stalk** is formed as the connection between the eye and the diencephalon. The axons from the ganglion cells of the inner layer of the retina meet at the base of the eye and travel down to the optic stalk. Initially, the optic stalk represents a narrow neck that connects the optic cup to the diencephalon. Once the axons reach the optic stalk, they grow into it forming the optic nerve (~E15.5 in the mouse) and relay the eye with the visual centers of the brain. The optic stalk is now referred to as the optic nerve (Carlson, 1994; Gilbert, 1994). In humans, beginning at the seventh month of gestation, the axons of the optic nerve become myelinated—a process that spreads out, back to the eye. At birth, the optic nerve is 3 mm thick but its diameter continues to increase for 6–8 years after birth (Hinrichsen, 1993).

One of the critical steps in this process is the formation of the optic disc, the interface between optic stalk and retina, where the visual fibers of the RGCs can exit and the hyaloid artery enters the developing eye chamber. Morcillo *et al.* (2006) demonstrated that in mouse embryos retinal fissure precursor cells can be recognized by the expression of netrin 1 (gene symbol: *Ntn1*) and the overlapping distribution of markers of the optic stalk (*Pax2, Vax1*) and the ventral neural retina (*Vax2, Raldh3*). They could also show that in *Bmp7*$^{-/-}$ fissure formation is not initiated leading eventually to a lack of the hyaloid artery, optic nerve aplasia, and intra-retinal misrouting of RGCs.

In humans, congenital optic nerve defects now account for about 15% of severe visual impairment or blindness (data for the UK): optic nerve hypoplasia (ONH) alone accounting for 12% (Taylor, 2007). In addition, colobomata occur as a rare congenital disorder with an incidence ranging from (per 10,000 births) 0.5 in Spain, 1.4 in France, and 2.6 in the USA to 7.5 in China. Coloboma has been reported in 0.6–1.9% of blind adults in Canada and 3.2–11.2% of blind children worldwide (Gregory-Evans *et al.*, 2004). A coloboma is a defect resulting from an abnormal closure of the fetal fissure in the inferonasal quadrant of the developing optic cup. Closure starts at the equator and extends anteriorly and posteriorly: incomplete closure creates a defect of any size from the margin of the pupil to the optic disc. Correspondingly, several forms are described; the GENEYE database (www.lmdatabases.com—with costs) lists 80 (partially overlapping) syndromes associated with optic disc coloboma and 104 with iris coloboma (Taylor, 2007).

There are two syndromes, which might be mentioned: the first is the **septo-optic dysplasia**, which is characterized by the absence of the *septum pellucidum* and ONH together with a variety of other structural abnormalities of the cerebral hemispheres and commisures, hypothalamus, visual system, the pituitary body, and stalk. It is caused by mutations in the gene *HESX1* (homeobox gene expressed in ES cells; OMIM 182230).

The second group is referred to as **renal-coloboma syndrome**; it is characterized by defects in eye (optic nerve coloboma) and kidney, which is caused by mutations in *PAX2*. A dominant mouse disease including defects in the optic nerve development, the retinal layer of the eye, and several defects of the kidney and brain was shown to be caused by an insertion of a G in a stretch of seven already existing G residues in the region coding for the paired domain of mouse Pax2 (Favor *et al.*, 1996). Since the characterization of the *Pax2* mouse mutants revealed defects in eye and kidney, corresponding mutational analyses of *PAX2* were conducted in several, independent human families with renal-coloboma syndrome, and, in two cases, a mutation identical to that in the *Pax2*^{1Neu} mouse was detected. Moreover, a deletion of one G or the insertion of even two Gs (probably as a result of slippage during replication) has been reported in humans at the same site in the gene confirming the stretch of 7 Gs as a mutation hotspot for spontaneous expansion or contracting mutations. In total, the OMIM database describes 12 alleles of *PAX2* associated with renal-coloboma syndrome (OMIM 167409, March 2010).

8.1. Outlook: the visual system

The optic nerve collects all RGCs and connects thereby the visual information from the external world with the visual processing centers and cognitive domains in the brain. The challenge for the biological system becomes obvious, if we view just the numbers: in mice, over 50,000 RGCs exit the retina, and in humans over a million RGC axons must be guided accurately during embryonic development. In mammals, RGC neurogenesis is limited to a particular period in embryogenesis, but in lower vertebrates (e.g., fishes and frogs), new RGCs are constantly added to the eyes because of their ongoing development (Oster *et al.*, 2004).

There is some evidence that the EphB and the ephrin-B families of transmembrane receptors and ligands are involved in guidance of RGC axons to the optic nerve head. The embryonic retina contains a number of Eph types and of ephrin molecules, showing specific expression patterns. Each subfamily of Eph molecules is preferentially distributed in a particular retinal quadrant, while its corresponding ephrin is present in the highest levels in the opposing quadrant. For example, EphB receptor proteins are generally found in a high ventral to low dorsal gradient, while ephrin-B proteins are present in an opposite high dorsal to low ventral pattern. Correspondingly, mice lacking both EphB2 and EphB3 function show optic nerve head targeting errors. There is further evidence that Sema5A acts as an additional guidance factor for optic nerve development (for a review, see Oster *et al.*, 2004; and references therein), and for BMPs regulating the expression of the EphB (Plas *et al.*, 2008).

Behind the eye, the optic nerve transports visual information to the brain. A part of the RGC axons (from the nasal hemiretina) cross the brain midline at the **optic chiasm** and project to visual targets, the **lateral geniculate nucleus** (LGN) and the **supperior colliculus** (SC), in the contralateral brain hemisphere. On the other hand, the fibers from the temporal retina do not cross the midline and project to their targets in the same side of the brain. The ratio of crossed to uncrossed parts of the RGC varies among species; in humans or nonhuman primates, the uncrossed component reaches about 40% of all RGCs, but cats, ferrets, horses, rabbits, or mice show a gradually decreased uncrossed component, which ranges from 30% in cats to about 5% in mice. The chiasma is not formed in *Pax2*-null mice and zebrafish mutants; further important transcription factors are FoxD1 and FoxG1. Later, the RGC axon expresses *Robo2*, one of the receptors for *Slit*. Moreover, gradients of several ephrins and the corresponding receptors act as guidance molecules. The key player for the control of the spatiotemporal specificity of expression of these axon guidance molecules seems to be Zic2, a zinc finger transcription factor, which is expressed in the optic chiasm (and some other brain regions). Zic2 is crucial for directing the ipsilateral retinal projection, since genetically modified mice expressing low levels of this protein (*Zic2* knockdown mice) show a severe reduction in the number of uncrossed axons. RGC axons project to the LGN and SC in a topographic manner; in the SC, the nasal-to-temporal axis of the retina is mapped along the posterior-to-anterior axis of the SC, and the dorsal-to-ventral retinal axis maps along the lateral-to-medial axis of the SC (for review, see Haupt and Huber-Brösamle, 2008; Herrera and García-Frigola, 2008; and references therein).

9. Conclusion and Perspectives

Mutations that lead to clinically relevant phenotypes highlight important steps in eye development: some affect genes that act at the top of the regulatory hierarchy and therefore at the initial stages of eye development, the formation of the eye field. Mutations in these genes (e.g., *Pax6, Sox2*) lead to anophthalmia, microphthalmia, or aniridia. Other genes (*FoxC1, FoxE3, Pitx3, Maf*) act downstream or later during development. Some are important just for one particular tissue, for example, the crystallins in the lens and *Pax2* for the optic nerve development. However, many genes involved in eye development are also active during development of other tissues and organs and, therefore, leading to pleiotropic or syndromic effects if mutated. Since the pathological events frequently occur in the eye first, eye disorders may serve as bioindicator for other disorders.

Another important aspect concerns the diagnosis of eye disorders. The overview presented here demonstrates that the same clinical phenotype might be caused by mutations in different genes; but in other cases, mutations in the same gene do not necessarily lead to the same phenotype. Therefore, further detailed investigations may lead to changes in the nomenclature and classification of eye disorders, which at present are solely based on the clinical phenotype and, from a genetic point of view, are sometimes confusing.

There is also some confusion coming from the comparison of eye development in different model systems, primarily in fishes, frogs, chicken, rodents, and humans. Each system has its own advantage and research history—and therefore its own nomenclature for genes (which is currently being harmonized). The orthologous genes have somewhat different expression patterns and therefore also distinct functions. Therefore, it would be helpful if a systematic comparison of spatial and temporal expression patterns of the relevant genes could be performed in the most important species to elaborate the basic principles, but also the diversity of eye development among vertebrates. The systematic comparison of corresponding mutants among these species might be helpful for a better understanding of human disorders or for therapies, in particular if the knowledge of tissue regeneration can be transferred from other organisms to human patients. Finally, major novel insights can be expected from epigenetics—the role of noncoding RNAs and the influence of chromatin modification were mentioned in this review only shortly. It may change our way for diagnosis and open also new therapeutic avenues for congenital and, clinically even more important, for age-related eye disorders (for a recent review, see Cvekl and Mitton, 2010).

ACKNOWLEDGMENTS

I would like to thank many friends and colleagues who have supported our work within the past years. Unfortunately, due to space limitations I could not refer to all papers dealing with molecular aspects in eye development, and I apologize to those colleagues, who have not been cited.

DATABASES USED

MGI database: Mouse Genome Informatics (http://www.informatics.jax.org/)
OMIM: Online Mendelian Inheritance in Man (http://www.ncbi.nlm.nih.gov/sites/entrez?db=omim)
ZFIN: The Zebrafish Model Organism Database (http://zfin.org)
mikro-RNA-Database: http://www.mirbase.org

REFERENCES

Adelmann, H. B. (1929). Experimental studies on the development of the eye. II. The eye-forming potencies of the urodelan neural plate (*triton teniatus* and *amblysoma punctatum*. *J. Exp. Zool.* **54**, 291–317.

Adler, R. (2000). A model of retinal cell differentiation in the chick embryo. *Prog. Retin. Eye Res.* 19, 529–557.

Adler, R., and Canto-Soler, M. V. (2007). Molecular mechanisms of optic vesicle development: Complexities, ambiguities and controversies. *Dev. Biol.* **305**, 1–13.

Alvarez, Y., Cederlund, M. L., Cottell, D. C., Bill, B. R., Ekker, S. C., Torres-Vazquez, J., Weinstein, B. M., Hyde, D. R., Vihtelic, T. S., and Kennedy, B. N. (2007). Genetic determinants of hyaloid and retinal vasculature in zebrafish. *BMC Dev. Biol.* **7**, 114.

Andreazzoli, M. (2009). Molecular regulation of vertebrate retina cell fate. *Birth Defects Res. (Part C)* **87**, 284–295.

Azuma, N., Hirakiyama, A., Inoue, T., Asaka, A., and Yamada, M. (2000). Mutations of a human homologue of the drosophila eyes absent gene (*EYA1*) detected in patients with congenital cataracts and ocular anterior segment anomalies. *Hum. Mol. Genet.* **9**, 363–366.

Beebe, D. C., and Coats, J. M. (2000). The lens organizes the anterior segment: Specification of neural crest cell differentiation in the avian eye. *Dev. Biol.* **220**, 424–431.

Belenkaya, T. Y., Han, C., Standley, H. J., Lin, X., Houston, D. W., and Heasman, J. (2002). Pygopus encodes a nuclear protein essential for wingless/wnt signaling. *Development* **129**, 4089–4101.

Blacque, O. E., and Leroux, M. R. (2006). Bardet-Biedl syndrome: An emerging pathomechanism of intracellular transport. *Cell Mol. Life Sci.* **63**, 2145–2161.

Blixt, A., Mahlapuu, M., Aitola, M., Pelto-Huikko, M., Enerbäck, S., and Carlsson, P. (2000). A forkhead gene, *FoxE3*, is essential for lens epithelial proliferation and closure of the lens vesicle. *Genes Dev.* **14**, 245–254.

Bonini, N. M., Leiserson, W. M., and Benzer, S. (1993). The *eyes absent* gene: Genetic control of cell survival and differentiation in the developing drosophila eye. *Cell* **72**, 379–395.

Bonini, N. M., Leiserson, W. M., and Benzer, S. (1998). Multiple roles of the *eyes absent* gene in *Drosophila*. *Dev. Biol.* **196**, 42–57.

Burmeister, M., Novak, J., Liang, M. Y., Basu, S., Ploder, L., Hawes, N. L., Vidgen, D., Hoover, F., Goldman, D., Kalnins, V. I., Roderick, T. H., Taylor, B. A., *et al.* (1996). Ocular retardation mouse caused by *chx10* homeobox null allele: Impaired retinal progenitor proliferatrion and bipolar cell differentiation. *Nat. Genet.* **12**, 376–384.

Carlson, B. M. (1994). *Human Embryology and Developmental Biology*. Mosby, St.Louis.

Chase, H. B. (1944). Studies on an anophthalmic strain of mice. IV. A second major gene for anophthalmia. *Genetics* **29**, 264–269.

Cho, S.-H., and Cepko, C. L. (2006). Wnt2b/β-catenin-mediated canonical wnt signaling determines the peripheral fates of the chick eye. *Development* **133**, 3167–3177.

Chow, R. L., Altmann, C. R., Lang, R. A., and Hemmati-Brivanlou, A. (1999). Pax6 induces ectopic eyes in a vertebrate. *Development* **126**, 4213–4222.

Ciulla, T. A., and Rosenfeld, P. J. (2009). Antivascular endothelial growth factor therapy for neovascular age-related macular degeneration. *Curr. Opin. Ophthalmol.* **20**, 158–165.

Crawford, S. E., Qi, C., Misra, P., Stellmach, V., Rao, M. S., Engel, J. D., Zhu, Y., and Reddy, J. K. (2002). Defects of the heart, eye, and megakaryocytes in peroxisome proliferator activator receptor-binding protein (PBP) null embryos implicate GATA family of transcription factors. *J. Biol. Chem.* **277**, 3585–3592.

Cvekl, A., and Mitton, K. P. (2010). Epigenetic regulatory mechanisms in vertebrate eye development and disease. *Heredity*. doi 10.1038/hdy.2010.16.

Cvekl, A., and Tamm, E. R. (2004). Anterior eye development and ocular mesenchyme: New insights from mouse models and human diseases. *BioEssays* **26**, 374–386.

Cvekl, A., and Wang, W.-L. (2009). Retinoic signaling in mammalian eye development. *Exp. Eye Res.* **89**, 280–291.

Cvekl, A., Yang, Y., Chauhan, B. K., and Cveklova, K. (2004). Regulation of gene expression by pax6 in ocular cells: A case of tissue-preferred expression of crystallins in lens. *Int. J. Dev. Biol.* **48**, 829–844.

Dahm, R., Schonthaler, H. B., Soehn, A. S., van Marle, J., and Vrensen, G.F.J.M. (2007). Development and adult morphology of the eye lens in the zebrafish. *Exp. Eye Res.* **85**, 74–89.

Danno, H., Michiue, T., Hitachi, K., Yukita, A., Ishiura, S., and Asashima, M. (2008). Molecular links among the causative genes for ocular malformation: Otx2 and sox2 coregulate *rax* expression. *Proc. Natl. Acad. Sci. USA* **105**, 5408–5413.

Davis, L. K., Meyer, K. J., Rudd, D. S., Librant, A. L., Epping, E. A., Sheffield, V. C., and Wassink, T. H. (2008). Pax6 3' deletion results in aniridia, autism and mental retardation. *Hum. Genet.* **123**, 371–378.

Davis-Silberman, N., and Ashery-Padan, R. (2008). Iris development in vertebrates; genetic and molecular considerations. *Brain Res.* **1192**, 17–28.

de Iongh, R. U., Abud, H. E., and Hime, G. R. (2006). WNT/frizzled signaling in eye development and disease. *Front. Biosci.* **11**, 2442–2464.

de Iongh, R., and McAvoy, J. W. (1993). Spatio-temporal distribution of acidic and basic FGF indicates a role for FGF in rat lens morphogenesis. *Dev. Dyn.* **198**, 190–202.

den Hollander, A. I., Roepman, R., Koenekoop, R. K., and Cremers, F. P. (2008). Leber congenital amaurosis: Genes, proteins and disease mechanisms. *Prog. Retin. Eye Res.* **27**, 391–419.

Dias da Silva, M. R., Tiffin, N., Mima, T., Mikawa, T., and Hyer, J. (2007). FGF-mediated induction of ciliary body tissue in the chick eye. *Dev. Biol.* **304**, 272–285.

Duester, G. (2009). Keeping an eye on retinoic acid signaling during eye development. *Chem.-Biol. Interact.* **178**, 178–181.

Dyer, M. A., Livesey, F. J., Cepko, C. L., and Oliver, G. (2003). Prox1 function controls progenitor cell proliferation and horizontal cell genesis in the mammalian retina. *Nat. Genet.* **34**, 53–58.

Eggert, T., Hauck, B., Hildebrandt, N., Gehring, W. J., and Walldorf, U. (1998). Isolation of a *Drosophila* homolog of the vertebrate homeobox gene Rx and its possible role in brain and eye development. *Proc. Natl. Acad. Sci. USA* **95**, 2343–2348.

Faber, S. C., Robinson, M. L., Makarenkova, H. P., and Lang, R. A. (2002). Bmp signaling is required for development of primary lens fiber cells. *Development* **129**, 3727–3737.

Fadool, J. M., and Dowling, J. E. (2008). Zebrafish: A model system for the study of eye genetics. *Prog. Retin. Eye Res.* **27**, 89–110.

Favor, J., Gloeckner, C. J., Neuhäuser-Klaus, A., Pretsch, W., Sandulache, R., Saule, S., and Zaus, I. (2008). Relationship of *pax6* activity levels to the extent of eye development in the mouse, *mus musculus*. *Genetics* **179**, 1345–1355.

Favor, J., Peters, H., Hermann, T., Schmahl, W., Chatterjee, B., Neuhäuser-Klaus, A., and Sandulache, R. (2001). Molecular characterization of $Pax6^{2Neu}$ through $Pax6^{10Neu}$. *Genetics* **159**, 1689–1700.

Favor, J., Sandulache, R., Neuhäuser-Klaus, A., Pretsch, W., Chatterjee, B., Senft, E., Wurst, W., Blanquet, V., Grimes, P., Spörle, R., and Schughart, K. (1996). The mouse $Pax2^{1Neu}$ mutation is identical to a human *PAX2* mutation in a family with renal-coloboma syndrome and results in developmental defects of the brain, ear, eye, and kidney. *Proc. Natl. Acad. Sci. USA* **93**, 13870–13875.

Fritz, A., Walch, A., Piotrowska, K., Rosemann, M., Schäffer, E., Weber, K., Timper, A., Wildner, G., Graw, J., Höfler, H., and Atkinson, M. J. (2002). Recessive transmission of multiple endocrine neoplasia syndrome in the rat. *Cancer Res.* **62**, 3048–3051.

Fuhrmann, S., Levine, E. M., and Reh, T. A. (2000). Extraocular mesenchyme patterns the optic vesicle during early eye development in the embryonic chick. *Development* **127**, 4599–4609.

Furuta, Y., and Hogan, B. L. (1998). BMP4 is essential for lens induction in the mouse embryo. *Genes Dev.* **12**, 3764–3775.

Gage, P. J., Qian, M., Wu, D., and Rosenberg, K. I. (2008). The canonical wnt signaling antagonist DKK2 is an essential effector of PITX2 function during normal eye development. *Dev. Biol.* **317**, 310–324.

Garcia, C. M., Yu, K., Zhao, H., Ashery-Padan, R., Ornitz, D. M., Robinson, M. L., and Beebe, D. C. (2005). Signaling through FGF receptor-2 is required for lens cell survival and for withdrawal from the cell cycle during lens fiber cell differentiation. *Dev. Dyn.* **233**, 516–527.

Geng, X., Speirs, C., Lagutin, O., Inbal, A., Liu, W., Solnica-Krezel, L., Jeong, Y., Epstein, D. J., and Oliver, G. (2008). Haploinsufficiency of *six3* fails to activate *sonic hedgehog* expression in the ventral forebrain and causes holoprosencephaly. *Dev. Cell* **15**, 236–247.

Gilbert, S. F. (1994). *Developmental Biology*, 4th edition. Sinauer, Sunderland.

Glass, A. S., and Dahm, R. (2004). The zebrafish as a model organism for eye development. *Ophthalmic Res.* **36**, 4–24.

Graw, J. (2003). Genetic and molecular basis of congenital eye defects. *Nat. Rev. Genet.* **4**, 876–888.

Graw, J. (2009a). Crystallins. Cataract and beyond. *Exp. Eye Res.* **88**, 173–189.

Graw, J. (2009b). Mouse models of cataract. *J. Genet.* **88**, 469–486.

Gregory-Evans, C. Y., Williams, M. J., Halford, S., and Gregory-Evans, K. (2004). Ocular coloboma: A reassessment in the age of molecular neuroscience. *J. Med. Genet.* **41**, 881–891.

Halder, G., Callaerts, P., and Gehring, W. J. (1995). Induction of ectopic eyes by targeted expression of the *eyeless* gene in *Drosophila. Science* **267**, 1788–1792.

Harrington, L., Klintworth, G. K., Secor, T. E., and Breitman, M. L. (1991). Developmental analysis of ocular morphogenesis in αA-crystallin/diphtheria toxin transgenic mice undergoing ablation of the lens. *Dev. Biol.* **148**, 508–516.

Haupt, C., and Huber-Brösamle, A. (2008). How axons see their way—axonal guidance in the visual system. *Front. Biosci.* **13**, 3136–3149.

Hay, E. D. (1979). Development of the vertebrate cornea. *Int. Rev. Cytol.* **63**, 263–322.

Herrera, E., and García-Frigola, C. (2008). Genetics and development of the optic chiasm. *Front. Biosci.* **13**, 1646–1653.

Hertwig, P. (1942). Neue mutationen und koppelungsgruppen bei der hausmaus. *Z. Indukt. Abstammungs. Vererbungsl.* **80**, 220–246.

Hill, R. E., Favor, J., Hogan, B.L.M., Ton, C.C.T., Saunders, G. F., Hanson, I. M., Prosser, J., Jordan, T., Hastie, N. D., and van Heyningen, V. (1991). Mouse *small eye* results from mutations in a paired-like homeobox-containing gene. *Nature* **354**, 522–525.

Hinrichsen, K. V. (1993). *Human-Embryologie.* Springer-Verlag, Heidelberg.

Hjortshøj, T. D., Grønskov, K., Philip, A. R., Nidshimura, D. Y., Riise, R., Sheffield, V. C., Rosenberg, T., and Brøndum-Nielsen, K. (2010). Bardet-Biedl syndrome in Denmark—report of 13 novel sequence variations in six genes. *Hum. Mutat.* **31**, 429–436.

Horsford, D. J., Nguyen, M. T., Sellar, G. C., Kothary, R., Arnheiter, H., and McInnes, R. R. (2005). *Chx10* repression of *mitf* is required for the maintenance of mammalian neuroretinal identity. *Development* **132**, 177–187.

Kamachi, Y., Uchikawa, M., Collignon, J., Lovell-Badge, R., and Kondoh, H. (1998). Involvement of sox1, 2 and 3 in the early and subsequent molecular events of lens induction. *Development* **125**, 2521–2532.

Kaufman, M. H. (1992). *The Atlas of Mouse Development.* Academic Press Ltd, London.

Kreslova, J., Machon, O., Ruzickova, J., Lachova, J., Wawrousek, E. F., Kemler, R., Krauss, S., Piatigorsky, J., and Kozmik, Z. (2007). Abnormal lens morphogenesis and ectopic lens formation in the absence of β-catenin function. *Genesis* **45**, 157–168.

Kriebel, M., Müller, F., and Hollemann, T. (2007). *Xeya3* regulates survival and proliferation of neural progenitor cells within the anterior neural plate of *xenopus* embryos. *Dev. Dyn.* **236**, 1526–1534.

Kurose, H., Bito, T., Adachi, T., Shimizu, M., Noji, S., and Ohuchi, H. (2004). Expression of *fibroblast growth factor 19* (*fgf19*) during chicken embryogenesis and eye development, compared with *fgf15* expression in the mouse. *Gene Expr. Patterns* **4**, 687–693.

Kurose, H., Okamoto, M., Shimizu, M., Bito, T., Marcelle, C., Noji, S., and Ohuchi, H. (2005). FGF19-FGFR4 signaling elaborates lens induction with the FGF8-L-maf cascade in the chick embryo. *Dev. Growth Differ.* **47**, 213–223.

Kuwabara, T., and Imaizumi, M. (1974). Denucleation process of the lens. *Invest. Ophthalmol.* 13, 973–981.

Lagutin, O. V., Zhu, C. C., Kobayashi, D., Topczewski, J., Shimamura, K., Puelles, L., Russell, H. R., McKinnon, P. J., Solnica-Krezel, L., and Oliver, G. (2003). Six3 repression of wnt signaling in the anterior neuroectoderm is essential for vertebrate forebrain development. *Genes Dev.* **17**, 368–379.

Lang, R. A. (2004). Pathways regulating lens induction in the mouse. *Int. J. Dev. Biol.* **48**, 783–791.

Lequeux, L., Rio, M., Vigouroux, A., Titeux, M., Etchevers, H., Malecaze, F., Chassaing, N., and Calvas, P. (2008). Confirmation of *RAX* gene involvement in human anophthalmia. *Clin. Genet.* 74, 392–395.

Liu, W., Lagutin, O. V., Mende, M., Streit, A., and Oliver, G. (2006). Six3 activation of *pax6* expression is essential for mammalian lens induction and specification. *EMBO J.* **25**, 5383–5395.

Loosli, F., Staub, W., Finger-Baier, K. C., Ober, E. A., Verkade, H., Wittbrodt, J., and Baier, H. (2003). Loss of eyes in zebrafish caused by mutation of *chokh/rx3*. *EMBO Rep.* **4**, 894–899.

Lwigale, P. Y., and Bronner-Fraser, M. (2007). Lens-derived Semaphorin3A regulates sensory innervation of the cornea. *Dev. Biol.* **306**, 750–759.

Lwigale, P. Y., and Bronner-Fraser, M. (2009). Semaphorin3A/neuropilin-1 signaling acts as a molecular switch regulating neural crest migration during cornea development. *Dev. Biol.* **336**, 257–265.

Lyon, M. F., Jamieson, R. V., Perveen, R., Glenister, P. H., Griffiths, R., Boyd, Y., Glimcher, L. H., Favor, J., Munier, F. L., and Black, G. (2003). A dominant mutation within the DNA-binding domain of the bZIP transcription factor maf causes murine cataract and results in selective alteration in DNA binding. *Hum. Mol. Genet.* **12**, 585–594.

Lyu, J., and Joo, C.-K. (2004). Wnt signaling enhances FGF2-triggered lens fiber cell differentiation. *Development* **131**, 1813–1824.

Machon, O., Kreslova, J., Ruzickova, J., Vacik, T., Klimova, L., Fujimura, N., Lachova, J., and Kozmik, Z. (2010). Lens morphogenesis is dependent on pax6-mediated inhibition of the canonical wnt/beta-catenin signaling in the lens surface ectoderm. *Genesis* **48**, 86–95.

Martínez-Morales, J. R., Rodrigo, I., and Bovolenta, P. (2004). Eye development: A view from the retina pigmented epithelium. *BioEssays* **26**, 766–777.

Massé, K., Bhamra, S., Eason, R., Dale, N., and Jones, E. A. (2007). Purine-mediated signalling triggers eye development. *Nature* **449**, 1058–1062.

Mathers, P. H., Grinberg, A., Mahon, K. A., and Jamrich, M. (1997). The *rx* homeobox gene is essential for vertebrate eye development. *Nature* **387**, 603–607.

McAvoy, J. W., and Chamberlain, C. G. (1989). Fibroblast growth factor (FGF) induces different responses in lens epithelial cells depending on its concentration. *Development* **107**, 221–228.

Mic, F. A., Molotkov, A., Molotkova, N., and Duester, G. (2004). *Raldh2* expression in optic vesicle generates a retinoic acid signal needed for invagination of retina during optic cup formation. *Dev. Dyn.* **231**, 270–277.

Miller, D. L., Ortega, S., Bashayan, O., Basch, R., and Basilico, C. (2000). Compensation by fibroblast growth factor 1 (FGF1) does not account for the mild phenotypic defects observed in FGF2 null mice. *Mol. Cell. Biol.* **20**, 2260–2268. *Erratum* in: *Mol. Cell. Biol.* **20**, 3752.

Molotkov, A., Molotkova, N., and Duester, G. (2006). Retinoic acid guides eye morphogenetic movements via paracrine signaling but is unnecessary for retinal dorsoventral patterning. *Development* **133**, 1901–1910.

Morcillo, J., Martínez-Morales, J. R., Trousse, F., Fermin, Y., Sowden, J. C., and Bovolenta, P. (2006). Proper patterning of the optic fissure requires the sequential activity of BMP7 and SHH. *Development* **133**, 3179–3190.

Moshiri, A., Gonzalez, E., Tagawa, K., Maeda, H., Wang, M., Frishman, L. J., and Wang, S. W. (2008). Near complete loss of retinal ganglion cells in the *math5/brn3b* double knockout elicits severe reductions of other cell types during retinal development. *Dev. Biol.* **316**, 214–227.

Muller, J., Stoetzel, C., Vincent, M. C., Leitch, C. C., Laurier, V., Danse, J. M., Hellé, S., Marion, V., Bennouna-Greene, V., Vicaire, S., Megarbane, A., Kaplan, J., *et al.* (2010). Identification of 28 novel mutations in the Bardet-Biedl syndrome genes: The burden of private mutations in an extensively heterogeneous disease. *Hum. Genet.* **127**, 583–593.

Murato, Y., and Hashimoto, C. (2009). *Xhairy2* functions in *xenopus* lens development by regulating $p27^{xic1}$ expression. *Dev. Dyn.* **238**, 2179–2192.

Ohsawa, R., and Kageyama, R. (2008). Regulation of retinal cell fate specification by multiple transcription factors. *Brain Res.* **1192**, 90–98.

Oliver, G., Loosli, F., Köster, R., Wittbrodt, J., and Gruss, P. (1996). Ectopic lens induction in fish in response to the murine homeobox gene *six3*. *Mech. Dev.* **60**, 233–239.

Ormestad, M., Blixt, A., Churchill, A., Martinsson, T., Enerbäck, S., and Carlsson, P. (2002). Foxe3 haploinsufficiency in mice: A model for Peters' anomaly. *Invest. Ophthalmol. Vis. Sci.* 43, 1350–1357.

Oster, S. F., Deiner, M., Birgbauer, E., and Sretavan, D. W. (2004). Ganglion cell axon pathfinding in the retina and optic nerve. *Sem. Cell Dev. Biol.* **15**, 125–136.

Panda, S., Provencio, I., Tu, D. C., Pires, S. S., Rollag, M. D., Castrucci, A. M., Pletcher, M. T., Sato, T. K., Wiltshire, T., Andahazy, M., Kay, S. A., van Gelder, R. N., *et al.* (2003). Melanopsin is required for non-image-forming photic responses in blind mice. *Science* **301**, 525–527.

Pearce, W. G., Mielke, B. C., and Walter, M. A. (1999). Histopathology and molecular basis of iridogoniodysgenesis syndrome. *Ophthalmic Genet.* **20**, 83–88.

Pei, W., and Feldman, B. (2009). Identification of common and unique modifiers of zebrafish midline bifurcation and cyclopia. *Dev. Biol.* **326**, 201–211.

Pellegata, N. S., Quintanilla-Martinez, L., Siggelkow, H., Samson, E., Bink, K., Höfler, H., Graw, J., and Atkinson, M. (2006). Germline mutations in *p27kip1* cause a multiple endocrine neoplasia syndrome in rats and man. *Proc. Natl. Acad. Sci. USA* **103**, 15558–15563.

Percin, F. E., Ploder, L. A., Yu, J. J., Arici, K., Horsford, D. J., Rutherford, A., Bapat, B., Cox, D. W., Duncan, A. M., Kalnins, V. I., Kocak-Altintas, A., Sowden, J. C., *et al.* (2000). Human microphthalmia associated with mutations in the retinal homeobox gene *CHX10*. *Nat. Genet.* **25**, 397–401.

Pereiro, I., Valverde, D., Piñeiro-Gallego, T., Baiget, M., Borrego, S., Ayuso, C., Searby, C., and Nishimura, D. (2010). New mutations in BBS genes in small consanguineous families with bardet-biedl syndrome: Detection of candidate regions by homozygosity mapping. *Mol. Vis.* **16**, 137–143.

Perveen, R., Favor, J., Jamieson, R. V., Ray, D. W., and Black, G.C.M. (2007). A heterozygous c-maf transactivation domain mutation causes congenital cataract and enhances target gene activation. *Hum. Mol. Genet.* **16**, 1030–1038.

Pézeron, G., Lambert, G., Dickmeis, T., Strähle, U., Rosa, F. M., and Mourrain, P. (2008). Rasl11b knock down in zebrafish suppresses *one-eyed-pinhead* mutant phenotype. *PLoS One* **16**, e1434.

Piatigorsky, J. (2001). Enigma of the abundant water-soluble cytoplasmic proteins of the cornea. *Cornea* **20**, 853–858.

Plas, D. T., Dhande, O. S., Lopez, J. E., Murali, D., Thaller, C., Henkemeyer, M., Furuta, Y., Overbeek, P., and Crair, M. C. (2008). Bone morphogenetic proteins, eye patterning, and retinocollicular map formation in the mouse. *J. Neurosci.* **28**, 7057–7067.

Puk, O., Löster, J., Dalke, C., Soewarto, D., Fuchs, H., Budde, B., Nürnberg, P., Wolf, E., Hrabé de Angelis, M., and Graw, J. (2008). Mutation in a novel connexin-like gene (*gjf1*) in the mouse affects early lens development and causes a variable small-eye phenotype. *Invest. Ophthalmol. Vis. Sci.* **49**, 1525–1532.

Qiu, R., Liu, Y., Wu, J. Y., Liu, K., Mo, W., and He, R. (2009). Misexpression of miR-196a induces eye anomaly in *Xenopus laevis*. *Brain Res. Bull.* **79**, 26–31.

Quiring, R., Walldorf, U., Kloter, U., and Gehring, W. J. (1994). Homology of the eyeless gene in drosophila to the small eye gene in mice and aniridia in humans. *Science* **265**, 785–789.

Ragge, N. K., Lorenz, B., Schneider, A., Bushby, K., de Sanctis, L., de Sanctis, U., Salt, A., Collin, J.R.O., Vivian, A. J., Free, S. L., Thompson, P., Williamson, K. A., *et al.* (2005). *SOX2* anophthalmia syndrome. *Am. J. Med. Genet.* **135A**, 1–7.

Rajagopal, R., Huang, J., Dattilo, L. K., Kaartinen, V., Mishina, Y., Deng, C.-X., Umans, L., Zwijsen, A., Roberts, A. B., and Beebe, D. C. (2009). The type i BMP receptors, bmpr1a and acvr1, activate multiple signaling pathways to regulate lens formation. *Dev. Biol.* **335**, 305–316.

Rebagliati, M. R., Toyama, R., Haffter, P., and Dawid, I. B. (1998). *Cyclops* encodes a nodal-related factor involved in midline signaling. *Proc. Natl. Acad. Sci. USA* **18**, 9932–9937.

Rieger, D. K., Reichenberger, E., McLean, W., Sidow, A., and Olsen, B. R. (2001). A double-deletion mutation in the *pitx3* gene causes arrested lens deveopment in *aphakia* mice. *Genomics* **72**, 61–72.

Riesenberg, A. N., Le, T. T., Willardsen, M. I., Blackburn, D. C., Vetter, M. L., and Brown, N. L. (2009). Pax6 regulation of math5 during mouse retinal neurogenesis. *Genesis* **47**, 175–187.

Rosemann, M., Ivashkevich, A., Favor, J., Dalke, C., Hölter, S. M., Becker, L., Rácz, I., Bolle, I., Klempt, M., Rathkolb, B., Kalaydjiev, S., Adler, T., *et al.* (2010). Microphthalmia, Parkinsonism and enhanced nociception in *Pitx3*^{416insG} mice. *Mamm. Genome.* **21**, 13–27.

Saint-Geniez, M., and D'Amore, P. A. (2004). Development and pathology of the hyaloid, choroidal and retinal vasculature. *Int. J. Dev. Biol.* **48**, 1045–1058.

Sander, M., Neubüser, A., Kalamaras, J., Ee, H. C., Martin, G. R., and German, M. S. (1997). Genetic analysis reveals that PAX6 is required for normal transcription of pancreatic hormone genes and islet development. *Genes Dev.* **11**, 1662–1673.

Semina, E. V., Ferrell, R. E., Mintz-Hittner, H. A., Bitoun, P., Alward, W.L.M., Reiter, R. S., Funkhauser, C., Daack-Hirsch, S., and Murray, J. C. (1998). A novel homeobox gene *PITX3* is mutated in families with autosomal-dominant cataracts and ASMD. *Nat. Genet.* **19**, 167–170.

Semina, E., Murray, J. C., Reiter, R., Hrstka, R. F., and Graw, J. (2000). Deletion in the promoter region and altered expression of pitx3 homeobox gene in aphakia mice. *Hum. Mol. Genet.* **9**, 1575–1585.

Shen, W., and Mardon, G. (1997). Ectopic eye development in *Drosophila* induced by directed *dachshund* expression. *Development* **124**, 45–52.

Simonelli, F., Maguire, A. M., Testa, F., Pierce, E. A., Mingozzi, F., Bennicelli, J. L., Rossi, S., Marshall, K., Banfi, S., Surace, E. M., Sun, J., Redmond, T. M., *et al.* (2010). Gene therapy for leber's congenital amaurosis is safe and effective through 1.5 Years after vector administration. *Mol. Ther.* **18**, 643–650.

Smith, A. N., Miller, L.-A., Radice, G., Ashery-Padan, R., and Lang, R. A. (2009). Stage-dependent modes of pax6-sox2 epistasis regulate lens development and eye morphogenesis. *Development* **136**, 2977–2985.

Smith, A. N., Miller, L.-A. D., Song, N., Taketo, M. M., and Lang, R. A. (2005). The duality of β-catenin function: a requirement in lens morphogenesis and signaling suppression of lens fate in periocular ectoderm. *Dev. Biol.* **285**, 477–489.

Smith, A. N., Radice, G., and Lang, R. A. (2010). Which FGF ligands are involved in lens induction? *Dev. Biol.* **337**, 195–198.

Smith, R. S., Zabaleta, A., Kume, T., Savinova, O. V., Kidson, S. H., Martin, J. E., Nishimura, D. Y., Alward, W. L., Hogan, B. L., and John, S. W. (2000).

Haploinsufficiency of the transcription factors FOXC1 and FOXC2 results in aberrant ocular development. *Hum. Mol. Genet.* **9**, 1021–1032.

Söker, T., Dalke, C., Puk, O., Floss, T., Becker, L., Bolle, I., Favor, J., Hans, W., Hölter, S. M., Horsch, M., Kallnik, M., Kling, E., *et al.* (2008). Pleiotropic effects in *eya3* knockout mice. *BMC Dev. Biol.* **8**, 118. doi:10.1186/1471-213X-8-118.

Song, N., Schwab, K. R., Patterson, L. T., Yamaguchi, T., Lin, X., Potter, S. S., and Lang, R. A. (2007). Pygopus 2 has a crucial, Wnt pathway-independent function in lens induction. *Development* **134**, 1873–1885.

Spemann, H. (1924). Über organisatoren in der tierischen entwicklung. *Naturwissenschaften* **48**, 1092–1094.

Stigloher, C., Ninkovic, J., Laplante, M., Geling, A., Tannhäuser, B., Topp, S., Kikuta, H., Becker, T. S., Houart, C., and Bally-Cuif, L. (2006). Segregation of telencephalic and eye-field identities inside the zebrafish forebrain territory is controlled by rx3. *Development* **133**, 2925–2935.

Streit, A. (2007). The preplacodal region: An ectodermal domain with multipotential progenitors that contribute to sense organs and cranial sensory ganglia. *Int. J. Dev. Biol.* **51**, 447–461.

Stump, R.J.W., Ang, S., Chen, Y., von Bahr, T., Lovicu, F. J., Pinson, K., de Iongh, R. U., Yamaguchi, T. P., Sassoon, D. A., and McAvoy, J. W. (2003). A role of wnt/β-catenin signaling in lens epithelial differentiation. *Dev. Biol.* **259**, 43–61.

Svoboda, K.K.H., Fischman, D. A., and Gordon, M. K. (2008). Embryonic chick corneal epithelium: A model system for exploring cell-matrix interactions. *Dev. Dyn.* **237**, 2667–2675.

Taylor, D. (2007). Developmental abnormalities of the optic nerve and chiasm. *Eye* **21**, 1271–1284.

Tomita, K., Moriyoshi, K., Nakanishi, S., Guillemot, F., and Kageyama, R. (2000). Mammalian achaete-scute and atonal homologs regulate neuronal versus glial fate determination in the central nervous system. *EMBO J.* **19**, 5460–5472.

Tsonis, P. A., and Fuentes, E. J. (2006). Focus on molecules: Pax-6, the eye master. *Exp. Eye Res.* **83**, 233–234.

Tucker, P., Laemle, L., Munson, A., Kanekar, S., Oliver, E. R., Brown, N., Schlecht, H., Vetter, M., and Glaser, T. (2001). The *eyeless* mouse mutation (*ey1*) removes an alternative start codon from the *rx/rax* homeobox gene. *Genesis* **31**, 43–53.

Varga, Z. M., Wegner, J., and Westerfield, M. (1999). Anterior movement of ventral diencephalic precursors separates the primordial eye field in the neural plate and requires *cyclops*. *Development* **126**, 5533–5546.

Vrensen, G.F.J.M., Graw, J., and de Wolf, A. (1991). Nuclear breakdown during terminal differentiation of primary lens fibers in mice. A transmission electron microscopic study. *Exp. Eye Res.* **52**, 647–659.

Wawersik, S., Purcell, P., Rauchman, M., Dudley, A. T., Robertson, E. J., and Maas, R. (1999). BMP7 acts in murine lens placode development. *Dev. Biol.* **207**, 176–188.

Weisschuh, N., Wolf, C., Wissinger, B., and Gramer, E. (2008). A novel mutation in the *FOXC1* gene in a family with Axenfeld-Rieger syndrome and Peters' anomaly. *Clin. Genet.* **74**, 476–480.

Westheimer, G. (2007). The ON-OFF dichotomy in visual processing: From receptors to perception. *Prog. Retin. Eye Res.* **26**, 636–648.

Winkler, S., Loosli, F., Henrich, T., Wakamatsu, Y., and Wittbrodt, J. (2000). The conditional medaka mutation *eyeless* uncouples patterning and morphogenesis of the eye. *Development* **127**, 1911–1919.

Wirth, D., Gama-Norton, L., Riemer, P., Sandhu, U., Schucht, R., and Hauser, H. (2007). Road to precision: Recombinase-based targeting technologies for genome engineering. *Curr. Opin. Biotech.* **18**, 411–419.

Yun, S., Saijoh, Y., Hirokawa, K. E., Kopinke, D., Murtaugh, L. C., Monuki, E. S., and Levine, E. M. (2009). *Lhx2* links the intrinsic and extrinsic factors that control optic cup formation. *Development* **136**, 3895–3906.

Zhang, J., Talbot, W. S., and Schier, A. F. (1998). Positional cloning identifies zebrafish *one-eyed pinhead* as a permissive EGF-related ligand required during gastrulation. *Cell* **92**, 241–251.

Zhao, S., Chen, Q., Hung, F. C., and Overbeek, P. A. (2002). BMP signaling is required for development of the ciliary body. *Development* **129**, 4435–4442.

Zhao, H., Yang, T., Madakashira, B. P., Thiels, C. A., Bechtle, C. A., Garcia, C. M., Zhang, H., Yu, K., Ornitz, D. M., Beebe, D. C., and Robinson, M. L. (2008). Fibroblast growth factor signaling is essential for lens fiber cell differentiation. *Dev. Biol.* **318**, 276–288.

Zieske, J. D. (2004). Corneal development associated with eyelid opening. *Int. J. Dev. Biol.* **48**, 903–911.

Zuber, M. E., Gestri, G., Viczian, A. S., Barsacchi, G., and Harris, W. A. (2003). Specification of the vertebrate eye by a network of eye field transcription factors. *Development* **130**, 5155–5167.

Subject Index

A

Activating transcription factor-4 (Atf4), 303*f*, 304–305
Airway peristalsis, 80, 113–116, 130
Alagille syndrome (AGS), 60, 219
Aldehyde dehydrogenase (Aldh1a2), 271, 352, 363
Alport syndrome, 217
Alveolar epithelial progenitors, 121–122
 bronchioalveolar junctions, 121
 glutathione depletion, 122
 mitochondrial failure, 122
 naphthaline injury, 121
Alveolarization, 75, 76*f*, 77*f*, 78, 87, 89*t*, 90*t*, 94*t*, 97, 100–101, 105, 107, 110
Anatomy of lung, 75–81
 abnormal development, 80
  ciliary dyskinesia, 81
  cystic adenomatoid malformation, 80
  gas exchange, 80
  kartagener syndrome, 81
  neonatal respiratory failure, 80
 bauplan, 75
Angiogenesis
 antiangiogenesis, 44
 embryonic, 50
 expansion of blood vessel networks, 47–48
  filopodia, 48
  phalanx phenotype, 48
  stalk cells, 47
 Jagged1-regulated, 59
 lymphangiogenesis, 52
 sprouting, 46*f*, 47, 56
Anterior intestinal portal, 165
Anterior–posterior (AP) axis, patterning along
 interactions between Shh and Gli3, 325–326
  Bmp antagonist gremlin (Grem1), 325
  cleavage of full-length Gli3 protein (Gli3FL), 325
 primary cilium and Gli3 function in limb, 326–327
  intraflagellar transport (IFT), 326
  mechanosensory and signaling function, 326
  Meckel syndrome gene (Mks1), 327
  pathway markers Ptch1 and Gli1, 327
  pleiotropic human disorders, 326
  polydactyly, 326
  Shh gain-of-function mutants, 327
 Shh-mediated digit specification and growth, 327–328
  digit formation, 328
  optical projection tomography analysis, 328
  Shh RNA, 327
  studies in *Drosophila,* 328
Anti-testis Z factor, 278
Aorta (Ao), 4*f*, 5, 12*f*, 16, 47, 54, 60
Apert syndrome, 84–85
Apical ectodermal ridge (AER), 320, 321*f*, 324*f*, 325, 328–334
Apoptosis, 50, 77, 86, 99, 101, 106, 122, 128, 161, 167, 175, 182, 197, 199–200, 206–207, 214, 242, 277, 357
Arterial pole (AP), 4*f*, 5–6, 8, 11, 12*f*, 13, 16–17, 19*f*, 20–28, 320, 322–323, 326–327
Arteriovenous malformations, 54
Arteriovenous shunts, 53–54
Articular cartilage (AC) degeneration, *See* Osteoarthritis
Atrioventricular canal (AVC) cells, 8
Autocrine signaling, 23, 333
Axenfeld–Rieger syndrome, 367

B

Basic Fgf (bFGF), 362
Biochemical factors in lung development, 107–111
 extracellular matrix, 108–110
  cell–cell/cell–ECM interaction, 108
  cell polarization, 109
  cross-shaped molecule, 108
  dexamethasone exposure, 110
  fibronectin (FN), 109
  laminins (LN), 108
  nidogen, 109
  proteoglycan sulfation, 109
  tissue inhibitors of metalloproteinases (TIMP), 109
 retinoic acid signaling, 110–111
  proximal–distal, 111
  RAR/RXR heterodimers, 110
  RA signaling, 110
  retinaldehyde dehydrogenase-2 (Raldh-2), 110
  retinoid X receptor (RXRα), 111
 small noncoding microRNA (miRNA), 107–108
  intranuclear primary transcript (pri-miRNA), 107

Biochemical factors in lung development (*Continued*)
  RNA-induced silencing complex (RISC), 108
  RNase III endonuclease, 107
Biochemical regulators, 87–111
  factors, 107–111
    extracellular matrix/lung development, 108–110
    retinoic acid signaling, 110–111
    small noncoding microRNA (miRNA)/ lung development, 107–108
  peptide growth factors, 97–107
    Bmp subfamily, 101–102
    epidermal growth factor (EGF), 105
    FGF family, 97–99
    insulin-like growth factors (IGF), 106
    platelet-derived growth factors (PDGF), 106
    roundabout (ROBO)/SLIT, 107
    sonic hedgehog (Shh) pathway, 102–103
    Tgf-β/Bmp family, 99
    Vegf isoforms/cognate receptors, 106–107
    Wnt/β-catenin pathway, 103–105
  transcription factors, 87–97
    forkhead box, 87
    Gli family zinc-finger, 97
    Hox family, 96
    Nkx and Hox homeodomain, 95–96
Blepharophimosis ptosis epicanthus inversus syndrome (BPES), 281*t*
Blood vessel formation, 45*f*
  molecular signaling and endothelial heterogeneity in, 46*f*
Bone cells (osteoblasts), 292, 294, 296, 301, 303–307, 303*f*
Bone development, 302–306
  elastic tissues (sutures), 304
  endochondral and intramembranous ossification, 302–303
    and osteoblast differentiation, 303*f*
  intramembranous ossification, 304
    bone homeostasis, 304
    bone-specific proteins (Bgp/Bsp), 304
    Runx2, master osteogenic factor, 304
  matrix metalloproteinases (Mmp13/Mmp14), 303
  osteoblasts and osteoclasts, 303
    Atf4 and Creb2, 305
    Bmp2 expression, 306
    craniosynostosis, 306
    Fgf signaling, 305–306
    forkhead factor FoxO1, 305
    osteoblastogenesis, 305
    Osx or Sp7, 305
    Wnt/beta-catenin pathway, 305
  periosteum or bone collar, 303
  primary ossification center, 304
Bone loss disease, *See* Osteoporosis
Bone morphogenetic protein (Bmp)
  Bmp-2, -4, -5, -6, and -7, 178
  Bmp signaling, 9*f*, 13–14, 18, 25, 27–28, 53–55, 57, 61, 84, 102, 119, 124–125, 172–173, 175, 178, 185–186, 299, 325, 332–333, 353*f*, 354, 366
  See also Vascular development, genetic mechanisms
Bone sialoprotein (Bsp), 294, 304–305
Brachydactyly (short digits), 297
Brain development, 293, 351
Branchio-oto-renal syndrome, 218
Bromodeoxyuridine (BrdU), 6, 119, 176, 183, 235
Bronchioalveolar stem cells (BASC), 104, 120–121
Bronchoalveolar junction, 118
Bronchopulmonary dysplasia (BPD), 80–81, 87, 100, 113, 122, 128

C

CADASIL, 45, 58–59
Cadherin
  E-cadherin expression, 108, 357
  N-cadherin, 298
  P-cadherin, 216
  VE-cadherin, 245, 250
Cancer, intestinal development pathways, 185–186
  AXIN1/AXIN2/Tcf4, 186
  β-catenin destruction complex, 186
  casein kinase 1 (CK1), 186
  glycogen synthase kinase (GSK), 186
  microsatellite-unstable colon cancers, 186
  nonsense mutations, 185
  proteasomal pathway, 186
  Wnt/β-catenin/Hh/Eph/notch/Bmp signaling, 185
Canonical Wnt pathway/signaling, 19*f*, 22–23, 25–26, 28, 30, 83, 101, 104, 204, 210, 275, 349, 362, 367
Cap mesenchyme (CM), 195*f*, 196, 198*f*, 203, 204*f*, 205–207, 209
Cardiac crescent (CC), 3, 4*f*, 7–8, 11, 18, 27
Cardiac neural crest, 4*f*, 5, 9*f*, 11, 25
Cardiac progenitors
  and cell fate determination, 7–8
  contributions to cell types of heart, 10–14
    cell types derived from SHF, 10–11
    neural crest, 11–13
    proepicardial organ (PEO), 13–14
  and distinction between FHF and SHF, 7–10
    cell fate determination, 7–8
    markers of heart fields, 8–10
Cardiac tube, 3, 4*f*, 6, 8, 10, 23, 27
Cardiogenesis
  early stages of, 19*f*
  mammalian, 3
  in zebrafish embryo, 27
Cardiovascular birth defects, 51

Cartilage anlagen, development of, 298–300
chondrogenesis, 298–299
chondrocyte early differentiation, 298–299
precartilaginous condensation, 298
See also Individual
Cartilage growth plates, development of, 300–302
cartilage GP, 302
and chondrocyte maturation, 300*f*
Fgf signaling, 302
growth plate chondrocytes, 301
growth plates, 300
Hdac4 histone deacetylase, 301–302
Ihh signaling, 302
MADS-box transcription factors (Mef2c/Mef2d), 301
pathways with roles in GP during gestation, 302
prehypertrophic chondrocytes, development of, 301
Pthrp signals, 302
receptor for parathyroid hormone
Ppr/Ihh and collagen-10 (Col10a1), 300
Runt-domain transcription factors (Runx2/Runx3), 301
stage-specific markers
matrilin-1 (Matn1) and FGF receptor 3 (Fgfr3), 300
tissue elongation
chondrocyte proliferation and hypertrophy, 300
TUNEL assay, 301
Wnt ligands and pathways, 302
Cartilaginous sleeve malformation, 84
Cataracts, 351, 356–357, 358, 361, 363, 369
β-Catenin destruction complex, 188
β-Catenin signaling, 9*f*, 23–24, 28–29, 82, 104–105, 116, 124, 174, 177–179, 180*f*, 181, 183–186, 241, 297–299, 308, 357
Cell adhesion, 23, 61, 108–109, 164, 195*f*, 335
Cell culture systems, 30
Cell fate determination, 293
and cardiac progenitors, 7–8
cardiovascular cell fates, 7
down-regulating pluripotency genes, 7
ectopic heart formation, 7
skeletal myogenesis, 7
T-box transcription factor, Tbx5 and Hand1/2, 7
transcription factors, Gata4/Nkx2-5, 7
Cell painting methods, 165
Cell types of heart, 10–14
derived from SHF, 10–11
endothelial/endocardial fate, 10
hematopoietic/vascular lineage, 10
myocardial/endocardial progenitors, 10
neural crest, 11–13
ectopic myocardial differentiation, 13
pharyngeal arch artery remodeling, 12*f*
proepicardial organ (PEO), 13–14
Centrosomal protein 290 (CEP290), 372
Chondrocytes
cartilage cells, 292
early differentiation, 298–299, 298*f*
campomelic dysplasia, *See* Chondrodysplasia
perichondrium, 299
Sox5/Sox6 and Sox9, differentiation, 299
Sox trio in chondrocytes, 299
Wnt/beta-catenin signaling, 299
growth plate, 301
hypertrophic, 300*f*, 301–302, 303*f*, 304–305
maturation, 301–302
and osteoblasts, 294
prechondrocyte, 298*f*
prehypertrophic, 300*f*, 301–302, 303*f*, 304–305
proliferating, 300–302, 300*f*
terminal, 300–302, 300*f*, 303*f*
Chondrodysplasia, 299
Chondrogenesis, 293, 297–299, 298*f*
Chronic obstructive pulmonary disease (COPD), 81, 130
Ciliary dyskinesia, 81
Ciliary process, 345*f*
Clavicle hypoplasia or aplasia, *See* Runx2 haploinsufficiency
Cleidocranial dysplasia, *See* Runx2 haploinsufficiency
Collagens
cartilage collagen network, 294
collagen-1, 294, 296, 298, 300, 304
collagen-2, 294
collagen-10, 300
Sox9 inducestype II collagen (Col2a1), 84
type I and III, 235
type I and IV, 243
type IV, 108–109, 217
Collecting ducts (CD), 196
Congenital aganglionic megacolon, *See* Hirschsprung's disease
Congenital diaphragmatic hernia (CDH), 81, 84, 112, 114, 210
Congenital dysmorphology syndromes, 322
Congenital heart defects, 3
Congenital/infantile glaucoma, 367
Congenital nephrotic syndrome, 216
Contractile oscillation, 80
Cornea, 363–365
axon-guidance protein semaphorin-3A (Sema3a), 365
corneal embryonic development in chick, 363
corneal endothelium, 363, 365
function, 365
corneal epithelium, 363
corneal stroma, 363, 365
extracellular matrix proteins, 365
formation of cornea, 364

Cornea (*Continued*)
- keratocytes, 365
- lens stalk, 363
- neuropilin-1 (Nrp1), 365
- periocular mesenchyme, 346
- thyroxine, influence of, 365

Corneal stroma, 346, 362–363, 364*f*, 365
Craniofacial skeletal structures, 295
Craniosynostosis syndromes, 84, 304, 306
- Crouzon syndrome, 84
- Goldenhar syndrome/Apert syndrome, 84
- Pfeiffer syndrome, 84

Crouzon syndrome, 84
Crypts of Lieberkühn, 161, 180*f*
Crypt–villus axis, 175–178
- generation of, 176*f*
- intestinal cell biology, 175
- pseudostratified endoderm, 175
- pseudostratified epithelium, 175
- radial asymmetry, 175
- role of Bmps, 178
  - Bmp antagonist Noggin, 178
  - crypt fission, 178
  - duplication of stem cells, 178
  - epithelial hyperplasia, 178
  - fatty acid binding protein (Fabp1), 178
  - phosphatidylinosital-3 kinase (PI3K), 178
  - stem cell/progenitor cell proliferation, 178
  - stromal hyperplasia, 178
  - villus morphogenesis, 178
- role of hedgehog genes, 176–177
  - duodenal obstruction, 176
  - Hh interacting protein (Hhip), 177
  - proliferation, 176
  - Shh–Ihh signal, 176
  - stem cell compartment, 176
  - *Villin–Hhip* mice, 177
- role of Wnt signaling, 177–178
  - Tcf4-deficient mice, 177
  - TOP-GAL (Tcf)-reporter gene, 177
- transmission electron microscopy (TEM), 175

Cyclic AMP-responsive element-binding protein 2 (Creb2), 305
Cyclopia, 346, 351–352
Cystic adenomatoid malformation, 80

D

Denys–Drash syndrome, 214, 219
Dichotomy, 233
DiGeorge syndrome, 20, 51
Disorders of sex development (DSD), 232, 237, 242, 254
Distal tip progenitors, 118
Di/trichotomous branching tree, 200–202
- angiotensinogen gene (Agt), 201
- Bmp/Fgf/Tgfβ and retinoic acid, 201
- CD network of kidney, 200
- Cxcr4/Myb/Met and Mmp14, role of, 201
- expression of Wnt11/Emx2 and Ret, 201–202
- Gdnf/Ret relationship, 201
- Hgf/Met signaling, 202
- high-affinity FGF receptors (Fgfr1 and Fgfr2), 201
- Hoxa11/Hoxd11 double mutant kidneys, 201
- MM-derived secreted factors, 202
- pattern of UB branching, 200
- recombinant FGF, 202
- transcription factors Etv4 and Etv5, 201

Drosophila
- airways, 197
- beyond Sox9, 241
- doublesex in, 242
- ectopic eye development, 350
- eyeless phenotype in, 354
- Fgf family, 97
- hairy and enhancer of split, 180*f*, 369
- Hh signaling pathway, 102
- lung organogenesis and airway morphogenesis, 80
- ommatidal eyes in, 354
- rhabdomeric eyes in, 345
- Shh-mediated digit specification and growth studies, 328
- Wnt pathway in, 358

Duplex collecting systems, 200
Dysplasia
- bronchopulmonary dysplasia (BPD), 80
- campomelic, 238, 299
- cleidocranial, 304
- and defects in Met in humans, 218–219
- goblet cell, 104
- Langer's mesomelic, 309
- lung, 80, 100
- osteochondrodysplasias, 293
- renal hypoplastic, 218
- septo-optic, 375
- skeletal, 311

E

Embryogenesis, 81, 194, 237, 376
Embryonic lethality, 50, 56, 99
Embryonic lung progenitors and proximal–distal patterning, 124–125
- Bmp antagonists gremlin, 125
- lung epithelial progenitor, 125

Embryonic stem (ES) cells, 7, 10–11, 13, 23, 25–26, 30, 118, 375
Endocardial cushions, 4*f*, 11
Endochondral ossification, 292, 301, 303, 303*f*
Endoderm, formation of definitive, 162–165
- blastomeres, 164
- chemokine signaling, 164
- colorectal cancer (CRC), 164
- Cxcl12–Cxcr4 signaling, 164
- early intestinal organogenesis, 163*f*
- embryonic gut, 165

epiboly, 164
fluorescence microscopy, 164
foregut/midgut/indgut, 165
gastrulation movements, 163
Hnf3β-deficient mice, 165
hypoblast, 164
microarray, 164
nodal signaling, 162
pluripotent cells, 162
somitogenesis, 164
Sry-box family TF, 163
T-box transcription factor (TF), 163
transgenic expression, 165
Xenopus, 163
zebrafish, 163
Endogenous epithelial progenitor cells, 118–122
alveolar epithelial progenitors, 121–122
bromodeoxyuridine (BrdU), 119
bronchoalveolar junction, 118
distal tip progenitors, 118
pulmonary neuroendocrine cell, 118
self-renewing progenitors, 118
tracheal and bronchial epithelial progenitors, 119–121
Endogenous mesenchymal progenitors, 122–123
smooth muscle progenitors, 122
localization of peripheral ASM progenitors, 122
vascular progenitors, 122–123
distal airspace, 122
Flk1 promoter, 123
heterogenous pulmonary endothelial cells, 122
laryngotracheal groove, 123
Endothelial apoptosis, 50
Endothelial cells, 249–251
coelomic vessel, 249
filopodia, 250
fluorescently labeled lectin, 249
gonad/mesonephros boundary, 249
perfusing embryos, 249
testicular vascular structures, 249
time-lapse confocal microscopy, 250
VegfA signaling, 250
wingless-related MMTV integration site 4 (Wnt4), 251
Enteric nervous system, formation of, 167–168
active Crohn's disease, 168
Claudin (Cldn)15, 167
endothelin-3 (ET-3), 167
enteric nervous system (ENS), 167
Glial-derived neurotrophic family (Gdnf), 167
GPI-anchored receptor (GFRα1), 167
Hirschsprung's disease, 167
inflammatory bowel disease (IBD), 166
ion-conducting pores, 166
megaintestine, 166
myenteric plexi, 167
Na^+/K^+-ATPase, 166
neural tube, 167
Ret signaling, 166
tight junctions, 166
tyrosine kinase Ret, 167
ulcerative colitis, 167
Eph–ephrin signaling, 176*f*, 180*f*, 181, 184
Epiboly, 164
Epididymis, 232, 246
Epithelial/mesenchymal transition (EMT), 3, 7, 13
Esophageal atresia (EA), 83–84, 83*f*, 102
Estrogen synthesis, 279
ET, *See* Transcription repression factor (Tbx3)
Extracellular matrix (ECM), 45*f*, 49, 81, 97, 108–109, 111, 162, 195*f*, 201, 215, 243, 244*f*, 250, 293–294, 295*f*, 299, 301, 307, 363, 365, 374
and lung development, 108–110
cell–cell/cell–ECM interaction, 108
cell polarization, 109
cross-shaped molecule, 108
dexamethasone exposure, 110
fibronectin (FN), 109
laminins (LN), 108
nidogen, 109
proteoglycan sulfation, 109
tissue inhibitors of metalloproteinases (TIMP), 109
Extracellular-regulated kinases (ERK) signaling, 53, 99, 105
Eye development, 343–378
adult human eye, schematic section through, 345*f*
cornea, 363–365
early stage, eye field
formation, 348–349
optic vesicle to optic cup, 352–354
patterning, 349–351
splitting, 351–352
iris and ciliary body, 366–367
from late gastrula to optic cup, 347*f*
lens development
formation of lens placode and Pax6, 354–356
from lens vesicle to mature lens, 358–363
signaling cascades, 357–358
optic nerve
visual system, 376–377
overview, 346
cornea, 346
lens, 346
lens placode, 346
ocular structures, 346
optic pits, 346
optic stalk, 346

Eye development (*Continued*)
optic vesicles, 346
retina, 346
retina
development of hyaloid and retinal vasculature, 373–374
neural retina, 368–373
retinal pigmented epithelium, 367–368
Eye field
cyclopia, 346–348
eyeless phenotypes (anophthalmia), 346
formation, 348–349
anophthalmia, 348
bilateral anophthalmia, 348
c.909C→G, nonsense mutation, 349
c.664delT, frame-shift deletion, 349
ectopic eye formation, 349
eye-field transcription factors (EFTF), 349
masterblind (affected gene axin) and headless, 349
of mouse and fish, 348
Otx2 and Sox2 transcription factors, 349
role of miRNA in eye development, 349
spontaneous mutant eyeless, 348
in zebrafish, 348
gastrulation, 346
optic vesicle to optic cup, 352–354
bone morphogenetic protein (Bmp) signaling, 354
formation of iris and ciliary body, 352
formation of retina/optic stalk/ presumptive optic nerve, 352
lens development, 352
model of Lhx2 function during mouse early eye organogenesis, 353*f*
paracrine RA signaling, 352
Raldh2/Aldh1a2, 352–353
retinoic acid (RA) signaling system, 352
transcription factor Lhx2 (LIM homeobox protein 2), 353
patterning, 349–351
dachshund genes, 351
eya function in *D. melanogaster,* 351
eye field (in *Xenopus*), 349–350
eye field transcription factors (EFTF), 350
genetic factors forming single eye field, 350*f*
mutations affecting Pax6 and Lhx2, 350
role of genes in eye development, 350
transcription factors, 349
splitting, 351–352
cyclopia, 351
function of Sonic hedgehog (Shh), 351
one-eyed pinhead (oep) gene, 352
Six3, single eye field, 351
Southpaw (Spaw) gene, 352
Squint (Sqt) gene, 352
start codon (ATG→ATA) in cyclops, 351
in zebrafish, 351–352

F

Female germline, establishment of, 268–272
Aldh1a2, 271
Cyp26b1 expression, 270*f*
dictyate stage, 268
diplotene stage, 269
DNA replication, 271
ectopic expression, 272
embryonic lung cells, 269
germ cell meiosis, 269, 270*f*
meiosis-inducing substance, 269
meiosis-inhibiting substance, 269
meiotic division, 268
meiotic oocytes, 268
meiotic prophase I, 268
mesoderm-derived tissues, 271
NANOS2, 272
neo-oogenesis, 268
oogonia, 268
P450 enzyme, 271
spermatogonia, 268
stray germ cells, 269
Y-bearing oocytes, 269
Fetal ovary development in mice and humans, 281*t*–282*t*
Fibroblast growth factor (Fgf), 47
Fgf–FgfR–sprouty signaling, 80
winged-helix, 87
Fgf receptor 3 (Fgfr3), 300
Fgf signaling, 28
fibroblast growth factor receptor (FgfR)2b, 85
First heart field (FHF), 3
and SHF, distinction between, 7–10
cardiac progenitors and cell fate determination, 7–8
markers of heart fields, 8–10
Forkhead box transcription factor family, 87–95
bone morphogenetic protein 4 (Bmp-4), 95
brachyury T-Box family (Tbx2–Tbx5), 95
CCSP, 95
Hnf-3α (Foxa1)/Hnf-3β (Foxa2), 87
lung Kruppel-like factor, 95
mutations in mouse, 88*t*–94*t*
neonatal respiratory distress, 96
Nkx homeodomain, 87
RA receptors, 87
SP-C promoter, 87
winged-helix, 87
Frasier syndrome, 219

G

Gastroesophageal reflux, 84
Gastrulation, 6, 10, 13, 108, 163*f*, 164–165, 168, 170, 185, 252, 295, 346, 349, 352
Germ cells, 251–253
gametes, 251
gene regulation, 252

meiotic cycle, 252
microRNA, 252
mitotic arrest/cycle, 252
pluripotency markers, 252
puberty, 252
retinoblastoma 1 (RB1), 252–253
transmembrane protein 184a (TMEM184a), 253
XY germ cell development, 253*f*
Germline stem cells, 268, 272
Glaucoma-related eye disease, 367
Glial cell line-derived neurotrophic factor (Gdnf), 197
and Ret signaling initiates and guides UB outgrowth, 197–199
Gdnf/Ret and Wnt11, positive feed back loop, 198
Gfrcα1, co-receptor for Gdnf, 198
IM/nephrogenic cord, 198
mesenchyme signaling, 197
receptor-tyrosine kinase ligand (Egf/Fgf/Tgfα), 197
Ret receptor tyrosine kinase, 198
Spry1, inhibitor of Ret-signaling, 199
ureteric bud outgrowth and branching morphogenesis, 198*f*
Gli family zinc-finger transcription factors, 97
normal proximal–distal differentiation, 97
Goblet-shaped invaginations, 161
Goldenhar syndrome, 84
Goldilocks hypothesis, 101
Gonadal dysgenesis, 367
Gonads, cell biology and morphology of, 233–234
coelomic epithelial cells, 233
genital ridges, 233
gonadal somatic cells, 233
interstitial cell precursors, 233
peritubular myoid (PM) cells, 234
primordial germ cells (PGC), 233
steroidogenic precursor cells, 234
testis development in mouse, 234*f*
Granulosa cells, differentiation of
β-catenin, 275
ectopic testis vasculature, 274
extracellular factors, 273
follistatin (Fst), 274*f*
Foxl2, 273
gain-of-function experiments, 275
ovary organogenesis, pathways of, 274*f*
R-spondin1 (Rspo1), 273
sex determination hypothesis, 276*f*
somatic cell environment, 272
Sox9 expression, 275
Sry-expressing Sertoli cells, 273
steroidogenic factor 1 (SF1), 275
transcriptional regulator, 273
transcription factors, 273
Wnt4 expression, 273
Granulosa-to-sertoli transdifferentiation, 279
Growth plates (GP), 300, 300*f*, 302, 304
Gut morphogenesis, 82
Gut, patterning early, 168
bilaterians/cnidarians, 168
epiblast cells, 168
Hox/ParaHox genes, 168
rostrocaudal patterning, 168
Gut tube closure, 82

H

Hairy and enhancer of split (Hes), 180*f*
in *Drosophila,* 180*f*, 369
Hes1/Hes5, 180*f*
Haploinsufficiency, 51, 59–60, 170, 304, 309
Heart fields and cardiac progenitor cell behavior, origin of, 5–7
cell labeling/grafting, 5
clonal analysis, 6
endodermal folding, 6
fate mapping analysis, 5
myocardial cell lineages, 6
OFT myocardium, 5
primitive streak (PS), 5
retrospective clonal analysis, 6
spatiotemporal fate mapping, 7
Heart/origin and regulation of cardiac progenitor cells, 1–30
cell types of heart, 10–14
derived from SHF, 10–11
neural crest, 11–13
proepicardial organ, 13–14
markers and distinction between FHF and SHF, 7–10
cardiac progenitors and cell fate determination, 7–8
heart fields, markers of, 8–10
multiple cellular contributions to heart, 4*f*
origin of heart fields and cardiac progenitor cell behavior, 5–7
SHF cell behavior, 18–29
anterior/posterior patterning, 20
heart tube and differentiation, 26–29
interactions with neural crest, 24–25
left/right patterning, 21–22
maintenance of proliferation in SHF, 22–24
prevention of differentiation, 25–26
See also Second heart field (SHF) cell behavior, 18–29
subdomains of SHF, 14–18
anterior SHF and contributions to AP of heart, 16–17
posterior SHF and formation of VP, 17–18
Hedgehog (Hh) signaling, 23
desert hedgehog (Dhh) signaling, 244*f*, 245
Indian hedgehog (Ihh), 23, 170
sonic hedgehog (Shh) signal, 9*f*, 16–17, 21–22, 23, 26, 82*f*, 102–103, 299, 327

Hemangioma, 45*f*, 51
Hematopoietic/vascular lineage, 10
Hepatocyte nuclear factor (Hnf), 82, 184
  Hnf-3α (Foxa1)/Hnf-3β (Foxa2), 87
  Hnf3β-deficient mice, 165
  Hnf-3/Foxa2β, 82
  mutations in Hnf1β/Tcf2/Pax2EYA1/Six1 and Sall1, 218
  role of Hnfα, 184–185
    epithelial apicobasal polarity, 185
    hepatocyte nuclear factor (Hnf)-4α, 184
    intestinal barrier function, 185
    single nucleotide polymorphism, 185
    ulcerative colitis, 184
Hereditary hemorrhagic telangiectasia (HHT), 45*f*, 54, 100
Hereditary lymphedema syndromes, 45*f*
High-mobility-group DNA-binding protein A2 (HMGA2) mutations, 310
Hirschsprung's disease, 167
Hox family transcription factors, 96
  alveolar wall thickening, 96
  Hoxa5/Hoxb2/Hoxb5/Hoxb3/Hoxb4, 96
  Hox and ParaHox genes, 168
  Hox clusters, 20
  proximal/distal mesenchyme, 96
  surfactant protein expression, 96
Human congenital dysmorphology syndromes, 322
Hydroureter and duplex kidney, 197
Hyperoxia, 87, 98, 119, 122
Hyperuricemic nephropathy, 212

I

Imaging technologies, 320
Indian Hedgehog (Ihh) signaling, 47, 163, 170, 173, 176, 178, 300–303, 305, 308
In situ hybridization (ISH), 170, 240, 321
Insulin-like growth factor (IGF) signaling, 106, 236
Intermediate mesoderm (IM), 194, 198–199, 219
Intervertebral discs (IVD), 292, 294–296
Intestinal sub-epithelial myofibroblasts (ISEMF), 161, 176–179, 181
Intraflagellar transport (IFT), 326–327
Intramembranous ossification, 292, 302–304, 306
Intraocular vascularization, 373
Intrapulmonary airway branching, 78
Iridogoniodysgenesis, 367
Iris and ciliary body, 366–367
  Bmp signaling, 366
  cytotoxic diphtheria toxin A, 366
  iridogoniodysgenesis, 367
    Dkk2/Pitx2, function of, 367
    nonsense mutation in mouse homolog of Foxc1, 367
    type II mutations in Pitx2, 367
    type I mutations in Foxc1, 367
  neural retina, 366
  optic cup, 366
  periocular mesenchymal cells, 366
  pigmented layer of retina, 366
  Wnt2b signaling, 366

J

Joints, 292
  cavitation, 307
  morphogenesis, 306–308
  structures, 292, 294, 306
    fibrous joints, 294
    synovial joints, 294
Juvenile, familial, 212
Juvenile glaucoma, 367

K

Kartagener syndrome, 81, 95
Kidney development, 193–220
  disruptions to kidney tubulogenesis in human disease
    consequences of defects in branching morphogenesis, 218
    dysplasia and defects in Met in humans, 218–219
    nephron number and renal disease, 219–220
  moving from structure to functional filter
    maturation of RC, 216–217
    patterning and vascularization of developing RC, 213–216
  origin, 194–197
    cap mesenchyme (CM), 196
    collecting ducts (CD), 196
    10 days post coitum (dpc) in mouse, 195
    epithelial renal vesicle (RV), 196
    mesenchyme-to-epithelial transition (Met), 196
    nephrons, 196
    paired epithelial buds, 196
    pretubular aggregate (PA) stage, 196
    primary nephric duct (ND) or Wolffian duct, 194
    pronephros/mesonephros and metanephros, 194
    ureteric tip (UT), 196
    Wnt ligands, 197
  patterning resulting tubules
    notch signaling in patterning of proximal nephron, 210–211
    patterning of LH and DT, 211–212
    patterning of UT *vs.* stalk and differentiation of ureter–pelvic junction, 212–213
  tubulogenesis, 195*f*
    CM as population of nephron progenitors, 205–207

di/trichotomous branching tree, 200–202
Gdnf/Ret signaling initiates and guides UB outgrowth, 197–199
nephron progenitors and cessation of nephrogenesis, 207
regulation of UB initiation/position/width and number, 199–200
via Met-tube one induces tube two, 203–207

L

Lamina propria, 160–161
Langer's mesomelic dysplasia, 309
Laryngotracheal groove, 76, 81–82, 123
Lateral geniculate nucleus (LGN), 376–377
Lateral plate mesoderm (LPM), 19, 294–295, 320, 322–323
Lens development, 352, 354–363, 373
formation of lens placode and Pax6, 354–356
aniridia, panocular disorder, 356
cell autonomous manner, 355
cornea epithelium, 354
Cre- or Flp-mediated recombinations, 355
Fgf- and Wnt-signaling cascades, 354
formation in surface ectoderm, 354
homozygous mutants Pax6^{3Neu} and Pax6^{7Neu}, 355
hypomorphic alleles Pax6$^{132\text{-}14Neu}$, 355
lens cup (lens pit), 354
lens/non-lens placode formation, 355
lens vesicle, 354
nasal dysfunction, 355
Pax6 and Sox2 expression, 356
preplacodal stage, 356
retina, transcription factors in formation of, 356
Six3/Sox2 expression, 356
transcription factors in, 356
Walter Gehring's group, 354
See also Pax6 mutations
from lens vesicle to mature lens, 358–363
anterior epithelial cells, 360
anterior segment mesenchymal dysgenesis (ASMD), 361
Bmp-receptor 1b (Bmpr1b) gene, 362
concentrations of Fgf2 (basic Fgf or bFGF), 362
co-receptors (Lrp5/Lrp6/Sfrp1–3/Dkk1–3) genes, 362
expression pattern of Pax6/Prox1/Six3 and Crygd, 361
formation of lens, 359*f*
Fzd receptors (Fzd1–8) gene, 362
mature lens of structural proteins, 362–363
in mice (Pitx3 and Foxe3) genes, 360
mutation in gene Gjf1, 361
opaque flecks in lens, 361
Parkinson's disease, 361
phenotype of aphakia (ak) or eyeless (eyl) mouse mutants, 360
Pparbp gene, knockout, 362
primary lens fiber cells, 360
secondary lens fibers, 360
in zebrafish, 360
signaling cascades in early, 357–358
BMP-mediated signaling, 358
Fgf signaling, 357
Wnt signaling, 357–358
Lens pit, 347*f*, 352, 354, 358
Leydig cells, 245–249
adrenal steroidogenic cells, 246
3-beta-hydroxy-delta-5- steroid dehydrogenase (3βHSD), 246
17beta-hydroxysteroid dehydrogenase (17βHSDIII), 246
cyclopamine, 248
cytochrome P450 17-hydroxylase (Cyp17), 246
embryonic testes, 245
fetal Leydig cells, 245
γ-secretase inhibitor, 248
insulin-like factor 3 (INSL3), 246
LIM homeobox gene 9 (Lhx9), 248
masculinization, 246
mesonephric cell migration, 249
notch signaling pathway, 248
platelet-derived growth factors (Pdgf), 248
P450 side chain cleavage (Cyp11A1/P450SCC/SCC), 246
regulation of Leydig cell development, 247*f*
transabdominal/inguinoscrotal, 246
Limb, early
limb induction, 322–323
Fgf8/Fgf10 expression, 322
forelimb budding, 323
limb budding, 322
mesenchyme–ectoderm signaling loop, 322
nested collinear pattern of Hox gene expression, 322
retinoic acid (RA), 323
T-box transcription factors Tbx5 and Tbx4, 322–323
pre-patterning/positioning of ZPA and asymmetry, 323–325
GLI3 and HAND2, genetic interactions between, 323
hedgehog receptor Patched1, 324
non-AER (DV) border ectoderm, 325
posterior restriction (POST) region, 324
Shh-independent and dependent genetic interactions, 324*f*
transcription factors Etv4 and Etv5, 325

Limb patterning, system biology approach to, 334–335
 gene-by-gene analyses, 334
 genomics-based screens, 334
 hypomorphic Bmp4 allele, 334
 molecular limb patterning, 334
 Shh–GREM1–FGF loop, 334
Limb skeletal elements (appendicular skeleton), 292, 295, 310
Lipid-filled interstitial fibroblasts (LFIF), 86
Live-cell imaging technique, 253
Lumenization, 47
Lung organogenesis, 74–130
 airways forming in sequential manner, 79*f*
 developmental anatomy, 75–81
  abnormal development, 80–81
  bauplan, 75
  branching morphogenesis, 78–80
 histological stages, 75–78
  canalicular stage, 75
  diagrammatic overview of, 75*f*
  mouse lung at characteristic stages of development, 77*f*
  ongoing differentiation of mesenchymal cells, 77
  respiratory tree, 77
 mechanobiology, 111–118
  and Ca2þ-sensing receptor (CaSR), 116–118
  embryonic airway peristalsis, 114–115
  hydraulic pressure, 113–114
  lung stretch transduction and PthrP, 115–116
  mechanobiology from human and *in vivo* studies, 112–113
 molecular embryology, 81–111
  biochemical regulators, 87–111
  extracellular matrix (ECM), 81
  process-driven, 81–87
  transcription factors, 87–97
 murine embryonic lung organ culture, 74
 postnatal and adult lung, 128–130
  lung aging and involution, 130
  transition to air breathing, 128–130
 stem/progenitor cell biology, 118–128
  embryonic lung progenitors and proximal–distal patterning, 124–125
  emergence of specific cell types, 125–127
  endogenous epithelial progenitor cells, 118–122
  endogenous mesenchymal progenitors, 122–123
  lung progenitor cell proliferation, control of, 123–124
  in postnatal respiratory system, 127–128
  potential strategies to protect lung progenitors, 128
Lung progenitors
 cell proliferation, control of, 123–124
  autocrine bmp signaling, 124
  fox transcription factors, 124
  goblet cell metaplasia, 124
  HMG box transcription factor, 123
 potential strategies to protect, 128
  glutathione repletion, 128
Lymphatic capillaries, 52, 162

M

Male sertoli cell, 233
Mammalian eggs (testicle of female), 264
Mammalian model systems, 87, 374
Mammalian RA synthesis, 110
Mammalian testis development, 231–254
 beyond Sox9, 241–243
 cell biology and morphology of gonads, 233–234
 disorders of sex development (DSD), 232
 endothelial cells, 249–251
 epididymis, 232
 female granulosa cell, 233
 germ cells, 251–253
 leydig cells, 245–249
 male sertoli cell, 233
 müllerian duct, 232
 origin of sertoli cells, 235
  5-bromo-2-deoxyuridine (BrdU), 235
  chicken/quail gonad/mesonephros grafts, 235
  gonadal soma, 235
  laminin, 235
  subepithelial basement membrane, 235
 ovotestes, 233
 peritubular myoid (PM) cells, 243–245
 post coitum (dpc), 232
 seminal vesicle, 232
 sexual differentiation, 232
 Sox9 and sertoli cell differentiation, 237–241
  Sox9 initiation, 238–239
  Sox9 maintenance, 239–241
  SRY-box containing gene 9 (Sox9), 237
  testis-specific enhancer of Sox9 core (TESCO), 238
Marfan's syndrome, 101
Markers of heart fields, 8–10
Mastermind-like (MAML), 24, 27
Matrilin-1 (Matn1), 300–301
Mature onset diabetes of the young—MODY5, 218
Mechanobiology of developing lung, 111–118
 and Ca^{2+}-sensing receptor (CaSR), 116–118
  -evoked inhibition and PI_3 kinase signaling, 117
  phosphoinositide 3 (PI_3), 116
 embryonic airway peristalsis, 114–115
  cholingergic agents, 114
  pacemaker-driven airway contractility, 114
  peristaltic contraction, 114

sequential photomicrographs of cultured embryonic lung, 115*f*
hydraulic pressure, 113–114
Fgf10–FgfR2b–sprouty, 114
lung stretch transduction and PthrP, 115–116
lipofibroblast lineage, 115
zeitgeber mechanism, 116
mechanobiology from human and *in vivo* studies, 112–113
airway smooth muscle (ASM), 113
ASM-led airway occlusion, 113
congenital laryngeal atresia, 112
diaphragmatic contraction, 112
fetal thorax compression, 112
lung distension, 113
lung liquid production, 112
nephrectomy, 112
postpneumonectomy, 113
potter's syndrome, 112
renal agenesis, 112
Rho–ROCK system, 112
tangential epithelial stress, 111
Meckel syndrome gene (Mks1), 327
Medullary cystic kidney disease type 2 (MCKD2), 212
Meiosis-inducing agent, 272
Mesenchyme–ectoderm signaling loop, 322
ectodermal (FgfR2b) receptor, 322
mesenchymal (FgfR2c) receptor, 322
Mesonephros, 194
Metabolic homeostasis, 160
Metanephric mesenchyme (MM), 194–195
Metanephros, 194
metanephric mesenchyme (MM), 194–195
ureteric bud (UB), 194
Missense mutation, 51–52, 58, 186
Molecular embryology of lung, 81–111
biochemical regulators, 87–111
factors, 107–111
peptide growth factors, 97–107
transcription factors, 87–97
process-driven, 81–87
alveolar septum formation, 86–87
branching morphogenesis of airway and vasculature, 85–86
induction of early lung anlagen, 81–83
tracheal cartilage formation, 84–85
tracheoesophageal septation, 83–84
transcription factors, 87–97
Molecular signaling, 46
Morphogenesis
of airway and vasculature, 85–86
gut, 82
joint, 306
of ovary, *See* Ovary
synovial joint, 308
ureteric bud outgrowth and branching, 198*f*
villus, 178
Mouse
and chick limbs, Shh transcription, 332
10 days post coitum (dpc) in mouse, 195
early eye organogenesis, 353*f*
eye field formation, 348
homolog of FoxC1, nonsense mutation, 367
lung at characteristic stages of development, 77*f*
mutations in, 88*t*–94*t*
ovary organogenesis in embryo, 263–282, 264
phenotype of aphakia (ak) or eyeless (eyl) mouse mutants, 360
signaling centers in 11.5 dpc mouse limb, 321*f*
testis development in, 234*f*
transgenic mouse technology, 87
Müllerian duct, 232, 242
Multi-layered organization of mature GI tract, 161*f*
Multiorgan syndrome, 218
Multipotential stem cells, 161
Murine embryonic lung organ culture, 74
Mutagenesis, 238
Myocardial/endocardial progenitors, 10–11
Myogenesis, skeletal, 7–8

N

Nail Patella syndrome, 216
Neonatal respiratory failure, 80, 102
Neovascular ocular diseases, 374
Nephron endowment, 204, 206, 219
Neural crest, 11–13
cells, 2, 4–5, 11, 21, 24, 167, 246, 295, 346, 363, 365
ectopic myocardial differentiation, 13
pharyngeal arch artery remodeling, 12*f*
Neurogenesis, 376
Neurulation, 346
Nkx/Hox homeodomain transcription factors, 95–96
congenital hypothyroidism, 96
eosinophil infiltration, 96
neonatal respiratory distress, 96
TTF-1 (thyroid-specific transcription factor), 95–96
type II pneumocyte proliferation, 96
Non-LFIF (NLFIF), 86
Notch/delta/jagged in vascular development, 55–60
delta mutations, 59
heterozygous Dll4þ/– embryos, 59
postnatal arteriogenesis, 59
endothelial crosstalk and heterogeneity, coordination of, 56–58
notch–BMP, 57–58
notch signaling, 57
notch–VEGF, 56–57
temporal oscillation, 57
vessel lengthening, 57

Notch/delta/jagged in vascular development (*Continued*)
jagged mutations, 59–60
AGS phenotype, 60
endothelial-specific manipulations, 59
notch2 hypomorphic allele, 60
notch/delta/jagged signaling pathway, 55–56
cell–cell communication, 55
delta-like (Dll)1/Dll3/Dll4, 55
embryonic lethality, 56
hematopoietic progenitor cells, 56
notch mutations, 58–59
cerebral autosomal dominant arteriopathy, 58
leukoencephalopathy, 58
vascular-specific gene manipulation, 58
signaling pathway, 55–56
Notch intracellular domain (NICD), 55, 58, 211, 248
Notch signaling pathway, 55, 210, 248, 369
Nuclear factor of activated T cells (NFAT), 10, 51

O

Ocular structures, 346
neural crest cells form central part of cornea, 346
neural ectoderm forms retina, 346
surface ectoderm gives rise to lens and part of cornea, 346
Oogenesis
female germline, 268
neo-oogenesis, 268
Optical projection tomography, application of, 203, 328
Optic chiasm, 347, 376–377
Optic cup, 346
inner layer gives rise to retina, 346
optic vesicle to, 352–354
outer layer form retinal pigment epithelium, 346
Optic nerve
coloboma, 375
congenital optic nerve defects, 375
GENEYE database, 375
mutations in PAX2, 375–376
optic disc, 375
optic nerve hypoplasia (ONH), 375
optic stalk, 374
renal-coloboma syndrome, 375
septo-optic dysplasia, 375
visual system, 376–377
chiasma, 376
ephrin molecules (Eph), 376
nasal hemiretina, 376
temporal retina, 376
Zic2, zinc finger transcription factor, 377
Optic stalk, *See* Optic nerve
Organizer of lens, *See* Optic cup
Organogenesis
early intestinal, 163*f*
germ layers (ectoderm/mesoderm/endoderm), 294
lung, *See* Lung organogenesis
ovary
in mammals, 264–282
in mouse embryo, 263–282
Orthogonal bifurcation, 79
Osteoarthritis, 293
Osteoblasts, 53, 292, 294, 296–297, 301, 303–307*f*
Osteocalcin (Bgp), 294, 304–305
Osteochondrodysplasias, 293
Osteochondroprogenitors, 296
Osteoporosis, 293
Osterix (Osx or Sp7)
Krüppel-like zinc finger domain, 305
Outflow tract (OFT) development, 5, 22
Ovarian follicles, 264, 266, 268
Ovary
-determining gene (or *od*) hypothesis, 265*f*
morphogenesis of, 266–268
basal lamina, 266
coelomic epithelium, 266
days post coitum (dpc), 266
genital ridge, 266
granulosa cells, 266
leydig cells, 267
mouse fetal ovary, 267*f*
ovarian surface epithelium, 267
primordial follicles, 266
rete ovarii, 267
sertoli cells, 267
small-molecular-weight proteins, 268
somatic cells, 266
steroidogenic tissue, 268
tubule structures, 266
organogenesis in mammals, hypotheses for, 264–266
embryonic development, 264
female semen, 264
germ cell meiosis, 266
gonadal primordium, 264
gonads/phenotypic sexual characteristics, 264
male organizer, 264
Od gene, 264
ovary differentiation, 266
sex determination, 265*f*
Tdy gene, 264
testis formation, 264
XO aneuploidy, 264
X/Y chromosome, 264
Z hypothesis, 265*f*
organogenesis in mouse embryo, 263–282
anti-testis Z factor, 278

estrogen synthesis, 279
female germline, establishment of, 268–272
female somatic environment and female germ cells, 276–278
fetal ovary development in mice and humans, 281*t*–282*t*
germline stem cells, 272
granulosa cell identity in fetal and adult ovaries, 280*f*
granulosa-to-sertoli transdifferentiation, 279
meiosis-inducing agent, 272
morphogenesis of ovary, 266–268
ovarian follicles, 264
ovary organogenesis in mammals, hypotheses for, 264–266
Rspo1 knockout model, 280
Ovotestes, 233, 239

P

Parathyroid hormone-related protein (PTHrP), 115–116
Parkinson's disease, 361
Passive diffusion, 44
Pax6 mutations, 354–356
aniridia, panocular disorder, 356
cataracts/macular hypoplasia/keratitis/Peters anomaly, 356
disorders of pancreas, 356
homozygous mutants $Pax6^{3Neu}$ and $Pax6^{7Neu}$, 355
hypomorphic alleles $Pax6^{132\text{-}14Neu}$, 355
ocular diseases behavioral and neurodevelopmental phenotypes, 356
See also Lens development
PAX2 (renal–coloboma syndrome), 199, 202, 206, 214, 218–219, 356, 361, 375–377
Peptide growth factors in lung development, 97–107
autocrine, 97
Bmp subfamily, 101–102
Bmp type II receptor (BmpRII), 102
neonatal atelectasis, 102
prenatal lung malformation, 102
primary pulmonary hypertension (PPH), 102
respiratory failure, 102
surfactant protein C (SP-C), 101
epidermal growth factor (Egf), 105
Egf receptor (EgfR), 105
matrix metalloprotease protein (MMP), 105
tumor necrosis factor-α (Tnfα)-converting enzyme (TACE), 105
Fgf family, 97–99
bud formation, 97
cell proliferation, 97
ciliated cells, 98
clara cells, 98
differentiation, 97
distal airway epithelial cell, 98
Drosophila, 97
ECM glycoprotein, 97
emphysema, 97
Erm/Pea3 expression, 98
extracellular-regulated kinases (Erk), 99
Fgf7 (KGF), 98
heparan sulfate proteoglycan, 97
hyperoxia, 98
lung hypoplasia, 98
MAP ERK kinase (MEK), 99
migration, 97
prenatal airway tubule formation, 97
protein sequence homology, 97
Ras–MAP kinase signaling, 99
signaling cascade, 99
Sp-C expression, 98
Spred-1/Spred-2, 99
insulin-like growth factors (Igf), 106
dwarfism, 106
leukemia inhibitory factor (Lif), 106
paracrine, 97
platelet-derived growth factors (Pdgf), 106
antisense oligodeoxynucleotide, 106
homodimers (AA or BB)/heterodimers (AB), 106
Pdgf-A/B, 106
roundabout (ROBO)/SLIT, 107
non-neuronal cell migration, 107
sonic hedgehog (Shh) pathway, 102–103
cubitus interruptus (Ci), 102
hedgehog interacting protein 1 (Hip1), 102
mesenchyme–epithelium interaction, 102
Tgf-β/Bmp family, 99
Tgf-β subfamily, 99–101
bleomycin/endotoxin, 100
Dermo1, 100
hereditary hemorrhagic telangiectasia, 100
interstitial fibrosis, 100
Marfan's syndrome, 101
neonatal alveolar hypoplasia, 100
pulmonary vasculogenesis, 100
Tgf-β activated kinase-1 binding protein-1 (TAB1), 100
vascular endothelial growth factor (Vegf) isoforms/cognate receptors, 106–107
endothelial crosstalk, 107
hypoxia inducible factor-1 (Hif1), 107
Wnt/β-catenin pathway, 103–105
basal cells (BC), 104
bronchioalveolar stem cells (BASC), 104
cartilaginous condensation, 103
Dermo1-cre/β-catenin CKO embryos, 105
Frizzled (Fzd) genes, 103
global gene expression, 105

Peptide growth factors in lung development (*Continued*)
JNK pathway, 103
PITX family transcription factors, 105
TOPGAL/BATGAL, 103
Pericardial edema, 52
Perichondrium, 299, 302–304
Pericytes, 44–45, 48, 61, 65, 214
Peripheral squamous cells, 77
Peristalsis, 80, 113–116, 130, 212–213
Peritubular myoid (PM) cells, 243–245
cell lineage tracing, 244
endothelin, 243
enhanced yellow fluorescent protein (EYFP), 244
fibronectin, 243
green fluorescent protein (GFP), 245
interstitial gonad compartments, 243
and other testicular cells, 245
Dax1 null mutations, 245
desert hedgehog (Dhh), 245
receptor Patched 1 (PTCH1), 245
prostaglandin F2α, 243
proteoglycans, 243
recombination, 244
seminiferous tubules, 243
and sertoli cells, interactions between, 244*f*
α-smooth muscle actin (αSma), 244
type I and IV collagens, 243
vasopressin, 243
VE-cadherin, 245
Persistent Müllerian duct syndrome, 242
Peters anomaly, 356, 361, 367
Pfeiffer syndrome, 84
Pharyngeal arch arteries (PAA), 11
remodeling, 12*f*
Pharyngeal arches (PA), 5
mesoderm, 26
Phenotypic heterogeneity, 44–45, 45, 47, 52, 56–57, 60–61
Phosphoinositide 3 (PI_3) kinase signaling, 117
Planar cell polarity (PCP) signaling, 174, 204
Platelet-derived growth factor (Pdgf)-BB, 48, 106, 173, 247*f*, 248
Polled intersex syndrome (PIS), 265
Polydactyly (extranumerary digits), 297, 325–327
Postnatal and adult lung
lung aging and involution, 130
amelioration, 130
DNA polymorphisms, 130
pre- or perinatal death, 130
transition to air breathing, 128–130
corticotrophin releasing hormone, 129
disaturated phosphatidylcholine (DSPC), 129
glucocorticoid receptors, 129
smoking and genetics synergize to degrade lung function, 129*f*
sodium/potassium ATPase, 129
Postnatal respiratory system, 127–128
BPD and asthma, 128
bronchoalveolar duct junction, 127
cholinergic-agonist, 128
nicotine, 128
syrian hamster fetal lung epithelial M3E3/C3 cells, 127
Postnatal steroid exposure, 87
Potter's syndrome, 112
Precartilaginous condensation, 298
in vitro studies, 298
in vivo data, 298
Primary nephric duct (ND) or Wolffian duct, 194
Primary ossification center, 304
Primitive gut tube/formation and regionalization of, 165–174
bioinformatics approach, 174
functional clustering, 174
pathway analysis tools, 174
quantitative RNA sequencing technologies, 174
role of Bmp signaling, 172–173
AIP formation, 173
Bmp2/Bmprlb/BmprII, 173
Bone morphogenetic proteins (Bmps), 172
morphogenesis, 172
platelet-derived growth factor-A (Pdgf-A), 173
pyloric sphincter, 173
Smad1/5/8 phosphorylation, 173
role of Cdx2, 168–171
benign hamartomas, 170
chromatin immunoprecipitation assays, 171
cis-regulatory elements, 170
foregut–midgut junction, 169
haploinsufficiency, 170
immunohistochemistry, 170
midgut–hindgut border, 169
pancreatic/duodenal homeobox factor-1 (Pdx1), 171
ParaHox gene family members
qRT-PCR analysis, 171
in situ hybridization, 170
Villin–Cre^{ER} system, 171
whole genome microarray analysis, 170
role of hedgehog genes, 173
epithelial–mesenchymal interactions, 173
Hh signaling, 173
role of Hox genes, 171–172
anal sphincters, 172
fibroblast growth factor (Fgf), 172
ileo-cecal, 172
posterior prevalence, 172
rostrocaudal axis, 171
role of Wnt/planar cell polarity (PCP) signaling, 174
anorectal malformation, 174
bifurcated lumen, 174
Primordial germ cells (PGC), 233, 234*f*, 266

Proepicardial organ (PEO), 3, 13–14
Bmp signaling, 13
coelomic mesenchyme, 13
coronary blood vessels, 13
dye-labeled epicardium, 14
fluorescent markers, 13
Pronephros, 194, 212
Protein delivery methods, 311
Proximal–distal patterning and outgrowth
Fgf signaling from Aer, 328–329
Aer–Fgfs, 329
limb outgrowth defects, 329
models of PD patterning, 329–331
Aer–Fgf signaling, 329–330
analysis of Raldh2–Raldh3, 331
classical progress zone (PZ) model of limb, 329
differentiation front, 331
internal clock-like mechanism, 329
proximal or distal identity, 329
PZ model of PD limb specification, 330*f*
two-signal model of PD specification, 331
undifferentiated zone, 331
X-irradiation-induced phocomelia in chick, 331
Pulmonary arterial hypertension (PAH), 45, 54
Pulmonary neuroendocrine cell, 104, 118
Pulmonary trunk (PT), 4–5, 5, 12, 15–17, 22, 25–26, 195*f*, 211
Pulmonary vein (PV), 4, 15, 18, 21, 86

Q

Quantitative reverse transcriptase polymerase chain reaction (qRT-PCR), 240

R

Radial asymmetry, 161–162, 175
Renal–coloboma syndrome, 218–219, 375–376
Renal corpuscle (RC)
maturation of RC, 216–217
alphaactinin-4 (ACTN4), 216
Alport syndrome, 217
cell cycle inhibiters ($p27^{Kip1}$ and $p57^{Kip2}$), 217
collagen IV and laminins, 217
collagen IV isoforms (COL4A3/COL4A4, or COL4A5), 217
endothelial cells/podocytes and GBM, 216
focal and segmental glomerular sclerosis (FSGS), 216
foot processes, 216
GBM matrix molecules, 216
Lmx1b, Lim-homeodomain transcription factor, 216
slit diaphragm proteins Nephrin (Nphs1)/ Podocin (Nphs2), 216
patterning and vascularization of developing RC, 213–216
Bowman's capsule, 214
comma-shape body, 213–214
Crim1 mutant, 215
Denys–Drash syndrome, 214
Foxc2, podocyte marker, 213
genetic ablation of Pdgfb or Pdgfrβ, 215
glomerular basement membrane (GBM), 214
glomerular filtration barrier (GFB), 215
intussusceptive angiogenesis, 214–215
kidney (filter of blood), 213
Lmx1b/Podxl/Nphs1/Nphs2, podocyte markers, 214
mesangial precursors, 214
podocyte-specific gain-/loss-of-function mutants, 215
receptor Pdgfrβ, 214
Vegfa receptors, 214–215
Renal hypoplastic dysplasia, 218
Retina
development of hyaloid and retinal vasculature, 373–374
intraocular vascularization, 373
retinal vascular network, 374
tunica vasculosa lentis (TVL), 373–374
Vegf (vascular endothelial growth factor) and its receptor Vegfr2, 374
neural retina, 368–373
amacrine cells, 370
Bardet–Biedl syndrome, 372
bipolar cells, differentiation of, 372
CEP290 mutation, 373
formation of retina, 370*f*
fovea centralis, 368
ganglion cell layer (GCL), 368
horizontal cells, 372
inner nuclear layer (INL), 368
Leber congenital amourosis (LCA), 372–373
Math5/Atoh7, 369
Math3/NeuroD, 370
Müller cells, 368
mutations in CHX10 gene, 369
outer nuclear layer (ONL), 368
photoreceptor cells, differentiation of, 372
recessive ocular retardation, 369
regulation of retinal cell fate specification, 371*t*
retinal development, 369
retinal precursor cells, 369
retinal progenitor cells, 369
somatic gene therapy, 373
types of neurons, 368
retinal pigmented epithelium (RPE), 367–368
activin A (Tgfβ family), 367
genes coding for transcription factors (Mitf/Otx1/Otx2/Pax6), 368
melanogenic enzyme Tyrosinase (Tyr), 368
microphthalmia (Mi), 368
second neural retina, 367–368
Retinaldehyde dehydrogenase (Raldh2), *See* Aldehyde dehydrogenase (Aldh1a2)

Retinal pigmented epithelium (RPE), 344–345, 348, 367–368, 370
Retinoic acid (RA), 14, 15, 19–20, 82–83, 87, 91, 110–111, 201, 252, 253*f*, 270–272, 274*f*, 279, 297, 302, 323, 352–353, 367
  signaling, 110–111
    proximal–distal, 111
    RAR/RXR heterodimers, 110
    RA signaling, 110
    retinaldehyde dehydrogenase-2 (Raldh-2), 110
    retinoid X receptor (RXRα), 111
Runx2 haploinsufficiency, 304

S

Second heart field (SHF) cell behavior, 18–29
  anterior/posterior patterning
    Mef2c SHF enhancer (Cre), 20
    Raldh2 mutant, 20
    retinoic acid signaling, 20
  -Cre driver lines, *Tbx1-Cre/Mef2c-Cre,* 8
  interactions with neural crest
    Fgf/notch/semaphorin signaling, 24–25
  left/right patterning
    nodal signaling through Pitx2c, 21–22
  leftsided view of heart region, 19*f*
  maintenance of proliferation
    Fgf/hedgehog/canonical Wnt signaling, 22–24
    zebrafish embryo, 23
  marker genes, 11
  prevention of differentiation
    canonical Wnt signaling and transcriptional repression, 25–26
    myocardin gene, 26
    stabilized β-catenin, 25
  proliferation, 23
  regulation of differentiation potential
    heart tube and differentiation, 26–29
    Shh/notch/Bmp and non-canonical Wnt signaling, 26–29
Self-renewing progenitors, 118, 120
Semaphorin signaling, 1, 24–25, 25
Seminal vesicle, 232, 246
SERKAL syndrome, 219, 282
Sertoli cells, origin of, 235
  5-bromo-2-deoxyuridine (BrdU), 235
  chicken/quail gonad/mesonephros grafts, 235
  gonadal soma formation, 235
  laminin, 235
  subepithelial basement membrane, 235
Sex determination hypothesis, 276*f*
Sexual differentiation, 232–233, 264
Shh signaling, 26
Short stature homeobox (SHOX) genes, 309
Short stature syndromes, 310
ShRNA knockdown, 54
Signaling pathways
  Aer-Fgf, 333–334
  autocrine, 23, 333
  β-catenin, *See* β-catenin signaling
  Bmp, *See* Bone morphogenetic protein (Bmp)
  canonical Wnt pathway, 19*f*, 22–23, 25–26, 28, 30, 83, 101, 104, 204, 210, 275, 349, 362, 367
  cascades in early lens development, 357–358
  center, 320–321, 325, 328, 332
  chemokine, 164
  Cxcl12-Cxcr4, 164
  delta/jagged/notch, 55–56
  and endothelial heterogeneity, 46*f*
  Eph-Ephrin, 176*f*, 180*f*, 181, 184
  extracellular-regulated kinases (ERK), 53, 99, 105
  Fgf, 28, 302, 305–306
    Fgf-FgfR-sprouty, 80
    Fgf/hedgehog/canonical Wnt, 22–24
    Fgf/notch/semaphorin, 24–25
  Gdnf/Ret, 197–199
    See also Glial cell line-derived neurotrophic factor (Gdnf)
  Hedgehog (Hh)
    desert hedgehog (DHH) signaling, 244*f*, 245
    Indian hedgehog (IHH), 23, 170
    sonic hedgehog (Shh) signal, 9*f*, 16–17, 21–22, 23, 26, 82*f*, 102–103, 299, 327
  Hgf/Met, 202
  Igf, 106, 236
  mesenchyme–ectoderm signaling loop, 322
  nodal, 21–22, 162
  notch, 55, 182–183, 210–211, 248, 308, 369
  paracrine RA, 352
  phosphoinositide 3 (PI_3) kinase, 117
  planar cell polarity (PCP), 174, 204
  prostaglandin, 240
  Pthrp/Ihh/Bmp and Tgf-beta, 308
  Ras-MAP kinase, 99
  Ret, 167
  retinoic acid, 20, 110–111
  semaphorin, 1, 24–25, 25
  Shh/notch/Bmp/non-canonical Wnt, 26–29
  sprouty, 80, 114, 116
  Tgf-β, 9, 24, 28–29
  Tgfbr2/Wnt/Gdf5, 307
  transcriptional networks and, 159–186
  VegfA, 49–50, 250
  Vegf-C/VegfR3, 52
  Wnt, 25–26, 177–178, 204
    planar cell polarity (PCP), 174
    Wnt2, 17
    Wnt4, 239
    Wnt2b, 366
    Wnt/β–catenin, 179–182, 298*f*, 300, 303*f*, 307*f*

Single nucleotide polymorphisms, 51
Sinus venosus, 4, 13, 18, 21
Skeletal cells, factors determining, 293
  differentiation factors, 293
  patterning factors, 293
Skeletal myogenesis, 7–8
Skeletogenesis, *See* Vertebrate skeletogenesis
Skeletogenic cells, development of, 294–297
  alanine stretch in Hoxd13
    brachydactyly (short digits), 297
    polydactyly (extranumerary digits), 297
    syndactyly (fusion of digits), 297
  craniofacial skeletal elements, 295
  dysostosis-causing mutations, 296
  fate and molecular control of skeletogenic mesenchymal cells, 296*f*
  homeodomain, 297
  Hox proteins, 297
  lineage potential of skeletogenic cells
    chondrogenic factor Sox9, 297
    osteogenic Runt-domain transcription factor Runx2, 297
  networks of morphogens, 297
    bone morphogenetic proteins (Bmps), 297
    fibroblast growth factors (Fgfs), 297
    retinoic acid, 297
    Sonic hedgehog (Shh), 297
    Wnt ligands, 297
  neural crest, 294
  neuronal cells/melanocytes and skeletogenic cells, 295
  nuclei pulposi (NP) of IVD, 296
  organogenesis
    germ layers (ectoderm/mesoderm/endoderm), 294
  origin of skeletal cells in vertebrate embryo, 295*f*
  osteochondroprogenitors, 296
  Sox4/Sox11 and Sox12, group C of Sry-related Hmg box transcription factors, 296
Skeleton degeneration diseases, 293
Skeleton malformation diseases, 293, 302, 310
Skeleton variation, 309–310
  HmgA2 mutations, 310
  homozygous mutations, 309
  protein sequence, 310
  Prx1 enhancer variation, 309–310
  Shox genes, 309
    in humans (Shox/Shox2), 309
    mutations in X-linked Shox gene, 309
    one in mice (Shox2), 309
  single nucleotide polymorphism, effect of, 310
  skeletogenesis, 309
Slit diaphragm, 216
SMAD, 53, 99
Small noncoding microRNA (miRNA)/lung development, 107–108
  intranuclear primary transcript (pri-miRNA), 107
  RNA-induced silencing complex (RISC), 108
  RNase III endonuclease, 107
Smooth muscle cell (SMC), 4, 11, 13, 48, 56, 58–59, 84, 109, 124
Smooth muscle progenitors, 122
  localization of peripheral ASM progenitors, 122
Somatic environment and germ cells, female, 276–278
  Acbb expression, 277
  activin βA and βB, 277
  adrenal cell lineage, 278
  androgen-producing cells, 278
  genetic models, 277
  germ cell loss phenotypes, 277
  leydig cells, 278
  masculinization, 278
  morphological differences, 276
  testis-specific vasculature, 277
  Wnt4 knockout ovary, 278
Somatic gene therapy, 373
Somitogenesis, 163*f*, 164, 169*f*, 173
Sonic Hedgehog (Shh) signaling, 9, 16, 17, 21, 23, 26, 82, 297, 299, 327
Sox9 and sertoli cell differentiation, 237–241
  autoregulatory loop, 238
  campomelic dysplasia, 238
  chromatin immunoprecipitation (ChIP) assays, 238
  downstream target, 238
  Hmg box, 237
  LacZ reporter, 238
  skeletal malformation, 238
  Sox9 initiation, 238–239
    adrenal hypoplasia congenital (AHC), 238
    dosage-sensitive sex reversal (DSS), 238
    Mus domesticus poschiavinus Y chromosome (Y^{POS}), 239
    ovarian development, 238
    testicular development, 238
    testis cord, 238
    WNT4 signaling, 239
  Sox9 maintenance, 239–241
    fibroblast growth factor 9 (Fgf9), 239
    glucose starvation, 240
    ovary differentiation, 241
    proliferation, 239
    prostaglandin D2 (PGD2), 240
    protein kinase A (PKA), 240
    protein-to-DNA interactions, 239
    testicular architecture, 241
  SRY-box containing gene 9 (Sox9), 237
  testis-specific enhancer of Sox9 core (TESCO), 238
Sox9, beyond, 241–243
  anti-Müllerian hormone, 242
  apoptosis, 242
  cerebellin 4 precursor, 242

Sox9, beyond (*Continued*)
cryptorchidism, 242
DMRT1 hemizygosity, 243
DNA-binding motif, 242
double-knockout analyses, 242
Drosophila, 242
dysgenic testes, 243
fallopian tubes, 242
feminization, 243
gonadogenesis, 241
human DSD phenotypes, 242
idiopathic weight loss, 241
infertility, 242
late spermatogenic defect, 241
reduced bone density, 241
testis differentiation, 242
transforming growth factor-β (Tgfβ), 242
uterus, 242
vanin-1, 242
XY gonads, 241
Sox9 inducestype II collagen (Col2a1), 84
Spatiotemporal SRY function, 232
Specific cell types, emergence of, 125–127
aquaporin 5, 127
calcitonin gene-related peptide (CGRP), 126
chronic airway injury, 126
cystic fibrosis transmembrane conductance regulator (Cftr), 126
end-expiratory atelectasis, 127
gastrin-releasing peptide (GRP), 126
goblet cell hyperplasia, 126
kulchitsky cells, 126
mucin markers (MUC5B, 5A, 5C), 126
mucin-positive cells, 126
Na^+ ion transporter, 126
pulmonary neurendocrine (PNE), 125
watery secretion, 126
Splanchnic mesoderm, 2–3, 3, 5, 95
Splicing, alternative, 49
Sprouty signaling, 80, 114, 116
Squamous type I pneumocytes, 77
SRY gene (sex-determining region on Y chromosome), 264–265, 275
Stem cell niche and homeostasis in intestinal epithelium, 179–185
crypt fission, 179
enteroendocrine cells, 179
notch and Eph/ephrin pathways, 179
Paneth cells, 179
role of eph–ephrin signaling, 184
c-Abl-dependent stabilization, 184
cell–cell communication, 184
cyclin D1 protein, 184
delta-notch signaling, 184
downward migration, 184
proliferative zone, 184
role of Hnfα, 184–185
epithelial apicobasal polarity, 185
hepatocyte nuclear factor (Hnf)-4α, 184
intestinal barrier function, 185
single nucleotide polymorphism, 185
ulcerative colitis, 184
role of notch signaling, 182–183
absorptive cells, 183
adenoma formation, 183
APC mutations, 183
C_2H_2 zinc-finger, 183
Kruppel-like factor (Klf)-4, 183
mucin-producing goblet cells, 182
notch receptor (N1ic), 183
pro-proliferative activity, 183
secretory cell, 183
transcriptional repressor, 183
role of Wnt/β–catenin signaling, 179–182
Achete scute-like 2 (Ascl2), 182
adenomatous polyposis coli (APC), 181
basic helix–loop–helix (bHLH), 182
β-napthoflavone, 182
crypt hyperplasia, 182
cryptidins/defensins, 181
diabetes/IBD, 182
ectopic crypts, 182
homeostasis, 181
lineage tracing experiments, 181
lysozyme, 181
organization of intestinal epithelium and crypts of Lieberkühn, 180*f*
stem cell niche, 181
Stem cell therapies, 3
Stem/progenitor cell biology of lung, 118–128
ASM stem cells, 118
embryonic lung progenitors and proximal–distal patterning, 124–125
emergence of specific cell types, 125–127
endogenous epithelial progenitor cells, 118–122
alveolar epithelial progenitors, 121–122
bromodeoxyuridine (BrdU), 119
bronchoalveolar junction, 118
distal tip progenitors, 118
pulmonary neuroendocrine cell, 118
self-renewing progenitors, 118
tracheal and bronchial epithelial progenitors, 119–121
endogenous mesenchymal progenitors, 122–123
smooth muscle progenitors, 122
vascular progenitors, 122–123
lung progenitor cell proliferation, control of, 123–124
in postnatal respiratory system, 127–128
potential strategies to protect lung progenitors, 128
Subdomains of SHF, 14–18, 15*f*
anterior SHF and contributions to AP of heart, 16–17

PT myocardium, 16
Semaphorin3c gene, 16
SHF enhancer, 16
sonic hedgehog (Shh) signal, 16
tetralogy of fallot, 16
posterior SHF and formation of VP, 17–18
caval vein myocardium, 18
feed-forward regulatory loop, 17
mesenchymal protusion, 17
pulmonary vein (PV), 18
sinus venosus, 18
Submucosa, 126, 160–162, 167, 177
Supperior colliculus (SC), 376–377
Syndactyly (fusion of digits), 297
Synovial joint formation, 306–309, 307*f*
articular chondrocytes, 308
cell death, 307
chondrocytes and osteoblasts, 307
Ets-domain transcription factor Erg, 308
expression of Sox trio, 308
intervertebral and cartilaginous joints, 306
interzones, 307
joint cavitation, 307–308
joint morphogenesis, 306
joint specification, 306
notch signaling, 308
Pthrp/Ihh/Bmp and Tgf-beta signaling, 308
sutures, fibrous joints, 306
synovial joint morphogenesis, 308
Tgfbr2/Wnt/Gdf5 signaling, 307
Wnt/beta-catenin signaling, 308

T

Testis cords, 234, 241, 243, 249–250, 252, 266, 270*f*, 275
Testis determination, Sry and sertoli cell specification, 235–237
comparative genomics, 236
evolutionary conserved elements, 236
Glu/Asp-rich carboxy-terminal domain, 237
human SRY promoter, 237
immunofluorescence, 236
mitogen-activated protein kinase kinase kinase (MAP3K4), 237
phosphorylation, 237
postulated molecular pathways, 236*f*
Sry transcription, 236
synergism, 237
transgenic mice, 235
Wilms' tumor suppressor 1 (WT1), 236
WT1(+KTS), 236
XY gonads, 235
Testis-specific enhancer of Sox9 core (TESCO), 238
TGF-β signaling, 9, 24, 28–29
TGF-β superfamily, 53, 99, 110, 163, 172, 201, 242, 351
Thyroid-specific transcription factor (TTF-1), 95, 123
Townes–Brocks syndrome, 218
Tracheal and bronchial epithelial progenitors, 119–121
adult homeostasis, 119
clonality assay, 119
endothelial progenitor cell (EPC), 119
hoechst dye, 119
interfollicular epidermis, 120
naphthalene injury, 120
oncogenic upregulation, 121
progenitor pool, 120
pseudostratified epithelium, 119
tamoxifen (TM), 120
transit amplifying (TA) cells, 120
Tracheoesophageal fistula (TEF), 83–84, 88, 91, 102
Tracheomalacia, 84
Transcriptional networks and signaling pathways, 159–186
alimentary/gastrointestinal (GI) tract, 160
apoptosis, 161
cell-fate determination, 161
collagenous extracellular matrix, 162
concentric layers of tissue, 160
crypts of Lieberkühn, 161
crypt–villus axis, 175–178
role of Bmps, 178
role of hedgehog genes, 176–177
role of Wnt signaling, 177–178
derivative organs, 162
elimination, 160
formation and regionalization of primitive gut tube, 165–174
formation of definitive endoderm, 162–165
formation of enteric nervous system, 167–168
patterning early gut, 168
Transcription factors
biochemical regulators, 87–97
forkhead box, 87
GLI family zinc-finger, 97
Hox family, 96
Nkx and Hox homeodomain, 95–96
Dax1/Foxl2 and β-catenin, intracellular factors, 273
Erm/Pea3, ETS domain, 98
Ets family of, 333
Etv4 and Etv5, 201, 325
eye-field transcription factors (EFTF), 349
in formation of retina, 356
Foxa1/Foxa2/Hfh8 and Hfh4, forkhead box family, 87
Foxc1/Foxc2, forkhead/winged helix, 199
FoxD1 and FoxG1, 376
Gata4/Nkx2-5, cardiac progenitors, 7, 87
GLI family zinc-finger, 97

Transcription factors (*Continued*)
- Hes1 and Hes5, helix-loop-helix, 369
- Hox family, 96
  - Hoxa5/Hoxb2/Hoxb5/Hoxb3/Hoxb4, 96
  - Hox and ParaHox genes
  - Hox clusters, 20
- in lung development, 87–97
- Mef2c/Mef2d, MADS-box, 301
- Mitf/Otx1/Otx2/Pax6, genes coding for, 368
- Nkx/Hox homeodomain, 95–96
  - TTF-1 (thyroid-specific transcription factor), 95–96
  - type II pneumocyte proliferation, 96
- Osr1/Sall1/Wt1/Hoxa11/Eya1, MM, 206
- Otx2 and Sox2, 349
- PITX family, 105
- Runx2/Runx3, Runt-domain, 301
- SMADS, 53
- Sox4/Sox11 and Sox12, group C of Sry-related Hmg box, 296
- Tbx5 and Hand1/2, T-box, 7
- TBX5 and TBX4, T-box, 322–323
- Tbx2–Tbx5, Brachyury T-Box family, 95

Transcription repression factor (Tbx3), 13, 325, 349

Transforming growth factor (Tgf)-β, 48, 87–88, 162, 242, 298, 351

Transgenic mouse technology, 87

Tube formation, 165–166
- anterior intestinal portal, 165
- caudal intestinal portal (CIP), 165
- plate-derived splanchnic (visceral) mesoderm, 166

Tubules, patterning resulting
- cyclinD1 (Ccnd1), cell cycle, 210
- Dkk1 expression, 210
- genes involved in nephron segmentation, 209*f*
- of LH and DT, 211–212
  - comma-shaped body (CB) stage, 211
  - gain- and loss-of-function studies in *Xenopus,* 212
  - iroquois (Irx) gene family (Irx1/Irx2/Irx3), 211
  - POU-domain transcription factor, 212
  - Pou3f3 (Brn1), 212
  - UMOD mutations, 212
  - Wnt9b-mediated canonical signal, 211
- Lhx1/Pou3f3/Dll1/Wt1, expression in RV, 208
- notch signaling in patterning of proximal nephron, 210–211
  - Jag1/Dll1/Lhx1 regional expression, 210–211
  - notch intracellular domain (NICD), 211
  - notch signaling pathway (notch1/2), 210
  - podocytes, development of, 210
  - presenilins (Psen1/Psen2), 211
  - proximal convoluted tubules (PCT), 210
- parietal (Bowman's capsule) and visceral (podocyte) layers, 208
- proximal–distal axis, 207
- S-shaped body (SB) stage, 208
- Tcf/Lef signaling, 210
- of UT *vs.* stalk and differentiation of ureter–pelvic junction, 212–213
  - BMP4 signaling, 213
  - in branching ureteric compartment, 212
  - Disc-large homolog 1 (Dlgh1), 213
  - expression of Ret/Gfrα1 and Wnt11, 213
  - hedgehog ligand Shh, 212
  - ureter function, 212
  - water channel aquaporin 2 (Aqp2), expression of, 212
- Wnt4/Pax8 and Fgf8, 209

Tubulogenesis (kidney) in human disease
- defects in branching morphogenesis, 218
  - congenital anomalies of kidney and urinary tract (CAKUT), 218
  - mutations in Hnf1β/Tcf2/Pax2EYA1/Six1 and Sall1, 218
  - renal hypodysplastic disease (RHD), 218
  - renin–angiontensin system, 218
- dysplasia and defects in MET in humans, 218–219
  - Alagille syndrome, 219
  - Denys–Drash syndrome, 219
  - Frasier syndrome, 219
  - notch2 (ALGS2) or Jag1 (ALGS1), mutation in, 219
  - renal–coloboma syndrome, 219
  - renal dysplasias, 218
  - renal hypoplasia, 219
  - unilateral renal agenesis, 219
  - WAGR(Wilms–Aniridia–Genital Anomoloies–Retardation), 219
  - Wilms' tumor, 219
- nephron number and renal disease, 219–220
  - Australian Aboriginal, 219
  - intrauterine growth retardation, 219
  - intrauterine insults, 220
  - low birth weight, 220
  - risk of renal disease and hypertension, 220
  - variability in nephron number, 219

Tubulogenesis VIA MET—Tube One Induces Tube Two
- CM as population of nephron progenitors, 205–207
  - FACS analysis, 206
  - MM transcription factors (Osr1/Sall1/Wt1/Hoxa11/Eya1), 206
  - renal hypoplasia, 206
  - role of Wt1 in MET, 207
  - Six2 gain-of-function, 206
- conditional β-catenin (Ctnnb1) gene deletion, 205

deletion of novel FGF receptor-like 1 (FGFrl1) gene, 205
embryonic brain and spinal cord, 203–204
Fgf8 gene deletion, 205
genes regulating nephron endowment via MET, 204*f*
heparan sulfate proteoglycans, 204
MM-derived CM cells, 203
nephron progenitors and cessation of nephrogenesis, 207
CM apoptosis, 207
$Six2^{+}cited1^{+}$ CM, 207
survival of CM phenotype, 207
non-canonical Wnt/PCP (planar cell polarity) pathway, 204
role of Wnt11, 205
transcription factor Emx2, expression of, 204
transcription factor gene Lhx1, 205
Wnt signaling, 204
Tumor endothelial marker-8 (TEM-8), 51
Tumorigenesis processes, 103
Twist1/2 homeodomain transcription factor, 297

U

Umbrella cell, 212
Ureteric bud (UB)
initiation/position/width and number, regulation of, 199–200
Bmp4 inhibits UB branching, 200
expression of Sall1, 200
Eya1 expression in IM, 199
forkhead/winged helix transcription factors (Foxc1/Foxc2), 199
Gata3 inactivation, 199
Gene ablation of Eya1 or Six1, 199
Gremlin (Grem1), 200
paralogous Hox11 genes (Hoxa11/Hoxc11/Hoxd11), 199
Pax2, 199
Wt1/Six2, 199
outgrowth, Gdnf/Ret signaling initiates and guides, 197–199
See also Glial cell line-derived neurotrophic factor (Gdnf)
Ureteric tree in 3D, 203

V

Vascular development, genetic mechanisms, 44–62
BMP, 53–55
BMP–VEGF crosstalk, 53–54
mutations, 54–55
signaling pathway, 53
notch/delta/jagged, 55–60
Vascular endothelial growth factor (Vegf), 44, 49–52, 86, 106–107, 250, 303, 374
human genetic disorders, 49
Vegf-A mutations, 50–51
Vegf-A signaling pathway, 49–50
spatial distribution, 50
vessel development, 49
Vegf-C/VegfR3 mutations, 52
abnormal fluid accumulation, 52
distal chromosome 5q, 52
lymphatic vessel formation, 52
lymphedema, 52
Vegf-C/VegfR3 signaling pathway, 52
Vegf receptor 2 (VegfR2), 49
Vascular progenitors, 122–123
distal airspace, 122
Flk1 promoter, 123
heterogenous pulmonary endothelial cells, 122
laryngotracheal groove, 123
Vascular smooth muscle cells (vSMC), 4, 48, 59, 214
Vasculogenesis, 47, 50, 95, 100, 102
Vas deferens, 232, 246
Venous pole (VP), 5–6, 8, 13, 17–19, 21, 23, 26
Vertebral column, 292, 294
Vertebrate endoderm organ development, 162
Vertebrate limb patterning, molecular regulation of, 319–335
axes
anterior–posterior (AP), 320
dorsal–ventral (DV), 320
proximal–distal (PD), 320
biology approach to limb patterning, 334–335
early limb
limb induction, 322–323
pre-patterning/positioning of ZPA and asymmetry, 323–325
interaction between Zpa and Aer
Shh–Grem1–Fgf feedback loop, 332–333
termination of limb outgrowth, 333–334
patterning along AP axis
interactions between Shh and Gli3, 325–326
primary cilium and Gli3 function in limb, 326–327
Shh-mediated digit specification and growth, 327–328
proximal–distal patterning and outgrowth
Fgf signaling from Aer, 328–329
models of PD patterning, 328–329
shape and length of mature limb, 321
autopod (hand–foot), 321
stylopod (upper arm–thigh), 321
zeugopod (forearm–lower leg), 321
signaling centers in 11.5 dpc mouse limb, 321*f*
Vertebrate skeletogenesis, 291–311
bone development, 302–306
development of cartilage
anlagen, 298–300
growth plates, 300–302
development of skeletogenic cells, 294–297
skeleton variation, 309–310

Vertebrate skeletogenesis (*Continued*)
  structural organization and advantages, 293–294
    airways and synovial joints, 294
    cartilage and bone, 294
    collagen-1 and collagen-2, 294
    invertebrate and vertebrate skeletons, 294
    joint structures, 294
    osteocalcin and bone sialoprotein, 294
  synovial joint formation, 306–309
Vesicoureteric reflux (VUR), 218–219
Vessel assembly, 46–47
  flow-mediated remodeling, 47
  somite VEGF production, 47
  vertebrate development, 46
Vessel remodeling and stabilization, 48
  hemodynamic stress, 48
  multicellular vessel networks, 48
Villi, 161, 170, 175–177, 180*f*, 182

W

Waardenburg syndrome, 368
  type 2, 368
Wnt signaling
  beta-catenin signaling, 297–299, 308
  ligands, 23, 197, 275, 297, 302
  Wnt2/2b, 82–83, 92, 124
  Wnt2 signaling, 17
Wolffian duct, 194, 232, 246

X

Xenopus laevis, 369
  effects of morpholino-dependent silencing of hairy2 in, 358
  embryo, 25, 28, 349, 351
  endoderm formation in, 163
  eye field in, 349
  gain- and loss-of-function studies in, 212
  pronephros, 212
  role of Wnt-signaling, 349
  SHF in, 8

Z

Zebrafish
  angiogenesis defects, 54
  anterior neural plate of, 349
  BMP signaling in blood vessel formation, 54
  cell fate map analysis in wild-type, 352
  cyclopic mutants, 351
  embryo, 20, 23, 27, 56, 164
  gastrulation, 164
  lumen formation in, 166
  mutations, 349
  myocardial differentiation in, 8
  neutrophils, 61
  notch signaling, 56
  notochord, 47
  retinal vasculature in, 374
  role of Wnt-signaling, 349
  Squint (Sqt) and Southpaw (Spaw) gene, 352
Z hypothesis, 265, 273
Zinc-finger transcription factor, 97, 297
Zone of polarizing activity (ZPA) and Aer
  Shh–Grem1–Fgf feedback loop, 332–333
    Bmp antagonist Grem1, 332
    Bmp–Shh negative feedback loop, 333
    Ets family of transcription factors, 333
    regulatory circuits controlling limb patterning and outgrowth, 332*f*
    Shh transcription in chick/mouse limbs, 332
  termination of limb outgrowth, 333–334
    Aer–Fgf signaling, 333–334
    bead implantation studies in chick, 333
    Fgf–Grem1 inhibitory loop, 334
    Grem1-expressing zone of cells, 333
    Shh expressing cells, 333
    Shh–Grem1–Fgf loop, 333

Contents of Previous Volumes

Volume 47

1. **Early Events of Somitogenesis in Higher Vertebrates: Allocation of Precursor Cells during Gastrulation and the Organization of a Moristic Pattern in the Paraxial Mesoderm**

   Patrick P. L. Tam, Devorah Goldman, Anne Camus, and Gary C. Shoenwolf

2. **Retrospective Tracing of the Developmental Lineage of the Mouse Myotome**

   Sophie Eloy-Trinquet, Luc Mathis, and Jean-François Nicolas

3. **Segmentation of the Paraxial Mesoderm and Vertebrate Somitogenesis**

   Olivier Pourqulé

4. **Segmentation: A View from the Border**

   Claudio D. Stern and Daniel Vasiliauskas

5. **Genetic Regulation of Somite Formation**

   Alan Rawls, Jeanne Wilson-Rawls, and Eric N. Olsen

6. **Hox Genes and the Global Patterning of the Somitic Mesoderm**

   Ann Campbell Burke

7. **The Origin and Morphogenesis of Amphibian Somites**

   Ray Keller

8. **Somitogenesis in Zebrafish**

   Scoff A. Halley and Christiana Nüsslain-Volhard

9. **Rostrocaudal Differences within the Somites Confer Segmental Pattern to Trunk Neural Crest Migration**

   Marianne Bronner-Fraser

Volume 48

1. **Evolution and Development of Distinct Cell Lineages Derived from Somites**
Beafe Brand-Saberi and Bodo Christ

2. **Duality of Molecular Signaling Involved in Vertebral Chondrogenesis**
Anne-Hélène Monsoro-Burq and Nicole Le Douarin

3. **Sclerotome Induction and Differentiation**
Jennifer L. Docker

4. **Genetics of Muscle Determination and Development**
Hans-Henning Arnold and Thomas Braun

5. **Multiple Tissue Interactions and Signal Transduction Pathways Control Somite Myogenesis**
Anne-Gaëlle Borycki and Charles P. Emerson, JR.

6. **The Birth of Muscle Progenitor Cells in the Mouse: Spatiotemporal Considerations**
Shahragim Tajbakhsh and Margaret Buckingham

7. **Mouse–Chick Chimera: An Experimental System for Study of Somite Development**
Josiane Fontaine-Pérus

8. **Transcriptional Regulation during Somitogenesis**
Dennis Summerbell and Peter W. J. Rigby

9. **Determination and Morphogenesis in Myogenic Progenitor Cells: An Experimental Embryological Approach**
Charles P. Ordahl, Brian A. Williams, and Wilfred Denetclaw

Volume 49

1. **The Centrosome and Parthenogenesis**
Thomas Küntziger and Michel Bornens

2. **γ-Tubulin**
Berl R. Oakley

3. **γ-Tubulin Complexes and Their Role in Microtubule Nucleation**
Ruwanthi N. Gunawardane, Sofia B. Lizarraga, Christiane Wiese, Andrew Wilde, and Yixian Zheng

4. **γ-Tubulin of Budding Yeast**
Jackie Vogel and Michael Snyder

5. **The Spindle Pole Body of Saccharomyces cerevisiae: Architecture and Assembly of the Core Components**
Susan E. Francis and Trisha N. Davis

6. **The Microtubule Organizing Centers of Schizosaccharomyces pombe**
Iain M. Hagan and Janni Petersen

7. **Comparative Structural, Molecular, and Functional Aspects of the Dictyostelium discoideum Centrosome**
Ralph Gräf, Nicole Brusis, Christine Daunderer, Ursula Euteneuer, Andrea Hestermann, Manfred Schliwa, and Masahiro Ueda

8. **Are There Nucleic Acids in the Centrosome?**
Wallace F. Marshall and Joel L. Rosenbaum

9. **Basal Bodies and Centrioles: Their Function and Structure**
Andrea M. Preble, Thomas M. Giddings, JR., and Susan K. Dutcher

10. **Centriole Duplication and Maturation in Animal Cells**
B. M. H. Lange, A. J. Faragher, P. March, and K. Gull

11. **Centrosome Replication in Somatic Cells: The Significance of the Gi Phase**
Ron Balczon

12. **The Coordination of Centrosome Reproduction with Nuclear Events during the Cell Cycle**
Greenfield Sluder and Edward H. Hinchcliffe

13. **Regulating Centrosomes by Protein Phosphorylation**
Andrew M. Fry, Thibault Mayor, and Erich A. Nigg

14. **The Role of the Centrosome in the Development of Malignant Tumors**
Wilma L. Lingle and Jeffrey L. Salisbury

15. **The Centrosome-Associated Aurora/Ipl–like Kinase Family**
T. M. Goepfert and B. R. Brinkley

16. **Centrosome Reduction during Mammalian Spermiogenesis**
G. Manandhar, C. Simerly, and G. Schatten

17. **The Centrosome of the Early C. elegans Embryo: Inheritance, Assembly, Replication, and Developmental Roles**
Kevin F. O'Connell

18. The Centrosome in Drosophila Oocyte Development

Timothy L. Megraw and Thomas C. Kaufman

19. The Centrosome in Early Drosophila Embryogenesis

W. F. Rothwell and W. Sullivan

20. Centrosome Maturation

Robert E. Palazzo, Jacalyn M. Vogel, Bradley J. Schnackenberg, Dawn R. Hull, and Xingyong Wu

Volume 50

1. Patterning the Early Sea Urchin Embryo

Charles A. Ettensohn and Hyla C. Sweet

2. Turning Mesoderm into Blood: The Formation of Hematopoietic Stem Cells during Embryogenesis

Alan J. Davidson and Leonard I. Zon

3. Mechanisms of Plant Embryo Development

Shunong Bai, Lingjing Chen, Mary Alice Yund, and Zinmay Rence Sung

4. Sperm-Mediated Gene Transfer

Anthony W. S. Chan, C. Marc Luetjens, and Gerald P. Schatten

5. Gonocyte–Sertoli Cell Interactions during Development of the Neonatal Rodent Testis

Joanne M. Orth, William F. Jester, Ling-Hong Li, and Andrew L. Laslett

6. Attributes and Dynamics of the Endoplasmic Reticulum in Mammalian Eggs

Douglas Kline

7. Germ Plasm and Molecular Determinants of Germ Cell Fate

Douglas W. Houston and Mary Lou King

Volume 51

1. Patterning and Lineage Specification in the Amphibian Embryo

Agnes P. Chan and Laurence D. Etkin

2. Transcriptional Programs Regulating Vascular Smooth Muscle Cell Development and Differentiation

Michael S. Parmacek

3. **Myofibroblasts: Molecular Crossdressers**

   Gennyne A. Walker, Ivan A. Guerrero, and Leslie A. Leinwand

4. **Checkpoint and DNA-Repair Proteins Are Associated with the Cores of Mammalian Meiotic Chromosomes**

   Madalena Tarsounas and Peter B. Moens

5. **Cytoskeletal and Ca^{2+} Regulation of Hyphal Tip Growth and Initiation**

   Sara Torralba and I. Brent Heath

6. **Pattern Formation during C. elegans Vulval Induction**

   Minqin Wang and Paul W. Sternberg

7. **A Molecular Clock Involved in Somite Segmentation**

   Miguel Maroto and Olivier Pourquié

Volume 52

1. **Mechanism and Control of Meiotic Recombination Initiation**

   Scott Keeney

2. **Osmoregulation and Cell Volume Regulation in the Preimplantation Embryo**

   Jay M. Baltz

3. **Cell–Cell Interactions in Vascular Development**

   Diane C. Darland and Patricia A. D'Amore

4. **Genetic Regulation of Preimplantation Embryo Survival**

   Carol M. Warner and Carol A. Brenner

Volume 53

1. **Developmental Roles and Clinical Significance of Hedgehog Signaling**

   Andrew P. McMahon, Philip W. Ingham, and Clifford j. Tabin

2. **Genomic Imprinting: Could the Chromatin Structure Be the Driving Force?**

   Andras Paldi

3. **Ontogeny of Hematopoiesis: Examining the Emergence of Hematopoietic Cells in the Vertebrate Embryo**

   Jenna L. Galloway and Leonard I. Zon

4. **Patterning the Sea Urchin Embryo: Gene Regulatory Networks, Signaling Pathways, and Cellular Interactions**
Lynne M. Angerer and Robert C. Angerer

Volume 54

1. **Membrane Type-Matrix Metalloproteinases (MT-MMP)**
Stanley Zucker, Duanqing Pei, Jian Cao, and Carlos Lopez-Otin

2. **Surface Association of Secreted Matrix Metalloproteinases**
Rafael Fridman

3. **Biochemical Properties and Functions of Membrane-Anchored Metalloprotease-Disintegrin Proteins (ADAMs)**
J. David Becherer and Carl P. Blobel

4. **Shedding of Plasma Membrane Proteins**
Joaquin Arribas and Anna Merlos-Suárez

5. **Expression of Meprins in Health and Disease**
Lourdes P. Norman, Gail L. Matters, Jacqueline M. Crisman, and Judith S. Bond

6. **Type II Transmembrane Serine Proteases**
Qingyu Wu

7. **DPPIV, Seprase, and Related Serine Peptidases in Multiple Cellular Functions**
Wen-Tien Chen, Thomas Kelly, and Giulio Ghersi

8. **The Secretases of Alzheimer's Disease**
Michael S. Wolfe

9. **Plasminogen Activation at the Cell Surface**
Vincent Ellis

10. **Cell-Surface Cathepsin B: Understanding Its Functional Significance**
Dora Cavallo-Medved and Bonnie F. Sloane

11. **Protease-Activated Receptors**
Wadie F. Bahou

12. **Emmprin (CD147), a Cell Surface Regulator of Matrix Metalloproteinase Production and Function**
Bryan P. Toole

13. The Evolving Roles of Cell Surface Proteases in Health and Disease: Implications for Developmental, Adaptive, Inflammatory, and Neoplastic Processes

Joseph A. Madri

14. Shed Membrane Vesicles and Clustering of Membrane-Bound Proteolytic Enzymes

M. Letizia Vittorelli

Volume 55

1. The Dynamics of Chromosome Replication in Yeast

Isabelle A. Lucas and M. K. Raghuraman

2. Micromechanical Studies of Mitotic Chromosomes

M. G. Poirier and John F. Marko

3. Patterning of the Zebrafish Embryo by Nodal Signals

Jennifer O. Liang and Amy L. Rubinstein

4. Folding Chromosomes in Bacteria: Examining the Role of Csp Proteins and Other Small Nucleic Acid-Binding Proteins

Nancy Trun and Danielle Johnston

Volume 56

1. Selfishness in Moderation: Evolutionary Success of the Yeast Plasmid

Soundarapandian Velmurugan, Shwetal Mehta, and Makkuni Jayaram

2. Nongenomic Actions of Androgen in Sertoli Cells

William H. Walker

3. Regulation of Chromatin Structure and Gene Activity by Poly(ADP-Ribose) Polymerases

Alexei Tulin, Yurli Chinenov, and Allan Spradling

4. Centrosomes and Kinetochores, Who needs 'Em? The Role of Noncentromeric Chromatin in Spindle Assembly

Priya Prakash Budde and Rebecca Heald

5. Modeling Cardiogenesis: The Challenges and Promises of 3D Reconstruction

Jeffrey O. Penetcost, Claudio Silva, Maurice Pesticelli, Jr., and Kent L. Thornburg

6. **Plasmid and Chromosome Traffic Control: How ParA and ParB Drive Partition**
Jennifer A. Surtees and Barbara E. Funnell

Volume 57

1. **Molecular Conservation and Novelties in Vertebrate Ear Development**
B. Fritzsch and K. W. Beisel

2. **Use of Mouse Genetics for Studying Inner Ear Development**
Elizabeth Quint and Karen P. Steel

3. **Formation of the Outer and Middle Ear, Molecular Mechanisms**
Moisés Mallo

4. **Molecular Basis of Inner Ear Induction**
Stephen T. Brown, Kareen Martin, and Andrew K. Groves

5. **Molecular Basis of Otic Commitment and Morphogenesis: A Role for Homeodomain-Containing Transcription Factors and Signaling Molecules**
Eva Bober, Silke Rinkwitz, and Heike Herbrand

6. **Growth Factors and Early Development of Otic Neurons: Interactions between Intrinsic and Extrinsic Signals**
Berta Alsina, Fernando Giraldez, and Isabel Varela-Nieto

7. **Neurotrophic Factors during Inner Ear Development**
Ulla Pirvola and Jukka Ylikoski

8. **FGF Signaling in Ear Development and Innervation**
Tracy J. Wright and Suzanne L. Mansour

9. **The Roles of Retinoic Acid during Inner Ear Development**
Raymond Romand

10. **Hair Cell Development in Higher Vertebrates**
Wei-Qiang Gao

11. **Cell Adhesion Molecules during Inner Ear and Hair Cell Development, Including Notch and Its Ligands**
Matthew W. Kelley

12. Genes Controlling the Development of the Zebrafish Inner Ear and Hair Cells

Bruce B. Riley

13. Functional Development of Hair Cells

Ruth Anne Eatock and Karen M. Hurley

14. The Cell Cycle and the Development and Regeneration of Hair Cells

Allen F. Ryan

Volume 58

1. A Role for Endogenous Electric Fields in Wound Healing

Richard Nuccitelli

2. The Role of Mitotic Checkpoint in Maintaining Genomic Stability

Song-Tao Liu, Jan M. van Deursen, and Tim J. Yen

3. The Regulation of Oocyte Maturation

Ekaterina Voronina and Gary M. Wessel

4. Stem Cells: A Promising Source of Pancreatic Islets for Transplantation in Type 1 Diabetes

Cale N. Street, Ray V. Rajotte, and Gregory S. Korbutt

5. Differentiation Potential of Adipose Derived Adult Stem (ADAS) Cells

Jeffrey M. Gimble and Farshid Guilak

Volume 59

1. The Balbiani Body and Germ Cell Determinants: 150 Years Later

Malgorzata Kloc, Szczepan Bilinski, and Laurence D. Etkin

2. Fetal–Maternal Interactions: Prenatal Psychobiological Precursors to Adaptive Infant Development

Matthew F. S. X. Novak

3. Paradoxical Role of Methyl-CpG-Binding Protein 2 in Rett Syndrome

Janine M. LaSalle

4. Genetic Approaches to Analyzing Mitochondrial Outer Membrane Permeability

Brett H. Graham and William J. Craigen

5. **Mitochondrial Dynamics in Mammals**
Hsiuchen Chen and David C. Chan

6. **Histone Modification in Corepressor Functions**
Judith K. Davie and Sharon Y. R. Dent

7. **Death by Abl: A Matter of Location**
Jiangyu Zhu and Jean Y. J. Wang

Volume 60

1. **Therapeutic Cloning and Tissue Engineering**
Chester J. Koh and Anthony Atala

2. **α-Synuclein: Normal Function and Role in Neurodegenerative Diseases**
Erin H. Norris, Benoit I. Giasson, and Virginia M.-Y. Lee

3. **Structure and Function of Eukaryotic DNA Methyltransferases**
Taiping Chen and En Li

4. **Mechanical Signals as Regulators of Stem Cell Fate**
Bradley T. Estes, Jeffrey M. Gimble, and Farshid Guilak

5. **Origins of Mammalian Hematopoiesis: In Vivo Paradigms and In Vitro Models**
M. William Lensch and George Q. Daley

6. **Regulation of Gene Activity and Repression: A Consideration of Unifying Themes**
Anne C. Ferguson-Smith, Shau-Ping Lin, and Neil Youngson

7. **Molecular Basis for the Chloride Channel Activity of Cystic Fibrosis Transmembrane Conductance Regulator and the Consequences of Disease-Causing Mutations**
Jackie F. Kidd, Ilana Kogan, and Christine E. Bear

Volume 61

1. **Hepatic Oval Cells: Helping Redefine a Paradigm in Stem Cell Biology**
P. N. Newsome, M. A. Hussain, and N. D. Theise

2. **Meiotic DNA Replication**
Randy Strich

3. **Pollen Tube Guidance: The Role of Adhesion and Chemotropic Molecules**
Sunran Kim, Juan Dong, and Elizabeth M. Lord

4. **The Biology and Diagnostic Applications of Fetal DNA and RNA in Maternal Plasma**
Rossa W. K. Chiu and Y. M. Dennis Lo

5. **Advances in Tissue Engineering**
Shulamit Levenberg and Robert Langer

6. **Directions in Cell Migration Along the Rostral Migratory Stream: The Pathway for Migration in the Brain**
Shin-ichi Murase and Alan F. Horwitz

7. **Retinoids in Lung Development and Regeneration**
Malcolm Maden

8. **Structural Organization and Functions of the Nucleus in Development, Aging, and Disease**
Leslie Mounkes and Colin L. Stewart

Volume 62

1. **Blood Vessel Signals During Development and Beyond**
Ondine Cleaver

2. **HIFs, Hypoxia, and Vascular Development**
Kelly L. Covello and M. Celeste Simon

3. **Blood Vessel Patterning at the Embryonic Midline**
Kelly A. Hogan and Victoria L. Bautch

4. **Wiring the Vascular Circuitry: From Growth Factors to Guidance Cues**
Lisa D. Urness and Dean Y. Li

5. **Vascular Endothelial Growth Factor and Its Receptors in Embryonic Zebrafish Blood Vessel Development**
Katsutoshi Goishi and Michael Klagsbrun

6. **Vascular Extracellular Matrix and Aortic Development**
Cassandra M. Kelleher, Sean E. McLean, and Robert P. Mecham

7. **Genetics in Zebrafish, Mice, and Humans to Dissect Congenital Heart Disease: Insights in the Role of VEGF**
Diether Lambrechts and Peter Carmeliet

8. **Development of Coronary Vessels**
Mark W. Majesky

9. **Identifying Early Vascular Genes Through Gene Trapping in Mouse Embryonic Stem Cells**
Frank Kuhnert and Heidi Stuhlmann

Volume 63

1. **Early Events in the DNA Damage Response**
Irene Ward and Junjie Chen

2. **Afrotherian Origins and Interrelationships: New Views and Future Prospects**
Terence J. Robinson and Erik R. Seiffert

3. **The Role of Antisense Transcription in the Regulation of X-Inactivation**
Claire Rougeulle and Philip Avner

4. **The Genetics of Hiding the Corpse: Engulfment and Degradation of Apoptotic Cells in C. elegans and D. melanogaster**
Zheng Zhou, Paolo M. Mangahas, and Xiaomeng Yu

5. **Beginning and Ending an Actin Filament: Control at the Barbed End**
Sally H. Zigmond

6. **Life Extension in the Dwarf Mouse**
Andrzej Bartke and Holly Brown-Borg

Volume 64

1. **Stem/Progenitor Cells in Lung Morphogenesis, Repair, and Regeneration**
David Warburton, Mary Anne Berberich, and Barbara Driscoll

2. **Lessons from a Canine Model of Compensatory Lung Growth**
Connie C. W. Hsia

3. **Airway Glandular Development and Stem Cells**
Xiaoming Liu, Ryan R. Driskell, and John F. Engelhardt

4. Gene Expression Studies in Lung Development and Lung Stem Cell Biology

Thomas J. Mariani and Naftali Kaminski

5. Mechanisms and Regulation of Lung Vascular Development

Michelle Haynes Pauling and Thiennu H. Vu

6. The Engineering of Tissues Using Progenitor Cells

Nancy L. Parenteau, Lawrence Rosenberg, and Janet Hardin-Young

7. Adult Bone Marrow-Derived Hemangioblasts, Endothelial Cell Progenitors, and EPCs

Gina C. Schatteman

8. Synthetic Extracellular Matrices for Tissue Engineering and Regeneration

Eduardo A. Silva and David J. Mooney

9. Integrins and Angiogenesis

D. G. Stupack and D. A. Cheresh

Volume 65

1. Tales of Cannibalism, Suicide, and Murder: Programmed Cell Death in C. elegans

Jason M. Kinchen and Michael O. Hengartner

2. From Guts to Brains: Using Zebrafish Genetics to Understand the Innards of Organogenesis

Carsten Stuckenholz, Paul E. Ulanch, and Nathan Bahary

3. Synaptic Vesicle Docking: A Putative Role for the Munc18/Sec1 Protein Family

Robby M. Weimer and Janet E. Richmond

4. ATP-Dependent Chromatin Remodeling

Corey L. Smith and Craig L. Peterson

5. Self-Destruct Programs in the Processes of Developing Neurons

David Shepherd and V. Hugh Perry

6. Multiple Roles of Vascular Endothelial Growth Factor (VEGF) in Skeletal Development, Growth, and Repair

Elazar Zelzer and Bjorn R. Olsen

7. **G-Protein Coupled Receptors and Calcium Signaling in Development**
Geoffrey E. Woodard and Juan A. Rosado

8. **Differential Functions of 14-3-3 Isoforms in Vertebrate Development**
Anthony J. Muslin and Jeffrey M. C. Lau

9. **Zebrafish Notochordal Basement Membrane: Signaling and Structure**
Annabelle Scott and Derek L. Stemple

10. **Sonic Hedgehog Signaling and the Developing Tooth**
Martyn T. Cobourne and Paul T. Sharpe

Volume 66

1. **Stepwise Commitment from Embryonic Stem to Hematopoietic and Endothelial Cells**
Changwon Park, Jesse J. Lugus, and Kyunghee Choi

2. **Fibroblast Growth Factor Signaling and the Function and Assembly of Basement Membranes**
Peter Lonai

3. **TGF-/β Superfamily and Mouse Craniofacial Development: Interplay of Morphogenetic Proteins and Receptor Signaling Controls Normal Formation of the Face**
Marek Dudas and Vesa Kaartinen

4. **The Colors of Autumn Leaves as Symptoms of Cellular Recycling and Defenses Against Environmental Stresses**
Helen J. Ougham, Phillip Morris, and Howard Thomas

5. **Extracellular Proteases: Biological and Behavioral Roles in the Mammalian Central Nervous System**
Yan Zhang, Kostas Pothakos, and Styliana-Anna (Stella) Tsirka

6. **The Genetic Architecture of House Fly Mating Behavior**
Lisa M. Meffert and Kara L. Hagenbuch

7. **Phototropins, Other Photoreceptors, and Associated Signaling: The Lead and Supporting Cast in the Control of Plant Movement Responses**
Bethany B. Stone, C. Alex Esmon, and Emmanuel Liscum

8. **Evolving Concepts in Bone Tissue Engineering**
Catherine M. Cowan, Chia Soo, Kang Ting, and Benjamin Wu

9. **Cranial Suture Biology**
Kelly A Lenton, Randall P. Nacamuli, Derrick C. Wan, Jill A. Helms, and Michael T. Longaker

Volume 67

1. **Deer Antlers as a Model of Mammalian Regeneration**
Joanna Price, Corrine Faucheux, and Steve Allen

2. **The Molecular and Genetic Control of Leaf Senescence and Longevity in Arabidopsis**
Pyung Ok Lim and Hong Gil Nam

3. **Cripto-1: An Oncofetal Gene with Many Faces**
Caterina Bianco, Luigi Strizzi, Nicola Normanno, Nadia Khan, and David S. Salomon

4. **Programmed Cell Death in Plant Embryogenesis**
Peter V. Bozhkov, Lada H. Filonova, and Maria F. Suarez

5. **Physiological Roles of Aquaporins in the Choroid Plexus**
Daniela Boassa and Andrea J. Yool

6. **Control of Food Intake Through Regulation of cAMP**
Allan Z. Zhao

7. **Factors Affecting Male Song Evolution in Drosophila montana**
Anneli Hoikkala, Kirsten Klappert, and Dominique Mazzi

8. **Prostanoids and Phosphodiesterase Inhibitors in Experimental Pulmonary Hypertension**
Ralph Theo Schermuly, Hossein Ardeschir Ghofrani, and Norbert Weissmann

9. **14-3-3 Protein Signaling in Development and Growth Factor Responses**
Daniel Thomas, Mark Guthridge, Jo Woodcock, and Angel Lopez

10. **Skeletal Stem Cells in Regenerative Medicine**
Wataru Sonoyama, Carolyn Coppe, Stan Gronthos, and Songtao Shi

Volume 68

1. **Prolactin and Growth Hormone Signaling**
Beverly Chilton and Aveline Hewetson

2. **Alterations in cAMP-Mediated Signaling and Their Role in the Pathophysiology of Dilated Cardiomyopathy**
Matthew A. Movsesian and Michael R. Bristow

3. **Corpus Luteum Development: Lessons from Genetic Models in Mice**
Anne Bachelot and Nadine Binart

4. **Comparative Developmental Biology of the Mammalian Uterus**
Thomas E. Spencer, Kanako Hayashi, Jianbo Hu, and Karen D. Carpenter

5. **Sarcopenia of Aging and Its Metabolic Impact**
Helen Karakelides and K. Sreekumaran Nair

6. **Chemokine Receptor CXCR3: An Unexpected Enigma**
Liping Liu, Melissa K. Callahan, DeRen Huang, and Richard M. Ransohoff

7. **Assembly and Signaling of Adhesion Complexes**
Jorge L. Sepulveda, Vasiliki Gkretsi, and Chuanyue Wu

8. **Signaling Mechanisms of Higher Plant Photoreceptors: A Structure-Function Perspective**
Haiyang Wang

9. **Initial Failure in Myoblast Transplantation Therapy Has Led the Way Toward the Isolation of Muscle Stem Cells: Potential for Tissue Regeneration**
Kenneth Urish, Yasunari Kanda, and Johnny Huard

10. **Role of 14-3-3 Proteins in Eukaryotic Signaling and Development**
Dawn L. Darling, Jessica Yingling, and Anthony Wynshaw-Boris

Volume 69

1. **Flipping Coins in the Fly Retina**
Tamara Mikeladze-Dvali, Claude Desplan, and Daniela Pistillo

2. **Unraveling the Molecular Pathways That Regulate Early Telencephalon Development**
Jean M. Hébert

3. **Glia–Neuron Interactions in Nervous System Function and Development**
Shai Shaham

4. **The Novel Roles of Glial Cells Revisited: The Contribution of Radial Glia and Astrocytes to Neurogenesis**
Tetsuji Mori, Annalisa Buffo, and Magdalena Götz

5. **Classical Embryological Studies and Modern Genetic Analysis of Midbrain and Cerebellum Development**
Mark Zervas, Sandra Blaess, and Alexandra L. Joyner

6. **Brain Development and Susceptibility to Damage; Ion Levels and Movements**
Maria Erecinska, Shobha Cherian, and Ian A. Silver

7. **Thinking about Visual Behavior; Learning about Photoreceptor Function**
Kwang-Min Choe and Thomas R. Clandinin

8. **Critical Period Mechanisms in Developing Visual Cortex**
Takao K. Hensch

9. **Brawn for Brains: The Role of MEF2 Proteins in the Developing Nervous System**
Aryaman K. Shalizi and Azad Bonni

10. **Mechanisms of Axon Guidance in the Developing Nervous System**
Céline Plachez and Linda J. Richards

Volume 70

1. **Magnetic Resonance Imaging: Utility as a Molecular Imaging Modality**
James P. Basilion, Susan Yeon, and Rene Botnar

2. **Magnetic Resonance Imaging Contrast Agents in the Study of Development**
Angelique Louie

3. **$^{1}H/^{19}F$ Magnetic Resonance Molecular Imaging with Perfluorocarbon Nanoparticles**
Gregory M. Lanza, Patrick M. Winter, Anne M. Neubauer, Shelton D. Caruthers, Franklin D. Hockett, and Samuel A. Wickline

4. **Loss of Cell Ion Homeostasis and Cell Viability in the Brain: What Sodium MRI Can Tell Us**
Fernando E. Boada, George LaVerde, Charles Jungreis, Edwin Nemoto, Costin Tanase, and Ileana Hancu

5. **Quantum Dot Surfaces for Use In Vivo and In Vitro**
Byron Ballou

6. **In Vivo Cell Biology of Cancer Cells Visualized with Fluorescent Proteins**
Robert M. Hoffman

7. **Modulation of Tracer Accumulation in Malignant Tumors: Gene Expression, Gene Transfer, and Phage Display**
Uwe Haberkorn

8. **Amyloid Imaging: From Benchtop to Bedside**
Chungying Wu, Victor W. Pike, and Yanming Wang

9. **In Vivo Imaging of Autoimmune Disease in Model Systems**
Eric T. Ahrens and Penelope A. Morel

Volume 71

1. **The Choroid Plexus-Cerebrospinal Fluid System: From Development to Aging**
Zoran B. Redzic, Jane E. Preston, John A. Duncan, Adam Chodobski, and Joanna Szmydynger-Chodobska

2. **Zebrafish Genetics and Formation of Embryonic Vasculature**
Tao P. Zhong

3. **Leaf Senescence: Signals, Execution, and Regulation**
Yongfeng Guo and Susheng Gan

4. **Muscle Stem Cells and Regenerative Myogenesis**
Iain W. McKinnell, Gianni Parise, and Michael A. Rudnicki

5. **Gene Regulation in Spermatogenesis**
James A. MacLean II and Miles F. Wilkinson

6. **Modeling Age-Related Diseases in Drosophila: Can this Fly?**
Kinga Michno, Diana van de Hoef, Hong Wu, and Gabrielle L. Boulianne

7. **Cell Death and Organ Development in Plants**
Hilary j. Rogers

8. **The Blood-Testis Barrier: Its Biology, Regulation, and Physiological Role in Spermatogenesis**
Ching-Hang Wong and C. Yan Cheng

9. **Angiogenic Factors in the Pathogenesis of Preeclampsia**
Hai-Tao Yuan, David Haig, and S. Ananth Karumanchi

Volume 72

1. **Defending the Zygote: Search for the Ancestral Animal Block to Polyspermy**
Julian L. Wong and Gary M. Wessel

2. **Dishevelled: A Mobile Scaffold Catalyzing Development**
Craig C. Malbon and Hsien-yu Wang

3. **Sensory Organs: Making and Breaking the Pre-Placodal Region**
Andrew P. Bailey and Andrea Streit

4. **Regulation of Hepatocyte Cell Cycle Progression and Differentiation by Type I Collagen Structure**
Linda K. Hansen, Joshua Wilhelm, and John T. Fassett

5. **Engineering Stem Cells into Organs: Topobiological Transformations Demonstrated by Beak, Feather, and Other Ectodermal Organ Morphogenesis**
Cheng-Ming Chuong, Ping Wu, Maksim Plikus, Ting-Xin Jiang, and Randall Bruce Widelitz

6. **Fur Seal Adaptations to Lactation: Insights into Mammary Gland Function**
Julie A. Sharp, Kylie N. Cane, Christophe Lefevre, John P. Y. Arnould, and Kevin R. Nicholas

Volume 73

1. **The Molecular Origins of Species-Specific Facial Pattern**
Samantha A. Brugmann, Minal D. Tapadia, and Jill A. Helms

2. **Molecular Bases of the Regulation of Bone Remodeling by the Canonical Wnt Signaling Pathway**
Donald A. Glass II and Gerard Karsenty

3. **Calcium Sensing Receptors and Calcium Oscillations: Calcium as a First Messenger**
Gerda E. Breitwieser

4. **Signal Relay During the Life Cycle of Dictyostelium**
Dana C. Mahadeo and Carole A. Parent

5. **Biological Principles for Ex Vivo Adult Stem Cell Expansion**
Jean-François Paré and James L. Sherley

6. **Histone Deacetylation as a Target for Radiosensitization**
David Cerna, Kevin Camphausen, and Philip J. Tofilon

7. **Chaperone-Mediated Autophagy in Aging and Disease**
Ashish C. Massey, Cong Zhang, and Ana Maria Cuervo

8. **Extracellular Matrix Macroassembly Dynamics in Early Vertebrate Embryos**
Andras Czirok, Evan A. Zamir, Michael B. Filla, Charles D. Little, and Brenda J. Rongish

Volume 74

1. **Membrane Origin for Autophagy**
Fulvio Reggiori

2. **Chromatin Assembly with H3 Histones: Full Throttle Down Multiple Pathways**
Brian E. Schwartz and Kami Ahmad

3. **Protein–Protein Interactions of the Developing Enamel Matrix**
John D. Bartlett, Bernhard Ganss, Michel Goldberg, Janet Moradian-Oldak, Michael L. Paine, Malcolm L. Snead, Xin Wen, Shane N. White, and Yan L. Zhou

4. **Stem and Progenitor Cells in the Formation of the Pulmonary Vasculature**
Kimberly A. Fisher and Ross S. Summer

5. **Mechanisms of Disordered Granulopoiesis in Congenital Neutropenia**
David S. Grenda and Daniel C. Link

6. **Social Dominance and Serotonin Receptor Genes in Crayfish**
Donald H. Edwards and Nadja Spitzer

7. Transplantation of Undifferentiated, Bone Marrow-Derived Stem Cells

Karen Ann Pauwelyn and Catherine M. Verfaillie

8. The Development and Evolution of Division of Labor and Foraging Specialization in a Social Insect (Apis mellifera L.)

Robert E. Page Jr., Ricarda Scheiner, Joachim Erber, and Gro V. Amdam

Volume 75

1. Dynamics of Assembly and Reorganization of Extracellular Matrix Proteins

Sarah L. Dallas, Qian Chen, and Pitchumani Sivakumar

2. Selective Neuronal Degeneration in Huntington's Disease

Catherine M. Cowan and Lynn A. Raymond

3. RNAi Therapy for Neurodegenerative Diseases

Ryan L. Boudreau and Beverly L. Davidson

4. Fibrillins: From Biogenesis of Microfibrils to Signaling Functions

Dirk Hubmacher, Kerstin Tiedemann, and Dieter P. Reinhardt

5. Proteasomes from Structure to Function: Perspectives from Archaea

Julie A. Maupin-Furlow, Matthew A. Humbard, P. Aaron Kirkland, Wei Li, Christopher J. Reuter, Amy J. Wright, and G. Zhou

6. The Cytomatrix as a Cooperative System of Macromolecular and Water Networks

V. A. Shepherd

7. Intracellular Targeting of Phosphodiesterase-4 Underpins Compartmentalized cAMP Signaling

Martin J. Lynch, Elaine V. Hill, and Miles D. Houslay

Volume 76

1. BMP Signaling in the Cartilage Growth Plate

Robert Pogue and Karen Lyons

2. The CLIP-170 Orthologue Bik1p and Positioning the Mitotic Spindle in Yeast

Rita K. Miller, Sonia D'Silva, Jeffrey K. Moore, and Holly V. Goodson

3. **Aggregate-Prone Proteins Are Cleared from the Cytosol by Autophagy: Therapeutic Implications**
Andrea Williams, Luca Jahreiss, Sovan Sarkar, Shinji Saiki, Fiona M. Menzies, Brinda Ravikumar, and David C. Rubinsztein

4. **Wnt Signaling: A Key Regulator of Bone Mass**
Roland Baron, Georges Rawadi, and Sergio Roman-Roman

5. **Eukaryotic DNA Replication in a Chromatin Context**
Angel P. Tabancay, Jr. and Susan L. Forsburg

6. **The Regulatory Network Controlling the Proliferation–Meiotic Entry Decision in the Caenorhabditis elegans Germ Line**
Dave Hansen and Tim Schedl

7. **Regulation of Angiogenesis by Hypoxia and Hypoxia-Inducible Factors**
Michele M. Hickey and M. Celeste Simon

Volume 77

1. **The Role of the Mitochondrion in Sperm Function: Is There a Place for Oxidative Phosphorylation or Is this a Purely Glycolytic Process?**
Eduardo Ruiz-Pesini, Carmen Díez-Sánchez, Manuel José López-Pérez, and José Antonio Enriquez

2. **The Role of Mitochondrial Function in the Oocyte and Embryo**
Rémi Dumollard, Michael Duchen, and John Carroll

3. **Mitochondrial DNA in the Oocyte and the Developing Embryo**
Pascale May-Panloup, Marie-Françoise Chretien, Yves Malthiery, and Pascal Reynier

4. **Mitochondrial DNA and the Mammalian Oocyte**
Eric A. Shoubridge and Timothy Wai

5. **Mitochondrial Disease—Its Impact, Etiology, and Pathology**
R. McFarland, R. W. Taylor, and D. M. Turnbull

6. **Cybrid Models of mtDNA Disease and Transmission, from Cells to Mice**
Ian A. Trounce and Carl A. Pinkert

7. **The Use of Micromanipulation Methods as a Tool to Prevention of Transmission of Mutated Mitochondrial DNA**
Helena Fulka and Josef Fulka, Jr.

8. **Difficulties and Possible Solutions in the Genetic Management of mtDNA Disease in the Preimplantation Embryo**
J. Poulton, P. Oakeshott, and S. Kennedy

9. **Impact of Assisted Reproductive Techniques: A Mitochondrial Perspective from the Cytoplasmic Transplantation**
A. J. Harvey, T. C. Gibson, T. M. Quebedeaux, and C. A. Brenner

10. **Nuclear Transfer: Preservation of a Nuclear Genome at the Expense of Its Associated mtDNA Genome(s)**
Emma J. Bowles, Keith H. S. Campbell, and Justin C. St. John

Volume 78

1. **Contribution of Membrane Mucins to Tumor Progression Through Modulation of Cellular Growth Signaling Pathways**
Kermit L. Carraway III, Melanie Funes, Heather C. Workman, and Colleen Sweeney

2. **Regulation of the Epithelial Na^+ Channel by Peptidases**
Carole Planés and George H. Caughey

3. **Advances in Defining Regulators of Cementum Development and Periodontal Regeneration**
Brian L. Foster, Tracy E. Popowics, Hanson K. Fong, and Martha J. Somerman

4. **Anabolic Agents and the Bone Morphogenetic Protein Pathway**
I. R. Garrett

5. **The Role of Mammalian Circadian Proteins in Normal Physiology and Genotoxic Stress Responses**
Roman V. Kondratov, Victoria Y. Gorbacheva, and Marina P. Antoch

6. **Autophagy and Cell Death**
Devrim Gozuacik and Adi Kimchi

Volume 79

1. **The Development of Synovial Joints**
   I. M. Khan, S. N. Redman, R. Williams, G. P. Dowthwaite, S. F. Oldfield, and C. W. Archer
2. **Development of a Sexually Differentiated Behavior and Its Underlying CNS Arousal Functions**
   Lee-Ming Kow, Cristina Florea, Marlene Schwanzel-Fukuda, Nino Devidze, Hosein Kami Kia, Anna Lee, Jin Zhou, David MacLaughlin, Patricia Donahoe, and Donald Pfaff
3. **Phosphodiesterases Regulate Airway Smooth Muscle Function in Health and Disease**
   Vera P. Krymskaya and Reynold A. Panettieri, Jr.
4. **Role of Astrocytes in Matching Blood Flow to Neuronal Activity**
   Danica Jakovcevic and David R. Harder
5. **Elastin-Elastases and Inflamm-Aging**
   Frank Antonicelli, Georges Bellon, Laurent Debelle, and William Hornebeck
6. **A Phylogenetic Approach to Mapping Cell Fate**
   Stephen J. Salipante and Marshall S. Horwitz

Volume 80

1. **Similarities Between Angiogenesis and Neural Development: What Small Animal Models Can Tell Us**
   Serena Zacchigna, Carmen Ruiz de Almodovar, and Peter Carmeliet
2. **Junction Restructuring and Spermatogenesis: The Biology, Regulation, and Implication in Male Contraceptive Development**
   Helen H. N. Yan, Dolores D. Mruk, and C. Yan Cheng
3. **Substrates of the Methionine Sulfoxide Reductase System and Their Physiological Relevance**
   Derek B. Oien and Jackob Moskovitz
4. **Organic Anion-Transporting Polypeptides at the Blood–Brain and Blood–Cerebrospinal Fluid Barriers**
   Daniel E. Westholm, Jon N. Rumbley, David R. Salo, Timothy P. Rich, and Grant W. Anderson

5. **Mechanisms and Evolution of Environmental Responses in Caenorhabditis elegans**
Christian Braendle, Josselin Milloz, and Marie-Anne Félix

6. **Molluscan Shell Proteins: Primary Structure, Origin, and Evolution**
Frédéric Marin, Gilles Luquet, Benjamin Marie, and Davorin Medakovic

7. **Pathophysiology of the Blood–Brain Barrier: Animal Models and Methods**
Brian T. Hawkins and Richard D. Egleton

8. **Genetic Manipulation of Megakaryocytes to Study Platelet Function**
Jun Liu, Jan DeNofrio, Weiping Yuan, Zhengyan Wang, Andrew W. McFadden, and Leslie V. Parise

9. **Genetics and Epigenetics of the Multifunctional Protein CTCF**
Galina N. Filippova

Volume 81

1. **Models of Biological Pattern Formation: From Elementary Steps to the Organization of Embryonic Axes**
Hans Meinhardt

2. **Robustness of Embryonic Spatial Patterning in Drosophila Melanogaster**
David Umulis, Michael B. O'Connor, and Hans G. Othmer

3. **Integrating Morphogenesis with Underlying Mechanics and Cell Biology**
Lance A. Davidson

4. **The Mechanisms Underlying Primitive Streak Formation in the Chick Embryo**
Manli Chuai and Cornelis J. Weijer

5. **Grid-Free Models of Multicellular Systems, with an Application to Large-Scale Vortices Accompanying Primitive Streak Formation**
T. J. Newman

6. **Mathematical Models for Somite Formation**
Ruth E. Baker, Santiago Schnell, and Philip K. Maini

7. Coordinated Action of N-CAM, N-cadherin, EphA4, and ephrinB2 Translates Genetic Prepatterns into Structure during Somitogenesis in Chick

James A. Glazier, Ying Zhang, Maciej Swat, Benjamin Zaitlen, and Santiago Schnell

8. Branched Organs: Mechanics of Morphogenesis by Multiple Mechanisms

Sharon R. Lubkin

9. Multicellular Sprouting during Vasculogenesis

Andras Czirok, Evan A. Zamir, Andras Szabo, and Charles D. Little

10. Modelling Lung Branching Morphogenesis

Takashi Miura

11. Multiscale Models for Vertebrate Limb Development

Stuart A. Newman, Scott Christley, Tilmann Glimm, H. G. E. Hentschel, Bogdan Kazmierczak, Yong-Tao Zhang, Jianfeng Zhu, and Mark Alber

12. Tooth Morphogenesis in vivo, in vitro and in silico

Isaac Salazar-Ciudad

13. Cell Mechanics with a 3D Kinetic and Dynamic Weighted Delaunay-Triangulation

Michael Meyer-Hermann

14. Cellular Automata as Microscopic Models of Cell Migration in Heterogeneous Environments

H. Hatzikirou and A. Deutsch

15. Multiscale Modeling of Biological Pattern Formation

Ramon Grima

16. Relating Biophysical Properties Across Scales

Elijah Flenner, Francoise Marga, Adrian Neagu, Ioan Kosztin, and Gabor Forgacs

17. Complex Multicellular Systems and Immune Competition: New Paradigms Looking for a Mathematical Theory

N. Bellomo and G. Forni

Volume 82

1. **Ontogeny of Erythropoiesis in the Mammalian Embryo**
Kathleen McGrath and James Palis

2. **The Erythroblastic Island**
Deepa Manwani and James J. Bieker

3. **Epigenetic Control of Complex Loci During Erythropoiesis**
Ryan J. Wozniak and Emery H. Bresnick

4. **The Role of the Epigenetic Signal, DNA Methylation, in Gene Regulation During Erythroid Development**
Gordon D. Ginder, Merlin N. Gnanapragasam, and Omar Y. Mian

5. **Three-Dimensional Organization of Gene Expression in Erythroid Cells**
Wouter de Laat, Petra Klous, Jurgen Kooren, Daan Noordermeer, Robert-Jan Palstra, Marieke Simonis, Erik Splinter, and Frank Grosveld

6. **Iron Homeostasis and Erythropoiesis**
Diedra M. Wrighting and Nancy C. Andrews

7. **Effects of Nitric Oxide on Red Blood Cell Development and Phenotype**
Vladan P. Čokić and Alan N. Schechter

8. **Diamond Blackfan Anemia: A Disorder of Red Blood Cell Development**
Steven R. Ellis and Jeffrey M. Lipton

Volume 83

1. **Somatic Sexual Differentiation in Caenorhabditis elegans**
Jennifer Ross Wolff and David Zarkower

2. **Sex Determination in the Caenorhabditis elegans Germ Line**
Ronald E. Ellis

3. **The Creation of Sexual Dimorphism in the Drosophila Soma**
Nicole Camara, Cale Whitworth, and Mark Van Doren

4. **Drosophila Germline Sex Determination: Integration of Germline Autonomous Cues and Somatic Signals**
Leonie U. Hempel, Rasika Kalamegham, John E. Smith III, and Brian Oliver

5. **Sexual Development of the Soma in the Mouse**
Danielle M. Maatouk and Blanche Capel

6. **Development of Germ Cells in the Mouse**
Gabriela Durcova-Hills and Blanche Capel

7. **The Neuroendocrine Control of Sex-Specific Behavior in Vertebrates: Lessons from Mammals and Birds**
Margaret M. McCarthy and Gregory F. Ball

Volume 84

1. **Modeling Neural Tube Defects in the Mouse**
Irene E. Zohn and Anjali A. Sarkar

2. **The Etiopathogenesis of Cleft Lip and Cleft Palate: Usefulness and Caveats of Mouse Models**
Amel Gritli-Linde

3. **Murine Models of Holoprosencephaly**
Karen A. Schachter and Robert S. Krauss

4. **Mouse Models of Congenital Cardiovascular Disease**
Anne Moon

5. **Modeling Ciliopathies: Primary Cilia in Development and Disease**
Robyn J. Quinlan, Jonathan L. Tobin, and Philip L. Beales

6. **Mouse Models of Polycystic Kidney Disease**
Patricia D. Wilson

7. **Fraying at the Edge: Mouse Models of Diseases Resulting from Defects at the Nuclear Periphery**
Tatiana V. Cohen and Colin L. Stewart

8. **Mouse Models for Human Hereditary Deafness**
Michel Leibovici, Saaid Safieddine, and Christine Petit

9. **The Value of Mammalian Models for Duchenne Muscular Dystrophy in Developing Therapeutic Strategies**
Glen B. Banks and Jeffrey S. Chamberlain

Volume 85

1. **Basal Bodies: Platforms for Building Cilia**
Wallace F. Marshall

2. **Intraflagellar Transport (IFT): Role in Ciliary Assembly, Resorption and Signalling**
Lotte B. Pedersen and Joel L. Rosenbaum

3. **How Did the Cilium Evolve?**
Peter Satir, David R. Mitchell, and Gáspár Jékely

4. **Ciliary Tubulin and Its Post-Translational Modifications**
Jacek Gaertig and Dorota Wloga

5. **Targeting Proteins to the Ciliary Membrane**
Gregory J. Pazour and Robert A. Bloodgood

6. **Cilia: Multifunctional Organelles at the Center of Vertebrate Left–Right Asymmetry**
Basudha Basu and Martina Brueckner

7. **Ciliary Function and Wnt Signal Modulation**
Jantje M. Gerdes and Nicholas Katsanis

8. **Primary Cilia in Planar Cell Polarity Regulation of the Inner Ear**
Chonnettia Jones and Ping Chen

9. **The Primary Cilium: At the Crossroads of Mammalian Hedgehog Signaling**
Sunny Y. Wong and Jeremy F. Reiter

10. **The Primary Cilium Coordinates Signaling Pathways in Cell Cycle Control and Migration During Development and Tissue Repair**
Søren T. Christensen, Stine F. Pedersen, Peter Satir, Iben R. Veland, and Linda Schneider

11. **Cilia Involvement in Patterning and Maintenance of the Skeleton**
Courtney J. Haycraft and Rosa Serra

12. **Olfactory Cilia: Our Direct Neuronal Connection to the External World**
Dyke P. McEwen, Paul M. Jenkins, and Jeffrey R. Martens

13. **Ciliary Dysfunction in Developmental Abnormalities and Diseases**
Neeraj Sharma, Nicolas F. Berbari, and Bradley K. Yoder

Volume 86

1. **Gene Regulatory Networks in Neural Crest Development and Evolution**
   Natalya Nikitina, Tatjana Sauka-Spengler, and Marianne Bronner-Fraser
2. **Evolution of Vertebrate Cartilage Development**
   GuangJun Zhang, B. Frank Eames, and Martin J. Cohn
3. **Caenorhabditis Nematodes as a Model for the Adaptive Evolution of Germ Cells**
   Eric S. Haag
4. **New Model Systems for the Study of Developmental Evolution in Plants**
   Elena M. Kramer
5. **Patterning the Spiralian Embryo: Insights from *Ilyanassa***
   J. David Lambert
6. **The Origin and Diversification of Complex Traits Through Micro- and Macroevolution of Development: Insights from Horned Beetles**
   Armin P. Moczek
7. **Axis Formation and the Rapid Evolutionary Transformation of Larval Form**
   Rudolf A. Raff and Margaret Snoke Smith
8. **Evolution and Development in the Cavefish *Astyanax***
   William R. Jeffery

Volume 87

1. **Theoretical Models of Neural Circuit Development**
   Hugh D. Simpson, Duncan Mortimer, and Geoffrey j. Goodhill
2. **Synapse Formation in Developing Neural Circuits**
   Daniel A. Colón-Ramos
3. **The Developmental Integration of Cortical Interneurons into a Functional Network**
   Renata Batista–Brito and Gord Fishell

4. **Transcriptional Networks in the Early Development of Sensory–Motor Circuits**
Jeremy S. Dasen

5. **Development of Neural Circuits in the Adult Hippocampus**
Yan Li, Yangling Mu, and Fred H. Gage

6. **Looking Beyond Development: Maintaining Nervous System Architecture**
Claire Bénard and Oliver Hobert

Volume 88

1. **The Bithorax Complex of Drosophila: An Exceptional Hox Cluster**
Robert K. Maeda and François Karch

2. **Evolution of the Hox Gene Complex from an Evolutionary Ground State**
Walter J. Gehring, Urs Kloter, and Hiroshi Suga

3. **Hox Specificity: Unique Roles for Cofactors and Collaborators**
Richard S. Mann, Katherine M. Lelli, and Rohit Joshi

4. **Hox Genes and Segmentation of the Vertebrate Hindbrain**
Stefan Tümpel, Leanne M. Wiedemann, and Robb Krumlauf

5. **Hox Genes in Neural Patterning and Circuit Formation in the Mouse Hindbrain**
Yuichi Narita and Filippo M. Rijli

6. **Hox Networks and the Origins of Motor Neuron Diversity**
Jeremy S. Dasen and Thomas M. Jessell

7. **Establishment of Hox Vertebral Identities in the Embryonic Spine Precursors**
Tadahiro Iimura, Nicolas Denans, and Olivier Pourquié

8. **Hox, Cdx, and Anteroposterior Patterning in the Mouse Embryo**
Teddy Young and Jacqueline Deschamps

9. **Hox Genes and Vertebrate Axial Pattern**
Deneen M. Wellik

Volume 89

1. **Intercellular Adhesion in Morphogenesis: Molecular and Biophysical Considerations**
Nicolas Borghi and W. James Nelson

2. **Remodeling of the Adherens Junctions During Morphogenesis**
Tamako Nishimura and Masatoshi Takeichi

3. **How the Cytoskeleton Helps Build the Embryonic Body Plan: Models of Morphogenesis from *Drosophila***
Tony J. C. Harris, Jessica K. Sawyer, and Mark Peifer

4. **Cell Topology, Geometry, and Morphogenesis in Proliferating Epithelia**
William T. Gibson and Matthew C. Gibson

5. **Principles of *Drosophila* Eye Differentiation**
Ross Cagan

6. **Cellular and Molecular Mechanisms Underlying the Formation of Biological Tubes**
Magdalena M. Baer, Helene Chanut-Delalande, and Markus Affolter

7. **Convergence and Extension Movements During Vertebrate Gastrulation**
Chunyue Yin, Brian Ciruna, and Lilianna Solnica-Krezel

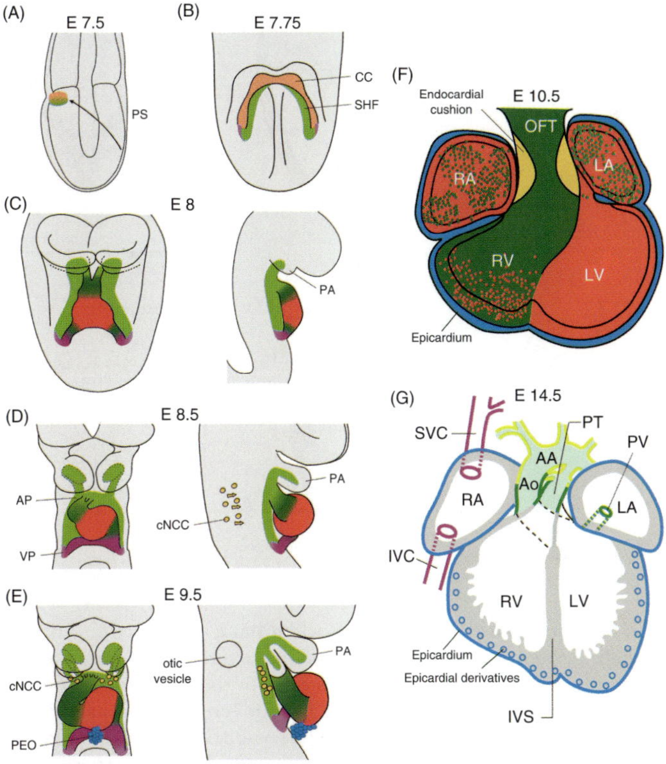

Stéphane D. Vincent and Margaret E. Buckingham, Figure 1.1 (A) Migration of cells anteriorly from the primitive streak (PS). (B) Formation of the cardiac crescent (CC), with the second heart field (SHF) lying medial to it. (C–E) Front (left) and lateral (right) views of the heart tube as it begins to loop with contributions of cardiac neural crest cells (cNCC), which migrate from the pharyngeal arches (PA) to the arterial pole (AP). The proepicardial organ (PEO) forms in the vicinity of the venous pole (VP). (F) The looped heart tube, with the cardiac compartments—OFT, outflow tract; RA, right atrium; LA, left atrium; RV, right ventricle; LV, left ventricle. (G) The mature heart which has undergone septation—IVS, interventricular septum; AA, aortic arch; Ao, aorta; PT, pulmonary trunk; PV, pulmonary vein; SVC, superior caval vein; IVC, inferior caval vein. The first heart field (FHF) and its myocardial contribution are shown in red, the SHF and its derivatives in dark green (myocardium) and pale green (vascular endothelial cells), cNCC in yellow (vascular smooth muscle of the AA, endocardial cushions), and PEO derivatives in blue.

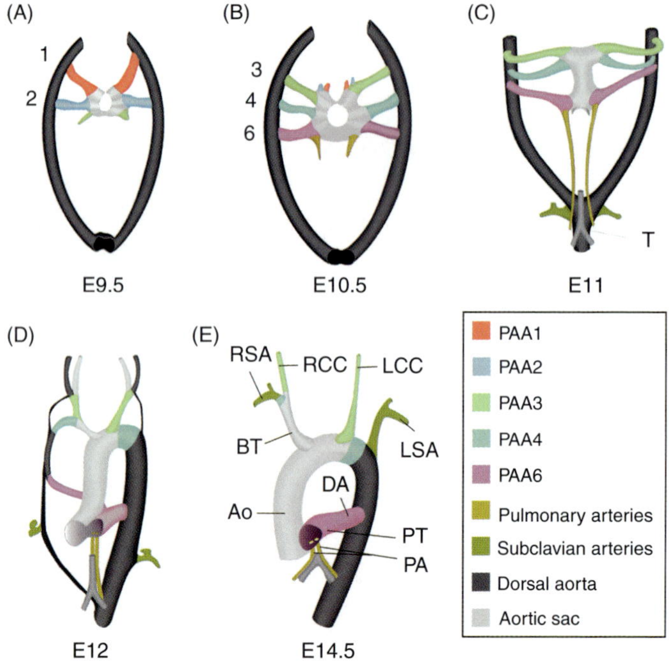

Stéphane D. Vincent and Margaret E. Buckingham, Figure 1.3 Pharyngeal arch artery remodeling. Ventral representation of the arterial network connected to the heart as development proceeds. The arterial pole of the heart (not represented here) feeds into the aortic sac (grey). The heart is connected to the paired dorsal aortas (black) via the artery of each of the five pharyngeal arches; the pharyngeal arch arteries (PAAs). During development, the pharyngeal arches and their arteries are symmetrically formed following a rostro-caudal, temporal gradient. (A) At E9.5, only the PAA1 (orange) and PAA2 (blue) are connected to the aortic sac. (B) At E10.5, the PAA1 and PAA2 are no longer connected directly to the heart but form the capillary beds of their respective pharyngeal arches. The heart is now connected via PAA3 (light green), PAA4 (dark green), and PAA6 (purple). (C) At E11, the network is still symmetric and the pulmonary arteries (dark yellow) are clearly visible. (E) From E11.5, remodeling of the PAAs is initiated via the increase of blood flow in the left PAA6. This results in the stabilization of the aortic arch on the left side at the expense of the right side. Segments of the dorsal aortas are degenerating, leading to the individualization of the future common carotid arteries (RCC, right common carotid and LCC, left common carotid). (D) At E14.5, the left PAA4 contributes to the segment of the aortic arch between the LCC and the left subclavian artery (in bronze), whereas the right PAA4 will form a segment that connects the RSA to the brachycephalic trunk (BT—remains of the right aortic arch). The BT is also connected to the derivative of the right PAA3, the right common carotid (RCC). The right PAA6 is not maintained, whereas the left PAA6 contributes to the *ductus arteriosus* (DA). The DA is an embryonic shunt that connects the (left) dorsal aorta to the pulmonary trunk. At birth, the DA closes, allowing the establishment of the pulmonary and systemic blood circulations (adapted from Kaufman and Bard (1999); Ao, aorta; PT, pulmonary trunk; T, trachea; PA, pulmonary arteries).

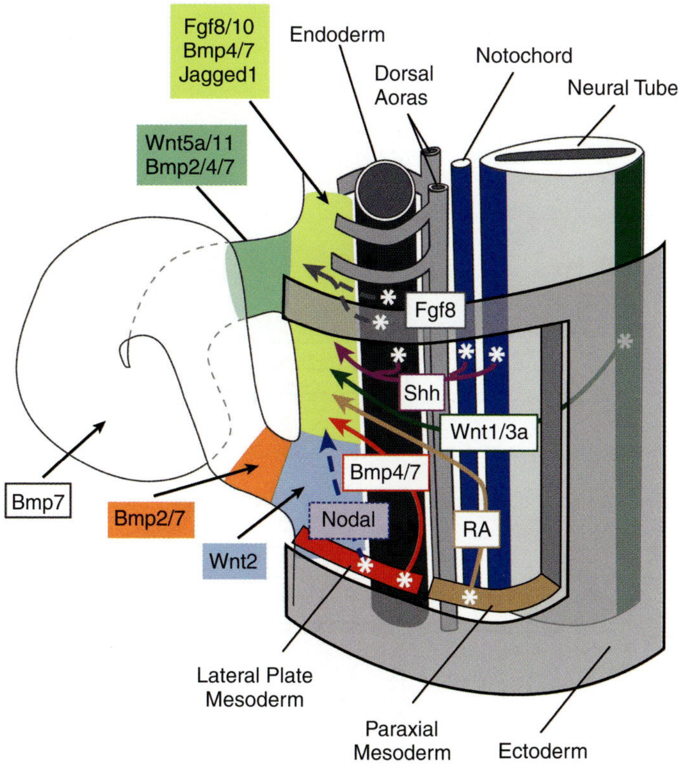

Stéphane D. Vincent and Margaret E. Buckingham, Figure 1.5 The SHF is a target of multiple signals. Schematic representation of a left-sided view of the heart region. The SHF itself is a source of signals: the Notch ligand Jagged1, is involved in the activation of *Fgf8*, but Notch signaling also interferes with canonical Wnt signaling, inhibiting progenitor proliferation. Fgf8 and Fgf10 are important for the promotion of SHF proliferation, whereas Bmp4/7 is required for the survival of the cNCCs. Bmp2, which is repressed by Nkx2-5 and also by canonical Wnt signaling, inhibits SHF proliferation. In the posterior SHF, Wnt2 controls the contribution of venous pole progenitors. Fgf8 also comes from the pharyngeal endoderm and ectoderm (grey arrow). The outflow tract is also a source of signals for the SHF: Bmp2 has been shown to be required for the deployment of SHF cells. Non-canonical Wnt11 and Wnt5a do not affect the SHF, but control outflow tract maturation. The midline is another signaling source. Shh (purple arrow), from the endoderm (in dark grey) (but also from the notochord and the floorplate, in blue), affects arterial pole formation, probably through an effect on SHF proliferation, and is required for the formation of the atrial septum at the venous pole. Also coming from the midline, canonical Wnts (Wnt1, Wnt3a, green arrow) from the dorsal neural tube (green domain) are important for the maintenance of proliferating progenitors within the SHF and inhibition of differentiation. On the contrary, lateral signals such as Bmp4 (red arrow) from the lateral plate mesoderm (red domain) promote cardiac differentiation. The paraxial mesoderm (posterior to the SHF, in brown) is a source of retinoic acid (RA, brown arrow) that is important in the early stages of cardiogenesis for anterior/posterior patterning and limiting the posterior boundary of the SHF. On the left side, and only between E8 and E8.5, Nodal (dashed blue arrow) activates *Pitx2c* asymmetrically on the left side of the SHF. Asymmetric *Pitx2c* expression is maintained independently of Nodal and affects both the arterial and venous poles of the heart (see Section 6.2).

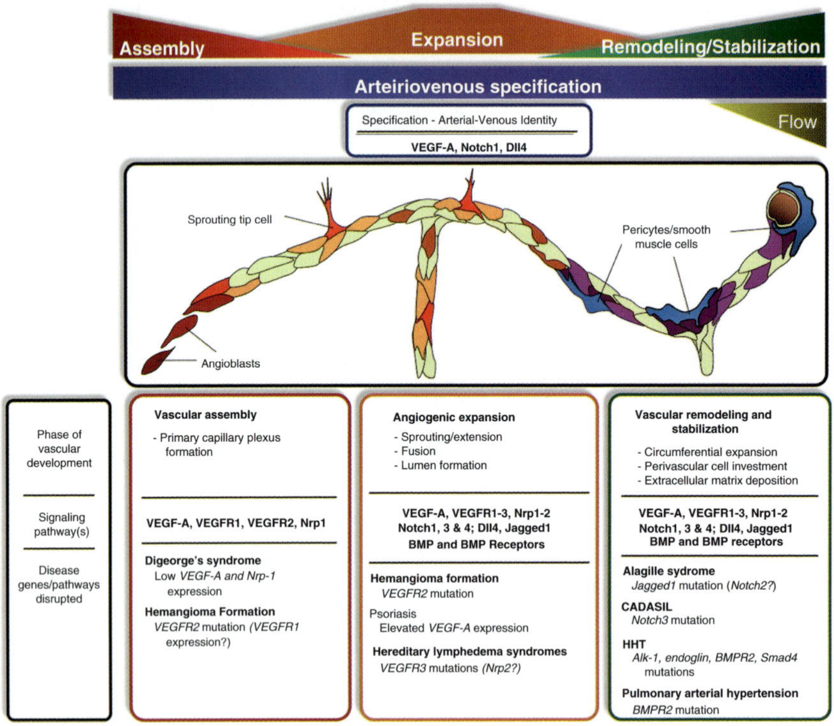

John C. Chappell and Victoria L. Bautch, Figure 2.1 Overview of blood vessel formation. The morphological events constituting vessel assembly, expansion, and remodeling intensify in activity and then attenuate during phase transitions, while arteriovenous specification events likely occur throughout the vascular development program. The different stages of vascular development utilize overlapping molecular pathways for these events, and endothelial phenotypic heterogeneity (represented schematically by color variation, i.e., shades of red, brown, and violet) is an essential aspect of each stage. Genetic lesions in human genes that disrupt these pathways are associated with specific vascular diseases.

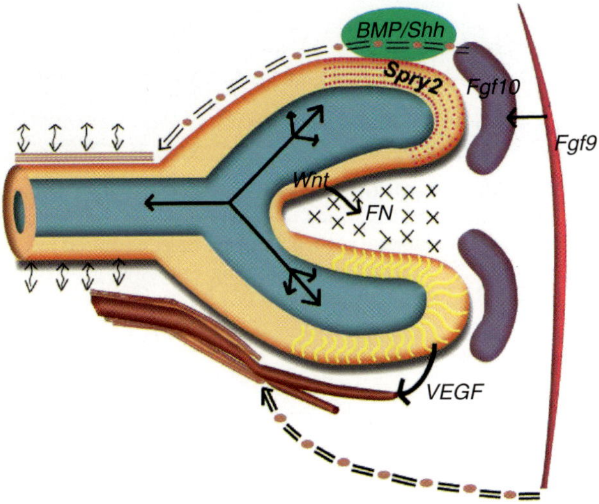

David Warburton *et al.*, Figure 3.4 Schematic illustration of the variety of biochemical and biomechanical regulators of lung growth. The epithelium of the end-bud (yellow) encloses a fluid filled lumen (blue-green) in which oscillatory fluid flows are depicted (solid bi-directional arrows) that result from periodic peristaltic contractions (bi-directional curvilinear arrows) of airway smooth muscle (ASM: sample shown in brown running parallel and above the main epithelial trunk). The ASM derives from the FGF10^{+} precursor pool (seen in purple on the right), and these ASM progenitors are seen becoming more proximal (double-lined "=" at the top of the epithelial outline) relative to the growing epithelium. Examples of key biochemical signaling are given: FGF9 from the mesothelium (far right) regulates the FGF10^{+} mesenchyme (purple), which in turn interacts locally with epithelial Sprouty (SPRY2), BMP4, and Sonic Hedgehog (SHH) signaling (the latter two epithelial signals are shown in green and, due to space constraint, adjacent to the schematized epithelium). Epithelial Wnt signaling regulates fibronectin (FN) elaboration (shown as "xx") in the cleft between epithelial branches. Epithelial VEGF signals to developing vasculature shown at the base of the figure. These vessels attract vascular smooth muscle precursors from the mesothelium (shown as "=" at the base of the figure).

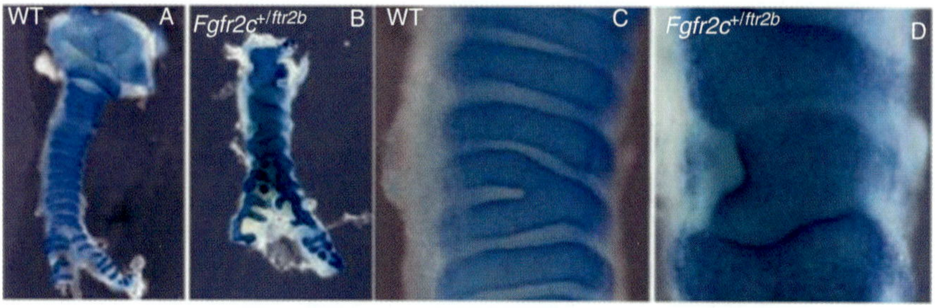

David Warburton *et al.*, Figure 3.6 Excessive mesenchymal FGF signaling leads to overgrowth of tracheal rings. Wild-type and mutant tracheas are stained with Alcian blue. (A) Wild-type trachea at P0 exhibiting regular cartilage rings separated by noncartilaginous mesenchyme; (B) *Fgfr2c*$^{+/Fgfr2b}$ trachea at P0 showing excessive growth of the cartilage with absence of noncartilaginous mesenchyme; (C, D) high magnification of A and B, respectively.

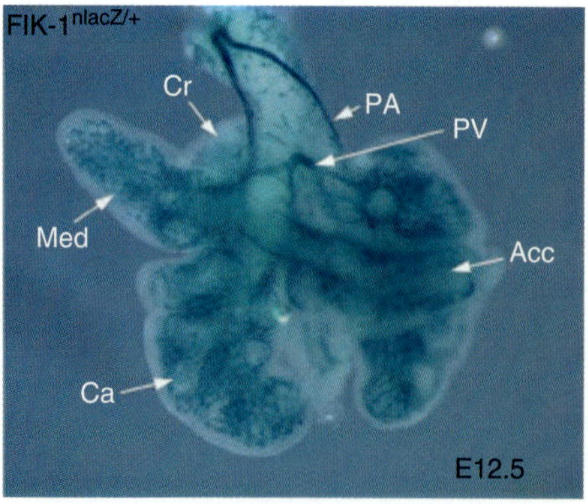

David Warburton *et al.*, Figure 3.7 Vascular endothelial development in E12.5 mouse lung is shown in whole mount as a blue signal resulting from transgenic expression of *Flk1*-β-galactosidase (Flk-1$^{nLacZ/+}$): pulmonary artery (PA); pulmonary vein (PV); cranial lobe (Cr); medial lobe (Med); caudal lobe (Ca); accessory lobe (Acc).

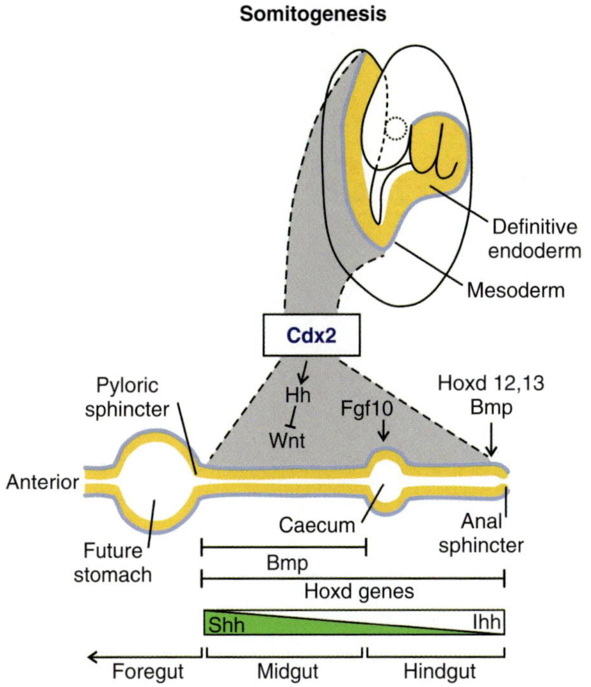

Joan K. Heath, Figure 4.3 Central role of Cdx2 in the regionalization of the primitive gut tube. During somitogenesis, the gut is patterned into regions: the foregut, midgut and hindgut, the later encompassing the cecum. The intestine-specific TF, Cdx2, is expressed throughout the intestinal endoderm, from the junction of the foregut and the midgut to the anus. The activity of Cdx2 choreographs complex patterns of *Hox, Hh*, and *Wnt* gene expression at different positions along the rostrocaudal axis of the developing gut. These patterns are in turn interpreted to stimulate the expression of growth factors such as Fgf10, which induce the formation of conspicuous anatomical features such as the cecum (for Fgf10), and the pyloric and anal sphincters.

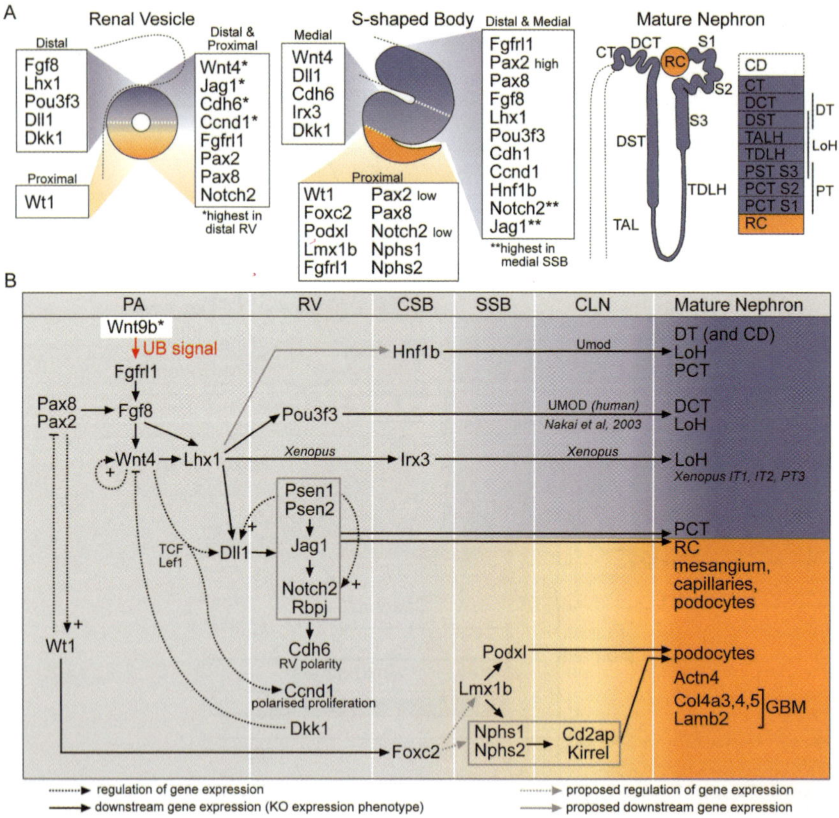

Melissa Little *et al.*, Figure 5.4 Summary of genes involved in nephron segmentation. (A) The expression patterns of polarized genes along the proximal–distal axis of the renal vesicle and SB. The ureteric epithelium is represented by a dotted black line. The mature nephron segments are shown (right); with CD derived from the ureteric epithelium and nephron segments derived from the CM; CT, DCT, distal straight tubule, thin ascending loop of Henle; thin descending loop of Henle; proximal straight tubule, PCT and RC. (B) The transcriptional hierarchy of genes governing nephron patterning. Genes are shown in order of their gene expression from left to right; in PA, RV, CSB, SSB, and capillary loop nephron. The heirarchy of specification within the mature nephron are indicated (right). This has been determined from the genetic analysis of mutant mice, see text (Section 4-5) for details.

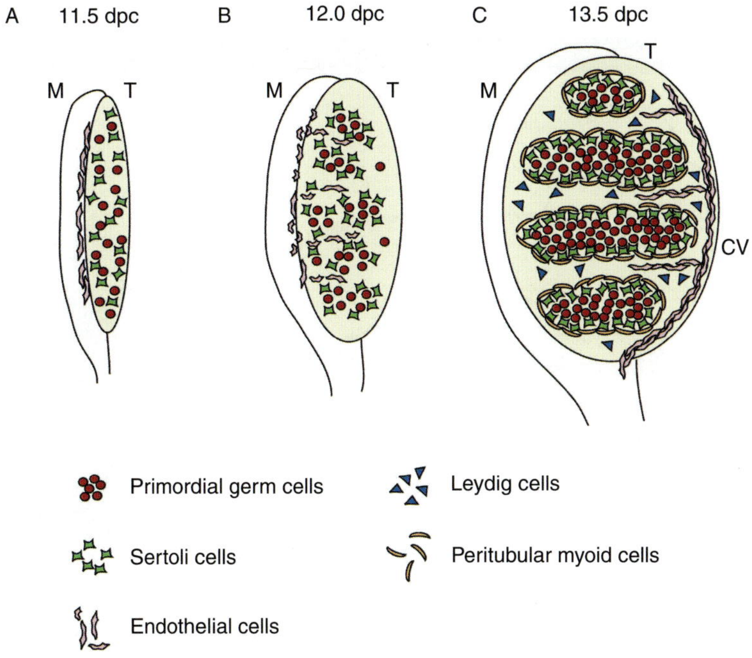

Elanor N. Wainwright and Dagmar Wilhelm, Figure 6.1 Schematic representation of testis development in mouse. (A) At 11.5 dpc SRY is expressed in pre-Sertoli cells (green) that are distributed evenly in between primordial germ cells (red) throughout the genital ridge. A vasculature network within the mesonephros disintegrates and endothelial cells (pink) start to migrate into the gonad. (B) By 12.0 dpc Sertoli cells become polarized and form epithelial aggregates around clusters of germ cells. Endothelial cells partition these clusters into the future testis cords. (C) At 13.5 dpc testis cords are defined with clusters of germ cells surrounded by one layer of Sertoli cells and one layer of peritubular myoid cells (yellow). In the interstitium, Leydig cells (blue) have differentiated to produce testosterone. T, testis; M, mesonephros, CV, coelomic vessel.

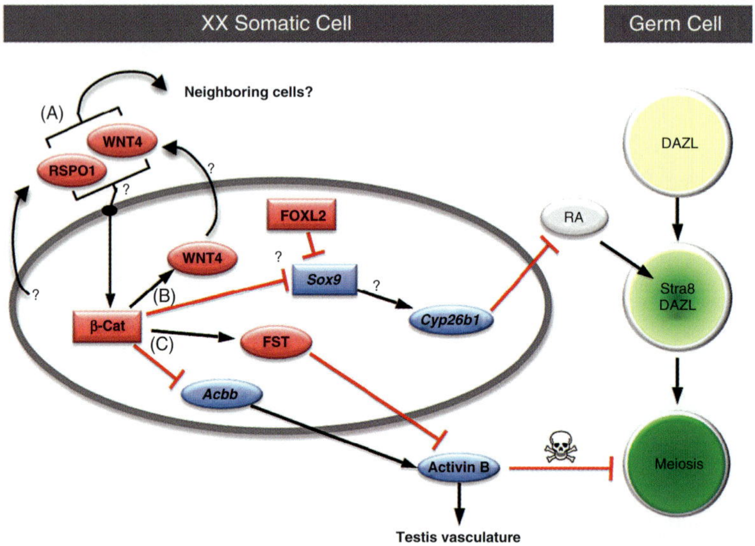

Chia-Feng Liu *et al.*, Figure 7.4 Putative pathways for ovary organogenesis in the mouse embryo. (A) Two somatic cell-derived factors, R-spondin1 (RSPO1) and WNT4, activate synergistically or independently the canonical **β**-catenin (**β**-cat) pathway in XX somatic cells in an autocrine or paracrine manner. (B) **β**-catenin then induces expression of *Wnt4* while suppresses expression of *Sox9* (probably through the action of FOXL2) and its putative downstream target *Cyp26b1*. Without the presence of *Cyp26b1*, RA is not degraded and therefore induces *Stra8* expression and meiosis in DAZL-positive germ cells. (C) **β**-catenin also induces expression of follistatin (*Fst*) and at the same time maintains a low expression of activin **β**B (*Acbb*). FST antagonizes the action of activin B, which is the protein product of *Acbb*. Lack of activin B ensures that no testis-specific vasculature forms and the survival of female germ cells is maintained. See text for more details.

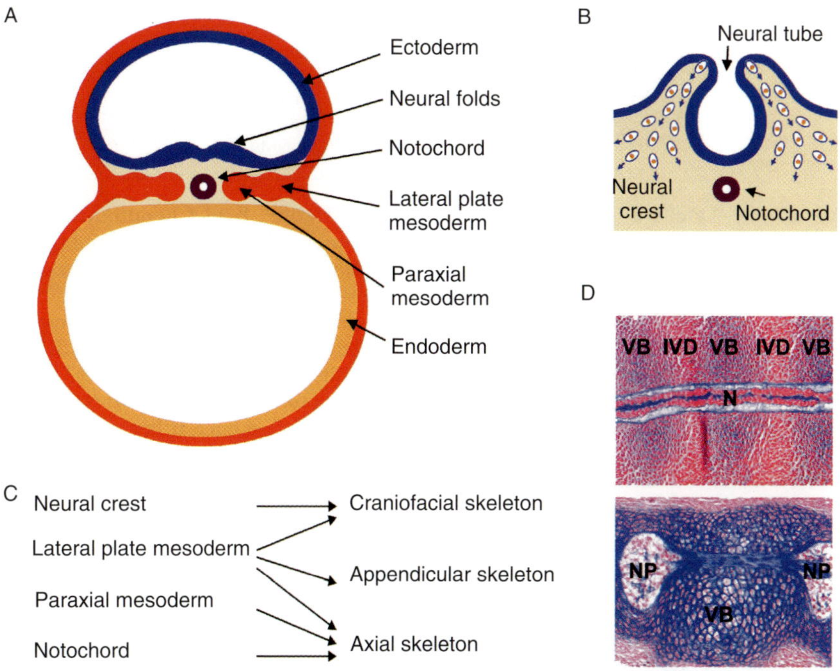

Véronique Lefebvre and Pallavi Bhattaram, Figure 8.1 Origin of skeletal cells in the vertebrate embryo. (A) Schematic of a cross-section through a mouse embryo soon after gastrulation at day 8 of development (equivalent to gestation day 17 in humans). The three germ layers are seen: ectoderm, endoderm, and mesoderm. Ectoderm-derived neural folds are rising. The mesoderm has formed the notochord and is starting to form lateral plate and paraxial derivatives on either sides of the midline. (B) Schematic showing the delamination of neural crest cells from the neural folds at the time of neural tube closure. These cells are starting to migrate inside the embryo (blue arrows), where they will participate in the formation of various structures. (C) Schematic showing the contribution of the neural crest, lateral plate mesoderm, paraxial mesoderm, and notochord to the three major parts of the skeleton. (D) Mid-sagittal sections through the notochord of mouse embryos at the gestation days 12.5 (E12.5, top) and E15.5 (bottom). The E12.5 notochord is a rod-like structure that becomes surrounded by the mesenchymal cell condensations of the prospective vertebral bodies (VB) and IVD. E15.5 VB are cartilaginous and notochord cells have migrated into the intervertebral disc spaces, where they have formed NP. Sections are stained with nuclear fast red and with Alcian blue, which is specific of the notochord and cartilage extracellular matrix.

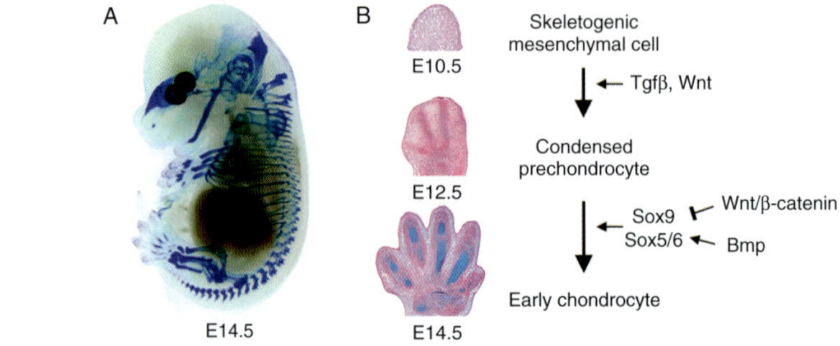

Véronique Lefebvre and Pallavi Bhattaram, Figure 8.3 Chondrocyte early differentiation and development of cartilage primordia. (A) Alcian blue staining of a mouse embryo at E14.5 demonstrates that chondrocyte differentiation of skeletogenic cells leads to the formation of a primary skeleton that is entirely cartilaginous. (B) Sections through the developing paws of mouse embryos illustrate the major steps of early chondrogenesis. At E10.5, the limb bud is filled with skeletogenic cells. By E12.5, some of these cells have formed precartilaginous condensations that prefigure the future digits. By E14.5, condensed prechondrocytes have undergone chondrocyte early differentiation. The sections are stained with Alcian blue and nuclear fast red.

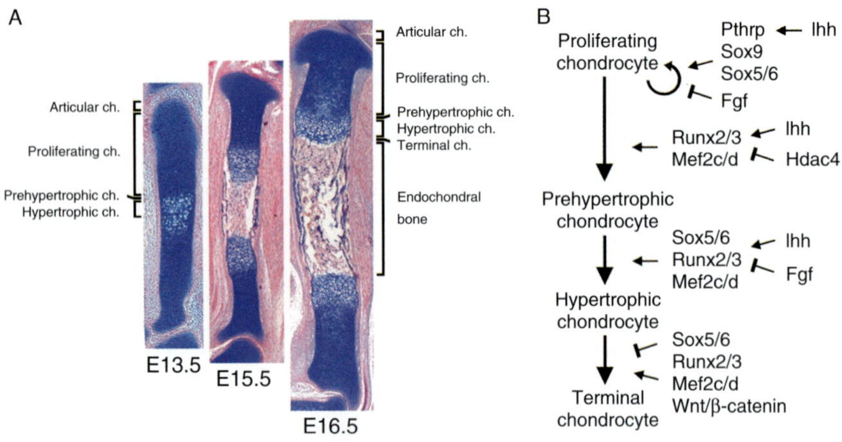

Véronique Lefebvre and Pallavi Bhattaram, Figure 8.4 Chondrocyte maturation and development of cartilage growth plates. (A) Sections through a mouse embryo tibia (T) illustrate the development of growth plates and endochondral bone. At E13.5, early chondrocytes in the center of cartilage primordia undergo prehypertrophic and hypertrophic maturation. They reach terminal maturation and are replaced by endochondral bone by E15.5. Later on, growth plates maintain themselves and elongate developing bones. Chondrocytes keep proliferating and give rise, layer by layer, to maturing chondrocytes. These cells, which eventually die, are replaced by bone. The sections are stained with Alcian blue and nuclear fast red. (B) Schematic of the molecular control of GP chondrocytes.

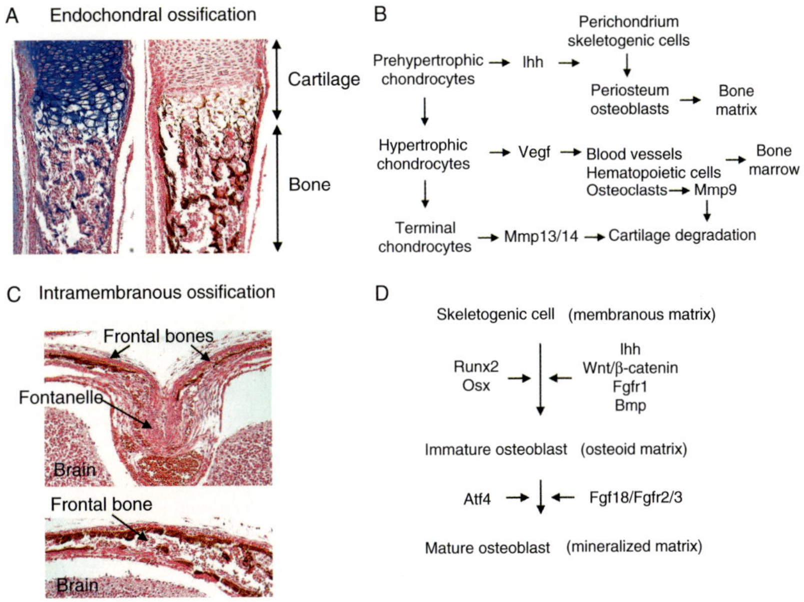

Véronique Lefebvre and Pallavi Bhattaram, Figure 8.5 Osteoblast differentiation and intramembranous and endochondral ossification. (A) Sections through an endochondral bone in a newborn mouse show the replacement of cartilage by bone. The left section is stained with Alcian blue and the right one with the von Kossa reagent, which leaves a brown precipitate on the mineralized bone matrix. (B) Schematic showing how GP chondrocytes and bone-forming cells interact with each other to achieve endochondral ossification. (C) Coronal sections of a newborn mouse head. In the suture linking the two frontal bones (top panel), osteoblast precursors are surrounded by an abundant collagenous matrix. Further away (bottom panel), osteoblasts mature and deposit a mineralized bone matrix. This matrix is stained with the von Kossa reagent. (D) Schematic of the molecular control of osteoblast differentiation.

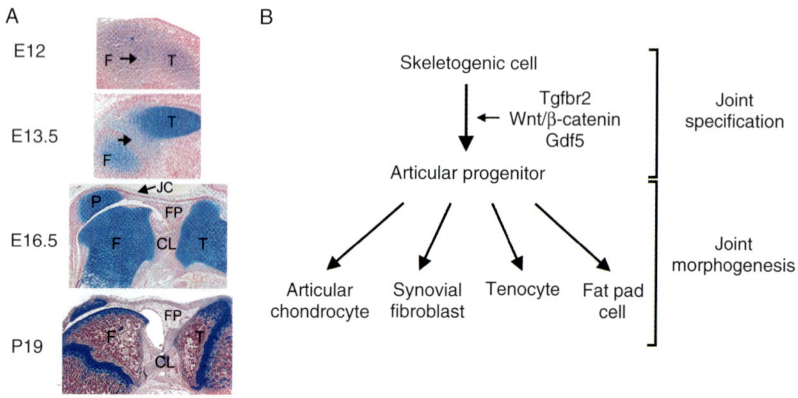

Véronique Lefebvre and Pallavi Bhattaram, Figure 8.6 Synovial joint development. (A) Sections through the mouse knee joint at various stages of development. At E12, the presumptive joint region (arrow) is not distinguishable from the femur (F) and T precartilaginous condensations. At E13.5, this region becomes distinguishable as surrounding cartilage primordia are overtly developing. At E16.5, joint morphogenesis is well advanced. The joint cavity has formed between the patella (P) and F. Cruciate ligaments and FP, lined with synovial tissue, are developed. At the postnatal day 19, the knee joint is mature. The AC is separated from the epiphyseal GP by a secondary center of ossification. The sections are stained with Alcian blue and nuclear fast red. (B) Schematic of the molecular control of synovial joint cell differentiation.

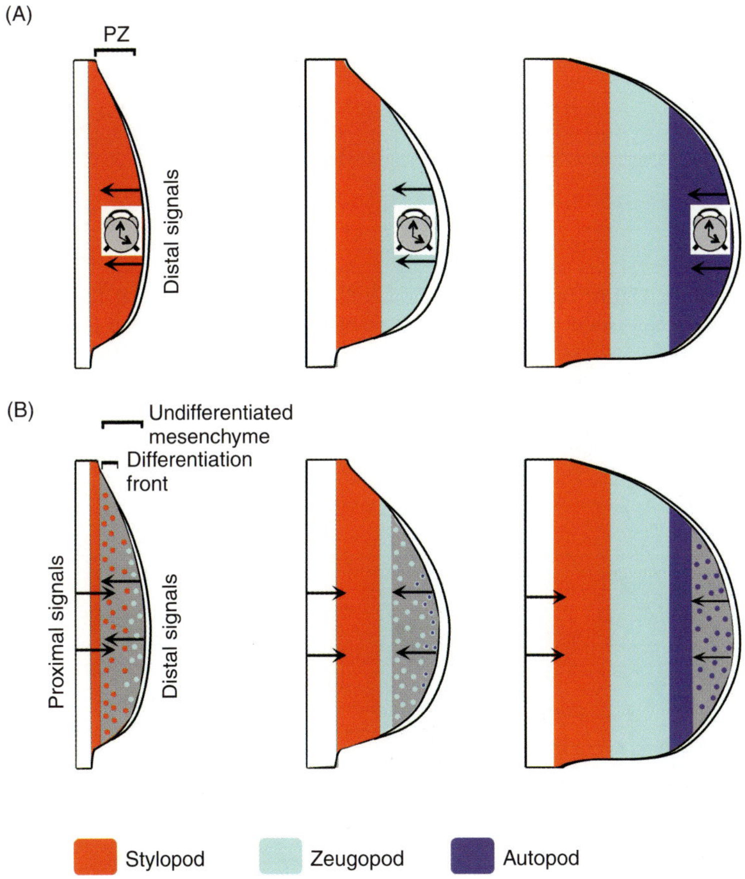

Natalie C. Butterfield *et al.*, Figure 9.3 Evolving models of PD limb patterning. (A) The PZ model of PD limb specification proposed that cells are maintained in a proliferative state in the PZ region immediately under the AER. Cells that are maintained in the PZ for longer acquire more distal information, in a process regulated by an internal clock mechanism. This model is thought to be inconsistent with available data, prompting the development of new theories to explain PD specification. (B) The two-signal or differentiation front theory suggests that PD specification is controlled by opposing gradients of proximalizing (possibly RA) and distalizing (likely AER–FGFs) factors. The proximal boundary of the zone of FGF influence is called the differentiation front, and once cells pass this front their fate is specified.

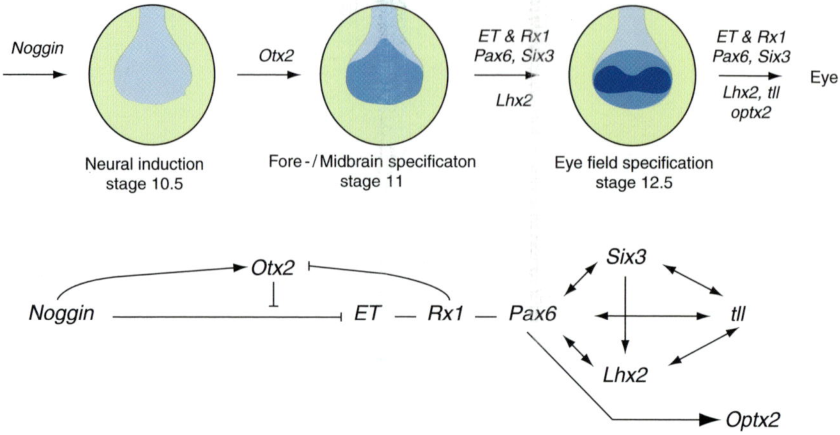

Jochen Graw, Figure 10.3 Genetic factors forming the single eye field. A simplified model of eye field induction in the anterior neural plate of vertebrates is given including the major transcription factors. Light blue indicates the neural plate, blue shows the area of *Otx2* expression, and dark blue represents the eye field (Zuber *et al.*, 2003; with permission from the Company of Biologists Ltd.).

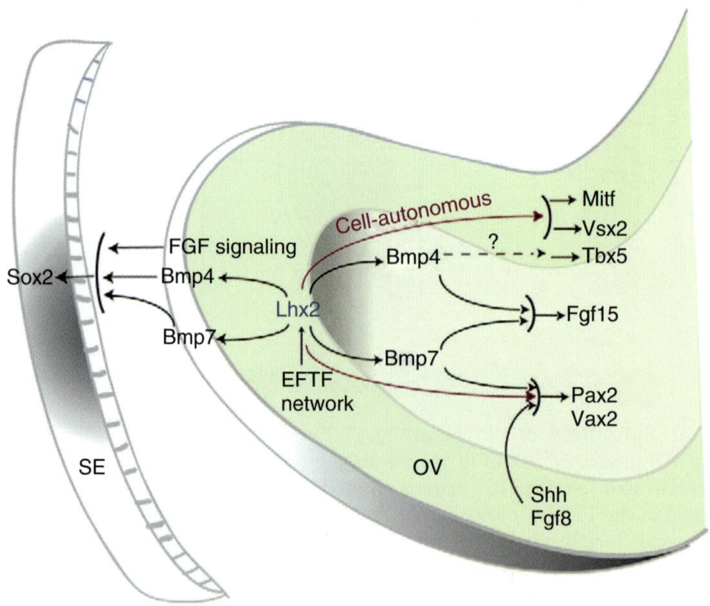

Jochen Graw, Figure 10.4 Model of *Lhx2* function during mouse early eye organogenesis. *Lhx2*, under the control of the EFTF network, links lens specification and optic vesicle patterning through the regulation of BMP signaling (black arrows). *Lhx2* also promotes optic vesicle patterning by cell-autonomous mechanisms (red arrows). Why Bmp4 fails to upregulate *Tbx5* expression is not resolved (dashed line). The timing of action and influence of *Lhx2* on several pathways suggest that it acts to coordinate the multiple patterning events necessary for optic cup formation (Yun *et al.*, 2009; with permission from the Company of Biologists Ltd.).